BASIC ALGEBRA

MILWAUKEE AREA TECHNICAL COLLEGE MATHEMATICS SERIES

BASIC ALGEBRA

THOMAS J. McHALE
PAUL T. WITZKE
Milwaukee Area Technical College, Milwaukee, Wisconsin

ADDISON-WESLEY PUBLISHING COMPANY
Reading, Massachusetts
Menlo Park, California • London • Amsterdam • Don Mills, Ontario • Sydney

Milwaukee Area Technical College Mathematics Series

BASIC ALGEBRA

CALCULATION AND SLIDE RULE

BASIC TRIGONOMETRY

ADVANCED ALGEBRA

Reproduced by Addison-Wesley from camera-ready copy prepared by the authors

Second printing, June 1971

ISBN 0-201-04625-3
LMNOPQ-MU-798

FOREWORD

The Milwaukee Area Technical College Mathematics Series is the product of a five-year project whose goal has been the development of a method of communicating the mathematics skills needed in basic science and technology to a wide range of students, including average and below-average students. This Series is not just a set of textbooks. It is a highly-organized, highly-assessed, and highly-successful system of instruction which has been designed to cope with the learning process of individual students by a combination of programmed instruction, continual diagnostic assessment, and tutoring. Though this system of instruction is different than conventional mathematics instruction, a deliberate effort has been made to keep it realistic in the following two ways:

1. It can be used in a regular classroom by regular teachers without any special training.
2. Its cost has been minimized by avoiding the use of educational hardware.

History. The five-year project was originally funded by the Carnegie Corporation of New York and subsequently funded by the Wisconsin State Board of Vocational, Technical, and Adult Education and the Milwaukee Area Technical College (MATC). The system of instruction offered in this Series was specifically developed for a two-semester Technical Mathematics course at MATC. It has been used in that capacity for the past five years with the following general results:

1. The dropout rate in the course has been reduced by 50%.
2. Average scores on equivalent final exams have increased from 55% to 85%.
3. The rate of absenteeism has decreased.
4. Student motivation and attitudes have been very favorable.

Besides its use in the Technical Mathematics course at MATC, parts or all of the system have been field tested in various other courses at MATC, in other technical institutes in Wisconsin and neighboring states, and in secondary schools in the Milwaukee area. These field tests with over 4,000 students have provided many constructive comments by teachers and students, plus a wealth of test data which has been item-analyzed and error-analyzed. On the basis of all this feedback, the textbooks and tests have been revised several times until a high level of learning is now achieved by a high percentage of students.

Uses of the System. Though specifically prepared for a two-semester Technical Mathematics course, the system of instruction can be used in a variety of contexts. At the college level, parts of it can be used in pre-technical, health occupations, apprentice, trade, and developmental programs, or in an intermediate algebra course. At the secondary school level, the whole system can be used in a two-year Technical Mathematics course in Grades 11 and 12; parts of it can be used in other courses or as supplementary materials in other courses. The system is highly flexible because it can be used either in a conventional classroom or in a learning center, with either a paced or self-paced schedule.

Written for students who have completed one year of algebra and one year of geometry in high school, the system has been used successfully with students without these prerequisites. Though not designed for students with serious deficiencies in arithmetic, it has been successfully used with students who were below average in arithmetic skills.

Three Main Elements in the System. The three main elements in the instructional system are programmed textbooks, diagnostic assessment, and a teacher. Though discussed in greater detail in the Teacher's Manual - MATC Mathematics Series, each main element is described briefly below:

1. Programmed Textbooks - The Series includes four programmed textbooks: BASIC ALGEBRA, CALCULATION AND SLIDE RULE, BASIC TRIGONOMETRY, and ADVANCED ALGEBRA. These textbooks include frequent short self-tests (with answers) which are an integral part of the instruction.

2. Diagnostic Assessment - Each textbook is accompanied by a test book which includes a diagnostic test for each assignment, chapter tests (three parallel forms), and a comprehensive final examination. Pre-tests in arithmetic and algebra are included in two of the test books, TESTS FOR BASIC ALGEBRA and TESTS FOR CALCULATION AND SLIDE RULE. A complete set of keys is provided in each test book for all tests in that book.

   Note: Test books are provided only to teachers, not to individual students. Copies of the tests for student use must be made by the Xerox process or some similar process.

3. A Teacher - A teacher is essential to the success of this system. Its success depends on the teacher's ability to diagnose learning difficulties and to remedy them by means of tutoring. Its success also depends on the teacher's skill in maintaining a suitable effort by each student. This new role for the teacher can be best described by calling him a "manager of the learning process". Teachers who use the system correctly find that this new role is a highly professional and satisfying one because it permits them to deal successfully with the wide range of individual differences which exist in any group of students.

Special Features. Some of the major features of the system of instruction are listed below:

Individual Attention - The system is designed so that the teacher can deal with the learning problems of individual students.

Learnability - When the system is used correctly, teachers can expect (and will obtain) a high level of learning in a high percentage of students.

Relevant Content - All of the content has been chosen on the basis of its relevance for basic science and technology. The content also provides a basis for the study of more advanced topics in mathematics.

Assessment - The many tests which are provided enable the teacher to maintain a constant assessment of each student's progress and the overall success of the system.

Student Motivation - Because students receive individual attention and have a high probability of success, motivation problems are minimized and student attitudes are generally quite positive.

Besides these main features, the system also provides a mechanism for dealing with absenteeism, and it lends itself quite easily to the use of paraprofessionals.

HOW TO USE THE SYSTEM

Though discussed in greater detail in the Teacher's Manual - MATC Mathematics Series, the procedure for using the system is briefly outlined below. All of the tests mentioned are available in the book Tests For Basic Algebra. Both books are available from the Addison-Wesley Publishing Company, Reading, Massachusetts 01867.

1. The entry diagnostic tests in arithmetic and algebra can be given to get an immediate assessment of the entry skills of the students. The scores on these tests quickly identify the students with very high or very low entry skills.

2. Each chapter is covered in a number of assignments. Each assignment can be assessed by the diagnostic test which is provided. The assignments for Basic Algebra are listed at the bottom of this page. Though the best results have been obtained with a paced schedule, whether daily or otherwise, the assignments can also be covered on a self-paced schedule. Since the diagnostic tests are designed to take only 15 to 25 minutes, ample time is left for correction and tutoring within a normal class period. The diagnostic tests need not be graded since they are simply a teaching tool.

3. After all assignments for a chapter are completed, one of the three equivalent forms of the chapter test can be administered. Ordinarily, these tests should be graded. If the diagnostic tests are used in conjunction with the assignments, high scores should be obtained on the chapter tests, and grades can be assigned on a percentage basis.

4. When all of the chapters in a book are completed in the manner above, the comprehensive test for that book can be given. Or, if the teacher prefers, the items in that comprehensive test can be included in a final exam at the end of the semester.

ASSIGNMENTS FOR BASIC ALGEBRA

Chapter	Assignment	Pages
Ch. 1:	#1	(pp. 1-14)
	#2	(pp. 15-36)
	#3	(pp. 37-53)
	#4	(pp. 54-66)
Ch. 2:	#5	(pp. 67-86)
	#6	(pp. 87-109)
	#7	(pp. 110-126)
Ch. 3:	#8	(pp. 127-142)
	#9	(pp. 143-162)
	#10	(pp. 163-181)
Ch. 4:	#11	(pp. 182-199)
	#12	(pp. 200-216)
	#13	(pp. 216-233)
Ch. 5:	#14	(pp. 234-249)
	#15	(pp. 250-267)
	#16	(pp. 267-281)
	#17	(pp. 281-292)
Ch. 6:	#18	(pp. 293-314)
	#19	(pp. 315-331)
	#20	(pp. 332-346)
	#21	(pp. 347-360)
Ch. 7:	#22	(pp. 361-392)
	#23	(pp. 393-414)
	#24	(pp. 415-436)
Ch. 8:	#25	(pp. 437-449)
	#26	(pp. 450-464)
	#27	(pp. 464-480)
Ch. 9:	#28	(pp. 481-499)
	#29	(pp. 500-520)
	#30	(pp. 521-533)
	#31	(pp. 534-548)
Ch. 10:	#32	(pp. 549-569)
	#33	(pp. 570-587)
	#34	(pp. 588-605)

BASIC ALGEBRA

PREREQUISITES: There are no prerequisites in the MATC Mathematics Series for Basic Algebra.

SEQUENCING: The chapters should be studied in their numerical order.

FEATURES: A complete review of fundamental algebra is included. The meaning and sensibleness of axioms and principles are informally justified by the use of numbers.

Manipulations of algebraic expressions and solutions of equations and formulas are based on an intuitive understanding of the relevant principles of modern mathematics.

The meaning of fractions and operations with fractions are heavily emphasized. Numerical fractions are used to show the basic principles which underlie operations with fractions. These principles are then generalized to the types of literal fractions which occur in formulas and derivations.

Fractional roots and fractional equations are not introduced until fractions are thoroughly reviewed.

Formal strategies for solving equations are an integral part of the instruction, and the strategies used in solving traditional equations are explicitly generalized to the rearrangement of literal equations and formulas.

When various solution-methods for a type of traditional equation exist, that method is taught which generalizes most readily to literal equations and formulas.

The methods of solving systems of equations are explicitly generalized to systems of formulas and formula derivation.

Both linear and curvilinear graphs of equations and formulas are treated. That is, the rectangular coordinate system is not restricted to functions in which the variables are "x" and "y".

ACKNOWLEDGMENTS

The Carnegie Corporation of New York made this Series possible by originally funding the project in 1965.

Dr. George A. Parkinson, now Director Emeritus of MATC, was instrumental in obtaining the original grant and totally supported the efforts of the project staff. Dr. William L. Ramsey, present Director of MATC, has continued this support.

Dr. Lawrence M. Stolurow, Professor and Director of Graduate Studies, Division of Educational Research and Development, State University of New York at Stonybrook, New York, served as the original chairman of the project.

Mr. Gail W. Davis of the project staff was largely responsible for the development of an efficient classroom procedure and the organization of a learning center. He also contributed many suggestions which helped to improve the instructional materials.

Mr. Keith J. Roberts, Mr. Allan A. Christenson, and Mr. Joseph A. Colla, members of the project staff, offered many constructive criticisms and helpful suggestions.

Many other teachers and administrators contributed to the implementation, field testing, and improvement of the instructional system.

Mrs. Arleen A. D'Amore typed the camera-ready copy of the textbooks and tests.

ABOUT THE AUTHORS

THOMAS J. McHALE

He has been director of the MATC Mathematics Project since its beginning in June, 1965. He received his doctorate in experimental psychology at the University of Illinois, with major emphasis on the psychology of learning. He is currently a part-time member of the Psychology Department at Marquette University. He has taught mathematics at the secondary-school level.

PAUL T. WITZKE

He has been a member of the MATC Mathematics Project staff since its beginning. He has taught mathematics, physics, electronics, and other technical courses at the Milwaukee Area Technical College for 25 years. He has written numerous instructional manuals and has developed several televised courses in mathematics.

CONTENTS

Chapter 1 SIGNED WHOLE NUMBERS

In this chapter, the following concepts and operations will be discussed:

(1) the number line and "signed" numbers
(2) addition, subtraction, and multiplication of "signed" numbers.

The discussion will be limited to "signed" whole numbers. "Signed" fractions and decimal numbers will be discussed in later chapters.

1-1 THE NUMBER LINE

Mathematicians have found it useful to arrange whole numbers on a line called the <u>number line</u>. On the number line, whole numbers are arranged according to their size. In this section, we will discuss the number line and the relative size of whole numbers.

1. Here is a part of the number line:

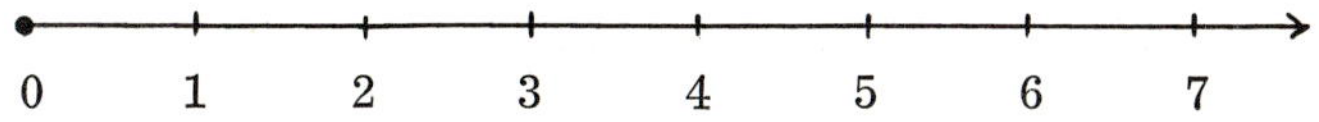

Notice these two features:

(1) Each point on the line stands for a different number.
(2) The arrowhead at the right means that the line goes on indefinitely in that direction.

Do the numbers get larger or smaller as you move <u>to the right</u> on the number line? __________

(Answer is below, to the right of Frame 2.)

Larger.

2. Here is a part of the number line again. It contains only counting numbers.

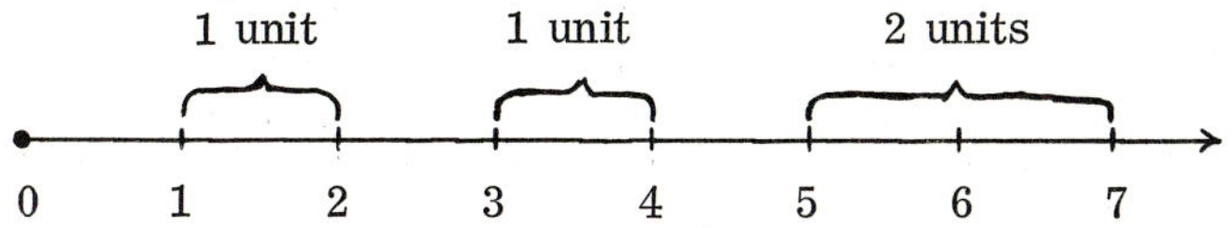

The distance between two consecutive whole numbers is called a "<u>unit distance</u>" or "<u>1 unit</u>."

There is 1 unit between "1" and "2".
There is 1 unit between "3" and "4".
There are 2 units between "5" and "7".

How many units are there between 1 and 6? __________

5 units

3. The part of the number line we are examining begins at "0". There is a special name for "point 0." It is called the <u>origin</u>.

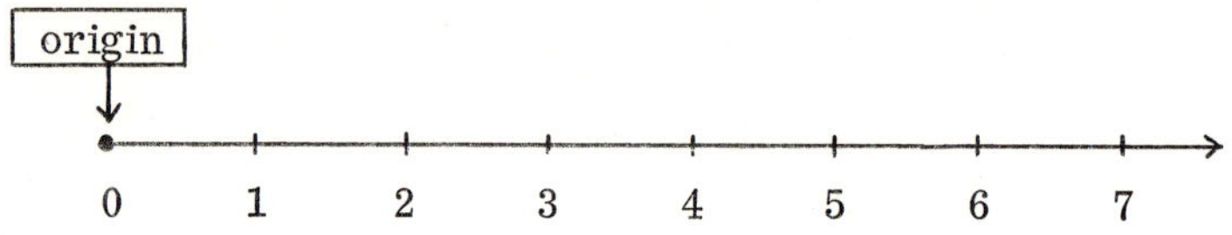

How many units are there:

(a) Between the origin and "5"? ________

(b) Between the origin and "7"? ________

---

4. On the number line, "3" is located 3 units to the right of the ________.

a) 5 units
b) 7 units

---

5. You can see how the number line is used to line up numbers according to their size. Larger numbers lie <u>to the right</u>; smaller numbers lie <u>to the left</u>.

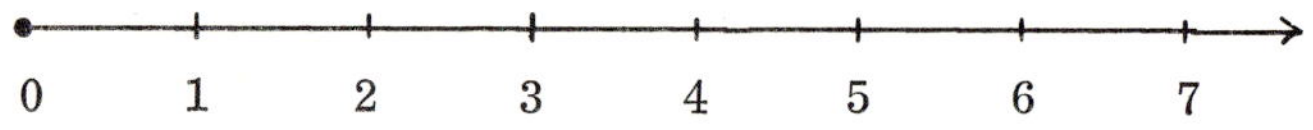

(a) "7" is greater than "4". Therefore, "7" is to the ________ (right/left) of "4" on the number line.

(b) "2" is less than "5". Therefore, "2" is to the ________ (right/left) of "5" on the number line.

origin

---

6. In mathematics, there are many statements we can make about numbers. These statements are called mathematical <u>sentences</u>.

You are all familiar with statements of equality. Here is an example:

5 + 2 is equal to 7

Mathematicians like to use symbols instead of words in order to make their sentences as short as possible. Therefore, instead of the words "is equal to," they use the symbol "=".

"=" means "is equal to."

They write the sentence above in this shorter way:

$5 + 2 = 7$

Here is another mathematical sentence:

7 - 3 is equal to 4

Write this sentence with a symbol instead of words: ____________

a) right
b) left

$7 - 3 = 4$

7. Not every mathematical sentence is a statement of equality. There are other types of statements which are based on the size of numbers. Since these statements are based on the size of numbers, they are also based on the position of numbers on the number line. Here is an example:

8 is greater than 4

Instead of the words ''is greater than,'' mathematicians use the symbol ''>''.

''>'' means ''is greater than.''

They write the sentence above in this shorter way:

$8 > 4$

Write the following sentence with a symbol instead of words:

15 is greater than 12 ______________

$15 > 12$

8. Here is another mathematical sentence based on the size of the numbers.

3 is less than 9

Instead of the words ''is less than,'' mathematicians use the symbol ''<''.

''<'' means ''is less than.''

They write the sentence above in this shorter way:

$3 < 9$

Write the following sentence with a symbol instead of words:

15 is less than 19 ______________

$15 < 19$

9. Any of the following three symbols can appear in a mathematical sentence:

''='', which means ''is equal to.''
''>'', which means ''is greater than.''
''<'', which means ''is less than.''

Write each of the following sentences with symbols instead of words:

(a) 7 is less than 9 ______________

(b) 14 is greater than 11 ______________

---

a) $7 < 9$

b) $14 > 11$

10. Write each of the following sentences with words instead of the symbols ''>'' or ''<''.

(a) $2 < 8$ ____________________________

(b) $11 > 5$ ____________________________

---

a) 2 is less than 8.

b) 11 is greater than 5.

11. (a) What does the symbol ">" mean in words?

____

(b) What does the symbol "<" mean in words?

____

---

12. In the blanks below, write either a ">" or a "<":

(a) 25 _____ 18 (c) 11 _____ 15

(b) 37 _____ 46 (d) 199 _____ 181

Answers to frame 11:

a) Is greater than.

b) Is less than.

---

13. (a) Since $49 > 36$, 49 is to the __________ (right/left) of 36 on the number line.

(b) Since $77 < 88$, 77 is to the __________ (right/left) of 88 on the number line.

Answers to frame 12:

a) > c) <

b) < d) >

Answers to frame 13:

a) right

b) left

SELF-TEST 1 (Frames 1-13)

On the number line:

1. What number is located at the origin? _______
2. What number is located 3 units to the right of the origin? _______
3. What number is located 5 units to the right of the number 2? _______
4. The number 6 is located to the __________ (right/left) of the number 12.
5. Since 28 is greater than 17, 28 is to the __________ (right/left) of 17 on the number line.

Write each of the following in symbol form:

6. 8 is greater than 3 __________
7. 5 + 9 is equal to 14 __________
8. 37 is less than 53 __________

ANSWERS:

1. 0
2. 3
3. 7
4. left
5. right
6. $8 > 3$
7. $5 + 9 = 14$
8. $37 < 53$

1-2 "SIGNED" NUMBERS AND THE NUMBER LINE

None of the numbers we have seen this far have had "positive" or "negative" signs attached to them. There are no "signed" numbers in simple arithmetic. However, "signed" numbers are necessary in algebra. In this section, we will extend the number line to include "signed" numbers.

There are various situations in the real world in which "signed" numbers are useful. Here are some examples:

(1) Temperatures above or below 0°
(2) Elevation above or below sea level
(3) Pressure above or below normal
(4) Increases or decreases

---

14. We will use a thermometer and temperature as an example to introduce the idea of "signed" numbers. Here is a picture of an ordinary thermometer turned sideways:

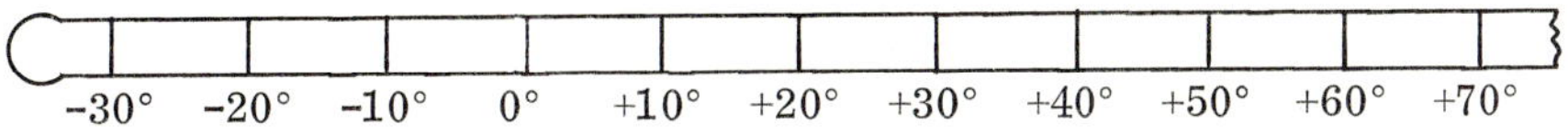

On an ordinary thermometer, some temperatures are "above 0°" and some temperatures are "below 0°." 0° is a reference point. For example, the temperature can be either:

20° above zero

or

20° below zero

In order to distinguish between "above zero" and "below zero" temperatures, mathematicians use the words "positive" and "negative," where:

"positive" means "above zero"
"negative" means "below zero"

Therefore:

"Positive 20°" means "20° above zero."
"Negative 20°" means "20° below zero."

As usual, mathematicians use a symbol instead of words because it is shorter. Instead of "positive," they use the symbol "+". Instead of "negative," they use the symbol "-".

"+" means "positive"
"-" means "negative"

Therefore:

+20° means "positive 20°" or "20° above zero."
-20° means "negative 20°" or "20° below zero."

---

15. Does "-30°" mean: (a) Positive 30° or negative 30°? ______________
(b) 30° above zero or 30° below zero? ______________

---

a) negative 30°
b) 30° below zero

16. The "+" or "-" in front of a number is called the "sign" of the number. Therefore, when a number has a "+" or "-" in front of it, it is called a "signed number."

What signed number could be used to represent each of the following temperatures?

(a) 47° above zero ________

(b) 32° below zero ________

---

17. When used as a sign of a number:

(a) The symbol "+" stands for what word? ____________

(b) The symbol "-" stands for what word? ____________

a) +47°

b) -32°

---

18. Here is another picture of a thermometer:

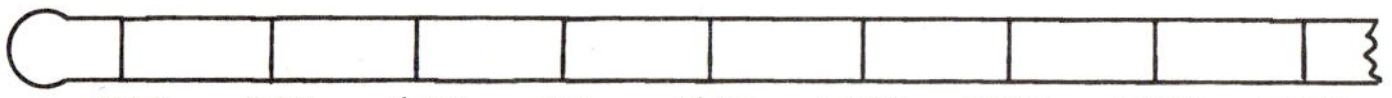

Notice these points:

(1) 0° is the reference point.
(2) Positive temperatures go to the right of 0°; negative temperatures go to the left of 0°.
(3) The numerical value of negative temperatures increases as you go to the left from 0°.

(a) -20° is to the ________ (right/left) of -10°.

(b) -30° is to the ________ (right/left) of -20°.

a) positive

b) negative

---

19. On a thermometer, for any positive temperature, there is a corresponding negative temperature which is the same distance from 0° on the scale.

For example: Both +20° and -20° are the same distance from 0°.

Complete: (a) +10° and ______ are the same distance from 0° on the scale.

(b) -30° and ______ are the same distance from 0° on the scale.

a) left

b) left

---

20. On the basis of your experience with temperatures, which temperature in each of the following pairs is higher?

(a) "20° above zero" or "40° above zero" ____________

(b) "10° below zero" or "10° above zero" ____________

(c) "20° below zero" or "10° below zero" ____________

a) -10°

b) +30°

21. If a thermometer is turned sideways with the positive values to the right, the "higher" of any two temperatures is <u>to the right</u>.

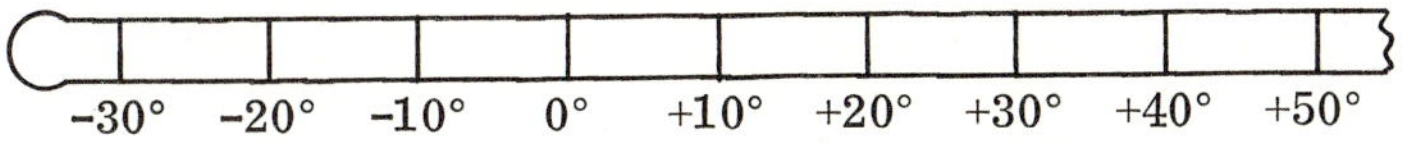

Complete the following statements:

(a) +20° is ___________ (higher/lower) than -20°.

(b) -10° is ___________ (higher/lower) than +10°.

(c) -20° is ___________ (higher/lower) than -30°.

(d) -40° is ___________ (higher/lower) than -30°.

a) 40° above zero
b) 10° above zero
c) 10° below zero

---

22. Instead of using the phrases "higher than" and "lower than," mathematicians use the phrases "greater than" and "less than."

Just as "higher than" and "lower than" refer to the relative <u>positions</u> of <u>two</u> temperatures on the scale, "greater than" and "less than" refer to the relative positions of two temperatures on the scale.

Here is a picture of a thermometer again:

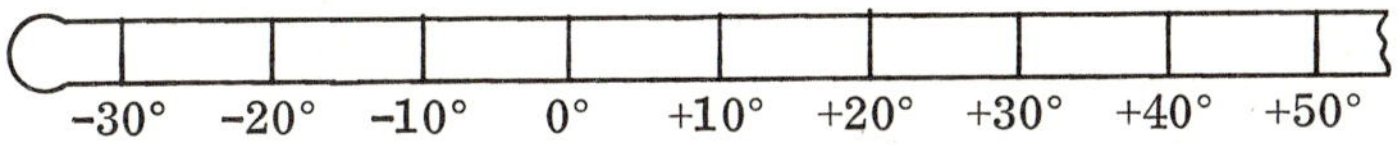

A thermometer is designed so that the "higher" or "greater" of any two temperatures is always <u>to the right</u>.

+50° is <u>to the right</u> of +40°, and +50° is greater than +40°.
+10° is <u>to the right</u> of -20°, and +10° is greater than -20°.
-10° is <u>to the right</u> of -30°, and -10° is greater than -30°.

When comparing any two signed temperatures:

(a) The "greater" is always the one to the ___________ (right/left).

(b) The "lesser" is always the one to the ___________ (right/left).

a) higher
b) lower
c) higher
d) lower

---

23. Complete these statements:

(a) -15° is ___________ (greater/less) than +5°.

(b) -5° is ___________ (greater/less) than -15°.

(c) -25° is ___________ (greater/less) than -15°.

a) right
b) left

---

24. Instead of using the phrases "higher than" or "lower than," we use the phrases "greater than" or "less than" in mathematics.

(a) What is the symbol for "is greater than?" ___________

(b) What is the symbol for "is less than?" ___________

a) less
b) greater
c) less

25. Complete these statements with either a ">" or "<".

(a) -7° _____ +5° (b) -17° _____ -13° (c) -12° _____ -19°

a) > b) <

---

26. Here is a part of the number line for signed numbers:

Notice these similarities with a thermometer:

(1) "0" is the reference point.

(2) Positive numbers go to the right of "0"; negative numbers go to the left of "0".

(3) The numerical value of negative numbers increases as you move to the left from "0".

What is the special name for "point 0"? ____________

a) < b) < c) >

---

27. On the number line:

(a) -5 is to the ____________ (right/left) of -1.

(b) -3 is to the ____________ (right/left) of -4.

origin

---

28. On the number line, signed numbers are arranged according to their size. Here are the definitions of the phrases "greater than" and "less than."

> A first number is greater than a second number if the first is located to the right of the second on the number line.

> A first number is less than a second number if the first is located to the left of the second on the number line.

On the number line, "+2" is to the right of "-3".

Therefore, we can say either:

(a) +2 is ____________ (greater/less) than -3.

(b) -3 is ____________ (greater/less) than +2.

a) left

b) right

---

29. On the number line, "-3" is to the left of "-1".

Therefore, we can say either:

(a) -3 is ____________ (greater/less) than -1.

(b) -1 is ____________ (greater/less) than -3.

a) greater

b) less

---

a) less

b) greater

30. Which number in each of the following pairs is the "greater" of the two?

(a) -4 or 0 _______ (d) +4 or +1 _______

(b) -2 or -4 _______ (e) +1 or 0 _______

(c) +3 or -2 _______ (f) -5 or -4 _______

---

31. Which number in each pair is the "lesser" of the two?

(a) +2 or +4 _______ (d) -6 or -3 _______

(b) 0 or +2 _______ (e) 0 or -1 _______

(c) -4 or -9 _______

a) 0 d) +4

b) -2 e) +1

c) +3 f) -4

---

32. Write each of the following with symbols instead of words:

(a) +17 is greater than +13. _______________

(b) -19 is less than -17. _______________

a) +2 d) -6

b) 0 e) -1

c) -9

---

33. In each blank below, write the symbol (either ">" or "<") which makes the statement true:

(a) +50 _____ +15 (c) -32 _____ -21

(b) +25 _____ +30 (d) -17 _____ -24

a) +17 > +13

b) -19 < -17

---

34. The sign of a number is determined by whether it is to the right or left of the origin ("0") on the number line.

Since "0" is neither to the right nor left of the origin, "0" is neither positive nor negative. "0" has no sign.

Would it make any sense to write "+0" or "-0"? __________

a) > c) <

b) < d) >

---

35. To help you remember what the symbols ">" and "<" mean, you might note this:

">" points to the right on the number line and this is where the "greater" numbers are. ">" means "is greater than."

"<" points to the left on the number line and this is where the "smaller" numbers are. "<" means "is less than."

Insert the correct symbol (> or <) in the blanks in the following examples. This concept is not difficult. There are ample problems to try here. When you are convinced that you know what you are doing, quit and go on to the next frame.

(1) +4 _____ -3 (5) -9 _____ -7 (9) -84 _____ -77

(2) -2 _____ -4 (6) -14 _____ -10 (10) +17 _____ +13

(3) +75 _____ +69 (7) +22 _____ +37 (11) -43 _____ -48

(4) +47 _____ +82 (8) -5 _____ -8 (12) +6 _____ -7

No.

36. On the number line, for any positive number there is a corresponding negative number which is the same distance from the origin. For example: Both +9 and -9 are the same distance from the origin. Complete: (a) +5 and _______ are the same distance from the origin. (b) -26 and _______ are the same distance from the origin.	1) > 5) < 9) < 2) > 6) < 10) > 3) > 7) < 11) > 4) < 8) > 12) >
	a) -5 b) +26

SELF-TEST 2 (Frames 14-36)

What signed number represents:

1. 72° above zero? _______
2. 12° below zero? _______
3. A point 17 units to the left of the origin? _______
4. A point 3 units to the right of the origin? _______

Write the symbol for: 5. negative _____ 6. positive _____

Complete:

7. -15° is ___________ (greater/less) than -4°.
8. 0° is ___________ (greater/less) than -12°.
9. -7 is ___________ (greater/less) than -20.
10. +2 is ___________ (greater/less) than -9.

Complete the following with either ">" or "<":

11. -19° _____ 0°
12. -12° _____ -11°
13. +7 _____ -8
14. -20 _____ -47

15. On the thermometer scale, +42° and _______ are the same distance from 0°.

16. On the number line, -6 and _______ are the same distance from the origin.

ANSWERS:

1. +72°	5. -	7. less	11. <	15. -42°
2. -12°	6. +	8. greater	12. <	16. +6
3. -17		9. greater	13. >	
4. +3		10. greater	14. >	

1-3 THE NUMBER LINE AND ABSOLUTE VALUE

There is a symmetry involved in the placement of signed numbers on the number line. That is, for each positive number to the right of the origin, there is a corresponding negative number to the left of the origin. Ignoring their signs, these two numbers have <u>the same numerical value</u>. Mathematicians use the term "absolute value" instead of "numerical value."

---

37. The <u>ABSOLUTE VALUE</u> of a signed number is the value of that number <u>when the sign is ignored</u>.

For example: The absolute value of +6 is "6".
The absolute value of -6 is "6".

Write the absolute values of each of the following signed numbers:

(a) +27 ________ (b) -65 ________

---

38. Both +10 and -10 have the same absolute value. What is it? ________

a) 27 b) 65

---

39. Name two signed numbers whose absolute value is "9".

________ and ________

10

---

40. What other signed number has the same absolute value as "-24"? ______

+9 and -9

---

41. Here is a picture of part of the number line:

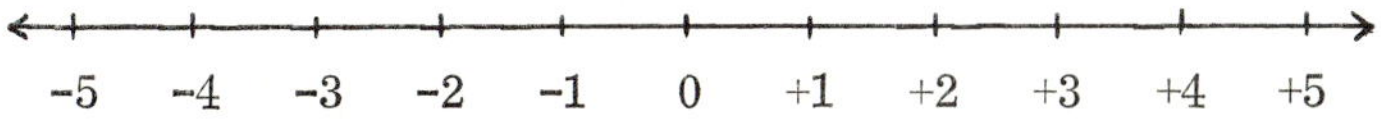

(a) As you move <u>to the right</u> from the origin, the absolute values of <u>positive</u> numbers ____________ (increase/decrease).

(b) As you move <u>to the left</u> from the origin, the absolute values of <u>negative</u> numbers ____________ (increase/decrease).

+24

---

42. The <u>absolute value</u> of any number tells you "how many units away from the origin" it is.

Since both +3 and -3 are "3 units away from the origin," both +3 and -3 have the same absolute value.

(a) The absolute value of +5 tells you that +5 is ______ units away from the origin.

(b) The absolute value of -7 tells you that -7 is ______ units away from the origin.

a) increase
b) increase

---

a) 5
b) 7

43. Of these two numbers: +4 and -10

(a) Which number has the larger absolute value? _______

(b) Therefore, which number is farther away from the origin on the number line? _______

---

44. Of these two numbers: -5 and -7

(a) Which number has the larger absolute value? _______

(b) Therefore, which number is farther away from the origin? _______

a) -10

b) -10

---

45. In each of the following pairs, which number lies <u>farther away</u> from the origin on the number line?

(a) +75 or -47 _____ (b) -65 or +59 _____ (c) -17 or -13 _____

a) -7

b) -7

---

46. The absolute value of any number, positive or negative tells you how far the number is from the origin.

In each of the following pairs, which number is <u>closer</u> to the origin on the number line?

(a) +3 or -4 _____ (b) +5 or -2 _____ (c) -4 or -5 _____

a) +75

b) -65

c) -17

---

47. Here is the part of the number line we were looking at before:

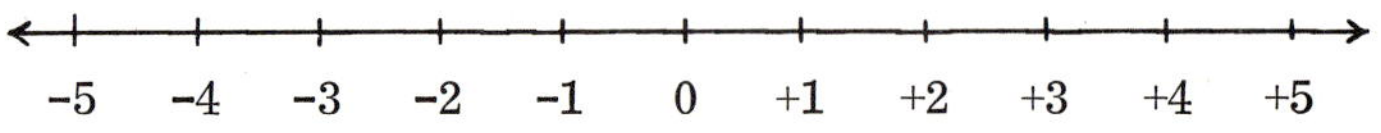

If you attempted to put the point for "+50" on the line above, it would be off this paper to the right. Similarly, the point for "-37" would be off this paper to the left. You can imagine the points for "+50" and "-37", even if we cannot draw them on the number line above.

But the number line is a tool, and we should feel free to use it as a tool. For example, there is no reason why we cannot draw that part of the number line which is around "+50". It would look like the line below.

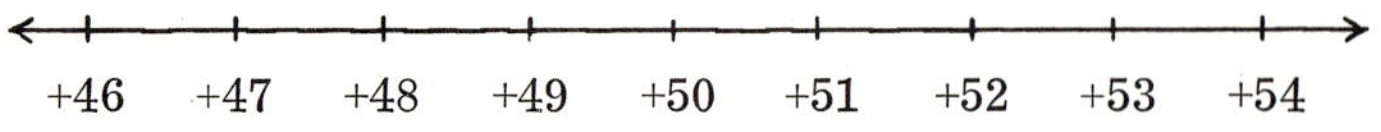

We can also draw the part of the number line around "-37". It would look like the line below.

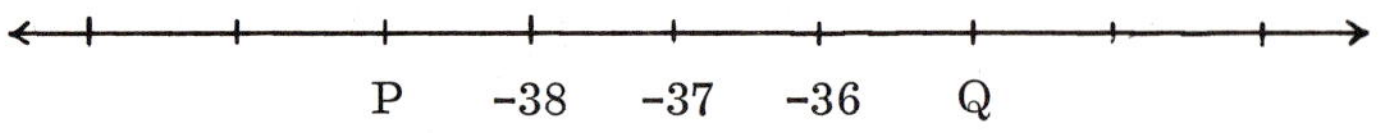

What signed numbers would you put at:

(a) The point labeled "P"? _________

(b) The point labeled "Q"? _________

a) +3

b) -2

c) -4

48. Remember that the absolute values of negative numbers increase as you move to the left on the number line.

Let's try another one. What numbers would you put at points "P" and "Q" on the number line below?

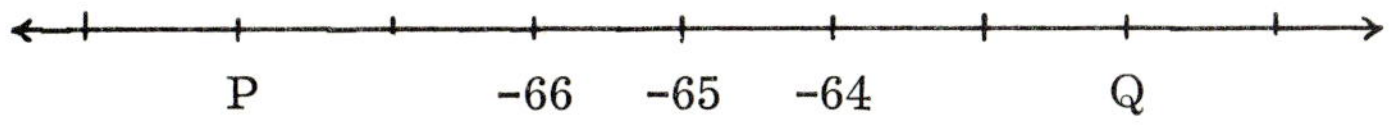

(a) P is __________ (b) Q is __________

a) P is -39

b) Q is -35

49.

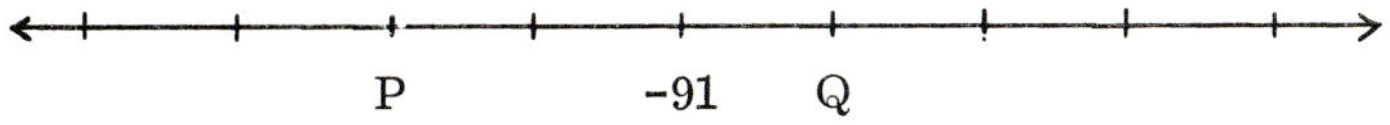

(a) What numbers would you put at points "P" and "Q" on the number line above?

P is __________ Q is __________

(b) Which number has the larger absolute value, the number at P or Q? __________

a) P is -68

b) Q is -62

---

50. Do not confuse the meaning of "greater than" and "less than" with that of absolute value.

<u>Absolute value</u> refers to the value of a signed number <u>with the sign ignored</u>.

<u>Greater than</u> and <u>less than</u> refer to relative position on the number line.

When comparing two signed numbers:

(1) Sometimes the one with the larger absolute value is "greater."

For example:

+5 has a larger absolute value than +4.
+5 also is "greater" than +4 since it is <u>to the right</u> of +4 on the number line.

(2) Sometimes the one with the larger absolute value is "less."

For example:

-3 has a larger absolute value than -2.
-3, however, is "less" than -2 since it is <u>to the left</u> of -2 on the number line.

Here are two signed numbers: -2 and +1

(a) Which has the larger absolute value? __________

(b) Which is "greater?" __________

a) P is -93
Q is -90

b) P, since 93 is larger than 90.

---

a) -2

b) +1

51. Even though the absolute value of "-5" is larger than that of +2, +2 is the "greater" of the two. Why? ______________________

____

Because +2 is to the right of -5 on the number line.

52. Here are three signed numbers: -35, -25, +15

(a) Which has the largest absolute value? __________

(b) Which of the three is "greatest?" __________

a) -35

b) +15, since it is farthest to the right on the number line.

SELF-TEST 3 (Frames 37-52)

1. What is the absolute value of -17? __________
2. A number lies seven units to the left of the origin. What is its absolute value? __________
3. Name two numbers whose absolute value is 24. __________ and __________
4. What signed number has the same absolute value as -81? __________

Here are three signed numbers: -7, -9, -14

5. Which number has the largest absolute value? __________
6. Which number is "greatest?" __________
7. Which number is "smallest?" __________
8. Which number has the smallest absolute value? __________

On the portion of the number line shown below, what number is at:

9. Point A? __________ 10. Point B? __________

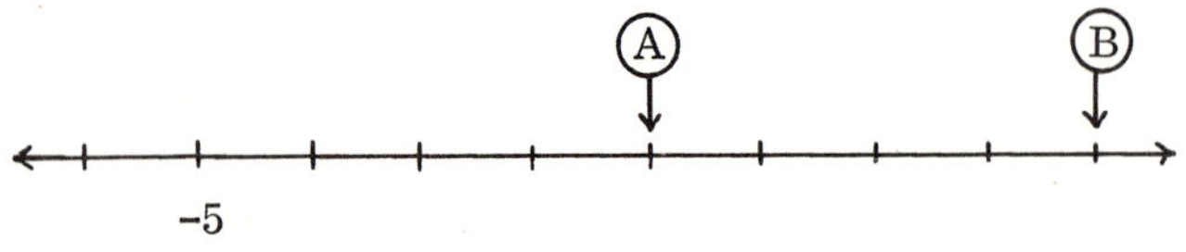

ANSWERS:

1. 17
2. 7
3. +24 and -24
4. +81
5. -14
6. -7
7. -14
8. -7
9. -1
10. +3

1-4 VECTORS AND "MOVES" ON THE NUMBER LINE

A "move" or "change" on the number line can be shown visually by means of an arrow. In mathematics, an arrow of this type is called a "vector." In this section, we will show that any vector can be represented by a signed number.

---

53. Before showing how vectors can be used to show a "move" or "change" on the number line, we will show how a vector can be used to show a "change" in temperature on a thermometer.

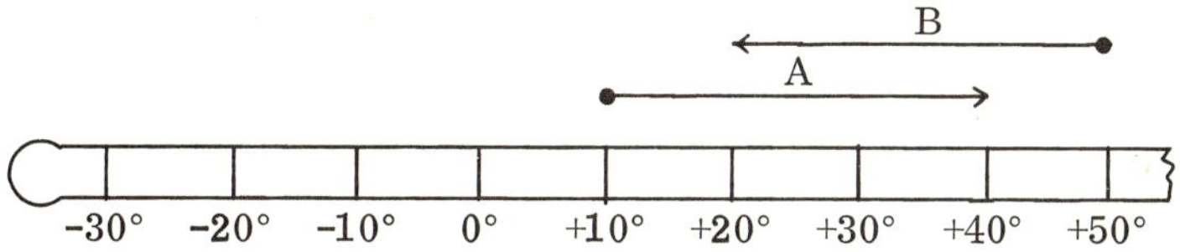

The arrow labeled "A" represents a change in temperature. In mathematics, we call this arrow "vector A." Since the arrowhead (→) points to the right, we know that the change is from +10° to +40°.

(a) The arrowhead for vector B points to the ________ (right/left).

(b) Therefore, vector B represents a change in temperature from ______ to ______.

54.

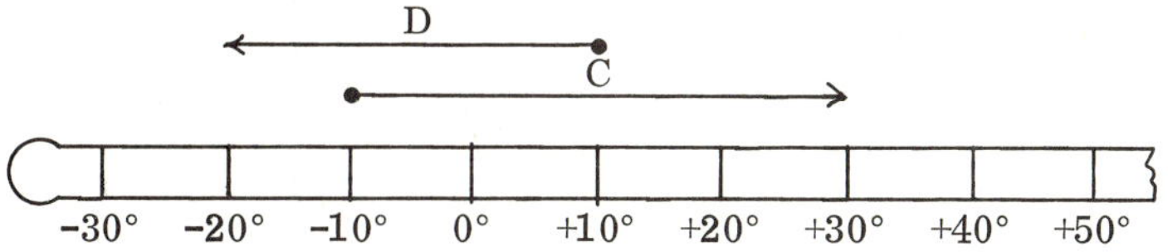

(a) Vector C represents a change in temperature from _____ to _____.

(b) Vector D represents a change in temperature from _____ to _____.

Answers to 53:

a) left

b) +50° to +20°

55.

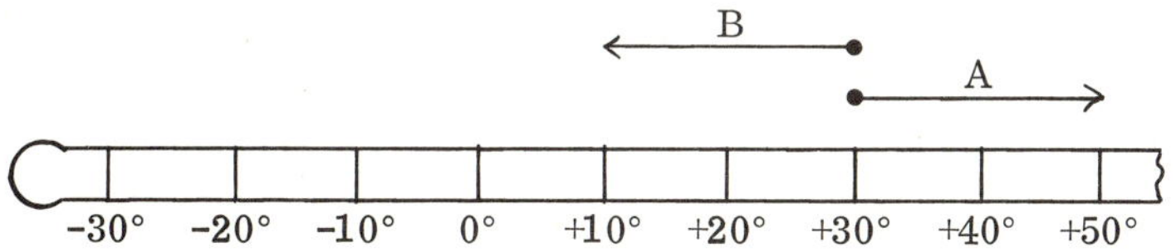

Signed numbers can be used to represent changes in temperature. If the temperature rises, we call the change "positive." If the temperature drops, we call the change "negative."

Vector A represents a change from +30° to +50°, or a "20° rise" in temperature. Since the change is a "rise," we can represent it by the positive signed number +20°.

Vector B represents a change from +30° to +10°, or a "20° drop" in temperature. Since the change is a "drop," we can represent it by the negative signed number -20°.

(Continued on following page.)

Answers to 54:

a) -10° to +30°

b) +10° to -20°

55. (Continued)
Represent each of the following temperature changes with a signed number:

(a) A change from +20° to +70° __________

(b) A change from −20° to −10° __________

(c) A change from +70° to +40° __________

---

a) +50°, since it is a "rise."
b) +10°, since it is a "rise."
c) −30°, since it is a "drop."

56. Both of the following temperature changes can be represented by the <u>same</u> signed number. What is it? __________

A change from −10° to +10°.
A change from +20° to +40°.

+20°

57. Both of the following temperature changes can be represented by the <u>same</u> signed number. What is it? __________

A change from +20° to −10°.
A change from +40° to +10°.

−30°

58. Mathematicians also use vectors to represent "moves" or "changes" on the number line. Here are two examples:

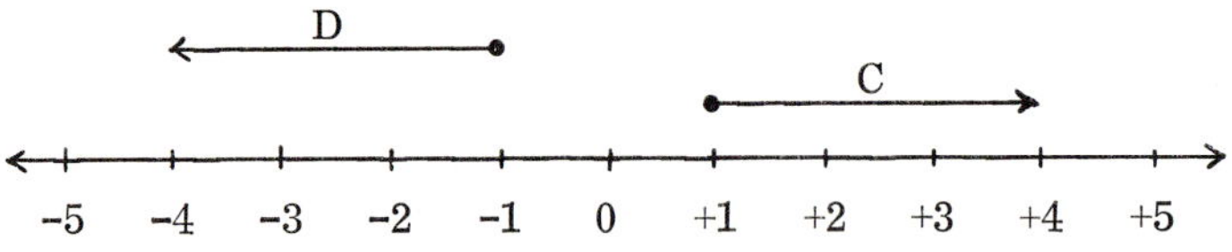

(a) Vector C represents a move from _______ to _______.

(b) Vector D represents a move from _______ to _______.

---

a) +1 to +4
b) −1 to −4

59. Here are two more vectors on the number line:

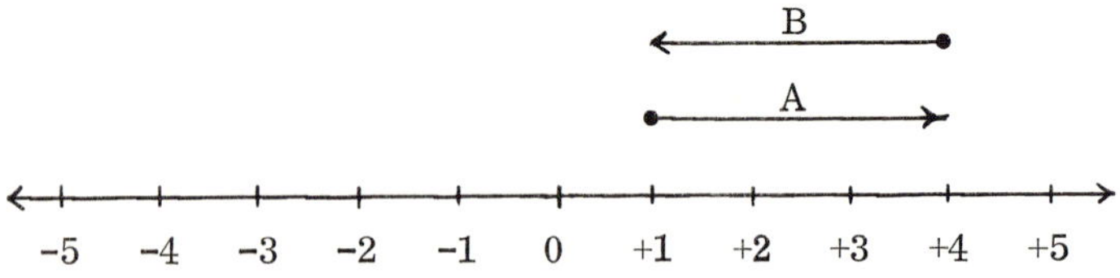

Vector A represents a move from +1 to +4.
Vector B represents a move from +4 to +1.

Though these moves are in opposite directions, both of them are __________ units long.

3

60. Here are the two moves we saw in the last frame:

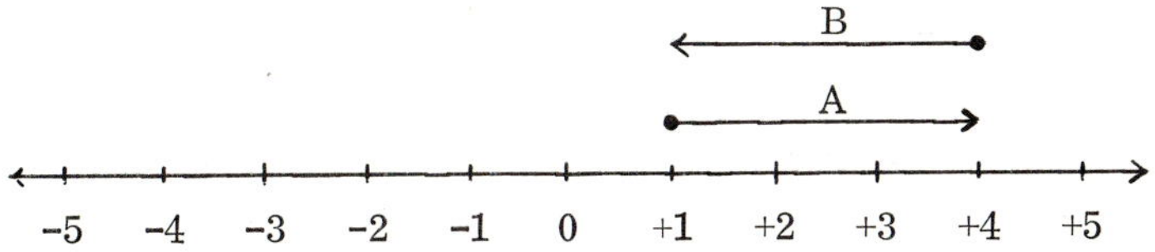

Move A is from "+1" to "+4".
Move B is from "+4" to "+1".

Both moves are 3-units long, but they are in <u>opposite</u> <u>directions</u>.

In order to show the <u>direction</u> of the moves, mathematicians call any move:

<u>Positive</u>, if it is "to the right."
<u>Negative</u>, if it is "to the left."

(a) Since move A is <u>to the right</u> or <u>positive</u>, we represent it with what signed number? ________

(b) Since move B is <u>to the left</u> or <u>negative</u>, we represent it with what signed number? ________

---

a) +3
b) -3

61. This vector represents a move from -1 to +3:

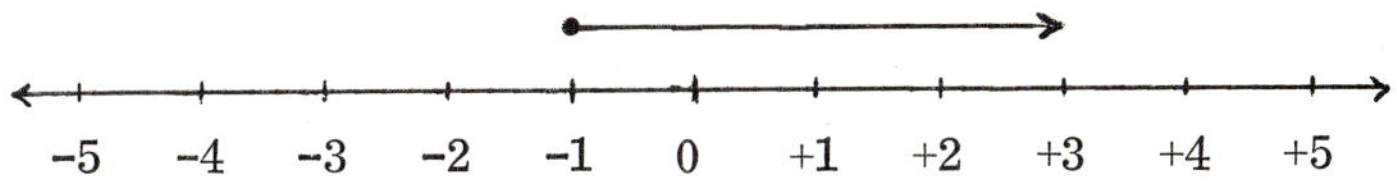

(a) The <u>length</u> of the move is ________ units.

(b) Since the move is <u>to the right</u>, its direction is ____________ (positive/negative).

(c) What signed number represents this move (or vector)? _____

---

a) 4
b) positive
c) +4

62. This vector represents a move from +2 to -3:

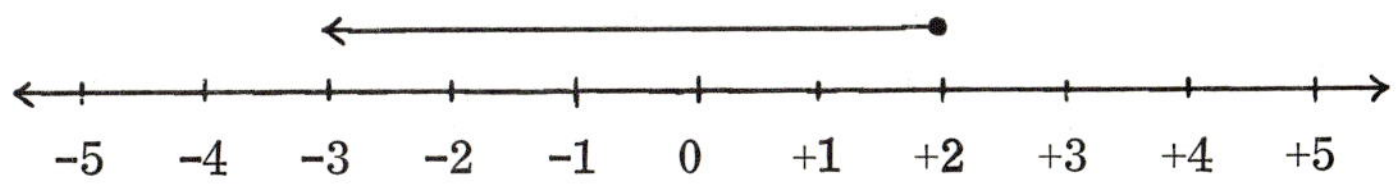

(a) The <u>length</u> of the move is ________ units.

(b) Since the move is <u>to the left</u>, its direction is ____________ (positive/negative).

(c) What signed number represents this move (or vector)? _____

---

a) 5
b) negative
c) -5

63. Here is a picture of a "move" from -3 to 0:

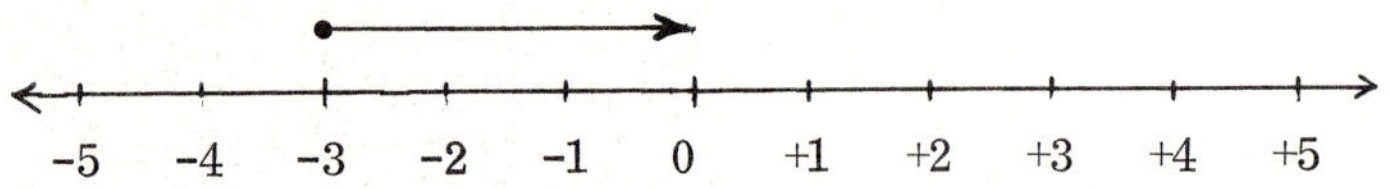

(a) The <u>length</u> of the vector is _______ units.

(b) Since the vector is <u>to the right</u>, we say that its direction is _______________ (positive/negative).

(c) What signed number is used to represent this vector? _______

---

a) 3

b) positive, since it is <u>to the right</u>

c) +3

64. Here is a "move" from "-1" to "-5". The tip of the vector gives the <u>direction</u> of the move.

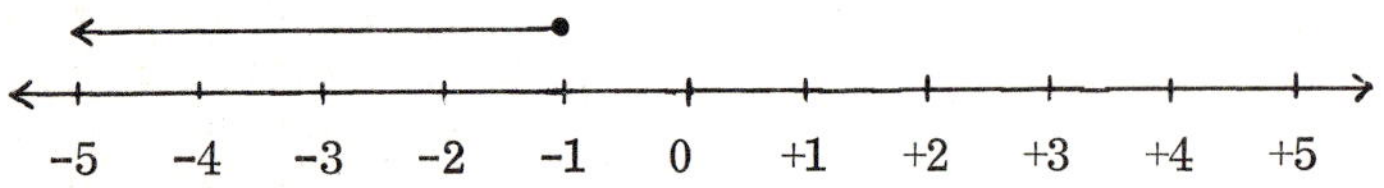

(a) The <u>length</u> of the vector is _______ units.

(b) The <u>direction</u> of the vector is _______________ (positive/negative).

(c) This vector is represented by what signed number? _______

---

a) 4

b) negative, since it is <u>to the left</u>

c) -4

65. Two <u>different</u> vectors of the <u>same length</u> can be drawn from the same starting point. Of course, they represent moves in <u>opposite</u> directions. Here is an example:

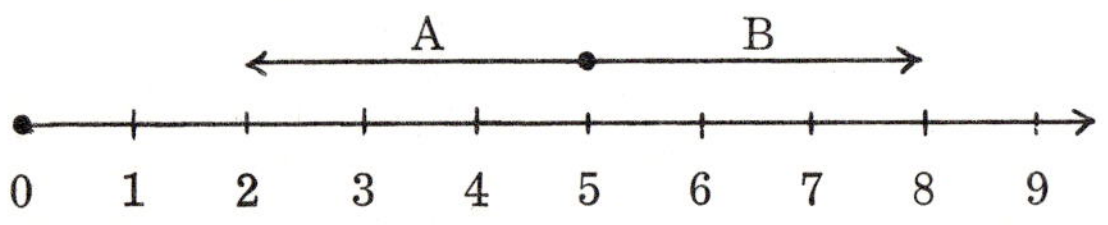

(a) What signed number represents vector A? _______

(b) What signed number represents vector B? _______

---

a) -3

b) +3

66. Many vectors are represented by the <u>same</u> signed number. It does not matter at which point they begin as long as they have the <u>same length and direction.</u>

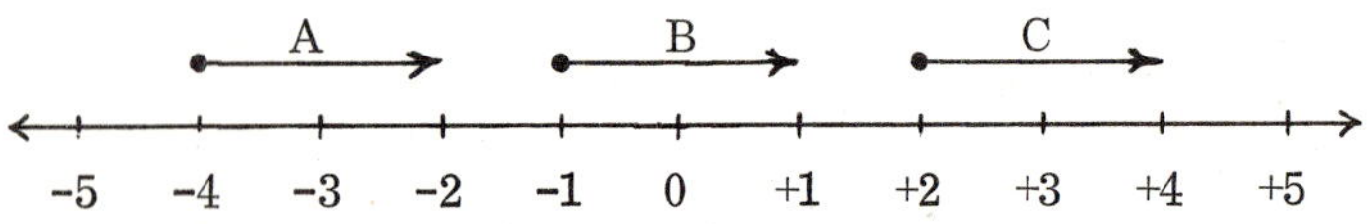

All three of these vectors are represented by what signed number? _______

+2

67.

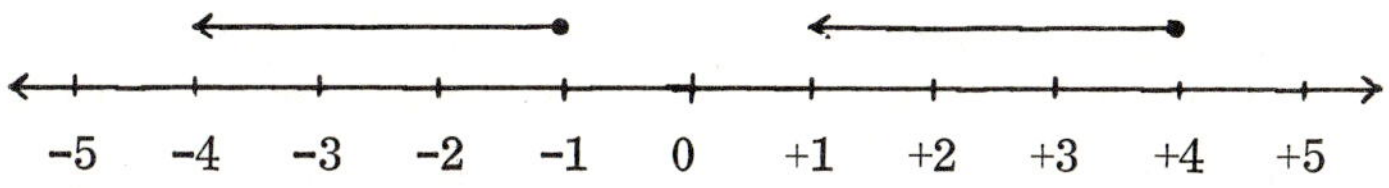

Both of these vectors are represented by what signed number? _______

---

68. The number "+8" represents any vector:

(a) Whose length is __________ units.

(b) Whose direction is ______________________________.

-3

---

69. The number "-4" represents any vector:

(a) Whose direction is ______________________________.

(b) Whose length is __________ units.

a) 8

b) to the right (or positive)

---

70. A "+3" vector and a "-3" vector:

(a) have the same ____________________ (lengths/directions),

(b) but opposite ______________________ (lengths/directions).

a) to the left (or negative)

b) 4

---

71. When signed numbers are used to represent vectors:

(1) The sign of the number gives the direction of the vector.
(2) The absolute value of the number gives the length of the vector.

Here are two signed numbers: "+4" and "-4"

(a) They have different signs. This fact means that they represent vectors whose directions are ____________________ (the same/opposite).

(b) They have the same absolute values. This fact means that they represent vectors whose lengths are ____________________ (the same/different).

a) lengths

b) directions

---

72. Any signed number can represent many vectors, as long as they have the same length and direction.

The absolute value of a signed number gives the ____________________ (length/direction) of all the vectors which it represents.

a) opposite

b) the same

---

73. When the signed number "-5" represents a vector:

(a) The sign "-" tells you the _______________ of the vector.

(b) The absolute value tells you the _______________ of the vector.

length

---

a) direction

b) length

74. 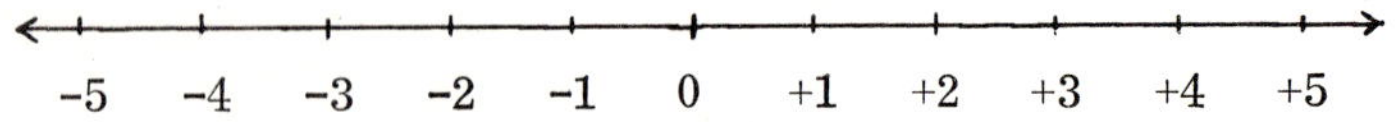

On the number line above:

(a) Draw a vector from "-1" to "+2". It is represented by what signed number? __________

(b) Draw a vector from "-1" to "-5". It is represented by what signed number? __________

---

75. Which of the following vectors are represented by "-2"? __________

(a) A vector from "+5" to "+2".
(b) A vector from "+3" to "+1".
(c) A vector from "-1" to "-4".

a) +3

b) -4

---

76. A "0" vector has no length. Since it has no length, if it begins at +3, it also ends at +3.

Would it make any sense to attach a <u>sign</u> to a "0" vector? __________

Only (b)

No, since it has no length, it has no direction.

<u>SELF-TEST 4 (Frames 53-76)</u>

Write the signed number which represents the vector:

1. Of 7 units length directed to the left. __________

2. Of 23 units length directed to the right. __________

3. A -4 vector has a length of __________ units. Its direction is to the __________ (right/left).

Referring to the diagram, write the signed number which represents:

4. Vector A __________ 5. Vector B __________ 6. Vector C __________ 7. Vector D __________

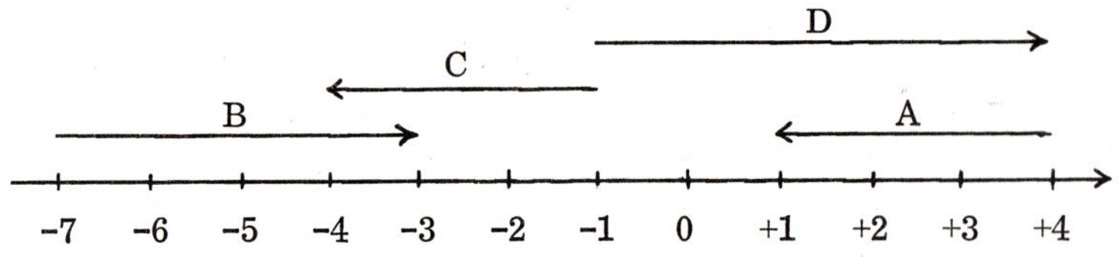

Write the signed number which represents the vector:

8. From +2 to +7 _____
9. From +3 to 0 _____
10. From 0 to +3 _____
11. From -15 to -7 _____
12. From -1 to -5 _____
13. From -35 to +35 _____
14. From +35 to -35 _____
15. From +80 to -20 _____

Answers to Self-Test 4:

1. -7	4. -3	7. +5	10. +3	13. +70
2. +23	5. +4	8. +5	11. +8	14. -70
3. 4, left	6. -3	9. -3	12. -4	15. -100

ADDITION OF SIGNED NUMBERS

"Addition of signed numbers" is a basic operation in algebra. It will be taught in the next few sections. After discussing the two different uses of the "+" symbol, we will teach how to add the following:

(a) Two positive numbers
(b) Two negative numbers
(c) One positive and one negative number
(d) Three or more signed numbers

The procedure for adding two signed numbers will be demonstrated by means of vectors on the number line. Then, rules for adding two signed numbers will be stated. Finally, the commutative property of addition and the concept of "opposites" will be discussed.

1-5 TWO DIFFERENT USES OF THE "+" SYMBOL

77. In algebra, the "+" symbol is used in two different ways:

(1) As the sign of a number.
(2) As the symbol for addition.

In (↓+5) + (↓+3):

(1) The two "+" symbols under the arrows are signs of numbers.
(2) The "+" symbol in the middle is the symbol for addition. It tells you to add the two numbers.

In (+10) + (+9), draw an arrow over the one "+" symbol which tells you to add.

Answer: (+10) ↓+ (+9)

78. When signed numbers are involved in an addition, we write them in parentheses. We do this so that the symbol for addition and the signs of numbers do not become confused.

Instead of +5 + +6 ,
we write (+5) + (+6).

Instead of -5 + -6 ,
we write (-5) + (-6).

Instead of -7 + +4 ,
we write ________.

Answer: (-7) + (+4)

79. Note: (1) We always write a negative sign (–) before each negative number.
(2) We frequently do not write a positive sign (+) before each positive number.

Both "+3" and "3" mean "positive 3."

Which two of the following symbols stand for "positive 7?" __________

+7 –7 7

---

80. Though it is not explicitly written, the sign of "147" is _______.

+7 and 7

---

81. 7 + 11 is a shorthand way of writing (+7) + (+11).

In 7 + 11, is the "+" symbol: (a) The sign of "11"?
or (b) The symbol for addition? __________

+

---

82. 6 + 12

Write the above expression so that the signs of the numbers are explicit: __________________

(b) The symbol for addition.

---

83. (–6) + 3

In the expression above, is the "+" the sign of "3" or the symbol for addition? __________________________

(+6) + (+12)

---

84. (–9) + 14

Write the expression above so that the sign of "14" is explicit:

____

The symbol for addition.

---

85. 7 + 3 2 + (–5)

(–9) + 7 (–4) + (–6)

In each of the four expressions above, only one "+" symbol is explicitly written. In each case, the "+" symbol is the symbol for ______________.

(–9) + (+14)

---

86. What are the two uses of the "+" symbol?

(1) ____

(2) ____

addition

---

1) The sign of positive numbers.

2) The symbol for addition.

1-6 ADDITION OF TWO POSITIVE OR TWO NEGATIVE NUMBERS

In this section, we will examine the addition of two positive or two negative numbers. First we will show some additions by means of vectors on the number line. Then, since using the number line is inefficient, we will develop "rules" which can be used instead.

---

87. In an addition: (1) The two original numbers are called "terms."
(2) The answer is called the "sum."

In 3 + 2 = 5: (a) "3" and "2" are called ____________.

(b) "5" is called the ____________.

---

a) terms

b) sum

88. Here is an addition of two positive numbers:

5 + 3 = 8

Here is the same addition by means of vectors on the number line:

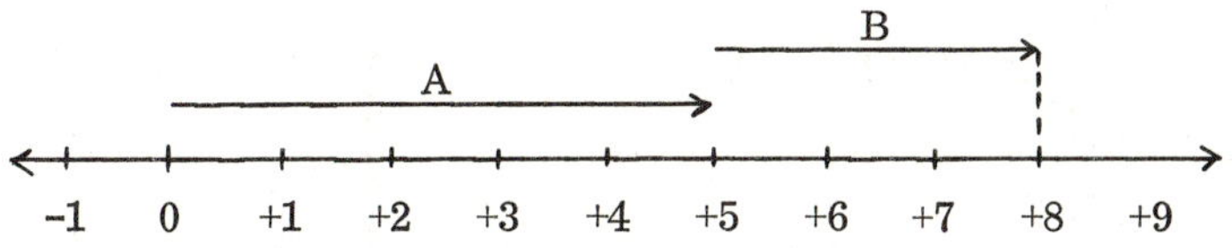

The procedure is:

(1) Vector A represents "+5", the first term. It begins at the origin and it is a 5-unit move to the right. (The first vector always begins at the origin.)

(2) Vector B represents "+3", the second term. It begins at the tip of vector A, and it is a 3-unit move to the right from that point.

(3) If you read down from the tip of vector B (see the dotted line), you find the sum, "+8". (Of course, this answer is not much of a surprise!)

---

89. Do the following addition by drawing vectors on the number line below:

2 + 4 = _____

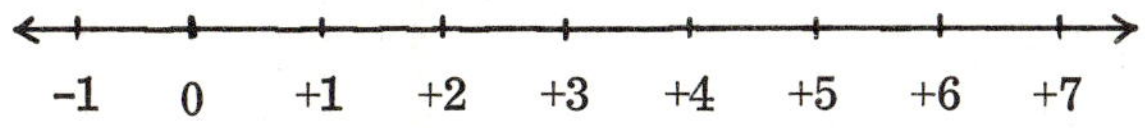

Answer to Frame 89:

The sum is "+6" or "6".
Your number line should look like this:

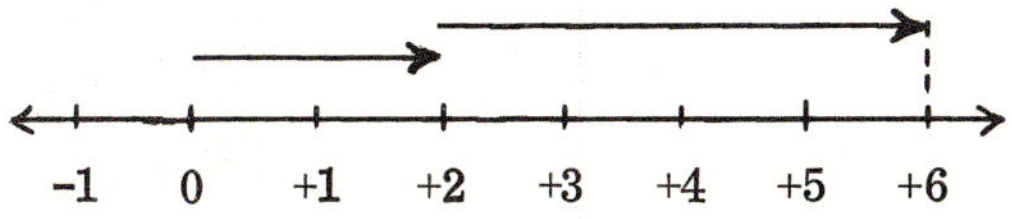

90. Here is a diagram of an addition of any two positive numbers:

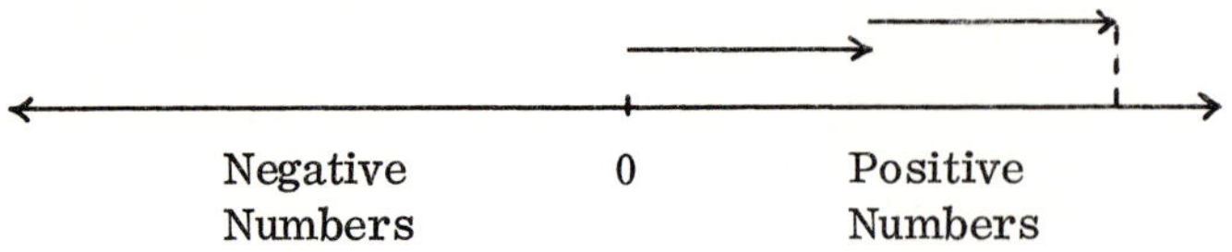

Since both terms are positive, both vectors are to the right, with the first one starting at the origin. Therefore, the "sign" of the sum must be ________________ (positive/negative).

---

91. Here are the additions we have done on the number line:

(+5) + (+3) = +8
(+2) + (+4) = +6

In each case, the absolute value of the sum is obtained by adding the absolute values of the two ____________.

Answer: positive (or "+")

---

92. When adding two positive numbers:

(1) The sign of the sum is always positive.
(2) The absolute value of the sum is obtained by adding the absolute values of the two terms.

Problem: (+10) + (+5) = ?

(a) Since both terms are positive, the sign of the sum must be ________________.

(b) The absolute value of the sum must be ______.

(c) The sum is ______.

Answer: terms

---

93. Here is an addition of two negative numbers. (Notice that the negative numbers being added are put in parentheses. It is not necessary that the sum "-5" be put in parentheses.)

(-3) + (-2) = -5

Here is the same addition by means of vectors on the number line:

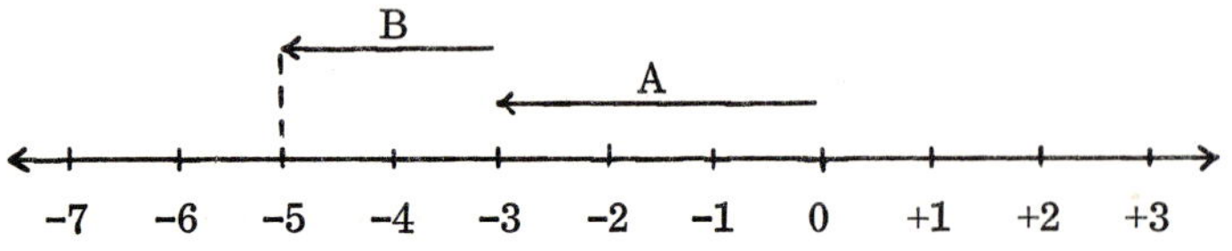

The procedure is:

(1) Vector A represents "-3", the first term. As usual, this first vector begins at the origin. It is a 3-unit move to the left since "-3" is negative.
(2) Vector B represents "-2", the second term. It begins at the tip of vector A, and it is a 2-unit move to the left from that point since "-2" is negative.
(3) If you read down from the tip of vector B (see the dotted line), you find the sum, "-5".

Answers:
a) positive (or "+")
b) 15
c) +15

94. Do this one on the number line below:

(−4) + (−3) = __________

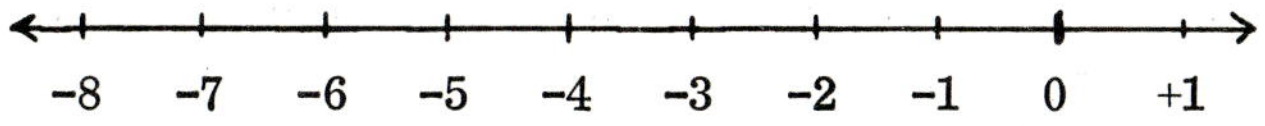

Answer to Frame 94:

The sum is "−7".
Your number line should look like this:

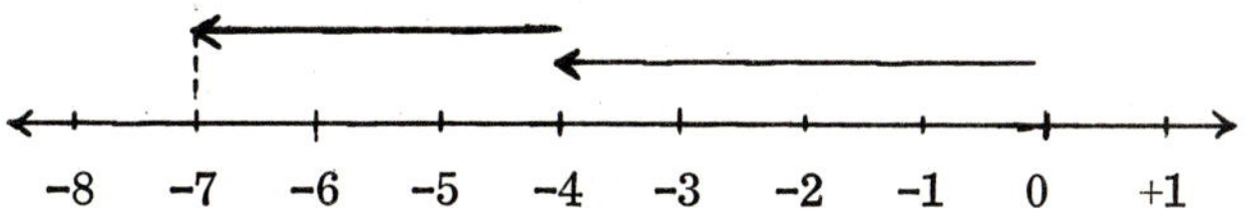

95. Do this one on the number line below:

(−1) + (−2) = __________

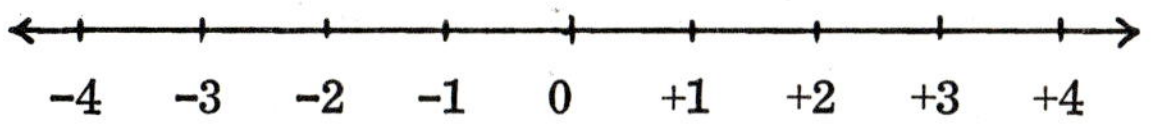

Answer to Frame 95:

The sum is "−3".
Your number line should look like this:

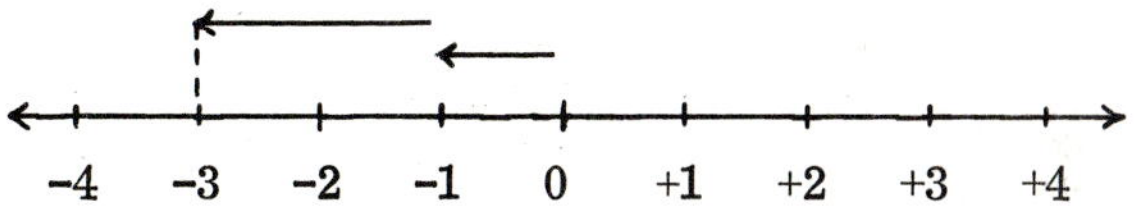

96. Here is a diagram of an addition of two negative numbers:

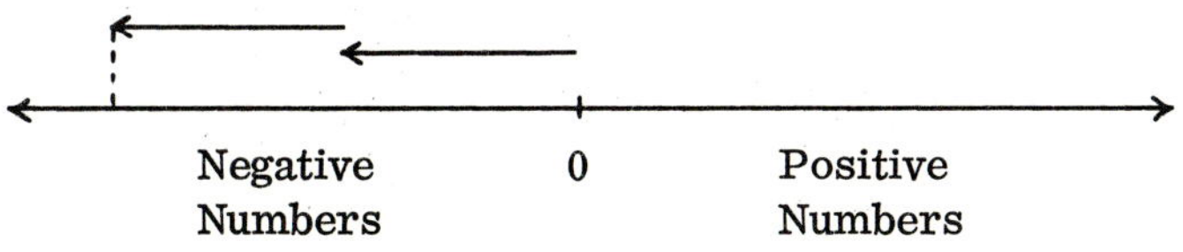

Since both terms are negative, both vectors are to the left, with the first one starting at the origin. Therefore, the "sign" of the sum must be ____________________ (positive/negative).

negative (or "−")

97. The three additions we have done on the number line are:

$$(-3) + (-2) = -5$$
$$(-4) + (-3) = -7$$
$$(-1) + (-2) = -3$$

In each case, the absolute value of the sum is obtained by adding the absolute values of the two __________.

98. When adding two negative numbers:

(1) The sign of the sum is always negative.
(2) The absolute value of the sum is obtained by adding the absolute values of the two terms.

Problem: (-7) + (-5) = ?

(a) The sign of the sum must be ________.

(b) The absolute value of the sum must be ________.

(c) The sum is ________.

Answer: terms

---

99. Do these: (a) (+7) + (+2) = ____ (c) (-1) + (-8) = ____

(b) (-7) + (-2) = ____ (d) (+4) + (+6) = ____

Answer:
a) negative (or "-")
b) 12
c) -12

Answer:
a) +9 c) -9
b) -9 d) +10

1-7 ADDITION OF ONE POSITIVE AND ONE NEGATIVE NUMBER

First we will show the addition of one positive and one negative number by means of vectors on the number line. Then we will use these examples to develop a "rule" which is more efficient.

---

100. In this problem, one term is positive and one term is negative.

5 + (-2) = ?

The vectors are drawn on the number line below:

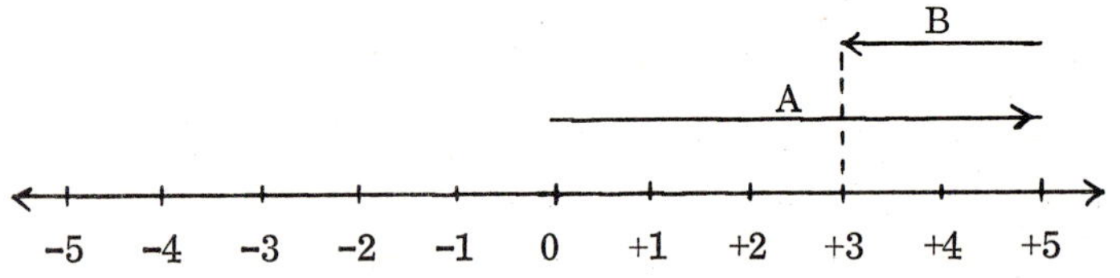

(a) Since "5" is positive, vector A begins at the origin and is a 5-unit move to the ________ (right/left).

(b) Since "-2" is negative, vector B begins at the tip of vector A and is a 2-unit move to the ________ (right/left).

(c) Reading down from the tip of vector B, the sum is ______.

Answer:
a) right
b) left
c) +3

101. In this problem, the first term is negative.

$$(-5) + 2 = ?$$

The vectors are drawn on the number line below:

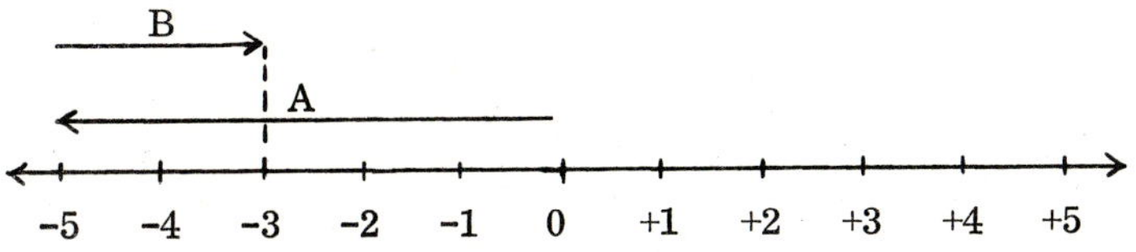

(a) Since "-5" is negative, vector A begins at the origin and is a 5-unit move to the ____________ (right/left).

(b) Since "2" is positive, vector B begins at the tip of vector A and is a 2-unit move to the ____________ (right/left).

(c) Reading down from the tip of vector B, the sum is _______.

---

102. Here are two more problems. A number line is provided for each. Do the additions on the number lines.

(a) $(-4) + 5 = ?$ The sum is _______.

(b) $1 + (-4) = ?$ The sum is _______.

(a)

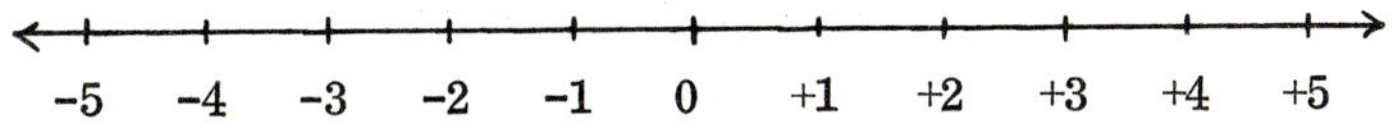

(b)

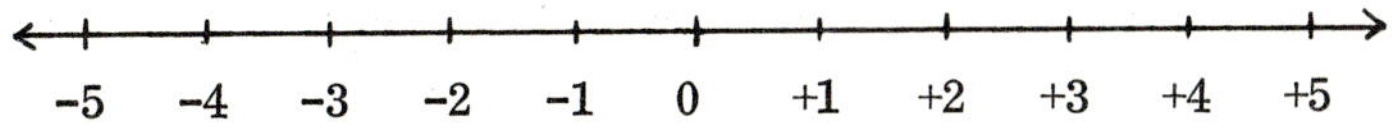

a) left

b) right

c) -3

Here are the correct vector diagrams for Frame 102:

(a) The sum is "+1".

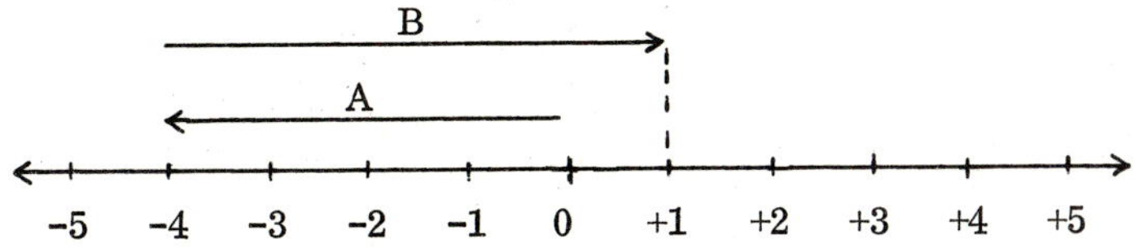

(b) The sum is "-3".

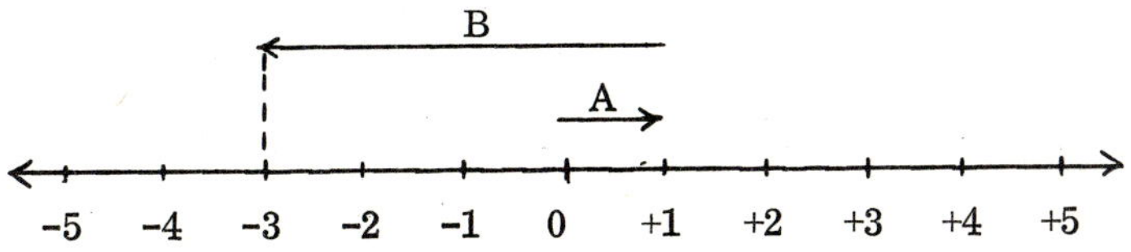

Answers to Frame 102 are shown at the left.

103. Here are the four problems we have done on the number line: $5 + (-2) = +3$ $(-5) + 2 = -3$ $(-4) + 5 = +1$ $1 + (-4) = -3$ In each case, the sign of the sum is the same as the sign of the term with the ______________ (larger/smaller) absolute value.	
104. $(-15) + 8$ For the addition above, the sum would be negative because the sign of the term with the larger absolute value is ______________.	larger
105. $27 + (-21)$ For the addition above, the sum would be positive because the sign of the term with the larger absolute value is ______________.	negative
106. For each of the following, state whether the sign of the sum would be "+" or "-". (a) $(-79) + (+42)$ _____ (c) $(+65) + (-59)$ _____ (b) $(-15) + 27$ _____ (d) $33 + (-37)$ _____	positive
107. Here are the same four problems: $5 + (-2) = +3$ $(-5) + 2 = -3$ $(-4) + 5 = +1$ $1 + (-4) = -3$ In each case, to find the absolute value of the sum, we subtract the smaller absolute value from the larger absolute value (of the two terms). $7 + (-9)$ (a) Which term has the larger absolute value? __________ (b) The absolute value of the sum must be __________.	a) - c) + b) + d) -
108. If one term is positive and one negative: (1) The sign of the sum is the same as the sign of the term with the larger absolute value. (2) The absolute value of the sum is obtained by subtracting the smaller absolute value from the larger absolute value (of the two terms). (Continued on following page.)	a) -9 b) 2, since $9 - 7 = 2$

108. (Continued) $4 + (-9) = ?$

(a) Since "-9" has the larger absolute value, the sign of the sum must be ___________.

(b) Subtracting the smaller absolute value from the larger, the absolute value of the sum must be _____.

(c) The sum is _____.

---

109. $8 + (-5) = ?$

(a) Since "8" has the larger absolute value, the sign of the sum must be ___________.

(b) The absolute value of the sum must be _____.

(c) The sum is _____.

Answers to 108:
a) negative (or "-")
b) 5, since $9 - 4 = 5$
c) -5

---

110. Complete these:

(a) $10 + (-7) =$ _____
(b) $(-6) + 12 =$ _____
(c) $7 + (-11) =$ _____
(d) $(-12) + 5 =$ _____

Answers to 109:
a) positive (or "+")
b) 3, since $8 - 5 = 3$
c) +3

---

111. Complete these:

(a) $8 + (-12) =$ _____
(b) $(-5) + 11 =$ _____
(c) $(-6) + (-5) =$ _____
(d) $3 + (-9) =$ _____

Answers to 110:
a) +3 c) -4
b) +6 d) -7

---

112. In the box below, we have listed the rules for adding two signed numbers. Memorize them.

(1) If both terms are positive:
  (a) The sign of the sum is positive.
  (b) The absolute value of the sum is obtained by adding the absolute values of the two terms.

(2) If both terms are negative:
  (a) The sign of the sum is negative.
  (b) The absolute value of the sum is obtained by adding the absolute values of the two terms.

(3) If one term is positive and one term negative:
  (a) The sign of the sum is the same as the sign of the term with the larger absolute value.
  (b) The absolute value of the sum is obtained by subtracting the smaller absolute value from the larger absolute value (of the two terms).

Answers to 111:
a) -4 c) -11
b) +6 d) -6

113. Do the following:

(1) 3 + (-9) = _______	(6) (-5) + (-2) = _______
(2) (-6) + (-3) = _______	(7) (-62) + 12 = _______
(3) (-2) + 4 = _______	(8) 12 + 30 = _______
(4) 8 + (-2) = _______	(9) 44 + (-24) = _______
(5) (-9) + 2 = _______	(10) (-23) + (-17) = _______

---

114. We want you to be able to do addition problems quickly and without errors. In this frame and Frame 115, we have given more practice problems. If you feel that you need more practice, do them. If you feel that you do not need more practice, go on to Frame 116.

(1) (-7) + (-2) = _______	(6) (-3) + (-2) = _______
(2) 2 + (-5) = _______	(7) 13 + (-33) = _______
(3) (-9) + 10 = _______	(8) (-12) + (-13) = _______
(4) 6 + (-5) = _______	(9) 24 + 20 = _______
(5) (-8) + 5 = _______	(10) (-19) + 49 = _______

1) -6	6) -7
2) -9	7) -50
3) 2	8) 42
4) 6	9) 20
5) -7	10) -40

115.

(1) (-3) + (-4) = _______	(6) (-6) + 5 = _______
(2) 4 + (-6) = _______	(7) 50 + 17 = _______
(3) 3 + (-1) = _______	(8) (-65) + (-15) = _______
(4) (-8) + 4 = _______	(9) (-42) + 12 = _______
(5) (-4) + (-5) = _______	(10) 60 + (-15) = _______

1) -9	6) -5
2) -3	7) -20
3) 1	8) -25
4) 1	9) 44
5) -3	10) 30

---

1) -7	6) -1
2) -2	7) 67
3) 2	8) -80
4) -4	9) -30
5) -9	10) 45

1-8 ADDITION OF THREE OR MORE TERMS

116. Only two signed numbers can be added at one time. To add three or more signed numbers, you can begin at the left and add "two at a time."

Here is an example:

Step 1: 3 + (-5) + 7

Step 2: (-2) + 7

Step 3: +5

In going from Step 1 to Step 2, we added "3" and "-5".
In going from Step 2 to Step 3, we added "-2" and "7".

Did we add more than two numbers at any one time? _________

No

117. Here is another example:

Step 1: (-9) + 3 + (-4) + 12

Step 2: (-6) + (-4) + 12

Step 3: (-10) + 12

Step 4: +2

In going from Step 1 to Step 2, we added "-9" and "3".
In going from Step 2 to Step 3, we added "-6" and "-4".
In going from Step 3 to Step 4, we added "-10" and "12".

Did we add more than two numbers at any one time? __________

No

118. Do each of the following:

(1) 3 + (-4) + 5 = _____

(2) 3 + 2 + 4 = _____

(3) (-9) + (-2) + 6 = _____

(4) (-7) + 5 + (-8) = _____

(5) (-6) + (-5) + (-2) = _____

(6) (-4) + (-3) + 5 + 2 = _____

(7) 7 + (-1) + 4 + (-6) = _____

(8) 1 + (-8) + 3 + (-8) + 4 = _____

(9) (-7) + 6 + (-7) + 2 + (-4) = _____

(10) (-9) + 2 + (-7) + 4 + (-6) = _____

---

1)	4	6)	0
2)	9	7)	4
3)	-5	8)	-8
4)	-10	9)	-10
5)	-13	10)	-16

119. If you feel that you need more practice, here are more problems. Otherwise, go on to the next frame.

(1) 6 + (-7) + 3 = _____

(2) (-4) + (-8) + 1 = _____

(3) 5 + 2 + 6 = _____

(4) (-1) + (-8) + (-3) = _____

(5) (-9) + 4 + (-2) = _____

(6) 2 + (-3) + 9 + (-5) = _____

(7) (-8) + 5 + (-3) + 6 = _____

(8) 8 + (-6) + 3 + (-7) + 5 = _____

(9) (-4) + 9 + (-7) + 6 + (-2) = _____

(10) (-9) + 8 + (-3) + 6 + (-7) = _____

---

1)	2	6)	3
2)	-11	7)	0
3)	13	8)	3
4)	-12	9)	2
5)	-7	10)	-5

1-9 THE COMMUTATIVE PROPERTY OF ADDITION

The point of this section is to show this fact: The order in which the terms of an addition are written makes no difference.

---

120. $(-7) + (-3) = -10$

If we interchange the two terms, we get:

$(-3) + (-7)$

(a) $(-3) + (-7) =$ ______

(b) Do we get the same sum when the terms are interchanged? ______

---

121. $7 + (-3) = +4$

Interchanging the two terms, we get:

$(-3) + 7$

(a) $(-3) + 7 =$ ______

(b) Do we get the same sum when the terms are interchanged? ______

Answers to 120:
a) -10
b) Yes

---

122. (a) $(-6) + 5 =$ ______

(b) $5 + (-6) =$ ______

(c) Does $(-6) + 5 = 5 + (-6)$? ______

Answers to 121:
a) +4
b) Yes

---

123. The word "commutative" means "to change around" or "interchange." In an addition, the sum is not changed even if the terms are interchanged. Therefore, mathematicians say that addition is "commutative." By the technical term "COMMUTATIVE PROPERTY OF ADDITION," they mean that the sum is not changed when the terms are interchanged.

(a) What does the word "commutative" mean?

______

(b) What does the technical term "commutative property of addition" mean? ______

______

Answers to 122:
a) -1
b) -1
c) Yes, since both equal "-1".

---

124. We know that: $1492 + 1973 = 1973 + 1492$

because of the ______ property of addition.

Answers to 123:
a) To change around or interchange.
b) The sum of an addition is not changed when the terms are interchanged.

125. The commutative property of addition is true even when more than two terms are involved.

$$(-6) + 3 + (-1) = -4$$

Let's interchange the three terms on the left in various ways and compute the sum to show that the commutative property still holds.

(a) $3 + (-1) + (-6) =$ _______

(b) $(-1) + (-6) + 3 =$ _______

(c) Do we get the same sum when the same terms are written in a different order? _______

Answer: commutative

(Watch the spelling!)

---

126. Since the order in which the terms are written makes no difference, there is a shortcut you can use to add three or more terms more quickly.

Here is such a problem:

$$(-2) + 5 + (-6) + 4 = ?$$

The shortcut involves writing all the negative numbers together and all the positive numbers together. We get:

$$\underbrace{(-2) + (-6)}_{(-8)} + \underbrace{5 + 4}_{9} = \text{\_\_\_\_\_\_\_}$$

Answers: a) -4 b) -4 c) Yes

---

127. Let's do another one:

$$8 + (-9) + (-3) + 7 + 5 + (-1) = ?$$

Grouping the positive and negative numbers, we get:

$$\underbrace{8 + 7 + 5}_{20} + \underbrace{(-9) + (-3) + (-1)}_{(-13)} = \text{\_\_\_\_\_\_\_}$$

Answer: +1

---

128. Of course, you can add the positive and negative numbers in your head without rewriting the order of the terms.

$$(-7) + 11 + 6 + (-10) + (-5) + 1 = ?$$

Do the following in your head:

(a) The sum of the positive numbers is _______.

(b) The sum of the negative numbers is _______.

(c) The total sum is _______.

Answer: +7

Answers: a) 18 b) -22 c) $18 + (-22) = -4$

129. Use the shortcut to improve your speed with these:

(1) $4 + (-2) + 5 =$ _______

(2) $1 + 5 + 3 =$ _______

(3) $(-6) + (-7) + 2 =$ _______

(4) $(-8) + 7 + (-9) =$ _______

(5) $(-3) + (-6) + (-5) =$ _______

(6) $(-2) + 5 + (-6) + 4 =$ _______

(7) $9 + (-8) + 7 + (-2) =$ _______

(8) $(-4) + 1 + (-6) + 3 + (-5) =$ _______

(9) $6 + (-2) + 8 + (-4) + 7 =$ _______

(10) $(-5) + (-4) + 8 + 2 + (-3) + 6 =$ _______

---

130. Here is another set of problems if you want more practice. Otherwise, go on to the next frame.

(1) $3 + (-4) + 2 =$ _______

(2) $(-8) + (-4) + 1 =$ _______

(3) $5 + (-6) + 2 =$ _______

(4) $(-4) + (-6) + 5 =$ _______

(5) $(-7) + 3 + (-9) =$ _______

(6) $2 + (-5) + 4 + (-6) =$ _______

(7) $(-8) + 1 + 6 + (-5) =$ _______

(8) $8 + (-4) + 3 + (-6) + 7 =$ _______

(9) $(-7) + 8 + (-6) + 3 + (-6) =$ _______

(10) $(-5) + 6 + 4 + (-3) + 8 + (-9) =$ _______

1) 7	6) 1
2) 9	7) 6
3) -11	8) -11
4) -10	9) 15
5) -14	10) 4

---

1) 1	6) -5
2) -11	7) -6
3) 1	8) 8
4) -5	9) -8
5) -13	10) 1

1-10 THE CONCEPT OF "OPPOSITES"

131. Add: (a) $3 + (-3) =$ _______ (c) $114 + (-114) =$ _______

(b) $(-7) + 7 =$ _______ (d) $(-1974) + 1974 =$ _______

The sum is "0" for all.

132. Put the number in the box which makes each statement true:

(a) 9 + ☐ = 0 (c) 5 + ☐ = 0

(b) (-11) + ☐ = 0 (d) (-2) + ☐ = 0

---

a) -9 c) -5

b) 11 d) 2

133. Definition: IF THE SUM OF TWO NUMBERS IS "0", THEY ARE CALLED A "PAIR OF OPPOSITES."

Since 7 + (-7) = 0,

(a) "7" and "-7" are called a pair of ________.

(b) The opposite of "7" is ________.

(c) The opposite of "-7" is ________.

---

a) opposites

b) -7

c) +7 or 7

134. (a) What number would you add to "-11" to get "0" as a sum? ________

(b) The opposite of "-11" is ________.

(c) The opposite of "11" is ________.

---

a) +11

b) +11

c) -11

135. Name the opposite of each of the following:

(a) -14 ________ (c) -1 ________

(b) 21 ________ (d) 1 ________

---

a) 14 c) 1

b) -21 d) -1

136. If two numbers are a pair of opposites:

(a) Are their signs the same or different? ________

(b) Are their absolute values the same or different? ________

---

a) Different

b) The same

137. (a) What number would you add to "0" to get "0" as a sum? ________

(b) Therefore, the opposite of "0" is ________.

---

a) 0

b) 0

138. One number is its own opposite. What number is it? ________

0

SELF-TEST 5 (Frames 77-138)

Find each sum: 1. (-8) + (-3) = _____ 2. (-12) + (+7) = _____ 3. (+5) + (-4) = _____

4. Using vectors, show this addition on the number line below: (+2) + (-5) = _____

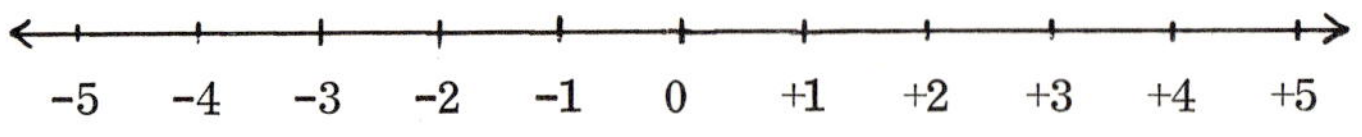

5. The sign of the sum of two negative numbers is _____________.
6. The sign of the sum of two positive numbers is _____________.
7. The sign of the sum of a positive and negative number is the same as the sign of the number having the greatest ______________________.

8. What number is the opposite of +17? _______
9. What number is the opposite of -17? _______
10. The sum of a pair of opposites is always _________.

Find each sum: 11. (+3) + (+7) + (-8) + (-4) = _____ 13. 17 + (-4) + 5 + 2 + (-9) = _____

12. (-5) + 9 + (-6) + 3 = _____ 14. 8 + (-8) + 6 + (-10) + 3 = _____

Complete each of the following using the commutative property of addition:

15. (+7) + (-4) = (-4) + (?) _____ 17. (-8) + 8 = (?) + (-8) _____

16. (-5) + (-9) = (-9) + (?) _____ 18. (-2) + (+3) + (-7) = (+3) + (-7) + (?) _____

ANSWERS:

1. -11	4. See below	7. absolute value	10. zero (or 0)	13. +11	16. -5
2. -5	5. negative (or -)	8. -17	11. -2	14. -1	17. +8
3. +1	6. positive (or +)	9. +17	12. +1	15. +7	18. -2

4. (+2) + (-5) = -3

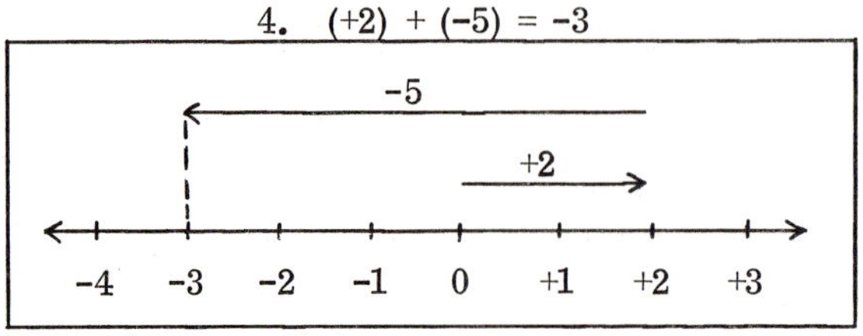

SUBTRACTION OF SIGNED NUMBERS

"Subtraction of signed numbers" will be taught in the next few sections. Of all the basic operations in algebra, subtraction is the one which causes the most difficulty. Be alert in this section. If you do not learn how to handle the subtraction operation, you will not be successful in further work in algebra. Our method for working any subtraction problem will involve two steps: (1) converting the subtraction problem to an equivalent addition problem and then (2) performing the addition.

1-11 TWO DIFFERENT USES OF THE "-" SYMBOL

139. Just as mathematicians use the "+" symbol in two different ways, they also use the "-" symbol in two different ways.

In $(\overset{\downarrow}{-}5) - (\overset{\downarrow}{-}7)$,

(1) The two "-" symbols with the arrows over them are signs of numbers.
(2) The "-" symbol in the middle is the symbol for subtraction.

In $7 \overset{\downarrow}{-} (-6)$, the "-" with the arrow over it is the symbol for __________.

---

140. In $10 - (\overset{\downarrow}{-}7)$, is the "-" with the arrow over it a symbol for subtraction? __________

Answer to 139: subtraction

---

141. Just as 5 + 3 is shorthand for (+5) + (+3),
5 - 3 is shorthand for (+5) - (+3).

In 5 - 3, the "-" is the symbol for __________.

Answer to 140: No. It is the sign of a number.

---

142. Write each of the following so that the signs of the positive numbers are explicit:

(a) 5 + 9 __________ (b) 5 - 9 __________

Answer to 141: subtraction

---

143. The point we are trying to make is this:

In 7 - 11,

(1) The "-" is the symbol for subtraction.
(2) The sign of "11" is "+". That is, "11" is a positive number.

In 6 - 10,

(a) The "-" is the symbol for __________.
(b) Is "10" a positive or negative number? __________

Answer to 142:
a) (+5) + (+9)
b) (+5) - (+9)

144. Similarly, in (-5) $\overset{\downarrow}{-}$ 7: (1) The "-" with the arrow over it is the symbol for subtraction. (2) "7" is a positive number. Though not written, its sign is "+". In (-10) $\overset{\downarrow}{-}$ 15, (a) The "-" with the arrow over it is the symbol for ________. (b) Is "15" a positive or negative number? ________	a) subtraction b) positive
145. Write each of the following so that the signs of the positive numbers are explicit: (a) 7 - 11 ________ (b) (-6) - 10 ________	a) subtraction b) positive
146. In both 7 - 10, and (-7) - 10, "10" is a ________ (positive/negative) number.	a) (+7) - (+11) b) (-6) - (+10)
147. Here are two expressions: (1) 3 - (-5) (2) 3 - 5 In (1), the number after the subtraction symbol is negative. In (2), the number after the subtraction symbol is positive. [After a subtraction symbol, the number is negative only if the negative sign is explicitly written.] In which of the following two cases does a negative number follow the subtraction sign? ________ (a) (-6) - 8 (b) (-6) - (-8)	positive
148. In 3 - 7, how do you know that "7" is a positive number? ________	Only in (b)
149. If you want "8" to be a negative number, do you write: (a) 5 - 8 ? or (b) 5 - (-8) ? ________	Because a negative sign is not explicitly written after the subtraction sign.
150. Does 7 - 11 mean: (a) 7 - (+11) ? or (b) 7 - (-11) ? ________	(b) 5 - (-8)
	(a) 7 - (+11)

151. Does (-5) - 9 mean:

(a) (-5) - (-9) ? or (b) (-5) - (+9) ? ____________

(b) (-5) - (+9)

152. Name the two different uses of the "-" symbol:

(1) ____________________________

(2) ____________________________

---

1) The <u>sign</u> <u>of</u> <u>negative</u> <u>numbers</u>

2) The symbol for <u>sub</u>-<u>traction</u>

153. Just as in addition, the two numbers involved in a subtraction are called "<u>terms</u>."

In (-6) - 9, the two terms are _______ and _______.

-6 and +9 (or 9)

154. We never write two "-" symbols in a row. When the second term is <u>negative</u>, we <u>always</u> put it in parentheses.

Instead of: 7 - -9,
we write: 7 - (-9).

Instead of: (-6) - -10,
we write: ____________

(-6) - (-10)

1-12 CONVERTING SUBTRACTION TO ADDITION

Any subtraction problem can be converted to an equivalent addition problem. In this section and the next, we will show how this conversion is made and used. Then, in a later section, we will show the sensibleness of this conversion by examining subtractions on the number line.

---

155. The conversion from subtraction to addition is based on the concept of <u>opposites</u>.

(a) Two numbers are a pair of opposites if their sum is _______.

(b) The opposite of "+9" is _______.

(c) The opposite of "-5" is _______.

---

a) 0

b) -9

c) +5

156. The following subtraction and addition are equivalent because both equal "+5".

7 - (+2) = 5
7 + (-2) = 5

Since they are equivalent, we can write this sentence:

7 - (+2) = 7 + (-2)

(Continued on following page.)

156. (Continued)

Let's examine the second term in each: $7 - (+2) = 7 + (-2)$

(a) The second term of the subtraction is ______.

(b) The second term of the addition is ______.

(c) These two second terms are a pair of ____________.

---

157. As the example in the last frame suggests:

Any "subtraction" can be converted to an equivalent "addition" by means of the following definition:

> The subtraction symbol "−" means:
>
> ADD THE OPPOSITE OF THE FOLLOWING TERM

That is: $11 - (+7) = 11 +$ (the opposite of "+7")
$= 11 + (-7)$

$7 - (+3) = 7 +$ (the opposite of "+3")
$= 7 + (\quad)$

Answers to 156: a) +2 from 7 − (+2) b) −2 c) opposites

---

158. $4 - (-3) = 4 +$ (the opposite of "−3")
$= 4 + (\quad)$

Answer to 157: 7 + (−3)

---

159. $(-6) - (+7) = (-6) +$ (the opposite of "+7")
$= (-6) + (\quad)$

Answer to 158: 4 + (+3)

---

160. $(-4) - (-10) = (-4) +$ (the opposite of "−10")
$= (-4) + (\quad)$

Answer to 159: (−6) + (−7)

---

161. The subtraction symbol means:

"Add the ____________ of the following number."

Answer to 160: (−4) + (+10)

---

162. (a) $6 - (+3) = 6 +$ (the ____________ of "+3")

(b) $(-4) - (-2) = (-4) +$ (the ____________ of "−2")

Answer to 161: opposite

---

163. When the number after the subtraction symbol is positive, its sign is ordinarily not written explicitly.

Example: $2 - 7$

In all such cases you must remember that the number after the subtraction symbol is positive. Then converting to "addition" is easy.

$2 - 7 = 2 - (+7) = 2 + (-7)$

(Continued on following page.)

Answers to 162: a) opposite b) opposite

163. (Continued)

Convert each of the following to addition:

(a) 7 - 1 = 7 - (+1) = ________

(b) (-4) - 5 = (-4) - (+5) = ________

(c) 2 - 14 = ________

(d) (-15) - 23 = ________

Answers: a) 7 + (-1) b) (-4) + (-5) c) 2 + (-14) d) (-15) + (-23)

164. Complete the following conversions to "addition."

(a) 4 - 2 = 4 + ☐

(b) 4 - (-3) = 4 + ☐

(c) (-3) - 1 = (-3) + ☐

(d) (-2) - (-4) = (-2) + ☐

Answers: a) -2 b) +3 c) -1 d) +4

165. Complete the following conversions to "addition."

(a) 3 - 17 = 3 + ☐

(b) (-1) - 16 = (-1) + ☐

(c) 27 - 36 = 27 + ☐

(d) (-100) - 75 = (-100) + ☐

Answers: a) -17 b) -16 c) -36 d) -75

1-13 SUBTRACTION BY MEANS OF ADDITION

166. Whereas the addition operation presents little difficulty, the subtraction operation is one of the major sources of errors in algebra. To avoid these errors, we use the following procedure with subtractions:

> (1) CONVERT TO THE EQUIVALENT "ADDITION."
> (2) THEN COMPUTE THE SUM OF THE ADDITION.

For example: 3 - 5 = 3 + (-5) = -2

(a) (-7) - 9 = (-7) + (-9) = ______

(b) 1 - 6 = 1 + (-6) = ______

Answers: a) -16 b) -5

167. Convert the following to "addition" and compute the sum:

		Addition		Sum
(a)	5 - 3 =	____________	=	______
(b)	(-4) - 9 =	____________	=	______
(c)	(-1) - (-3) =	____________	=	______
(d)	3 - (-7) =	____________	=	______
(e)	2 - 6 =	____________	=	______

---

a) 5 + (-3) = 2

b) (-4) + (-9) = -13

c) (-1) + 3 = 2

d) 3 + 7 = 10

e) 2 + (-6) = -4

168. Convert the following to "addition" and compute the sum:

		Addition		Sum
(a)	3 - 9 =	____________	=	______
(b)	7 - (-7) =	____________	=	______
(c)	(-3) - 4 =	____________	=	______
(d)	(-8) - (-7) =	____________	=	______
(e)	(-9) - (-3) =	____________	=	______

---

a) 3 + (-9) = -6

b) 7 + 7 = 14

c) (-3) + (-4) = -7

d) (-8) + 7 = -1

e) (-9) + 3 = -6

169. <u>In this course, we will convert all subtractions to additions</u>. By doing so, difficulties with "sign" errors will be avoided. Do the following:

(1) 6 - 7 = __________ (5) (-2) - (-4) = __________

(2) (-4) - 3 = __________ (6) 7 - (-8) = __________

(3) 6 - (-2) = __________ (7) (-4) - 7 = __________

(4) 8 - 4 = __________ (8) (-7) - (-3) = __________

---

1) -1 5) 2

2) -7 6) 15

3) 8 7) -11

4) 4 8) -4

170. This frame contains more problems. Do them if you need more practice. Otherwise, go on to the self-test.

(1) 4 - 8 = __________ (5) (-6) - (-5) = __________

(2) 4 - (-5) = __________ (6) (-2) - 5 = __________

(3) 5 - 1 = __________ (7) (-2) - (-8) = __________

(4) (-3) - 2 = __________ (8) 3 - (-9) = __________

---

1) -4 5) -1

2) 9 6) -7

3) 4 7) 6

4) -5 8) 12

SELF-TEST 6 (Frames 139-170)

Complete the following:

1. To subtract a quantity, add its ______________.

2. 5 - (-2) = 5 + (the ______________ of -2)

3. (-3) - 7 = (-3) + (the opposite of _______)

Convert the following subtractions to additions by filling in each box:

4. (-5) - (-3) = (-5) + ☐

5. 7 - (-4) = 7 + ☐

6. (-3) - (+9) = (-3) + ☐

7. 6 - 14 = 6 + ☐

Do the following subtractions:

8. (-7) - (-5) = _______

9. 6 - (-4) = _______

10. 3 - (+8) = _______

11. 3 - 8 = _______

12. (-1) - 6 = _______

13. (-9) - 4 = _______

14. (-7) - 2 = _______

15. (-3) - (-10) = _______

16. 12 - 9 = _______

ANSWERS:

1. opposite	4. +3	6. -9	8. -2	11. -5	14. -9
2. opposite	5. +4	7. -14	9. +10	12. -7	15. +7
3. +7			10. -5	13. -13	16. +3

1-14 SUBTRACTION ON THE NUMBER LINE

In this section, we will reinforce the fact that any subtraction can be converted to an equivalent addition. We will do so by contrasting subtraction and addition on the number line. We want to show that addition and subtraction are opposite operations by definition. By opposite operations, we mean that the addition symbol ("+") and the subtraction symbol ("-") give opposite instructions about the direction of the vector for the second term.

171. The following addition and subtraction have the same terms:

3 + 2 = 3 + (+2) = 5
3 - 2 = 3 - (+2) = 1

By doing each problem on the number line, the difference between addition and subtraction becomes obvious.

(Continued on following page.)

171. ADDITION: 3 + (+2) = 5

(Continued)

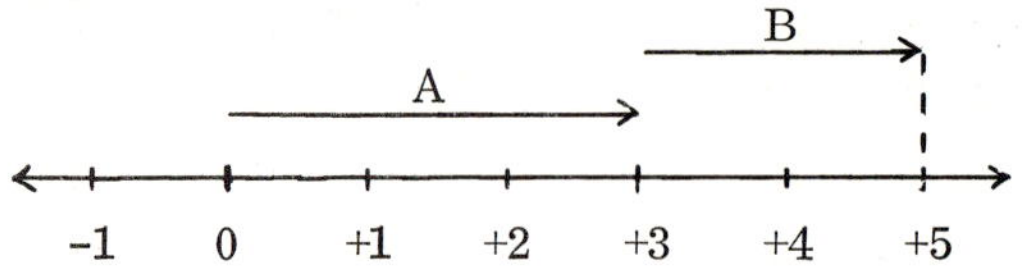

SUBTRACTION: 3 - (+2) = 1

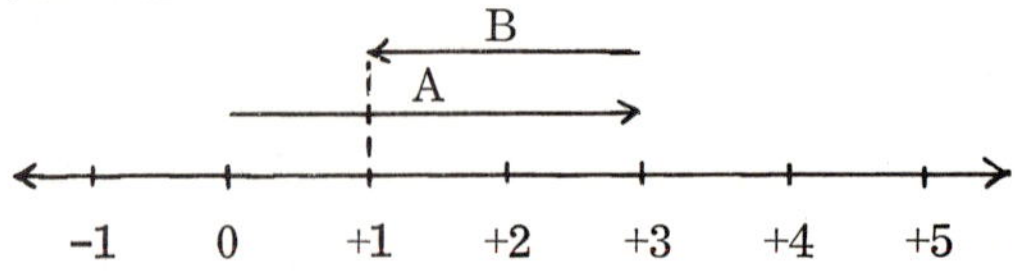

Notice these points:

> Vector A for the first term ("+3"):
>
> In both cases, the first term is handled in the same way. Vector A is a 3-unit move to the right from the origin for both the addition and the subtraction.
>
> Vector B for the second term ("+2"):
>
> In the addition, vector B is a 2-unit move to the right.
> In the subtraction, vector B is a 2-unit move to the left.
> THAT IS, VECTOR B GOES IN THE OPPOSITE DIRECTION FOR SUBTRACTION.

Go to next frame.

172. Let's show one more comparison of an addition and subtraction which have the same terms:

$$5 + (-3) = 2$$
$$5 - (-3) = 8$$

We will do each on a number line.

ADDITION: 5 + (-3) = 2

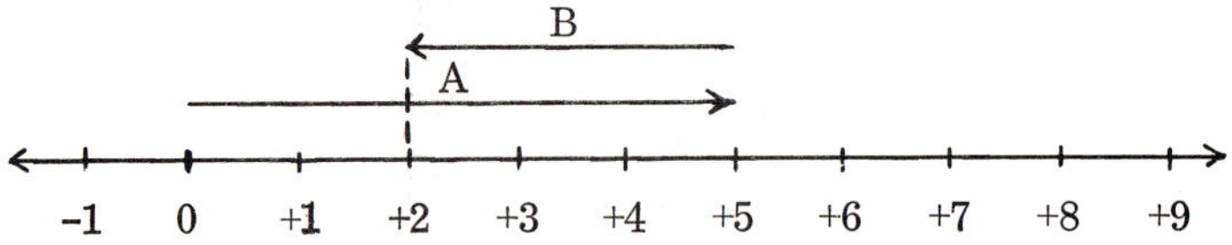

SUBTRACTION: 5 - (-3) = 8

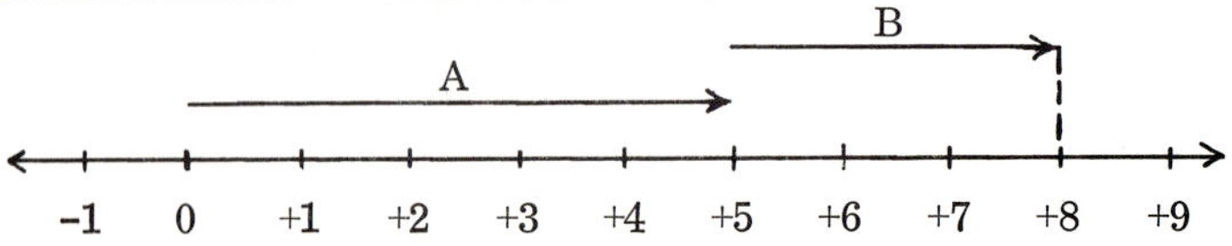

In each case:

(1) Vector A, representing the first term, is a 5-unit move to the right from the origin.

(2) Vector B, representing the second term, is a 3-unit move.

Though both vector B's have the same length, in what way do they differ? ______________________________

173. In terms of vectors on the number line, the addition symbol ("+") and the subtraction symbol ("-") give opposite instructions about the direction of the vector for the second term.

> Where you would go to the right in addition, you go to the left in subtraction.
>
> Where you would go to the left in addition, you go to the right in subtraction.

The following addition and subtraction have the same terms.

(-3) + (+2)
(-3) - (+2)

If the addition were performed on the number line, the vector for the second term ("+2") would go to the right since +2 is positive.

If the subtraction were performed on the number line, the vector for the second term ("+2") would go in the direction opposite to that of addition. Therefore, it would go to the ____________ (right/left).

Answer: Their directions are opposite.

---

174. The subtraction symbol means: "Whatever direction you would use for the vector of the second term in an addition, use the opposite direction for the same term in a subtraction."

(a) If "-5" is the second term in an addition, the vector representing it would go to the ____________ (right/left).

(b) If "-5" is the second term in a subtraction, the vector representing it would go to the ____________ (right/left).

Answer: left

---

175. To show the difference between addition and subtraction on the number line, we used additions and subtractions with the same terms. Here are the two pairs we used:

Pair 1	Pair 2
3 + (+2) = 5	5 + (-3) = 2
3 - (+2) = 1	5 - (-3) = 8

(a) In Pair 1, are the addition and subtraction equivalent? ____________

(b) In Pair 2, are the addition and subtraction equivalent? ____________

Answer:
a) left
b) right

Answer:
a) No, since the answers 5 and 1 are not equal.
b) No, since the answers 2 and 8 are not equal.

176. Whenever an addition and a subtraction have the same terms, they are not equivalent. However, for any subtraction, there is an equivalent addition. In the next few frames, we will use the number line to find additions which are equivalent to various subtractions.

When any subtraction is done on the number line, the vectors also represent an addition. We have done the following subtraction on the number line below:

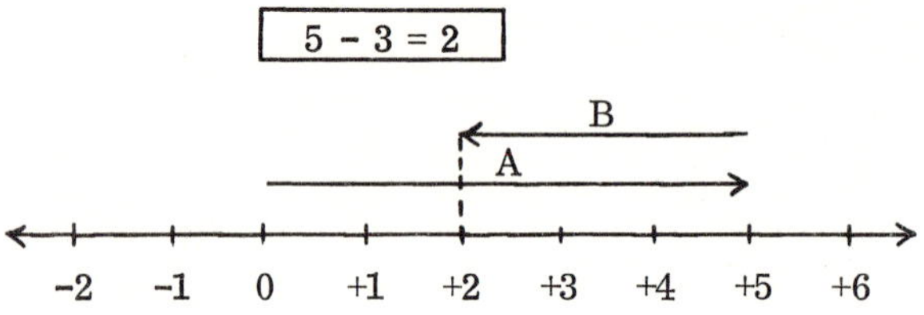

Let's see what equivalent addition the vectors represent.

(a) Vector A is a 5-unit move to the right. Therefore, the first term of the addition must be _______.

(b) Vector B is a 3-unit move to the left. Therefore, the second term of the addition must be _______.

(c) What addition do the vectors represent? _______________

Answers to 176:

a) +5 or 5

b) -3

c) (+5) + (-3) = +2

or

5 + (-3) = 2

---

177. We have done the following subtraction on the number line below:

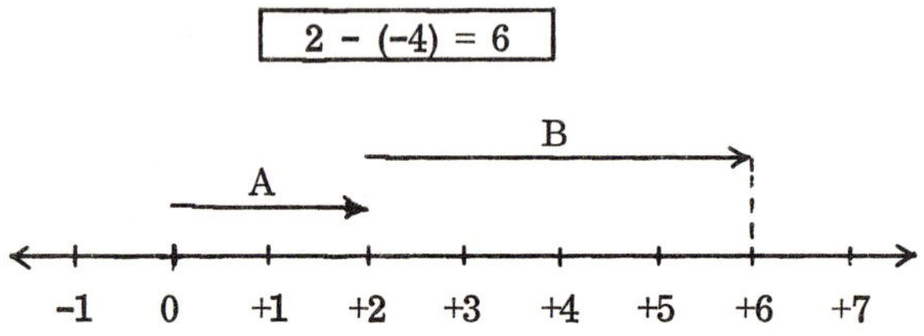

Let's see what equivalent addition the vectors represent.

(a) Vector A shows that the first term of the addition must be _____.

(b) Vector B shows that the second term of the addition must be _____.

(c) What addition do the vectors represent? _______________

Answers to 177:

a) +2 or 2

b) +4 or 4

c) (+2) + (+4) = +6

or

2 + 4 = 6

178. We have done the following subtraction on the number line below:

(-1) - 4 = -5

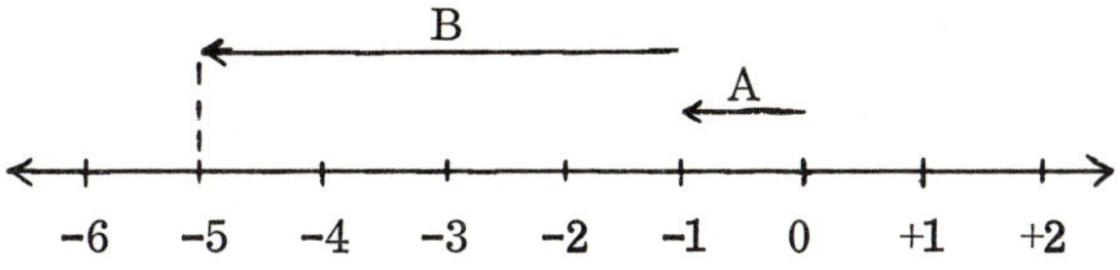

Let's see what equivalent addition these vectors represent:

(a) The first term of the addition must be _______.

(b) The second term of the addition must be _______.

(c) The addition represented is __________________.

---

a) -1 (see vector A)

b) -4 (see vector B)

c) (-1) + (-4) = -5

---

179. Here is another subtraction. The number line solution is shown below:

(-5) - (-4) = -1

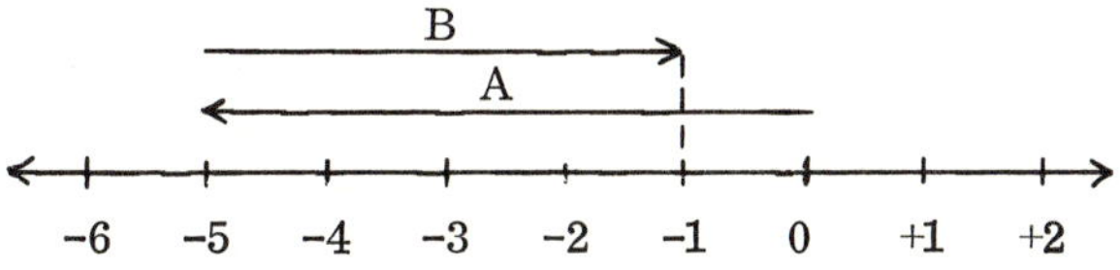

Let's see what equivalent addition these vectors represent:

(a) The first term of the addition must be _______.

(b) The second term of the addition must be _______.

(c) The addition is __________________.

---

a) -5

b) +4

c) (-5) + (+4) = -1

or

(-5) + 4 = -1

---

180. Here are the four sets of equivalent problems we saw on the number lines in the last four frames. In each case, we have made the sign of the second term explicit.

Set 1	Set 3
5 - (+3) = 2 5 + (-3) = 2	(-1) - (+4) = -5 (-1) + (-4) = -5
Set 2	**Set 4**
2 - (-4) = 6 2 + (+4) = 6	(-5) - (-4) = -1 (-5) + (+4) = -1

In each set, the first terms are identical.

In each set, the second terms are a pair of __________________.

---

opposites

181. In practice, we do not use the number line to perform additions and subtractions. However, using the number line does help to see the relationship between these two operations. From the preceding frames, it should be clear that:

> (1) For any subtraction, there is an equivalent addition.
>
> (2) We can convert the subtraction to addition by:
>
> ADDING THE OPPOSITE OF THE SECOND TERM

> IN THIS COURSE, WE WILL INSIST THAT YOU CONVERT EVERY SUBTRACTION TO AN ADDITION BECAUSE ADDITION IS AN EASIER OPERATION.

Go to next frame.

---

182. Convert each subtraction to an addition, and then write the final answer:

(a) 5 - 13 = 5 + [ ] = _______

(b) 4 - (-3) = 4 + [ ] = _______

(c) (-3) - 7 = (-3) + [ ] = _______

(d) (-6) - (-9) = (-6) + [ ] = _______

a) 5 + [-13] = -8 b) 4 + [+3] = +7 c) (-3) + [-7] = -10 d) (-6) + [+9] = +3

SELF-TEST 7 (Frames 171-182)

1. The vector solution of the addition [(-1) + (-3) = -4] is shown below. On the same diagram, draw the vector solution of the subtraction [(-1) - (-3) = +2].

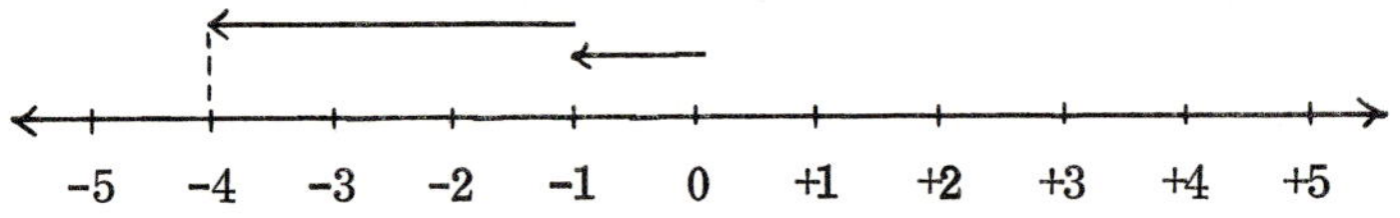

2. The vector solution of the following subtraction is shown below: [6 - 4 = 2]

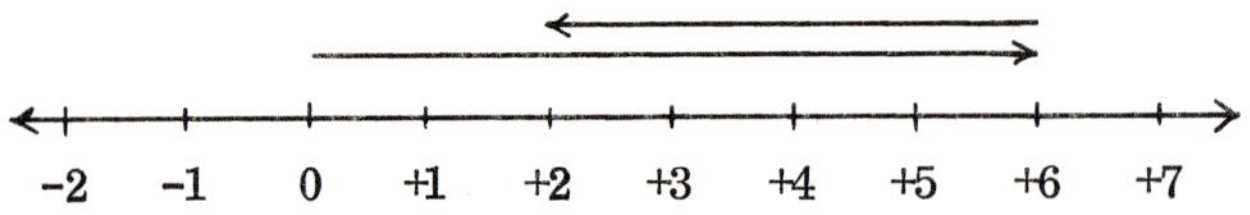

Write the equivalent addition which these vectors represent. (That is, convert the subtraction to an addition.) _______________

Without referring to the number line, do these problems:

3. 18 - (-12) = _____ 4. (-9) - 2 = _____ 5. 14 - 23 = _____ 6. (-7) - (-10) = _____

ANSWERS:

1.

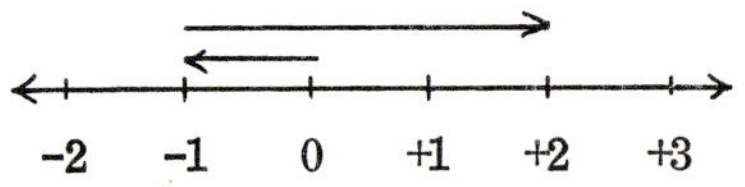

2. 6 + (-4) = 2

3. 30

4. -11

5. -9

6. 3

1-15 COMBINED ADDITION AND SUBTRACTION

183. Here is a problem which contains both an addition and a subtraction:

$$7 + 8 - (-6)$$

To find the answer, we convert the subtraction to addition first. Do so and find the sum: ______________

---

184. In this case, there are two subtractions: $9 - 7 - (-3)$

(a) Convert both subtractions to additions: ______________

(b) The sum is ______.

Answer to 183: $7 + 8 + (+6) = +21$

---

185. In this case, there are two subtractions and an addition: $14 - 9 + (-7) - (-3)$

(a) Convert both subtractions to additions: ______________

(b) The sum is ______.

Answers to 184: a) $9 + (-7) + (+3)$ b) $+5$

---

186. Do the following. As the first step in each problem, convert subtractions to additions:

(1) $(-1) - 6 + 7 =$ ______________

(2) $(-3) + (-8) - (-1) =$ ______________

(3) $6 + 5 - (-9) =$ ______________

(4) $(-3) - (-7) - (-6) =$ ______________

(5) $8 + 3 - 6 =$ ______________

(6) $2 + (-3) - 7 =$ ______________

(7) $(-3) - (-5) - 2 =$ ______________

(8) $(-4) - 6 + (-3) =$ ______________

(9) $(-1) + 7 - (-8) - 5 =$ ______________

(10) $6 - (-4) + (-1) - 5 =$ ______________

Answers to 185: a) $14 + (-9) + (-7) + (+3)$ b) $+1$

Answers to 186:

1) 0
2) -10
3) 20
4) 10
5) 5
6) -8
7) 0
8) -13
9) 9
10) 4

187. This frame and the next contain more problems. Do them if you need more practice. Otherwise, go on to Frame 189.

(1) 8 + 5 - (-1) = ______

(2) (-6) + (-7) - (-7) = ______

(3) (-7) - 6 + 5 = ______

(4) 1 - (-5) - 9 = ______

(5) 2 + 3 - 9 = ______

(6) (-5) - (-9) + (-1) = ______

(7) (-2) - 5 + (-3) = ______

(8) (-8) - (-1) - 6 = ______

(9) 9 - (-2) + (-8) - 1 = ______

(10) 4 + (-7) - 2 - (-3) = ______

---

188. (1) (-6) + (-2) - (-3) = ______

(2) 7 + 8 - (-2) = ______

(3) 9 + (-3) - 6 = ______

(4) (-6) - (-4) + (-9) = ______

(5) 2 - 8 - 1 = ______

(6) (-6) - 1 + (-9) = ______

(7) (-4) - (-8) + 1 = ______

(8) (-7) - 2 - 5 = ______

(9) (-3) + 6 - (-8) - 1 = ______

(10) 8 - (-2) + (-9) - 7 = ______

Answers to Frame 187:

1) 14
2) -6
3) -8
4) -3
5) -4
6) 3
7) -10
8) -13
9) 2
10) -2

Answers to Frame 188:

1) -5	3) 0	5) -7	7) 5	9) 10
2) 17	4) -11	6) -16	8) -14	10) -6

1-16 THE COMMUTATIVE PROPERTY AND SUBTRACTION

In an earlier section, we introduced the commutative property of addition. This property says: If the two terms of an addition are interchanged, the sum does not change. Here is an example:

$$(-5) + 3 = 3 + (-5)$$

(since both equal -2)

In this section, we will show:

(1) That subtraction itself is not commutative.
(2) That after a subtraction is converted to addition, the addition is commutative.

189. Subtraction itself is not commutative. Here is an example:

7 - 2 = +5
2 - 7 = -5

Does 7 - 2 = 2 - 7 ? ______

---

190. Here is a subtraction: (+4) - (-3)

Commuting the two terms, we get: (-3) - (+4)

We can show that subtraction is not commutative by showing that the two subtractions above are unequal.

(a) (+4) - (-3) = (+4) + ☐ = ______

(b) (-3) - (+4) = (-3) + ☐ = ______

(c) Are the two subtractions equal? ______

No, since +5 and -5 are not equal.

---

191. When an expression contains more than two terms and each operation is an addition, we can rearrange the order of the terms without changing the value of the expression.

Here is a three-term addition:

(+7) + (-4) + (-1) = +2

No matter how the above terms are rearranged, the sum is always "+2".

(-4) + (+7) + (-1) = +2
(-4) + (-1) + (+7) = +2
(-1) + (+7) + (-4) = +2

However, when an expression contains more than two terms and one or more of the operations is a subtraction, we cannot rearrange the terms in any order without changing the value of the expression.

In the expressions below, which contain subtractions, we have simply rearranged the same three terms. Compute the value of each expression to see whether they are equal when the terms are rearranged.

(a) (+10) - (+7) - (-3) = ______

(b) (+7) - (+10) - (-3) = ______

(c) (-3) - (+7) - (+10) = ______

(d) Are the three expressions equal? ______

a) [+3] = 7
b) [-4] = -7
c) No

---

192. Here is another example of the same type. In each case, the same terms are rearranged. Compute the value of each expression.

(a) (+7) - (+5) + (-3) - (-4) = ______

(b) (+5) - (+7) + (-3) - (-4) = ______

(c) (+5) - (+7) + (-4) - (-3) = ______

(d) (-3) - (+7) + (-4) - (+5) = ______

(e) Are the expressions equal when the terms are rearranged? ______

a) +6
b) 0
c) -20
d) No

193. We have seen these two facts:

(1) Simple subtractions are not commutative.
(2) More complex expressions which contain a subtraction are also not commutative.

After a simple subtraction is converted to addition, the addition (like any addition) is commutative.

Here is a conversion of a subtraction to addition:

$$(-5) - 7 = (-5) + (-7)$$

If we commute the addition on the right, we get:

$$(-7) + (-5)$$

Does $(-5) + (-7) = (-7) + (-5)$? _______

Answers: a) +3 b) -1 c) -3 d) -19 e) No

---

194. Similarly, when all of the subtractions in a complex expression are converted to additions, the new expression can be rearranged in any order.

Here is a combined addition and subtraction problem:

$$(-5) - 7 + (-2) - (-10) = -4$$

Converting the subtractions to additions, we get:

$$(-5) + (-7) + (-2) + (+10) = -4$$

Here are various rearrangements of the four terms in the complex additions:

$$(-7) + (-5) + (+10) + (-2)$$
$$(-5) + (+10) + (-7) + (-2)$$
$$(-2) + (+10) + (-5) + (-7)$$

Do all of the rearranged expressions equal "-4"? _______

Answer: Yes, since both equal "-12".

---

195. When working a combined addition and subtraction problem:

(1) We convert all subtractions to additions.
(2) Then we can rearrange (or commute) the terms. Frequently we do this to get a shortcut.

Here is an example:

$$(-3) + 4 - 9 - (-5)$$

Converting the subtractions to additions, we get:

$$(-3) + 4 + (-9) + (+5)$$

Collecting the positive and negative terms for a shortcut, we get:

$$\underbrace{(-3) + (-9)}_{\downarrow} + \underbrace{4 + (+5)}_{\downarrow}$$
$$(-12) \quad + \quad (+9) \quad = \_\_\_\_\_\_$$

Answer: Yes.

---

196. Before rearranging (or commuting) the terms in a combined addition and subtraction problem, what must be done first? _______________________

____

Answer: -3

197. Of the two operations, addition and subtraction: (a) Which one <u>is</u> commutative? ____________ (b) Which one <u>is</u> <u>not</u> commutative? ____________	All subtractions must be converted to additions.
	a) Addition b) Subtraction

SELF-TEST 8 (Frames 183-197)

Convert the following subtractions to additions:

1. 14 - (-3) - 20 = 14 + ( ) + ( )
2. (-7) - 5 - (-2) - 3 = (-7) + ( ) + ( ) + ( )

Do these:

3. (-8) + 5 - (-3) = ______
4. 19 - 23 - 7 = ______
5. 3 - (-1) - 7 + (-5) = ______
6. (-11) - 2 + 19 - (-4) = ______

Complete:

7. Addition __________ (is/is not) commutative.
8. Subtraction __________ (is/is not) commutative.

True or False:

__________ 9. 2 - 7 = 7 - 2

__________ 10. (-2) + (+5) - (+3) = (+5) + (-2) - (+3)

__________ 11. 9 - (-6) + (-2) = 9 - (-2) + (-6)

__________ 12. (+4) + (-3) + (+7) + (-2) = (+4) + (+7) + (-2) + (-3)

ANSWERS:

1. 14 + (+3) + (-20)
2. (-7) + (-5) + (+2) + (-3)
3. 0
4. -11
5. -8
6. 10
7. is
8. is not
9. False
10. True
11. False
12. True

MULTIPLICATION OF SIGNED NUMBERS

In the last section you saw that subtraction and addition are closely related operations. In fact, you learned how to change any subtraction to an equivalent addition. Multiplication is also closely related to addition.

When you learned to multiply (a few years ago), you were probably shown that multiplication is really a short-cut method of adding a group of like numbers. You then memorized the multiplication table. In this section, we will review the basic meaning of multiplication and how to multiply signed numbers. Since we are taking for granted that you can multiply, we are mainly interested in reviewing the "sign" of the product. (You probably remember that the answer in multiplication is called the product.)

First of all, let's review the different symbols that mathematicians use to tell you to multiply. Here are six of them:

(1) 2 times 4	(4) 2(4)
(2) 2 x 4	(5) (2)4
(3) 2 · 4	(6) (2)(4)

All six of these expressions mean exactly the same thing: Multiply 2 and 4. If any of the six are unfamiliar to you, memorize them now.

1-17 MULTIPLICATION WITH TWO POSITIVE FACTORS

198. Here is a multiplication: (4)(7) = 28

The two numbers on the left are called factors.
The answer ("28" in this case) is called the product.

In (9)(7) = 63: (a) The product is _______.

(b) The two factors are _______ and _______.

Answers to 198: a) 63 b) 9 and 7

199. Any multiplication is a short-cut way of writing an addition of a series of identical terms.

Instead of writing: 4 + 4 + 4,
we can write: 3(4).

(Notice that the first factor "3" tells how many times the second factor "4" should be added.)

(a) Instead of writing the addition: 6 + 6 + 6 + 6,
we can write the multiplication: _________.

(b) Instead of writing the addition: 2 + 2 + 2 + 2 + 2,
we can write the multiplication: _________.

Answers to 199: a) 4(6) b) 5(2)

200. Write each of the following multiplications as an addition:

(a) 3(7) _____________ (b) 4(9) _____________

Answers to 200: a) 7 + 7 + 7 b) 9 + 9 + 9 + 9

201. Let's examine the case when both factors are positive. By writing the multiplication as an addition, we can determine the sign of the product because it must be the same as the sign of the sum of the addition.

(3)(7) or (+3)(+7) means: (+7) + (+7) + (+7)
(4)(9) or (+4)(+9) means: (+9) + (+9) + (+9) + (+9)

(a) In each case above, the sign of the sum must be ____________ (positive/negative).

(b) Therefore, in each case above, the sign of the product must be ____________ (positive/negative).

---

202. Whenever both factors are positive, the product must be ____________ (positive/negative).

a) positive
b) positive

---

203. It is easy to see that multiplication is "commutative." That is, we get the same product when the factors are interchanged.

4(2) = 2 + 2 + 2 + 2 += +8
2(4) = 4 + 4 = +8

Use the commutative property of multiplication to fill in the blanks in each of these:

(a) 7(8) = 8( ) (b) 6(5) = ( )(6)

positive

---

204. Of the three operations: Addition, subtraction, and multiplication:

(a) Which are commutative? ____________

(b) Which are not commutative? ____________

a) (7)
b) (5)

---

a) Addition and multiplication b) Subtraction

1-18 MULTIPLICATIONS WITH ONE POSITIVE AND ONE NEGATIVE FACTOR

In this section, we will examine multiplications with one negative factor and one positive factor. First we will examine cases in which the second factor is negative. Then, we will examine cases in which the first factor is negative.

---

205. Here is a multiplication in which the first factor is positive but the second factor is negative: 3(−2)

Just as 3(2) means 2 + 2 + 2,
3(−2) means (−2) + (−2) + (−2).

(a) Write 2(−9) as an addition: ____________

(b) Write (−5) + (−5) + (−5) + (−5) as a multiplication: ____________

---

a) (−9) + (−9)
b) 4(−5)

206. When the first factor is positive and the second factor is negative, the multiplication is shorthand for "an addition of a series of identical terms which are negative."

(a) 2(−9) means (−9) + (−9) = _______

(b) 4(−5) means (−5) + (−5) + (−5) + (−5) = _______

---

207. To determine the sign of the product when the second factor is negative, we can examine the sign of the sum of the corresponding addition.

3(−4) or (+3)(−4) = (−4) + (−4) + (−4) = −12
4(−2) or (+4)(−2) = (−2) + (−2) + (−2) + (−2) = −8

(a) In all cases of this type, the sign of the sum must be _______________ (positive/negative).

(b) Therefore, in all cases of this type, the sign of the PRODUCT must be _______________ (positive/negative).

Answers to 206: a) −18 b) −20

---

208. Find these products:

(a) 5(−6) = _____ (b) 10(−7) = _____ (c) 3(−1) = _____

Answers to 207: a) negative b) negative

---

209. In the following cases, the first factor is negative and the second factor is positive.

(−2)(+5)
(−6)(+2)

We can use the commutative property of multiplication to determine the sign of the products.

(−2)(+5) = (+5)(−2)
(−6)(+2) = (+2)(−6)

(a) Since (+5)(−2) = −10, (−2)(+5) = _______

(b) Since (+2)(−6) = −12, (−6)(+2) = _______

Answers to 208: a) −30 b) −70 c) −3

---

210. Complete these: (a) (−3)(4) = 4(−3) = _______

(b) (−10)(5) = 5(−10) = _______

Answers to 209: a) −10 b) −12

---

211. When the first factor is negative and the second is positive, the multiplication can be commuted so that the one negative factor is second.

(−7)(+4) = (+4)(−7)

In all such cases, therefore, the sign of the product must be negative.

Find these products:

(a) (−8)(7) = _______ (b) (−5)(9) = _______ (c) (−9)(6) = _______

Answers to 210: a) −12 b) −50

Answers to 211: a) −56 b) −45 c) −54

Frame	Answer
212. Whenever one of the factors is negative and the other is positive (no matter whether the negative factor is first or second), the product must be _______________ (positive/negative).	
213. Which of the following products will be negative? _______________ (a) (7)(8) (c) (4)(-1) (b) (-3)(4) (d) (-2)(8)	negative
	(b), (c), and (d)

1-19 MULTIPLICATIONS IN WHICH "0" IS A FACTOR

Multiplications in which "0" is a factor are a special case. In this section, we will show that the product in multiplications of this type is always "0".

Frame	Answer
214. (a) Since (5)(0) means: 0 + 0 + 0 + 0 + 0, (5)(0) = _______. (b) Since (3)(0) means: 0 + 0 + 0, (3)(0) = _______.	
215. By the commutative property of multiplication: (a) (0)(5) = (5)(0) = _______ (b) (0)(7) = (7)(0) = _______	a) 0 b) 0
216. In each multiplication below, the product is _______. (a) (4)(0) (b) (0)(3)	a) 0 b) 0
217. Let's examine (0)(5) more closely. Just as (3)(5) means: "Add three 5's," (0)(5) means: "Add zero 5's." When you "add zero 5's," the sum is 0. Therefore, (0)(5) = 0. Let's examine (0)(-4). Just as (2)(-4) means: "Add two -4's," (0)(-4) means: "Add zero -4's." When you "add zero -4's," the sum is 0. Therefore, (0)(-4) = _______.	0
	0

218. The product for each of the following is _______. (a) (0)(−7) (b) (0)(−100)	
219. By the commutative property of multiplication: (a) (−3)(0) = (0)(−3) = _______ (b) (−9)(0) = (0)(−9) = _______	0
220. In each multiplication below, the product is _______. (6)(0) (0)(−6) (0)(6) (−6)(0)	a) 0 b) 0
221. In any multiplication in which one of the factors is "0", the product is always _______.	0
	0

1-20 MULTIPLICATION WITH TWO NEGATIVE FACTORS

In this section, we will examine the cases in which both factors are negative. We will do so by examining patterns of products.

222. In the table at the right, the second factor is always (−5). Examine the pattern of products. Then, complete the table so that this pattern is not broken.

(4)(−5) = −20
(3)(−5) = −15
(2)(−5) = −10
(1)(−5) = −5
(0)(−5) = 0
(−1)(−5) = _______
(−2)(−5) = _______
(−3)(−5) = _______

223. In the table at the right, the second factor is always (−3). Complete the table so that the pattern of products is not broken.

(4)(−3) = −12
(3)(−3) = −9
(2)(−3) = −6
(1)(−3) = −3
(0)(−3) = 0
(−1)(−3) = _______
(−2)(−3) = _______
(−3)(−3) = _______

Answers to 222:
(−1)(−5) = +5
(−2)(−5) = +10
(−3)(−5) = +15

Answers to 223:
(−1)(−3) = +3
(−2)(−3) = +6
(−3)(−3) = +9

224. In the last two frames, we have seen this fact: When both factors are negative, the sign of the product must be positive.

Therefore: (a) (-5)(-10) = ______ (b) (-3)(-7) = ______

---

225. Here are the rules for multiplying two signed numbers:

IF BOTH FACTORS ARE POSITIVE, THE PRODUCT IS POSITIVE.
IF BOTH FACTORS ARE NEGATIVE, THE PRODUCT IS POSITIVE.
IF ONE FACTOR IS POSITIVE AND ONE FACTOR IS NEGATIVE, THE PRODUCT IS NEGATIVE.

(a) If we multiply two factors with like signs, the product is always ______ (positive/negative).

(b) If we multiply two factors with unlike signs, the product is always ______ (positive/negative).

Answers to 224: a) +50 b) +21

---

226. Which of the following products will be positive? ______

(a) (-6)(-5) (b) (7)(8) (c) (-6)(5) (d) (-2)(-5)

Answers to 225: a) positive b) negative

---

227. Do not confuse the rules for adding signed numbers with the rules for multiplying signed numbers.

(a) When two negative numbers are ADDED, the sum is always ______ (positive/negative).

(b) When two negative numbers are MULTIPLIED, the product is always ______ (positive/negative).

Answer to 226: (a), (b), and (d)

---

228. Watch the signs in these:

(a) (-6) + (-5) = ______ (b) (-6)(-5) = ______

Answers to 227: a) negative b) positive

---

229. (a) Examine the following two additions:

(+7) + (-3) = +4
(-6) + (+5) = -1

When one positive and one negative number are added, is the sum "always positive," "always negative," or "sometimes positive and sometimes negative?" ______

(b) Examine the following two multiplications:

(+9)(-5) = -45
(-7)(3) = -21

When one positive and one negative number are multiplied, is the product "always positive," "always negative," or "sometimes positive and sometimes negative?" ______

Answers to 228: a) -11 b) +30

230. Watch the signs in these:

(a) 10 + (-5) = _______ (c) (-10)(5) = _______

(b) (-10) + 5 = _______ (d) 10(-5) = _______

a) Sometimes positive and sometimes negative

b) <u>Always</u> negative

---

231. Watch the signs in these:

(a) (-10) + (-5) = _______ (b) (-10)(-5) = _______

a) +5 c) -50

b) -5 d) -50

---

a) -15 b) +50

1-21 MULTIPLICATION WITH MORE THAN TWO FACTORS

232. Here is a multiplication in which there are <u>three</u> factors:

(2)(4)(5)

To perform such a multiplication, we begin at the left and multiply two factors at a time.

(2)(4)(5)
↓
(8) (5) = 40

Complete this one: (-4)(5)(3)
↓
( ) (3) = ( )

---

233. Complete this one: (-3)(-2)(-4)
↓
( ) (-4) = ( )

(-20)(3) = (-60)

---

234. Complete this one: (-4)(3)(-1)(2)
↓
(-12) (-1)(2)
↓
( ) (2) = ( )

(+6)(-4) = (-24)

---

235. Do these: (a) (-3)(2)(7) = _______

(b) (-4)(3)(-2) = _______

(c) (-1)(+5)(-3)(-2) = _______

(+12)(2) = 24

---

a) -42

b) +24

c) -30

236. Watch what happens when one of the factors is "0". (a) (9)(3)(0) → (27) (0) = _______ (b) (0)(-4)(3)(-7) → (0) (3)(-7) → (0) (-7) = _______	
237. When multiplying a series of factors, if one of the factors is "0", the product is always _______.	a) 0 b) 0
238. Always look for "0" factors in a multiplication problem, since you then know that the product is "0". In which cases below is the product "0"? _______________ (a) (5)(8)(0)(-5)(-2) (b) (75)(1,287)(-692) (c) (128,569)(-67,888)(0)(54,683)(234,569)	0
239. (-2)(3)(4) = -24 Let's rearrange the factors in various ways and see whether the product changes: (a) (3)(-2)(4) = _______ (b) (4)(-2)(3) = _______ (c) (4)(3)(-2) = _______ (d) Does the product change when the factors are rearranged? _____	The product is "0" in both (a) and (c).
240. The point of the last frame was to show you that the <u>commutative</u> property of multiplication <u>also holds for more than two factors.</u> Using this fact, complete the following: (a) (-3)(4)(6) = (4)( )(6) (b) (-2)(-5)(-8) = (-8)( )(-5) (c) (0)(-17)(-35) = (-35)(-17)( )	a) -24 b) -24 c) -24 d) No
241. Do these: (a) (-1)(-6) = _______ (b) (-7)(2) = _______ (c) (-5)(-1)(-2) = _______ (d) (8)(0)(-7)(-3) = _______	a) -3 b) -2 c) 0
242. Do these: (a) 5(-9) = _______ (b) (-4)(-6) = _______ (c) (-2)(-3)(10) = _______ (d) (5)(-4)(-9)(0) = _______	a) +6 c) -10 b) -14 d) 0

a) -45	c) +60
b) +24	d) 0

SELF-TEST 9 (Frames 198-242)

In the multiplication (7)(9) = 63:

1. The "63" is called the ______________.
2. The "7" and "9" are called ______________.

3. Which of the following operations are commutative? ______________

(a) Addition (b) Subtraction (c) Multiplication

4. If two numbers having the <u>same</u> sign are multiplied, the product is always ______________ (positive/negative).
5. If two numbers having <u>different</u> signs are multiplied, the product is always ______________ (positive/negative).

Multiply: 6. (-3)(9) = ______ 7. (-8)(-7) = ______ 8. (6)(-4) = ______

Do the following:

9. (-4) + (+8) = ______ 11. (-2) + (-5) = ______ 13. (-2) + 0 = ______

10. (-4)(+8) = ______ 12. (-2)(-5) = ______ 14. (-2)(0) = ______

Multiply:

15. (-1)(4)(-3) = ______ 18. (-2)(-4)(3)(-1) = ______

16. (5)(-2)(6) = ______ 19. (-1)(-1)(-5)(-2)(1) = ______

17. (-2)(-1)(7)(0) = ______ 20. (2)(-2)(-1)(1)(-3) = ______

ANSWERS:

1. product	6. -27	11. -7	16. -60
2. factors	7. +56	12. +10	17. 0
3. (a) and (c)	8. -24	13. -2	18. -24
4. positive	9. +4	14. 0	19. +10
5. negative	10. -32	15. +12	20. -12

1-22 COMBINED ADDITION, SUBTRACTION, AND MULTIPLICATION

Many mathematical expressions require additions, subtractions, and multiplications. In these more complex expressions, it is important:

(1) That you be able <u>to identify the terms</u>.
(2) That you know <u>the order in which the operations should be performed</u>.

We will examine these more complex expressions in this section.

243. In simple additions and subtractions, each of the signed numbers is called a "term."

Draw a box around each term in the following expressions:

(a) 10 + (-7) (b) (-6) - 5

---

244. When addition and subtraction are combined, each signed number is still called a "term."

How many terms are there in each of the following:

(a) 10 + (-4) - 9 ________ (b) 7 - 11 - 6 - (-5) ________

Answer to 243: Your answers should look like this:

a) [10] + [(-7)]

b) [(-6)] - [5]

---

245. Some terms are more complicated than single numbers. For example, <u>any multiplication is one term.</u>

(5)(-4) + 7

In the expression above, there are two terms.

The <u>first</u> is (5)(-4).
The <u>second</u> is 7 .

In the expression: 5 - (6)(4), there are <u>two</u> terms.

List them: ________ and ________

Answer to 244: a) three b) four

---

246. In the expression: (5)(7) + (-3)(2),

the <u>two</u> terms are ________ and ________.

Answer to 245: 5 and (6)(4)

---

247. As you can see, any mathematical expression is composed of terms. The terms are separated by either an addition symbol or a subtraction symbol.

Draw a box around each term in each of the following expressions:

(a) (7)(8) - 5 (b) 10 + (-7)(-3) (c) (-3)(4) - (-2)(-3)

Answer to 246: (5)(7) and (-3)(2)

---

248. In a simple addition or subtraction, the two terms are <u>single</u> numbers. By performing the indicated operation, the two terms are <u>reduced</u> to one.

In (-5) + 2 = -3
and 11 - 6 = 5

the two terms on the left are <u>reduced to one</u> on the right.

Additions and subtractions are used to reduce the number of terms in an expression. Multiplication, however, is an operation which is called for <u>within a term.</u> For example, a multiplication is called for <u>within the first term</u> below:

(5)(-4) + 7

(Continued on following page.)

Answer to 247: Your answers should look like this:

a) [(7)(8)] - [5]

b) [10] + [(-7)(-3)]

c) [(-3)(4)] - [(-2)(-3)]

248. (Continued)

When a multiplication is called for within a term, <u>we perform the multiplication first.</u> By performing the multiplication, <u>we replace the two factors with their product</u> which is a single number.

$$\underbrace{(5)(-4)} + 7$$
$$\downarrow$$
$$(-20) + 7$$

In this case we replaced (5)(-4) with -20. Of course, now we can perform a simple addition with the two single numbers, "-20" and "+7".

In 10 + (-7)(-3), what operation would you perform first? __________
____

---

249. In this one, watch how we perform the multiplication <u>within a term</u> first:

$$\underbrace{(5)(-4)} + 7$$
$$\downarrow$$
$$(-20) + 7 = -13$$

Complete this one: $\underbrace{(-6)(3)} + (-9)$
$\downarrow$
( ) + (-9) = ( )

> Multiply (-7)(-3), and then substitute +21 for the two factors.

---

250. Here is another example:

$$12 + \underbrace{(-9)(-3)}$$
$$\downarrow$$
$$12 + (+27) = 39$$

Complete this one: $(-6) + \underbrace{(-2)(-4)}$
$\downarrow$
(-6) + ( ) = _______

> (-18) + (-9) = (-27)

---

251. Evaluate each of the following:

(a) (7)(5) + (-20) = _______
(b) (-3)(4) + 14 = _______
(c) 100 + (6)(-5) = _______
(d) 100 + (-7)(-10) = _______

> (-6) + (+8) = 2

---

252. Evaluate each of these:

(a) (-7) + (-5)(2) = _______
(b) (-50) + (-6)(-10) = _______
(c) (-5)(-100) + 300 = _______
(d) (-10)(7) + (-20) = _______

> a) 15 c) 70
> b) 2 d) 170

> a) -17 c) +800
> b) +10 d) -90

253. In this case, the two terms are connected by a subtraction symbol.

$$(5)(-8) - 20$$
$$\downarrow$$
$$(-40) - 20 = (-40) + (-20) = -60$$

Notice the steps:

(1) We performed the multiplication first.
(2) Then, we converted from subtraction to addition by "adding the opposite of +20."

Complete this one: $(-7)(5) - 15$

$\downarrow$

( ) - 15 = ( ) + ( ) = _______

Answer: (-35) - 15 = (-35) + (-15) = -50

---

254. Here is another example which contains a <u>subtraction</u>:

$$14 - (7)(-3)$$
$$\downarrow$$
$$14 - (-21) = 14 + (+21) = 35$$

Notice the steps again:

(1) We performed the multiplication.
(2) Then, we converted to addition by "adding the opposite of -21."

Complete this one: $10 - (-4)(3)$

$\downarrow$

10 - ( ) = 10 + ( ) = _______

Answer: 10 - (-12) = 10 + (+12) = 22

---

255. Complete the following:

$$25 - (-6)(-4)$$
$$\downarrow$$

25 - ( ) = 25 + ( ) = _______

Answer: 25 - (+24) = 25 + (-24) = 1

---

256. Evaluate each of the following:

(a) (6)(5) - 21 = _______
(b) 35 - (5)(-3) = _______
(c) (-9)(7) - (-3) = _______
(d) (-50) - (-5)(-10) = _______

Answers: a) 9 b) 50 c) -60 d) -100

---

257. Evaluate each of the following:

(a) (-7)(-2) - (-6) = _______
(b) 20 + (-3)(5) = _______
(c) 30 - (-5)(-2) = _______
(d) (-4)(5) + (-10) = _______

Answers: a) 20 b) 5 c) 20 d) -30

258. Here is a case in which both terms call for a multiplication. We perform both multiplications before adding.

$$\underline{(-5)(8)} + \underline{(-3)(-10)}$$
$$\downarrow \qquad \downarrow$$
$$(-40) + (+30) = -10$$

Complete the following: $\underline{(-6)(-5)} + \underline{(-4)(6)}$

$$(\quad) + (\quad) = \_\_\_\_\_$$

---

259. Notice how we change subtraction to addition by "adding the opposite" in this one:

$$\underline{(-3)(2)} - \underline{(-4)(-1)}$$
$$\downarrow \qquad \downarrow$$
$$(-6) - (+4) = (-6) + (-4) = -10$$

Complete the following: $\underline{(-7)(-4)} - \underline{(-3)(5)}$

$$(\quad) - (\quad) = (\quad) + (\quad) = \_\_\_\_\_$$

Answer to 258: 30 + (-24) = 6

---

260. Evaluate each of the following:

(a) (-9)(6) + (8)(7) = _______

(b) (9)(-3) - (6)(-8) = _______

(c) (-11)(-5) - (-1)(-10) = _______

(d) (7)(11) + (-7)(9) = _______

Answer to 259: (28) - (-15) = (28) + (15) = 43

---

a) 2 b) 21 c) 45 d) 14

SELF-TEST 10 (Frames 243-260)

Draw a box around each term in the following expressions:

1. 3 + (-8) + 5
2. (-2)(-4) - 7
3. (4)(-6) + (-1)(-4)

Evaluate each:

4. (-9) + (5)(-4) = _______
5. (-8)(6) + 35 = _______
6. (-5)(-3) + (-10)(-1) = _______
7. (-7)(6) + (8)(-4) = _______

Evaluate each:

8. 14 - (3)(-2) = _______
9. (-11) - (-4)(2) = _______
10. (-9)(-6) - (-1)(14) = _______
11. (-10)(-5) - 1 = _______

Evaluate each:

12. (-5)(0) - (-5)(1) = _______
13. (-18) + (-4)(-7) = _______
14. (-1)(-3) - (-6)(-2) = _______
15. (8)(-1) + (-7)(0) = _______

ANSWERS:

1. [3] + [(-8)] + [5]
2. [(-2)(-4)] - [7]
3. [(4)(-6)] + [(-1)(-4)]
4. -29
5. -13
6. +25
7. -74
8. +20
9. -3
10. +68
11. +49
12. +5
13. +10
14. -9
15. -8

Chapter 2 NON-FRACTIONAL EQUATIONS

Algebra has a reputation of being mysterious and confusing. It has gained a bad reputation because it is frequently presented as a bag of tricks, rules, and short-cuts. We feel that this reputation is false, and we will convince you that we are right. If presented properly, algebra is quite simple and logical.

In this chapter, we will introduce the meaning of simple equations and the algebraic principles which are needed to solve them. There are only a few principles which you must understand to solve the equations in this chapter, and these principles are easy to understand.

We know that some of you have already learned some algebra. But whether you know a lot or very little or nothing about it, we will insist that you learn the principles which we teach. These principles will be used consistently throughout the course, and a thorough understanding of them will become more and more useful as the material in the course becomes more complicated.

2-1 MATHEMATICAL SENTENCES

In this section, we will introduce the idea of mathematical sentences. We will show these facts:

(1) A mathematical sentence may be true or false.
(2) Some mathematical sentences are "open" because one of the numbers is missing.
(3) An open sentence is solved when we write in the number which makes the sentence true.

---

1. Some ordinary sentences are true and some are false. For example:

"George Washington was the first president of the United States" is a true sentence.

"Abraham Lincoln was the first president of the United States" is a false sentence.

Here are two mathematical sentences: $7 + 3 = 10$
$9 - 7 = 5$

A sentence containing numbers may also be either true or false.

(a) "$7 + 3 = 10$" is a ____________ (true/false) sentence.

(b) "$9 - 7 = 5$" is a ____________ (true/false) sentence.

---

a) true
b) false

2. In the blank below each sentence, write either "True" or "False."

(a) $11 - 7 = 4$ ____________ (b) $3 + 6 = 10$ ____________ (c) $7(5) = 40$ ____________

---

a) True
b) False
c) False

3. "=" and "≠" are two symbols which can appear in a mathematical sentence. Each of these symbols is shorthand for some words.

"=" means "is equal to"
"≠" means "is not equal to"

Fill in the blanks with words:

(a) $3 + 9 = 12$ means: $3 + 9$ ________________ 12

(b) $7 + 2 \neq 8$ means: $7 + 2$ ________________ 8

---

4. If a sentence contains the symbol "≠", it may be either true or false. In the blank below each sentence, write either "True" or "False."

(a) $5 + 4 \neq 6$ (b) $7 - 2 \neq 5$ (c) $3(5) \neq 15$

________ ________ ________

a) is equal to

b) is not equal to

---

5. In each blank below, write either "=" or "≠", using the one which makes the sentence true.

(a) $5(4)$ _____ 20 (c) $5 + 6$ _____ 13

(b) $3(6)$ _____ 12 (d) $10 - 3$ _____ 7

a) True

b) False, since $7 - 2 = 5$

c) False, since $3(5) = 15$

---

6. Here is an example of a mathematical sentence in which one of the numbers is missing. A circle is written in the place of the missing number.

$$7 + \bigcirc = 9$$

As it stands, the sentence is neither true nor false. But as soon as we write a number in the circle, the sentence immediately becomes either true or false.

(a) If we write a 10 in the circle, we get:

$$7 + 10 = 9$$

Is this new sentence true or false? ________

(b) If we write a 2 in the circle, we get:

$$7 + 2 = 9$$

Is this new sentence true or false? ________

a) = c) ≠

b) ≠ d) =

---

a) False

b) True

7. When one of the numbers in a sentence is missing, the mathematical sentence is called an *open* sentence. We can use geometrical figures to show that a number is missing.

Here is another *open* sentence. A triangle is written in the place of the missing number.

$$7 + 2 = \triangle$$

(a) If you wrote an "8" in the triangle, would the new sentence be true? ________

(b) If you wrote a "9" in the triangle, would the new sentence be true? ________

---

8. In this open sentence, a square is written in the place of the missing number.

$$\square - 2 = 7$$

(a) If you wrote a "10" in the square, would the new sentence be true? ________

(b) If you wrote a "9" in the square, would the new sentence be true? ________

a) No

b) Yes

---

9. When a sentence is *open*, you can write any number you want in the geometrical figure. For example, you can write any number in the circle below.

$$\bigcirc + 2 = 5$$

However, if you want the new sentence to be true, there is only *one* number which you can write in the circle. What number is it? ______

a) No

b) Yes

---

10. When writing numbers in open sentences, the number which makes the sentence true is said to *satisfy* that sentence.

$$3 + \square = 9$$

What number satisfies the sentence above? ______

3

---

11. When we tell you to *solve a sentence*, we want you to find the number that *satisfies* it.

$$5 - 3 = \square$$

What number would you write in the square to solve the sentence above? ______

6

---

12. Solve the following sentences:

(a) $9 + 7 = \triangle$ (b) $9 - 7 = \bigcirc$ (c) $9(7) = \square$

2

---

a) 16 b) 2 c) 63

13. Solve these sentences:

(a) 7 + ○ = 11 (c) △ + 5 = 15

(b) 10 − □ = 5 (d) ○ − 4 = 5

---

14. When <u>a</u> <u>number</u> and <u>a</u> <u>figure</u> are written next to each other, multiplication <u>is</u> called for.

"3□" means "3 times □"

Of course, any of the other symbols for multiplication can be used.

(3)(□)
3(□)
(3)□
3 x □

→ All mean "3 times □".

Plug in a "4" in the box and write the product:

3□ = ______

a) 4 c) 10
b) 5 d) 9

---

15. For each of the following, plug a "6" in the box and write the product:

(a) 4□ = ______ (b) 6□ = ______ (c) 8□ = ______

3(4) = <u>12</u>

---

16. Solve these sentences:

(a) 3□ = 12 (b) 7□ = 21 (c) 6□ = 54

a) 24 b) 36 c) 48

---

17. Sometimes the number which <u>satisfies</u> a sentence is <u>negative</u>.

(−3) + □ = (−7)

(a) Does +4 satisfy this sentence? ______

(b) Does −4 satisfy this sentence? ______

a) 4 b) 3 c) 9

---

18. Solve the following sentences:

(a) (−6) + (−2) = △ (b) 3 − 7 = ○ (c) 6(−7) = □

a) No
c) Yes

---

19. 2□ = −16

(a) Does "+8" satisfy the sentence above? ______

(b) Does "−8" satisfy the sentence above? ______

a) −8 b) −4 c) −42

---

20. −4△ = 20

(a) Does "+5" satisfy this sentence? ______

(b) Does "−5" satisfy this sentence? ______

a) No
b) Yes

21. Solve the following sentences:

(a) $3\square = 3$ (c) $2\square = 0$

(b) $5 + \triangle = 5$ (d) $4 + \triangle = 0$

a) No
b) Yes

---

22. Solve these: (a) $1975 + \bigcirc = 1492 + 1975$

(b) $103\square = (98)(103)$

(c) $(9)(2)(3) = (3)(9)(\square)$

a) 1 c) 0
b) 0 d) -4

---

a) 1492 b) 98 c) 2

2-2 THE USE OF LETTERS IN OPEN SENTENCES

Up to this point, we have seen open sentences with geometrical figures in them, such as:

$$7 + \square = 9 \qquad \bigcirc - 3 = 4 \qquad 5\triangle = 30$$

In mathematics, we ordinarily use letters instead of geometrical figures in open sentences. Instead of the sentences above, we would write:

$$7 + x = 9 \qquad y - 3 = 4 \qquad 5z = 30$$

Though mathematicians traditionally use letters like "x", "y", and "z" in open sentences, using these particular letters is purely arbitrary. <u>Any letter</u> can be used <u>in an open sentence.</u> In this course, we will not limit ourselves to the use of the traditional letters.

---

23. In open sentences, letters are used in exactly the same way that we have been using geometrical figures. <u>Letters are written in place of a missing number.</u>

$$7 + x = 9$$

(a) If you plug in a "5" for the "x" in the sentence above, is the new sentence true? _______

(b) What number must you plug in for "x" to make the sentence true? _______

---

24. If you do not care whether the new sentence is true or false, you can plug in any number for a letter in an open sentence. You can, for example, plug in any number for the "t" in this sentence:

$$(-5) + (-4) = t$$

But if you want the new sentence to be true, what number must you plug in for "t"? _______

a) No
b) 2

-9

25. The sentence below is true if we plug in a "-5" for "y".

$$3 - 8 = y$$

Therefore, we say that "-5" ______________ this sentence.

---

26. What number satisfies each of the following sentences?

(a) $5 + (-7) = m$ (b) $4 - (-3) = D$ (c) $(-6)(-8) = r$

_______ _______ _______

satisfies

---

27. "<u>Solving</u> <u>an</u> <u>open</u> <u>sentence</u>" means: Finding the number which satisfies it. Solve each of the following:

(a) $S + 6 = 13$ (b) $14 - y = 9$ (c) $x - 3 = 6$

_______ _______ _______

a) -2 b) +7 c) +48

---

28. Just as $3\triangle$ means: 3 <u>times</u> $\triangle$,

$3x$ means: 3 <u>times</u> x.

In each case below, plug in a "+5" for "x" and find the product:

(a) $3x =$ _______ (b) $-5x =$ _______ (c) $-10x =$ _______

a) 7 b) 5 c) 9

---

29. Solve each of the following:

(a) $3t = 27$ (b) $7m = 56$ (c) $8y = 72$

_______ _______ _______

a) 15 b) -25 c) -50

---

30. $$4b = -24$$

(a) Does +6 satisfy this sentence? _______

(b) Does -6 satisfy this sentence? _______

a) 9 b) 8 c) 9

---

31. $$-7x = 35$$

(a) Does +5 satisfy this sentence? _______

(b) Does -5 satisfy this sentence? _______

a) No

b) Yes

---

a) No

b) Yes

2-3 EQUATIONS AND ROOTS

32. Sentences with an equal sign ("=") in them are called equations. Therefore, an equation is an equal sign ("=") with an expression on each side of it.

Here are two examples of equations: $12 + \square = 19$

$x + 3 = 7$

Which of the following is an equation? _______

(a) $3x + 9$ (b) $7 + t = 15$

Only (b). There is no "=" in (a).

33. Any equation consists of a left side, an equal sign ("="), and a right side. Here is the general pattern:

$\boxed{\text{Left Side}} = \boxed{\text{Right Side}}$

In the equation: $x + 7 = 9$

(a) The left side is _________.

(b) The right side is _________.

a) $x + 7$

b) 9

34. Either side of an equation can contain one or more terms. If a side of an equation contains more than one term, the terms are separated by an addition symbol ("+") or a subtraction symbol ("−").

In an equation: (1) Any single number is a term.
(2) Any single letter is a term.

We have drawn a box around each term in the following equation:

$\boxed{7} + \boxed{y} = \boxed{13}$

In the equation above, how many terms are there:

(a) On the left side? _________

(b) On the right side? _________

a) Two

b) One

35. (a) Draw a box around each term on both sides of this equation:

$14 = t - 7$

(b) How many terms are there on the left side? _________

(c) How many terms are there on the right side? _________

a) $\boxed{14} = \boxed{t} - \boxed{7}$

b) One

c) Two

36. Earlier in our course, we saw that any multiplication is one term.

For example: In 5(6) + 7,
5(6) is one term.

Since "3x" means: 3 times x,
and "7t" means: 7 times t,

"3x" and "7t" are really multiplications.

Therefore, "3x" is one term,
and "7t" is one term.

We have drawn a box around each term in the following equation:

$$\boxed{2x} + \boxed{7} = \boxed{15}$$

How many terms are there: (a) On the left side? _______

(b) On the right side? _______

---

37. (a) Draw a box around each term in this equation:

$$5 = 4d - 3$$

(b) There is _______ term on the left side.

(c) There are _______ terms on the right side.

a) Two
b) One

---

38. (a) Draw a box around each term in this equation:

$$7x = 56$$

(b) How many terms are there on each side? __________

a) $\boxed{5} = \boxed{4d} - \boxed{3}$
b) one
c) two

---

39. A NUMBER WHICH SATISFIES AN EQUATION IS CALLED THE SOLUTION OF THE EQUATION. A SOLUTION IS ALSO CALLED A ROOT.

What is the root or solution of this equation?

$7 + D = 12$ _______

a) $\boxed{7x} = \boxed{56}$
b) One

---

40. Find the solution of each of these equations:

(a) $(-9) + 3 = t$ (b) $(-4) - 7 = a$ (c) $(-5)(-1) = V$

_______ _______ _______

5

---

41. Find the root of each of these equations:

(a) $x + 9 = 17$ (b) $y - 9 = 1$ (c) $7b = 49$

_______ _______ _______

a) −6 b) −11 c) +5

---

a) 8 b) 10 c) 7

42. $7x = -21$

(a) Is "+3" the root of this equation? _______

(b) Is "-3" the root of this equation? _______

---

43. $-5x = 30$

(a) Is "+6" the root of this equation? _______

(b) Is "-6" the root of this equation? _______

a) No
[Since $7(3) = +21$]

b) Yes

---

44. $-7t = -35$

(a) Is "-5" the root of this equation? _______

(b) Is "+5" the root of this equation? _______

a) No
[Since $(-5)(+6) = -30$]

b) Yes

---

45. We can make the following general statements about the use of "letters" in equations:

(1) Letters are used in equations to fill in a place where a number is missing.
(2) Though any number can be plugged in for a letter, we are ordinarily interested in the number which makes the sentence or equation true. That is, we are ordinarily interested in solving an equation.
(3) The number which satisfies an equation is called its solution or root.

a) No

b) Yes

SELF-TEST 1 (Frames 1-45)

1. Write the symbol which means "is not equal to." _______

True or False: 2. $3 = 9 - (-6)$ _________ (3) $(-2) + (-3) \neq -5$ _________

Solve these sentences:

4. $12 - \square = 5$ 5. $4\triangle = 32$ 6. $5 + (-2) = (-2) + \bigcirc$ 7. $8\square = 0$

Find the number which satisfies each of the following sentences:

8. $4 - (-2) = w$ _____ 9. $15 - x = 10$ _____ 10. $9y = 54$ _____ 11. $3t = 3$ _____

12. In algebra, letters stand for _______________.

In this equation: $3d - 2 = 10$

13. How many terms are there on the left side? _______

14. How many terms are there on the right side? _______

15. State whether +2 or -2 is the root of: $5r = -10$ _______

16. State whether +9 or -9 is the solution of: $-4h = 36$ _______

17. State whether +5 or -5 is the root of: $-7a = -35$ _______

Find the root of each equation:

18. $t - 1 = 8$ _____ 19. $4p = 0$ _____ 20. $(-12) - 20 = s$ _____

ANSWERS:						
	1. ≠	4. 7	8. 6	12. numbers	15. -2	18. 9
	2. False	5. 8	9. 5	13. Two	16. -9	19. 0
	3. False	6. 5	10. 6	14. One	17. +5	20. -32
		7. 0	11. 1			

2-4 SOLVING INSTANCES OF THE BASIC EQUATION

There is one type of equation which we call the "basic" equation. We call it the "basic" equation because it occurs as the final step in the solution of any more complicated equation. In this section, we will examine and solve instances of the "basic" equation.

46. Here are instances of what we call the basic equation:

$$3x = 21$$
$$5y = -40$$
$$-6m = 42$$

Notice these points about the equations:

(1) There is one term on each side.
(2) The term on the right side is a signed number.
(3) The term on the left side contains both a signed number and a letter.

Which of the following are instances of the basic equation? ________

(a) $x + 7 = 15$ (c) $-3S = -9$

(b) $2t = 18$ (d) $4k + 1 = 13$

---

47. Let's examine the following instance of the basic equation:

(b) and (c)

$$7d = 28$$

The term on the left side calls for a multiplication.

That is: 7d means: (7)(d)

Therefore: "7" and "d" are factors.
"28" is the product.

In $-5t = -40$: (a) The two factors are _______ and _______.

(b) The product is _______.

---

48. When a term contains both a number and a letter, the number is called the "numerical coefficient" of the letter.

a) (-5) and (t)

b) -40

In 7x, "7" is called the numerical coefficient of "x".

(a) In -3y, the numerical coefficient of "y" is _______.

(b) In 5t, "5" is called the _______________ of "t".

---

a) -3

b) numerical coefficient

49. In an instance of the basic equation, the term which contains both a number and a letter calls for a multiplication.

In 7x = 42, "7x" means "7 times x."

To show that it is a multiplication, we could write parentheses around each factor:

$$7x = (7)(x)$$

And since multiplication is commutative, we could write the letter first:

$$7x = x7$$
$$\text{or } 7x = (x)(7)$$

However, when a term contains a letter and its numerical coefficient:

(1) We write the numerical coefficient first.
(2) We ordinarily do not use parentheses.

(a) Instead of (-5)(m), we would write _______.

(b) Instead of (R)(-3), we would write _______.

Answers: a) -5m b) -3R

50. In an instance of the basic equation, the letter stands in the place of a missing factor. Therefore, "solving the equation" means "finding the missing factor."

$$5m = 35$$

In this case: The solution or root is +7.
Therefore, the missing factor is _______.

Answer: +7

51. When there is no negative sign on either side, instances of the basic equation are easy to solve.

Find the root of each of these equations:

(a) 9r = 54 _______ (b) 8v = 48 _______ (c) 7d = 63 _______

Answers: a) 6 b) 6 c) 9

52. The basic equation is more difficult to solve when a negative sign appears on one or both sides. Here are some examples of this type:

(1) 5t = -40
(2) -9t = 54
(3) -8t = -56

In these equations:

(1) It is easy to find the absolute value of the root.
(2) It is more difficult to find the sign of the root.

To find the absolute value of the root, divide the absolute value of the product by the absolute value of the known factor.

(Continued on following page.)

52. (Continued)

(1) $5t = -40$
(2) $-9t = 54$
(3) $-8t = -56$

The absolute value of the root for equation (1) is 8, since:

$$\frac{40}{5} = 8$$

Find the absolute value of the root:

(a) For equation (2). _______ (b) For equation (3). _______

---

a) 6, since $\frac{54}{9} = 6$

b) 7, since $\frac{56}{8} = 7$

53. Find the absolute value of the root for each of these:

(a) $-10x = 60$ (b) $-11d = -77$ (c) $9y = -72$

_______ _______ _______

---

a) 6, since $\frac{60}{10} = 6$

b) 7, since $\frac{77}{11} = 7$

c) 8, since $\frac{72}{9} = 8$

54. After determining the absolute value of the root, the root must be either the positive or the negative number which has this absolute value. For example:

$$9t = -45$$

The absolute value of the root is 5, since $\frac{45}{9} = 5$.

Therefore, the root is either +5 or -5.

To determine the sign of the root, plug each of the two possible values into the equation and see which one satisfies it.

(a) Does +5 satisfy the equation? _______

(b) Does -5 satisfy the equation? _______

(c) The root is _______.

---

a) No, since $(9)(5) = +45$

b) Yes, since $9(-5) = -45$

c) -5

55. Here is another example: $-6x = 54$

The absolute value of the root is 9, since $\frac{54}{6} = 9$.

The root must be either +9 or -9.

(a) Does +9 satisfy the equation? _______

(b) Does -9 satisfy the equation? _______

(c) The root is _______.

---

a) No, since $(-6)(+9) = -54$

b) Yes, since $(-6)(-9) = +54$

c) -9

56. Here is another example: $-7q = -28$

The absolute value of the root is 4, since $\frac{28}{7} = 4$.

Therefore, the root is either +4 or -4.

(a) Does +4 satisfy the equation? _______

(b) Does -4 satisfy the equation? _______

(c) The root is _______.

Answers to 56:

a) Yes, since $(-7)(+4) = -28$

b) No, since $(-7)(-4) = +28$

c) +4

---

57. For each of the following, find the root by:

(1) Determining its absolute value.
(2) Then checking to see which of the two signed alternatives satisfies the equation.

(a) $3t = -12$ The root is _______.

(b) $6t = -30$ The root is _______.

(c) $5t = -50$ The root is _______.

Answers to 57:

a) -4

b) -5

c) -10

---

58. Use the same strategy for these:

(a) $-10s = 70$ The root is _______.

(b) $-5s = 55$ The root is _______.

(c) $-2s = 18$ The root is _______.

Answers to 58:

a) -7

b) -11

c) -9

---

59. Use the same strategy for these:

(a) $-5v = -35$ The root is _______.

(b) $-6v = -42$ The root is _______.

(c) $-7v = -63$ The root is _______.

Answers to 59:

a) +7

b) +7

c) +9

---

60. Do these:

(a) $-11m = 66$ The root is _______.

(b) $-9r = -36$ The root is _______.

(c) $7t = -49$ The root is _______.

Answers to 60:

a) -6

b) +4

c) -7

SELF-TEST 2 (Frames 46-60)

1. Which of the following are instances of the basic equation? ________

   (a) $-8w = 48$ (b) $5x = 2x - 6$ (c) $t - 8 = 3$ (d) $5r = -40$

What is the numerical coefficient of each of the following terms?

2. $12s$ ______ 3. $-6b$ ______ 4. $(9)(t)$ ______

5. Write the term $(p)(-8)$ in a simpler, preferred way: ______

Find the root of each equation:

6. $9y = 63$ ______ 7. $12k = -36$ ______ 8. $-5w = 55$ ______ 9. $-8t = -56$ ______

Solve each equation: 10. $-7v = 42$ ______ 11. $-6a = -54$ ______

ANSWERS:

1. (a) and (d)
2. 12
3. -6
4. 9
5. -8p
6. 7
7. -3
8. -11
9. 7
10. -6
11. 9

2-5 SIMPLIFYING EQUATIONS BY COMBINING NUMBER-TERMS

In the last section, you learned how to solve instances of the basic equation. Finding the roots for equations of that type is fairly easy. However, most equations which you must learn to solve are more complex than the basic equation. Here are some examples:

$$4x + 7 = 31$$

$$2(3x + 9) = 45 + 2x$$

$$10 - 3(5x - 2) = 25 - (2x - 1)$$

When confronted with a more complex equation, you can attempt to solve it by trial-and-error. That is, you can plug in one number after another for the letter until you find the one which satisfies the equation. However, this method is a very laborious one, and at times you would "give up" before finding the root.

To avoid the difficulties of the trial-and-error method, mathematicians have developed a set of procedures which can be used to simplify complex equations so that the root can be found more easily. These procedures are designed to reduce any complex equation to an instance of the basic equation (which, of course, is easy to solve).

In this section, we will introduce and justify one of these procedures for simplifying an equation. The procedure is called "combining number-terms."

61. Here is an instance of the basic equation:

$$9x = 36$$

In the basic equation, there is one term on each side. In the equation above:

"9x" is called a letter-term (even though there is a numerical coefficient).

"36" is called a number-term.

The following equation is not an instance of the basic equation because there are two terms on the right side.

$$3y = 9 + 12$$

In this equation:

(a) "3y" is called a ____________-term.

(b) Both "9" and "12" are called ____________-terms.

Answers: a) letter b) number

---

62. Here is the same equation. We can simplify it by adding (or combining) the two number-terms on the right.

$$3y = \underbrace{9 + 12}$$
$$\downarrow$$
$$3y = 21$$

(a) When the two terms on the right are combined (or added), how many terms are there on the right side of the new equation? _____

(b) Is the new equation an instance of the basic equation? _______

Answers: a) One b) Yes

---

63. Here is the same simplification again:

$$3y = \underbrace{9 + 12}$$
$$\downarrow$$
$$3y = 21$$

The new equation is an instance of the basic equation. Its root is "+7".

We can show that "+7" is also the root of the original equation by plugging "+7" in for "y" and showing that it makes the original sentence true.

$$3y = 9 + 12$$

$$\underbrace{3(+7)} = \underbrace{9 + 12}$$
$$\downarrow \qquad \downarrow$$
$$21 = 21$$

<u>When two equations have the same root, we call them "equivalent" equations.</u>

Since both $3y = 9 + 12$ and $3y = 21$ have the same root (+7), we call them ______________ equations.

Answer: equivalent

64. Whenever we combine number-terms in an equation, the new equation is equivalent to the original one.

$$4t = \underbrace{30 + 6}$$
$$\downarrow$$
$$4t = 36$$

The root of the new equation is "+9".

Show that +9 is also the root of the original equation. That is, plug in +9 for "t" and show that both sides of the equation are equal.

$$\underbrace{4(+9)} = \underbrace{30 + 6}$$
$$\downarrow \qquad \downarrow$$
$$36 = 36$$

65. Though the examples we have been using are trivial, they do show our general strategy for solving an equation.

(1) We reduce more complicated equations to an instance of the basic equation.
(2) Since the equations are equivalent, the root of the basic equation is also the root of the more complicated equation.

Here is another example: $5x = 25 + (-10)$

(a) Reduce this equation to an instance of the basic equation by combining the number-terms. ________

(b) The root of the basic equation is ______.

(c) Therefore, the root of the original equation must be ______.

---

a) $5x = \underbrace{25 + (-10)}$
$$\downarrow$$
$$5x = 15$$

b) +3

c) +3

66. Here is another one: $3x = (-19) + 7$

(a) Reduce this equation to an instance of the basic equation. ________

(b) The root of the basic equation is ______.

(c) The root of the original equation must be ______.

---

a) $3x = \underbrace{(-19) + 7}$
$$\downarrow$$
$$3x = -12$$

b) -4

c) -4

67. In the equation below, there is a subtraction on the right side. Our first step is to convert this subtraction to addition. Then we proceed in the usual way.

$$5b = 5 - 25$$
$$5b = \underbrace{5 + (-25)}$$
$$\downarrow$$
$$5b = -20$$

(a) The root of the last equation is ______.

(b) Therefore, the root of the original equation must be ______.

---

a) -4

b) -4

68. $8a = (-35) - 5$

(a) Show all the steps in reducing this equation to an instance of the basic equation:

(b) The root of the basic equation is ______.

(c) Therefore, the root of the original equation must be ______.

---

a) $8a = (-35) - 5$
$8a = (-35) + (-5)$
$8a = -40$

b) -5

c) -5

SELF-TEST 3 (Frames 61-68)

1. If two equations are equivalent, they have the same ________.

Solve the following equations:

2. $7w = 25 + 17$ ______

3. $4d = (-8) - 12$ ______

4. $-5h = 10 - 25$ ______

ANSWERS: 1. root 2. 6 3. -5 4. 3

2-6 GROUPINGS AND GROUPING SYMBOLS

In the last section, we showed how equations can be simplified by combining number-terms. Equations can also be simplified by combining letter-terms. Letter-terms are combined by means of the distributive principle. However, as a basis for understanding the distributive principle, we must introduce the idea of a grouping. Therefore, we will introduce groupings and grouping symbols in this section before discussing the distributive principle in the next section.

---

69. Earlier, we saw that multiplications are performed before additions or subtractions. Here is an example:

$$\underbrace{5(8)} + 9$$
$$\downarrow$$
$$40 + 9 = 49$$

The multiplication called for within the first term was performed before the addition.

Evaluate this one: $5(2) + 4(3)$ ________

70. There is one case in which addition or subtraction is performed before multiplication. You can recognize this case because it always includes a "grouping." Here are two examples:

$$5(3+1) \to 5\ (4) = 20 \qquad\qquad 5(3-1) \to 5\ (2) = 10$$

The parentheses ( ) around "3 + 1" and "3 - 1" indicate that they are groupings. A grouping always contains an addition or subtraction, and the operation within the grouping is always performed first. By performing the operation within the grouping, we reduce the grouping to a single number before multiplying.

Both of these expressions contain a grouping. Evaluate each of them.

(a) 7(5 + 4) _______ (b) 9(8 - 3) _______

Answer: 22, since

$$5(2) + 4(3) \to 10 + 12 = 22$$

---

71. Not every use of parentheses indicates a grouping. To indicate a grouping, the parentheses must contain either an addition or a subtraction.

In 7(8) + 3, the parentheses do not indicate a grouping since they do not enclose an addition or subtraction.

In 7(8 + 3), the parentheses do indicate a grouping since they do enclose an addition.

Which of the following expressions contain a grouping? __________

(a) 4(3 + 2) (c) 4(3) - 2

(b) 4(3) + 2 (d) 4(3 - 2)

Answer:
a) 7(9) = 63
b) 9(5) = 45

---

72. Remember, additions are only performed before multiplications when a grouping is involved.

Evaluate each of these:

(a) 5(6) + 2 _______ (b) 5(6 + 2) _______

Answer: Only (a) and (d)

---

73. Evaluate: (a) 7(8) - 3 _______ (b) 7(8 - 3) _______

Answer:
a) 30 + 2 = 32
b) 5(8) = 40

---

74. We have usually enclosed negative numbers in parentheses. However, when a grouping contains negative numbers, we would get a confusing series of parentheses if parentheses were also used to indicate the grouping. Here is an example:

$$3((-8) + (-7))$$

(Continued on following page.)

Answer:
a) 56 - 3 = 53
b) 7(5) = 35

74. (Continued)

To avoid this series of parentheses, mathematicians use brackets [ ] to indicate the grouping in this case. They write:

3[(-8) + (-7)]

Complete this evaluation: 3[(-8) + (-7)]

↓ ↓

3 ( ) = _______

---

75. Evaluate: (a) 7[(-9) + 7] = _______

(b) (-10)[(-1) + (-2)] = _______

Answer: 3(-15) = -45

---

76. When a grouping contains a subtraction, we convert the subtraction to addition <u>before</u> reducing the grouping to a single number.

5(3 - 9) = 5[3 + (-9)]

↓

= 5 (-6) = -30

Evaluate each of these:

(a) 7[(-3) - (-5)] = _______ (b) (-5)[(-7) - 3] = _______

Answer:
a) 7(-2) = <u>-14</u>
b) (-10)(-3) = <u>+30</u>

---

77. <u>Any multiplication is only one term. Even though one of the factors is a grouping, the multiplication is still only one term.</u>

For example, there are <u>two</u> terms in the following expression:

10 + 3(7 + 1)

The two terms are 10 and 3(7 + 1).

Draw a box around each term in these:

(a) 5[(-3) - 1] + 7

(b) 8 - 9(2 - 8) + 5(-3)

Answer:
a) 7(+2) = <u>14</u>
b) (-5)(-10) = <u>+50</u>

---

78. When a more complicated expression contains a term with a grouping, we perform the operation <u>within</u> this term first.

5(3 + 4) + 8

↓

5 (7) + 8

↓

35 + 8 = 43

Evaluate this one: 10 + 3[(-1) + (-2)] = _______

Answer:
a) $\boxed{5[(-3) - 1]} + \boxed{7}$
b) $\boxed{8} - \boxed{9(2 - 8)} + \boxed{5(-3)}$

Answer: 10 + 3(-3) = 10 + (-9) = <u>+1</u>

79. In this one, we reduce the complex term to a single number before converting the subtraction to addition.

$$25 - 4(7+3)$$
$$\downarrow$$
$$25 - 4\ (10)$$
$$\downarrow$$
$$25 - 40 \quad = 25 + (-40) = -15$$

Evaluate this one: $50 - 5[(-2) + (-1)]$

---

80. The following expression contains <u>three</u> terms. When evaluating it, we perform the operations <u>within</u> the terms before combining the terms.

$$20 + 10(8-4) + 5(7)$$
$$\downarrow \qquad \downarrow$$
$$20 + 10\ (4) \ + 35$$
$$\downarrow$$
$$20 + 40 \ + 35 = 95$$

Evaluate this one: $5(-4) + 30 + 10[(-2) + (-3)] =$ _______

Answer to 79: $50 - (-15) = 50 + (+15)$
$= \underline{65}$

Answer to 80: $(-20) + 30 + (-50) = \underline{-40}$

SELF-TEST 4 (Frames 69-80)

1. Underline those expressions which contain a <u>grouping</u>:

(a) $5(4) + 3$ (c) $2(8 - 3)$ (e) $10(2 - 5)$

(b) $5(4 + 3)$ (d) $2(8) - 3$ (f) $(10)(2) - 5$

2. The expression $4(7 + 2)$ is: (a) one term (b) two terms (c) three terms _______

Evaluate each of the following:

3. $7(6 - 2) =$ _______

4. $7(6) - 2 =$ _______

5. $4(3 - 8) =$ _______

6. $(-2)[(-7) + 4] =$ _______

7. $5 + 3[4 - (-1)] =$ _______

8. $7 + 9(1 - 3) - 4(-2) =$ _______

<u>ANSWERS:</u>
1. (b), (c), and (e)
2. (a) one term
3. 28
4. 40
5. -20
6. 6
7. 20
8. -3

2-7 THE DISTRIBUTIVE PRINCIPLE

In the last section, we introduced groupings and showed that either parentheses or brackets are used to indicate a grouping. We are now ready to introduce the distributive principle in this section.

---

81. We have seen that terms like "3x" or "10t" are really multiplications with two factors.

$$3x = (3)(x)$$
$$10t = (10)(t)$$

In these multiplications, the factors are:

(1) either single numbers like "3" or "10",
(2) or single letters like "x" or "t".

In the last section, we introduced terms which contain a grouping. These terms are also multiplications with two factors. However, one of the factors is either an addition or a subtraction.

In 4(7 + 3): The first factor is "4".
The second factor is "(7 + 3)".

Here is a similar expression: 7(6 + 2)

(a) How many factors are there in this expression? ______

(b) The factors are ______ and ______.

---

82. In 10(9 - 3), the factors are ______ and ______.

Answer to 81: a) Two b) 7 and (6 + 2)

---

83. When one of the factors is a grouping, we have reduced this factor to a single number before performing the multiplication.

$$10\underbrace{(4+3)}$$
$$\downarrow$$
$$10\ (7) = 70$$

However, there is a second way of evaluating expressions which contain a grouping. This second way is based on the distributive principle for multiplication over addition. Here is the method:

$$10(4+3) = \underbrace{10(4)} + \underbrace{10(3)}$$
$$= 40 + 30 = 70$$

Complete this one: $5(8+4) = \underbrace{5(8)} + \underbrace{5(4)}$

$$= (\quad) + (\quad) = \_\_\_\_\_\_$$

Answer to 82: 10 and (9 - 3)

Answer to 83: 40 + 20 = 60

84. This new method of evaluating a grouping is based on the distributive principle for multiplication over addition. Here is an example:

$$10(3+2) = 10(3) + 10(2)$$

The two arrows show both "3" and "2" are multiplied by "10". That is, the "multiplication by 10" is applied or distributed to both terms in the addition-factor (3 + 2).

Complete these:
(a) $7(3+9) = 7(3) + 7(\;\;)$
(b) $8(5+1) = 8(5) + (\;\;)(1)$
(c) $10(4+6) = 10(\;\;) + 10(6)$
(d) $5(9+3) = (\;\;)(9) + 5(3)$

---

a) 7(9)
b) (8)(1)
c) 10(4)
d) (5)(9)

85. Here is a general statement of THE DISTRIBUTIVE PRINCIPLE FOR MULTIPLICATION OVER ADDITION:

$$\triangle(\bigcirc + \square) = \triangle(\bigcirc) + \triangle(\square)$$

Using this principle, complete these:
(a) $10[7+(-3)] = 10(7) + 10(\;\;\;)$
(b) $3[(-2)+(-1)] = 3(-2) + (\;\;\;)(-1)$
(c) $7[(-5)+2] = 7(\;\;\;) + 7(2)$
(d) $8[(-4)+(-6)] = (\;\;\;)(-4) + 8(-6)$

---

a) 10(-3)
b) (3)(-1)
c) 7(-5)
d) (8)(-4)

86. By means of the distributive principle, we can multiply a grouping by a simple factor. The result is a product, even though it is a rather complicated one. Here is an example:

Two Factors		Product
$5(2+8)$	=	$5(2) + 5(8)$

On the left side, 5 and (2 + 8) are the two factors in a multiplication.

On the right side, 5(2) + 5(8) is a complicated product.

In $4(7+3) = 4(7) + 4(3)$:
(a) The factors on the left are _______ and _______.
(b) The complicated product is _______________.

---

a) 4 and (7 + 3)
b) 4(7) + 4(3)

87. When we multiply a grouping by a simple factor by means of the distributive principle, we increase the number of terms. In this example, we have drawn a box around each term on both sides.

$$\boxed{10(7 + 3)} = \boxed{10(7)} + \boxed{10(3)}$$

Notice these points:

(1) There is only one term on the left.
(2) There are two terms in the product on the right.
(3) Each term on the right calls for a multiplication with 10 as the first factor.

Write the complicated product on the right for each of these:

(a) 3(5 + 2) = ( )( ) + ( )( )

(b) 7[(-6) + (-4)] = ( )( ) + ( )( )

---

88. We have seen that multiplication is commutative. That is, the two factors can be interchanged without changing the product. For example:

$$10(5) = 5(10)$$

When one of the factors is a grouping, we can still interchange the factors without changing the product.

$$10(5 + 3) = (5 + 3)10$$

(Both sides equal 80.)

When the grouping is written as the first factor, the distributive principle looks like this:

$$(5 + 3)(10) = 5(10) + 3(10)$$

Notice in this case that 10 is the second factor in each multiplication on the right.

Write the complicated product on the right for each of these:

(a) (8 + 2)(7) = ( )( ) + ( )( )

(b) [(-5) + (-4)](3) = ( )( ) + ( )( )

Answers to 87:

a) (3)(5) + (3)(2)

b) (7)(-6) + (7)(-4)

---

89. Using the distributive principle to go from the factors to the complicated product is called "multiplying by the distributive principle."

If the grouping is the second factor on the left, the common factor on the right will be the first factor in each term.

$$8(9 + 1) = \underset{\downarrow}{8}(9) + \underset{\downarrow}{8}(1)$$

If the grouping is the first factor on the left, the common factor on the right will be the second factor in each term.

$$(9 + 1)(8) = 9\underset{\downarrow}{(8)} + 1\underset{\downarrow}{(8)}$$

Write the product for each of these:

(a) 5[(-6) + (-2)] = ( )( ) + ( )( )

(b) [(-3) + (-4)]9 = ( )( ) + ( )( )

Answers to 88:

a) (8)(7) + (2)(7)

b) (-5)(3) + (-4)(3)

90. Write the product for each of these. Don't let the letters bother you.

(a) $(5 + 4)x = (\quad)(\quad) + (\quad)(\quad)$

(b) $(7 + 3)t = (\quad)(\quad) + (\quad)(\quad)$

a) $(5)(-6) + (5)(-2)$

b) $(-3)(9) + (-4)(9)$

---

91. Write the product for each of these:

(a) $[7 + (-3)]m =$ ________ + ________

(b) $[(-5) + (-2)]R =$ ________ + ________

a) $(5)(x) + (4)(x)$

b) $(7)(t) + (3)(t)$

---

a) $7m + (-3m)$ b) $(-5R) + (-2R)$

2-8 FACTORING AND THE DISTRIBUTIVE PRINCIPLE

In the last section, we introduced the distributive principle. In this section, we will introduce the general idea of "factoring." Then we will discuss "factoring by means of the distributive principle," which is a special type of factoring. This later type of factoring is the basis for combining letter-terms.

---

92. "Breaking up a product into factors" is called factoring.

If we break up "20" into "(5)(4)", we have factored "20" into "5" and "4". The "5" and "4" are called factors.

Write in the missing factors in each of the following:

(a) $21 = 7\square$ (c) $-25 = 5\triangle$

(b) $24 = 4\bigcirc$ (d) $-28 = (-4)\square$

---

93. Write in the missing factor in each of these:

(a) $100 = 50\square$ (c) $42 = \square(7)$

(b) $75 = (-25)\triangle$ (d) $56 = \triangle(-8)$

a) 3 b) 6 c) −5 d) 7

---

a) 2 b) −3 c) 6 d) −7

---

94. Just as any product can be factored, the product in the distributive principle can be factored.

Here are two multiplications by means of the distributive principle:

$$8(7 + 3) = 8(7) + 8(3)$$

$$(5 + 9)(4) = 5(4) + 9(4)$$

If we reverse the procedure, we can begin with the complicated product and obtain the original factors. This procedure is called "factoring by the distributive principle."

Example: $8(7) + 8(3) = (8)(7 + 3)$ We have factored $8(7) + 8(3)$ into 8 and $(7 + 3)$.

Example: $5(4) + 9(4) = (5 + 9)(4)$ We have factored $5(4) + 9(4)$ into $(5 + 9)$ and 4.

95. Complete the following. In each case, you are factoring by the distributive principle.

(a) (3)(7) + (3)(6) = ☐(7 + 6)

(b) (5)(9) + (2)(9) = (5 + ☐)(9)

(c) (2)(3) + (2)(6) = 2(____ + ____)

(d) (4)(8) + (7)(8) = (____ + ____)8

---

96. Complete the following:

(a) 6(4) + 3(4) = (6 + 3)☐

(b) 7(5) + 2(5) = (____ + ____)(5)

(c) 9(y) + 8(y) = (9 + 8)☐

(d) 7(x) + 6(x) = (____ + ____)(x)

a) 3
b) 2
c) (3 + 6)
d) (4 + 7)

---

97. Any expression is a "product in the distributive principle" only if each term contains a common factor.

(6)(5) + (7)(5) is an instance of such a product because each term contains a "5".

(7)(3) + (4)(5) is not an instance of such a product because there is no common factor in each term.

Which of the following are instances of "products in the distributive principle?" ____________

(a) 7(4) + 2(4)   (c) 9(1) + 4(7)

(b) 11 (5) + 7(8)   (d) 8(3) + 8(2)

a) 4
b) (7 + 2)
c) y
d) (7 + 6)

---

98. 5x + 3x

Why is the above expression an instance of a "product in the distributive principle?" ____________________________

Only (a) and (d)

---

99. 5x + 3

Is the above expression an instance of a "product in the distributive principle?" ______

Because each term contains the common factor "x".

---

100. In each of the following, name the common factor if there is one.

(a) 4x + 5x ______   (b) 7x + 9 ______   (c) 5t + 3t ______

No. There is no common factor.

---

a) x
b) None
c) t

101. Factor each of the following if factoring is possible: (a) 3t + 5t = (+)( ) (b) 4a + 7 = (+)( ) (c) 10m + 9m = (+)( )	
102. "Factoring by the distributive principle" can be used to combine two letter-terms into one term. Here is an example: $2x + 9x = (2 + 9)x$ $= 11 \quad x$ Complete these: (a) $7x + 4x = (7 + 4)x =$ ______ (b) $5R + 2R = (5 + 2)R =$ ______	a) (3 + 5)(t) b) Cannot be factored. c) (10 + 9)(m)
103. If any of the following can be factored by the distributive principle and combined into one letter-term, do so: (a) 11t + 15t ______ (c) 5y + 4y ______ (b) 14x + 3 ______ (d) 7a + 5 ______	a) 11x b) 7R
104. Two letter-terms can be combined even if their numerical coefficients are <u>negative</u>. $7c + (-5c) = [7 + (-5)]c$ $= 2c$ Combine each of the following pairs of letter-terms into one letter-term: (a) $10x + (-7x) =$ ______ (b) $(-8y) + 3y =$ ______ (c) $(-10m) + (-11m) =$ ______	a) (11 + 15)t = 26t b) Cannot be done. c) (5 + 4)y = 9y d) Cannot be done.
105. When we combine the following two terms, the numerical coefficient of the sum is "0". $3x + (-3x) = [3 + (-3)]x = 0x$ 0x means: 0 times x. No matter what number we plug in for "x", 0 times x = 0. Therefore, 0x = ______	a) [10 + (-7)]x = 3x b) [(-8) + 3]y = -5y c) [(-10) + (-11)]m = -21m
106. Two quantities are a pair of opposites if their sum is 0. Since 3x + (-3x) = 0x or 0: (a) The opposite of 3x is ______. (b) The opposite of -3x is ______.	0 (When one of the factors is "0", the product is <u>always</u> "0".)

107. It should be easy to see that two terms which contain the same letters are a pair of opposites if their numerical coefficients are a pair of opposites. For example:

7t and -7t are a pair of opposites, since 7 and -7 are a pair of opposites.

$$7t + (-7t) = [7 + (-7)]t = 0t = 0$$

Write the opposites of each of these:

(a) 10m ______ (c) -6R ______

(b) -7S ______ (d) +14d ______

Answers: a) -3x b) +3x or 3x

---

108. We can perform the following subtraction by converting to addition.

$$7y - 10y = 7y + (\text{the opposite of } 10y)$$
$$= 7y + (-10y)$$
$$= [7 + (-10)]y = \_\_\_\_\_\_$$

Answers: a) -10m c) 6R b) 7S d) -14d

---

109. Convert the following subtractions to additions:

(a) 10x - 7x = 10x + ☐ (c) (-6v) - 11v = (-6v) + ☐

(b) 9t - 14t = 9t + ☐ (d) (-5m) - (-7m) = (-5m) + ☐

Answer: -3y

---

110. Combine the following terms by converting to addition first:

(a) 3z - 8z = 3z + (-8z) = ______

(b) (-5d) - 4d = (-5d) + (-4d) = ______

(c) 5h - (-6h) = _____ + _____ = ______

(d) (-9x) - (-7x) = _____ + _____ = ______

Answers: a) -7x c) -11v b) -14t d) +7m

---

111. Up to this point, we have used two steps to combine letter-terms: (1) factoring by means of the distributive principle, and then (2) simplifying the grouping. For example:

$$9y + 6y = (9 + 6)y$$
$$= 15 \quad y$$

You should be able to see that a shortcut is possible: Two letter-terms can be combined by simply adding their numerical coefficients. That is:

$$9y + 6y = 15y, \text{ since } 9 + 6 = 15.$$

Use the shortcut for these:

(a) (-6t) + 9t = ______

(b) 5R + (-7R) = ______

(c) (-2d) + (-3d) = ______

Answers: a) -5z b) -9d c) 5h + 6h = 11h d) (-9x) + 7x = -2x

a) 3t

b) -2R

c) -5d

112. When a subtraction is involved, the shortcut of "adding numerical coefficients" cannot be used until the subtraction is converted to addition.

$$3y - 7y = 3y + (-7y) = -4y$$

$$(\text{Since } 3 + (-7) = -4)$$

Convert the following to addition and then use the shortcut:

(a) $11x - (-2x) = 11x + 2x =$ ______

(b) $(-7t) - (-10t) =$ ______ + ______ = ______

(c) $(-4b) - 3b =$ ______ + ______ = ______

a) $13x$ b) $(-7t) + 10t = 3t$ c) $(-4b) + (-3b) = -7b$

SELF-TEST 5 (Frames 81-112)

Using the distributive principle, complete the following:

1. $4(3 + 7) = 4(\quad) + 4(\quad)$
2. $(3 + 7)4 = (3)(4) + (\quad)(\quad)$
3. $(-2)[(-3) + 7] = (\quad)(-3) + (\quad)(7)$
4. $[6 + (-1)](-5) = (\quad)(\quad) + (\quad)(\quad)$
5. $(2 + 7)w = (\quad)(\quad) + (\quad)(\quad)$
6. $[4 + (-6)]t = (\quad)(\quad) + (\quad)(\quad)$

Complete:

7. $\square(\triangle + \bigcirc) = (\quad)(\quad) + (\quad)(\quad)$
8. $(\square + \triangle)\bigcirc = (\quad)(\quad) + (\quad)(\quad)$

Factor:

9. $27 = (-3)\square$
10. $-50 = \triangle(-25)$

Factor by the distributive principle:

11. $3(7) + 3(2) = (\quad)(7 + 2)$
12. $5(2) + 8(2) = (\quad + \quad)(2)$
13. $6v + 2v = (\quad + \quad)v$
14. $(-5d) + 3d = (\quad + \quad)(\quad)$

15. The expression $2x + 7$ cannot be factored by the distributive principle. Why not? ______________________

Using the shortcut, combine the following:

16. $3w + 12w =$ ______
17. $9a + (-14a) =$ ______
18. $(-7r) + 7r =$ ______
19. $5g - 18g =$ ______
20. $(-16p) - (-11p) =$ ______
21. $(-4k) - (-4k) =$ ______

ANSWERS:

1. $4(3) + 4(7)$
2. $(3)(4) + (7)(4)$
3. $(-2)(-3) + (-2)(7)$
4. $(6)(-5) + (-1)(-5)$
5. $(2)(w) + (7)(w)$
6. $(4)(t) + (-6)(t)$
7. $(\square)(\triangle) + (\square)(\bigcirc)$
8. $(\square)(\bigcirc) + (\triangle)(\bigcirc)$
9. -9
10. 2
11. $(3)(7 + 2)$
12. $(5 + 8)(2)$
13. $(6 + 2)v$
14. $[(-5) + 3]d$
15. Because the second term does not contain "x" as a factor.
16. $15w$
17. $-5a$
18. 0
19. $-13g$
20. $-5p$
21. 0

2-9 SIMPLIFYING EQUATIONS BY COMBINING LETTER-TERMS

In this section, we will solve equations in which two letter-terms appear on the same side of the equation. To solve equations of this type, we can reduce the more complicated equation to an instance of the basic equation by combining the two letter-terms.

---

113. Here is an equation with two letter-terms on the left:

$$7x + 2x = 27$$

When the same letter appears in two terms in an equation, <u>the same number must be plugged in for it in both places.</u>

(a) Plug in +4 for both "x's" in the equation above.
Is +4 the root of the equation? _______

(b) Plug in +3 for both "x's" in the equation above.
Is +3 the root of the equation? _______

---

114. When a letter appears in two terms in an equation, it is difficult to solve the equation by trial-and-error. <u>Therefore, we combine the two letter-terms into one to reduce the equation to an instance of the basic equation which is easier to solve.</u>

$$5y + 10y = 30$$
$$15y = 30$$

The root of the new equation is +2.
The root of the original equation is also +2.

Therefore, we say that the two equations are _______________.

Answers to 113: a) No, since $28 + 8 \neq 27$ b) Yes, since $21 + 6 = 27$

---

115. When two letter-terms are combined into one, <u>the new equation is always equivalent to the original equation.</u>

$$3x + 2x = -30$$
$$5x = -30$$

(a) The root of $5x = -30$ is _______.

(b) Therefore, the root of the original equation must be _______.

Answer to 114: equivalent

116.

$$9m + (-3m) = 42$$

(a) Reduce this equation to an instance of the basic equation:

_______ = _______

(b) The root of the original equation must be _______.

Answers to 115: a) -6 b) -6

Answers to 116: a) $6m = 42$ b) +7

117. $(-4t) + 9t = -45$

(a) Reduce this equation to an instance of the basic equation:

_______ = _______

(b) The root of the new equation is _______.

(c) Show that this root satisfies the original equation:

---

a) $5t = -45$

b) -9

c) $(-4)(-9) + 9(-9) = -45$
$(+36) + (-81) = -45$
$-45 = -45$

118. $7x - 3x = 23 + 17$

(a) To reduce the equation above to an instance of the basic equation, we must combine the two letter-terms on the left and the two number-terms on the right. Do so:

_______ = _______

(b) The root of the original equation must be _______.

---

a) $4x = 40$

b) 10

119. $(-2m) + 8m = (-20) + (-4)$

(a) Reduce the equation above to an instance of the basic equation:

_______ = _______

(b) The root of the new equation is _______.

(c) Show that this root satisfies the original equation:

---

a) $6m = -24$

b) -4

c) $(-2)(-4) + 8(-4) = (-20) + (-4)$
$(+8) + (-32) = (-24)$
$-24 = -24$

120. $5d - (-3d) = 40$

(a) Reduce this equation to an instance of the basic equation:

_______ = _______

(b) The root of the new equation is _______.

(c) Show that this root satisfies the original equation:

---

a) $8d = 40$

b) 5

c) $5(5) - [(-3)(5)] = 40$
$25 - (-15) = 40$
$25 + (+15) = 40$
$40 = 40$

121. Find the root of each of these:

(a) $(-4x) - (-9x) = -35$

The root is _______.

(b) $2t - (-5t) = (-25) + (-3)$

The root is _______.

---

a) -7 b) -4

2-10 THE INTERCHANGE PRINCIPLE FOR EQUATIONS

Since we read from left to right, most of us feel more comfortable if the letter-term appears on the left side of an equation. In this section, we will introduce the "interchange principle for equations." This principle can be used to interchange the sides of an equation so that the letter-term is moved from the right side to the left side.

---

122. The root of the following equation is "5".

$$35 = 7x$$

If we interchange the two sides of the equation, we get this new equation:

$$7x = 35$$

Is "5" the root of this new equation? _______

123.

$$11 = 2x + 3$$

Yes

The root of this equation is "+4". If we interchange the two sides, we get:

$$2x + 3 = 11$$

(a) Is "+4" the root of this new equation? _______

(b) When we interchange the two sides of an equation, is the new equation <u>equivalent</u> to the original one? That is, does the new equation have the same root as the original equation? _______

---

124. The <u>INTERCHANGE</u> <u>PRINCIPLE</u> <u>FOR</u> <u>EQUATIONS</u> says this:

a) Yes

b) Yes

> IF THE TWO SIDES OF AN EQUATION ARE INTERCHANGED, THE NEW EQUATION IS EQUIVALENT TO THE ORIGINAL EQUATION.
>
> That is: If □ = ○ ,
>
> then ○ = □ .

If the root of $65 = 4x + 29$ is "9",

the root of $4x + 29 = 65$ must be "9".

We know these two equations are equivalent because of the ____________________ principle for equations.

---

interchange

125. The interchange principle for equations is especially useful with instances of the basic equation when the letter-term is on the right. (Most people like to have the letter-term on the left.)

Instead of finding the roots of:

$$27 = 9t \quad \text{and} \quad -60 = 10m$$

You can interchange the sides and get:

$$9t = 27 \quad \text{and} \quad 10m = -60$$

Since the root of $10m = -60$ is "-6",

the root of $-60 = 10m$ is _______.

Answer: -6

126. If you feel more comfortable with the letter-term on the left, use the "interchange principle" to find the roots for these:

(a) $56 = 7x$ The root is _______.

(b) $-36 = 3d$ The root is _______.

(c) $-45 = 9y$ The root is _______.

Answers: a) 8 b) -12 c) -5

127. The interchange principle can be used with any equation. For example:

If $35 + (-15) = 3t + 2t$,
then $3t + 2t = 35 + (-15)$.

However, rewriting a complicated equation is tedious. It is easier to reduce the equation as it stands to an instance of the basic equation and then interchange sides.

$$35 + (-15) = 3t + 2t$$
$$20 = 5t$$
$$5t = 20$$

(a) The root of the last equation is _______.

(b) The root of the original equation must be _______.

Answers: a) 4 b) 4

128. Solve each of these:

(a) $(-40) + 10 = 7m + (-2m)$ (b) $10 - 26 = 2x - (-2x)$

The root is _______. The root is _______.

Answers: a) -6 b) -4

2-11 THE OPPOSITING PRINCIPLE FOR EQUATIONS

Many people find it difficult to solve instances of the basic equation in which the numerical coefficient of the letter is negative. In this section, we will introduce the "oppositing principle for equations." By using this principle, we can avoid solving basic equations which have negative numerical coefficients.

---

129. Here is an equation in which the numerical coefficient of the letter is negative. The root is "+3".

$$-7x = -21$$

The opposite of $-7x$ is $7x$.
The opposite of -21 is 21.

If we replace each side of the original equation by its opposite, we get:

$$7x = 21$$

(a) Is "+3" the root of this new equation? _______

(b) Is the new equation equivalent to the original one? _______

---

130. The root of the following equation is "-4".

$$-9m = 36$$

(a) Write a new equation by replacing each side with its opposite:

_______ = _______

(b) The root of this new equation is _______.

(c) Is the new equation equivalent to the original one? _______

Answers to 129: a) Yes b) Yes

---

131. The OPPOSITING PRINCIPLE FOR EQUATIONS says this:

> IF EACH SIDE OF AN EQUATION IS REPLACED BY ITS OPPOSITE, THE NEW EQUATION IS EQUIVALENT TO THE ORIGINAL ONE.

Since the root of $5y = -10$ is "-2",

the root of $-5y = 10$ must be _______.

Answers to 130: a) $9m = -36$ b) -4 c) Yes

---

132. In each of the following:

(1) Write a new equation by applying the "oppositing principle."
(2) Then write the root of the original equation. (You may need the "interchange principle" also.)

(a) $-5x = 55$ _____ = _____ The root is _____.

(b) $-54 = -6m$ _____ = _____ The root is _____.

(c) $48 = -6t$ _____ = _____ The root is _____.

Answer to 131: -2

133. When you meet a basic equation like:

$$-5d = 20$$

you can find its root in either of two ways:

(1) Solve it as it stands.

The root is "-4" since $(-5)(-4) = 20$.

(2) Apply the oppositing principle before solving.

$$5d = -20$$

The root is "-4" since $5(-4) = -20$.

> You can do it either way. If you prefer to avoid negative numerical coefficients, use the oppositing principle first.

a) $5x = -55$
The root is -11.

b) $54 = 6m$
The root is 9.

c) $-48 = 6t$
The root is -8.

---

134. Solve each:

(a) $-12x = 48$ The root is ______.

(b) $36 = -6t$ The root is ______.

(c) $-56 = -7a$ The root is ______.

Go to next frame.

---

135. When equations are reduced to instances of the basic equation, the basic equation sometimes has a negative numerical coefficient. You can use the "oppositing principle" with the basic equation in these cases.

Solve each of these:

(a) $7b - 11b = 44$ The root is ______.

(b) $3R + (-8R) = (-30) + (-10)$ The root is ______.

a) -4

b) -6

c) +8

---

136. Solve each of these:

(a) $-32 = (-8x) - (-4x)$

The root is ______.

(b) $80 + (-8) = 7g - 16g$

The root is ______.

a) -11

b) +8

---

a) +8 b) -8

SELF-TEST 6 (Frames 113-136)

Find the root of each equation:

1. $9t - 5t = 2 - 10$

The root is _______.

2. $7h - (-2h) = (-3) + 21$

The root is _______.

3. Apply the "interchange principle" to the equation $15 = -3w$.

The resulting equation is ____________.

4. Apply the "oppositing principle" to the equation $-8p = -24$.

The resulting equation is ____________.

5. Apply both the "interchange principle" and the "oppositing principle" to the equation $28 = -7b$.

The resulting equation is ____________.

Solve these equations:

6. $(-10) - 25 = 9r - 4r$

The root is _______.

7. $(-5) - 3 = 7w - 9w$

The root is _______.

ANSWERS:

1. -2
2. 2
3. $-3w = 15$
4. $8p = 24$
5. $7b = -28$
6. -7
7. 4

2-12 THE ADDITION AXIOM FOR EQUATIONS

Our general strategy in solving more complicated equations is this: Simplify the equation by reducing it to an equivalent instance of the basic equation. Up to this point, we have simplified equations by:

(1) combining number-terms,

or (2) combining letter-terms.

Here are some examples of equations which cannot be reduced to instances of the basic equation by combining terms:

$$3x + 7 = 34$$

$$5x = 2x + 9$$

$$9x + 8 = 4x - 32$$

The first equation has number-terms on both sides.
The second equation has letter-terms on both sides.
The third equation has both number-terms and letter-terms on both sides.

To reduce equations of this type to an instance of the basic equation, we need the "addition axiom for equations." In this section, we will introduce the "addition axiom" and use it to solve equations with number-terms on both sides.

137. We can combine two letter-terms into one term because we can factor by means of the distributive principle and simplify the grouping:

$$3x + 7x = (3 + 7)x$$
$$= 10 \quad x$$

We cannot combine a letter-term and a number-term into one term because we cannot factor by means of the distributive principle. Here is an example:

$$3x + 7$$

We cannot factor by the distributive principle because the two terms do not contain a common factor.

Which of the following expressions cannot be factored by means of the distributive principle? _______

(a) $7y + 8y$ (b) $7y + 8$

---

138. Which of the following two-term expressions cannot be combined into one term? _______

(a) $5m + 7$ (b) $5m + 7m$

Answer (137): (b)

---

139. The following equation can be reduced to an instance of the basic equation by combining the two terms on the left side.

$$2x + 3x = 15$$
$$5x = 15$$

The following equation cannot be reduced to an instance of the basic equation by combining the two terms on the left side, since these two terms cannot be combined.

$$2x + 3 = 15$$

Therefore, we need another way to reduce this equation to an instance of the basic equation. This "other" way is called the "addition axiom for equations." We will begin to introduce this axiom in the next frame.

Answer (138): (a)

---

140. Here is a simple equation: $3x = 12$ (The root is "+4".)

We want to show this fact: If we add the same number to both sides of this equation, no matter what number it is, the new equation will be equivalent to $3x = 12$.

(a) Let's add "+1" to each side of $3x = 12$:

$$3x + 1 = 12 + 1$$

or $3x + 1 = 13$

Is "+4" the root of this new equation? _______

(Continued on following page.)

Answer (139): Go to next frame.

140. (Continued)

(b) Let's add "+5" to each side of $3x = 12$:

$$3x + 5 = 12 + 5$$

or $3x + 5 = 17$

Is "+4" the root of this new equation? _______

(c) Let's add "-2" to each side of $3x = 12$:

$$3x + (-2) = 12 + (-2)$$

or $3x + (-2) = 10$

Is "+4" the root of this new equation? _______

---

141. In the last frame, we saw that the following equations all have the same root (namely, "+4"):

$$3x = 12$$

$$3x + 1 = 13$$

$$3x + 5 = 17$$

$$3x + (-2) = 10$$

Since these equations have the same root, we call them _____________ equations.

Answers to frame 140:

a) Yes, since $12 + 1 = 13$

b) Yes, since $12 + 5 = 17$

c) Yes, since $12 + (-2) = 10$

---

142. Let's add the same number to both sides of this equation:

$$2x + 5 = 11 \qquad \text{(The root is "+3".)}$$

(a) Adding "+10" to both sides, we get:

$$2x + 5 + 10 = 11 + 10$$

or $2x + 15 = 21$

Is "+3" the root of this new equation? _______

(b) Adding "+50" to both sides, we get:

$$2x + 5 + 50 = 11 + 50$$

or $2x + 55 = 61$

Is "+3" the root of this new equation? _______

(c) Adding "-20" to both sides, we get:

$$2x + 5 + (-20) = 11 + (-20)$$

or $2x + (-15) = -9$

Is "+3" the root of this new equation? _______

Answer to frame 141: equivalent

143. The ADDITION AXIOM FOR EQUATIONS says this:

> IF YOU ADD THE SAME QUANTITY TO BOTH SIDES OF AN EQUATION, THE NEW EQUATION IS EQUIVALENT TO THE ORIGINAL EQUATION.

That is: Any equation in this form:

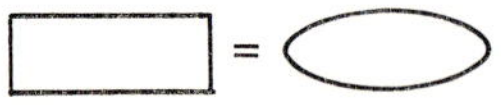

has the same root as one in this form:

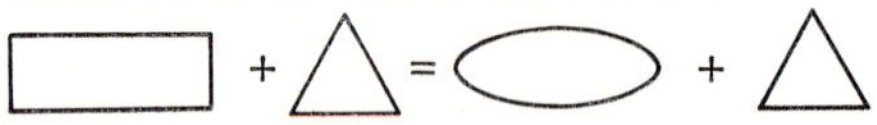

(The two triangles show symbolically that the same quantity is added to both sides of the equation.)

For example, the following two equations have the same root:

$$\boxed{3x + 9} = 21$$

and $\boxed{3x + 9} + 40 = 21 + 40$

a) Yes, since $6 + 15 = 21$

b) Yes, since $6 + 55 = 61$

c) Yes, since $6 + (-15) = -9$

---

144. The "addition axiom" says that we can add any number to both sides of an equation and obtain an equivalent equation. However, if we want to use the addition axiom to obtain an equivalent equation which is an instance of the basic equation, adding any number to both sides will not lead to an instance of the basic equation.

Here is an example: $3x + 5 = 17$ (The root is "+4".)

Let's add "+7" to both sides. We get:

$$3x + 5 + 7 = 17 + 7$$

Combining the number-terms on each side, we get:

$$3x + 12 = 24$$

The new equation is equivalent to the original one since its root is also "+4".

However, is the new equation an instance of the basic equation? ______

Go to next frame.

---

145. In the last frame, we began with the equation:

$$3x + 5 = 17$$

To obtain an instance of the basic equation, we must get rid of the number-term on the left side. By adding +7 to both sides, we obtained this equation:

$$3x + 12 = 24$$

Adding +7 to both sides changed the number-term on the left side from "5" to "12", but it did not get rid of the number-term. Therefore, the new equation is still not an instance of the basic equation.

(Continued on following page.)

No, there are still two terms on the left side.

145. (Continued) $3x + 5 = 24$

Let's add "-5" to both sides of the original equation. "-5" is the opposite of the number-term "5". We get:

$$3x + 5 + (-5) = 17 + (-5)$$

Combining number-terms, we get:

$$3x + 0 = 12$$

or

$$3x = 12$$

In this case, is the new equation an instance of the basic equation? _____

---

146. When using the addition axiom to obtain an equivalent instance of the basic equation, we ADD THE OPPOSITE OF THE TERM WE WANT TO ELIMINATE.

Adding the opposite of "5" worked in the last frame because:

(1) The sum of a pair of opposites is "0".

$$5 + (-5) = 0$$

(2) If "0" is added to any other term, the sum is the other term.

$$3x + 0 = 3x$$

Here is another example: $5x + 9 = 44$

We can eliminate the number-term on the left by adding its opposite "-9". We get:

$$5x + \underbrace{9 + (-9)} = \underbrace{44 + (-9)}$$
$$5x + \quad 0 \quad = \quad 35$$
$$5x = 35$$

(a) The new equation is an instance of the basic equation. Its root is _______.

(b) Show that this root is also the root of the original equation:

Answer (145): Yes

---

147. In this case, there are two terms on the right side of the equation.

$$83 = 7t + 69$$

To eliminate the number-term on the right, we add its opposite to both sides. We get:

$$\underbrace{83 + (-69)} = 7t + \underbrace{69 + (-69)}$$
$$14 \quad = 7t + \quad 0$$
$$14 = 7t$$

(a) The root of the basic equation is _______.

(b) Show that this root is also the root of the original equation:

Answers (146):

a) +7

b)
$$5x + 9 = 44$$
$$5(7) + 9 = 44$$
$$35 + 9 = 44$$
$$44 = 44$$

148. $4m + 23 = 71$

(a) To eliminate the number-term from the left side, what number must be added to both sides? _______

(b) Do so and write the equivalent basic equation:

_______ = _______

(c) The root of the basic equation is _______.

(d) The root of the original equation must be _______.

Answers
a) 2
b) $83 = 7(2) + 69$ $83 = 14 + 69$ $83 = 83$

---

149. $15 = 5R + (-10)$

(a) To eliminate the number-term on the right, what number should be added to both sides? _______

(b) Do so and write the basic equation:

_______ = _______

(c) The root of the basic equation is _______.

(d) Therefore, the root of the original equation must be _______.

Answers
a) "-23", the opposite of 23
b) $4m = 48$
c) 12
d) 12

---

150. Before applying the addition axiom to an equation, we must convert any subtraction to addition. Here is an example: $4d - 13 = 15$

Converting the subtraction to addition, we get: $4d + (-13) = 15$

(a) What should you add to both sides to eliminate the number-term on the left? _______

(b) Do so and write the basic equation:

_______ = _______

(c) The root of the original equation is _______.

Answers
a) +10, the opposite of -10
b) $25 = 5R$
c) 5
d) 5

---

151. (a) Convert the subtraction to addition: $59 = 8b - 29$

_____ = _______________

(b) Now use the addition axiom to find the root:

The root is _______.

Answers
a) +13, the opposite of "-13"
b) $4d = 28$
c) 7

---

152. (a) Convert the subtraction to addition: $6y - 45 = -81$

_______________ = _____

(b) Use the addition axiom to find the root:

The root is _______.

Answers
a) $59 = 8b + (-29)$
b) The root is "11". (From $8b = 88$)

Answers
a) $6y + (-45) = -81$
b) The root is "-6". (From $6y = -36$)

153. "Checking a root" means plugging it into the original equation to make sure that it satisfies it.

In the last frame, we found that the root of the original equation (shown at the right) is "-6". Check the root in the original equation:

$$6y - 45 = -81$$

Answer to 153:

$$6y - 45 = -81$$
$$6(-6) - 45 = -81$$
$$(-36) - 45 = -81$$
$$(-36) + (-45) = -81$$
$$-81 = -81$$

154. Before using the addition axiom, we always combine terms on each side of the equation if this is possible. Here is such a case:

$$\underbrace{2x + 3x}_{5x} + \underbrace{7 + 8}_{15} = \underbrace{20 + 30}_{50}$$

$$5x + 15 = 50$$

Combine terms on each side in this one:

$$7x + (-3x) + 20 + (-9) = 22 + (-3)$$

________ + ________ = ________

Answer to 154: $4x + 11 = 19$

155. Here is a case in which the letter-terms and number-terms on the left side are separated:

$$9d + 12 + (-4d) + (-7) = 30 + (-10)$$

In this case, we can proceed in either of two ways:

(1) Commute the terms on the left side so that the letter-terms and number-terms are together, and then combine:

$$\underbrace{9d + (-4d)}_{5d} + \underbrace{12 + (-7)}_{5} = \underbrace{30 + (-10)}_{20}$$

$$5d + 5 = 20$$

(2) Or, combine the letter-terms and number-terms on the left side in your head without commuting them. You can then immediately write:

$$5d + 5 = 20$$

Combine terms on each side in this one:

$$40 + (-15) = 10t + 9 + (-2t) + (-8)$$

________ = ________ + ________

Answer to 155: $25 = 8t + 1$

156. When combining terms, it does not make any difference whether you write the letter-term or the number-term first.

$$50 + 5a + (-30) + (-2a) = 25 + 7$$

When combining, we can write either: $20 + 3a = 32$

or: $3a + 20 = 32$

By the commutative property of addition: $20 + 3a = 3a + 20$

(Continued on following page.)

156. (Continued)

Combine terms in the following. Write the simpler equation in two possible ways:

$$4 + (-12) = 30 + 7y + (-8) + 3y$$

_______ = _______ + _______

or

_______ = _______ + _______

Answer: $-8 = 10y + 22$ or $-8 = 22 + 10y$

157. Before combining terms, all subtractions must be converted to additions.

$$10m - 15 - 3m + 20 = 2 - 11$$

(a) Convert all subtractions to additions:

_______________ = _______

(b) Combine terms on each side:

_______ + _______ = _______

(c) Now use the addition axiom to obtain an instance of the basic equation:

_______ = _______

(d) The root of the original equation must be _______.

Answers:

a) $10m + (-15) + (-3m) + 20 = 2 + (-11)$

b) $7m + 5 = -9$

c) $7m = -14$

d) -2

158. In this case, the number-term appears first on the left side:

$$10 + 5x = 50$$

To eliminate the number-term on the left, we add "-10" to both sides:

$$10 + 5x + (-10) = 50 + (-10)$$

(a) Combining terms on each side, we get: _______ = _______

(b) The root of the basic equation is _______.

(c) The root of the original equation must be _______.

Answers:

a) $5x = 40$

b) 8

c) 8

159. In this case, a letter-term appears after the subtraction symbol:

$$7 - 3x = 22$$

We first convert the subtraction to addition by adding the opposite of 3x:

$$7 + (-3x) = 22$$

Now we can use the addition axiom to eliminate the number-term from the left side:

$$7 + (-3x) + (-7) = 22 + (-7)$$

$$-3x + 0 = 15$$

$$-3x = 15$$

(Continued on following page.)

159. (Continued) $-3x = 15$ (a) Apply the <u>oppositing principle</u> to this last equation: _____ = _____ (b) The root of the basic equation is _______. (c) The root of the original equation must be _______.	
160. Let's solve this one: $30 = 8 - 2x$ (a) Convert the subtraction to addition. We get: _____ = _____ (b) What basic equation do you obtain after using the addition axiom to eliminate <u>the number-term</u> on the right? _____ = _____ (c) Apply both the interchange and oppositing principles to the last equation. You get: _____ = _____ (d) The root of the original equation must be _______.	a) $3x = -15$ b) -5 c) -5
161. In the last frame, we found that "-11" is the root of: $30 = 8 - 2x$ Check to see that this root satisfies the equation:	a) $30 = 8 + (-2x)$ b) $22 = -2x$ c) $2x = -22$ d) -11
162. Solve each of these: (a) $10 - 4d = 30$ (b) $60 + (-10) = 20 - 15q + 10q$ The root is _______. The root is _______.	$30 = 8 - \underbrace{2(-11)}$ ↓ $30 = 8 - (-22)$ $30 = 8 + (+22)$ $30 = 30$ Yes, -11 is the root.
	a) -5 b) -6

<u>SELF-TEST 7 (Frames 137-162)</u>

Solve the following equations. Show all steps.

1. $7r - 9 = 12$

   The root is _______.

2. $2 + 3t - 12 = 5 - 9$

   The root is _______.

3. $29 = 5 - 8e$

   The root is _______.

4. $22 + (-6) = 4h - 7 - 9h + 3$

   The root is _______.

<u>ANSWERS:</u> 1. 3 2. 2 3. -3 4. -4

2-13 EQUATIONS WITH LETTER-TERMS ON BOTH SIDES

In the last section, we solved equations with number-terms on both sides. In doing so, we used the addition axiom to obtain an equivalent instance of the basic equation.

In this section, we will show that the addition axiom can also be used to solve equations with a letter-term on each side.

---

163. We have shown that we obtain an equivalent equation if we add the same number-term to both sides of an equation. We also obtain an equivalent equation if we add the same letter-term to both sides of an equation. Here is an example:

$$7x = 2x + 30 \qquad \text{(The root is "6".)}$$

(a) If we add "3x" to each side of the original equation,

we get: $7x + 3x = 2x + 30 + 3x$

or $10x = 5x + 30$

Is "6" the root of this new equation? _______

(b) If we add "5x" to each side of the original equation,

we get: $7x + 5x = 2x + 30 + 5x$

or $12x = 7x + 30$

Is "6" the root of this new equation? _______

(c) If we add "-4x" to each side of the original equation,

we get: $7x + (-4x) = 2x + 30 + (-4x)$

or $3x = (-2x) + 30$

Is "6" the root of this new equation? _______

---

164. Just as we obtain an equivalent equation by adding any number-term to both sides of an equation, we obtain an equivalent equation by adding any letter-term to both sides of an equation.

However, our purpose in using the addition axiom is to obtain an equivalent instance of the basic equation. Therefore, we must be selective in the letter-term we add to each side. We add the opposite of the letter-term we want to eliminate.

To obtain an instance of the basic equation from the equation below, eliminate the letter-term on the right side. That is, we add "-2x" and get:

$$7x = 2x + 30$$
$$7x + (-2x) = 2x + 30 + (-2x)$$
$$5x = \underbrace{2x + (-2x)} + 30$$
$$5x = (0) + 30$$
$$5x = 30$$

(a) The root of the basic equation is _______.

(b) Therefore, the root of the original equation must be _______.

Answers to 163:

a) Yes, since: $60 = 30 + 30$

b) Yes, since: $72 = 42 + 30$

c) Yes, since: $18 = (-12) + 30$

165. $5y + 36 = 9y$

To obtain an instance of the basic equation, we must eliminate the letter-term on the left side.

(a) What letter-term do we add to both sides to eliminate "5y"? _______

(b) Do so and write the resulting basic equation:

_______ = _______

(c) The root of the original equation must be _______.

Answers: a) 6 b) 6

---

166. When an equation has a number-term on each side, we eliminate the number-term from the side which contains two terms. (Then we immediately get an instance of the basic equation.)

In $5x + 15 = 60$, we eliminate the "15" from the left side.

Similarly, when an equation has a letter-term on each side, we eliminate the letter-term from the side which contains two terms. (Then we immediately get an instance of the basic equation.)

In $6t = 2t + 28$, we eliminate the "2t" from the right side.

When using the addition axiom, which term should be eliminated in each of these?

(a) $10m + 27 = 7m$ _______ (c) $56 + 5d = 13d$ _______

(b) $8R = 2R + 48$ _______ (d) $15v = 49 + 8v$ _______

Answers: a) -5y, its opposite b) $36 = 4y$ c) 9

---

167. $3t = 45 + (-6t)$

(a) To obtain an instance of the basic equation, what term should we eliminate? _______

(b) Do so and write the resulting basic equation:

_______ = _______

(c) The root of the original equation must be _______.

Answers: a) 10m b) 2R c) 5d d) 8v

---

168. In this one, we must convert the subtraction to addition first:

$7m = 40 - 3m$

(a) Convert the subtraction to addition:

_______ = _______________

(b) Use the addition axiom to obtain an instance of the basic equation:

_______ = _______

(c) The root of the original equation must be _______.

Answers: a) (-6t) b) $9t = 45$ c) 5

169. (a) Solve: $24 - 3m = -9m$

(b) Check your root in the original equation:

The root is _______.

a) $7m = 40 + (-3m)$

b) $10m = 40$

c) 4

---

a) -4

b) $24 - (3)(-4) = (-9)(-4)$

$24 - (-12) = +36$

$24 + 12 = 36$

$36 = 36$

2-14 EQUATIONS WITH LETTER-TERMS AND NUMBER-TERMS ON BOTH SIDES

In this section, we will solve equations with letter-terms and number-terms on both sides. To solve them, we must use the addition axiom twice in order to obtain an instance of the basic equation.

---

170. This equation has a letter-term and a number-term on each side:

$$9x + 35 = 4x + 70$$

To obtain an instance of the basic equation, we must eliminate a letter-term from one side and a number-term from the other side. We have two choices:

(1) eliminating "9x" and "70",

or (2) eliminating "4x" and "35".

With either choice, we must use the addition axiom twice.

Let's eliminate "4x" and "35".

(1) To eliminate the "4x" on the right, we add "-4x" to both sides:

$$9x + (-4x) + 35 = 4x + (-4x) + 70$$

$$5x + 35 = 0 + 70$$

$$5x + 35 = 70$$

(2) The last equation still has number-terms on each side. To eliminate the "35", we add "-35" to both sides:

$$5x + 35 + (-35) = 70 + (-35)$$

$$5x + 0 = 35$$

$$5x = 35$$

Now we have an instance of the basic equation.

(a) The root of the basic equation is _______.

(b) The root of the original equation must be _______.

171. In the last frame, we decided that "7" is the root of the following equation: $9x + 35 = 4x + 70$ Check to see that this root satisfies this equation:	a) 7 b) 7
172. Here is another equation of the same type:	$9(7) + 35 = 4(7) + 70$ $63 + 35 = 28 + 70$ $98 = 98$

$$3x + 28 = 7x + 12$$

To obtain an instance of the basic equation, we must eliminate either:

(1) "3x" and "12",
or (2) "7x" and "28".

No matter which pair we eliminate, we will obtain basic equations with the same root.

(1) Eliminating "3x" and "12":

(a) $3x + (-3x) + 28 = 7x + (-3x) + 12$
$28 = 4x + 12$

(b) $28 + (-12) = 4x + 12 + (-12)$
$16 = 4x$

(2) Eliminating "7x" and "28":

(a) $3x + (-7x) + 28 = 7x + (-7x) + 12$
$(-4x) + 28 = 12$

(b) $(-4x) + 28 + (-28) = 12 + (-28)$
$-4x = -16$

We obtained these two basic equations:

(1) $16 = 4x$ and (2) $-4x = -16$

(a) The root of each of these equations is _______.

(b) Therefore, the root of the original equation must be _______.

---

173. In the last frame, we began with this equation:

a) +4 or 4 b) +4 or 4

$$3x + 28 = 7x + 12$$

Eliminating "3x" and "12", we obtained: $16 = 4x$

Eliminating "7x" and "28", we obtained: $-4x = -16$

The numerical coefficient of the letter is positive in the first basic equation and negative in the second. When choosing a pair of terms to eliminate, one choice will always lead to a positive numerical coefficient and the other choice to a negative numerical coefficient. Since we have a choice, we try to avoid the negative numerical coefficient because its equation is more difficult to solve.

(Continued on following page.)

173. (Continued)

Here is a new equation: $3b + 8 = 5b + 2$

We can eliminate either of these pairs:

(1) "5b" and "8",
or (2) "3b' and "2".

You can tell which pair will lead to a negative coefficient by beginning to eliminate the letter-terms. Which pair is it? ______________

---

174. In each of the following equations, which pair of terms should be eliminated to obtain a positive coefficient for the letter in the basic equation?

(a) $7y + 9 = 10y + 18$ _______ and _______

(b) $8d + 6 = 5d + 21$ _______ and _______

Answer (173): (1) "5b" and "8"

---

175. In each of the following, which pair of terms should be eliminated to obtain a positive coefficient in the basic equation? (Don't let the negative coefficients bother you. Just try them and see.)

(a) $10 + (-4b) = 40 + 6b$ _______ and _______

(b) $7 + (-8x) = 11 + (-10x)$ _______ and _______

Answer (174):
a) 7y and 18
b) 5d and 6

---

176. Find the root of: $3x + 8 = 5x + 2$

The root is _______.

Answer (175):
a) -4b and 40
b) -10x and 7

---

177. Find the root of: $25 + (-4y) = 5 + 6y$

The root is _______.

Answer (176): The root is 3. (From $6 = 2x$)

---

178. In this one, you must convert the subtractions to additions first. Find the root:

$70 - 5f = 30 - 9f$

The root is _______.

Answer (177): The root is 2. (From $20 = 10y$)

Answer (178): The root is -10. (From $4f = -40$)

179. NEVER USE THE ADDITION AXIOM BEFORE COMBINING TERMS ON EACH SIDE OF THE EQUATION. ALWAYS SIMPLIFY BY COMBINING TERMS FIRST.

The following equation should be simplified before using the addition axiom. Simplify it by combining terms. (Do not solve!)

$$10x + 12 + (-3x) + (-8) = 12x + 9 + (-15)$$

_______ + _______ = _______ + _______

Answer: $7x + 4 = 12x + (-6)$

180. Which of the following equations are simplified to the point where the addition axiom should be used? __________

(a) $12y + 9 = 33$

(b) $7d + 8 + (-10) = 9d + (-4d) + 10$

(c) $5m = 20 + 3m$

(d) $4t + (-7) = 14 + (-3t)$

(e) $2b + (-20) + 5b = 40 + (-3b) + 30$

Answer: Only (a), (c), and (d)

SELF-TEST 8 (Frames 163-180)

Solve the following equations. Show all steps:

1. $5y - 12 = 3y$ The root is _______.	2. $-7w = 40 - 2w$ The root is _______.	3. $9a - 4 = 3a + 8$ The root is _______.
4. $20 - 2h = 3h + 40$ The root is _______.	5. $8 - 2t + 3 = 4t - 13 + 2t$ The root is _______.	6. $7 - 3r - 5r = 35 - 2r - 10$ The root is _______.

ANSWERS: 1. 6 2. -8 3. 2 4. -4 5. 3 6. -3

2-15 FORMAL STRATEGIES FOR SOLVING EQUATIONS

We could attempt to solve any equation by trial-and-error. That is, we could plug in one number after another for the letter in the equation until we found its root. With complicated equations, however, this "guessing" procedure becomes tedious, and we would frequently "give up" before finding the root.

Therefore, mathematicians have developed strategies for solving equations. For complicated equations, any strategy includes a number of steps. Each step involves the use of an axiom or principle to obtain a simpler but equivalent equation which is easier to solve. We are convinced that a student must have a formal knowledge of these strategies in order to become skilled in equation solving. That is, the student must:

(1) Understand the axioms and principles.
(2) Know the order in which these axioms and principles are used.
(3) Be able to state at any time what is being done and why it is being done.

In the following sections, we will formally review the meaning of solving equations and the strategies for solving them. We will begin by reviewing the use of the following axioms and principles:

Converting subtraction to addition
Combining terms
Addition axiom
Interchange principle
Oppositing principle
Solution of basic equations

No new axioms or principles will be introduced at this time.

2-16 A REVIEW OF THE INDIVIDUAL STEPS USED IN SOLVING EQUATIONS

181. Any equation contains one or more terms on each side. The terms are connected by addition or subtraction symbols.

The simplest equations to solve are instances of the basic equation. In any instance of the basic equation, there is only one term on each side: A letter-term on one side, and a number-term on the other side.

Which of the following are instances of the basic equation? ________

(a) $10d = 60$ (c) $4x + 30 = 10x$

(b) $14 = 2 - 6m$ (d) $-42 = 6y$

182. Any equation which contains more than one term on either side is more complicated than the basic equation. When an equation is more complicated than the basic equation, there are various procedures which we use to solve it. The goal of all these procedures is to eventually obtain an equivalent instance of the basic equation. The latter equation is easy to solve, and its root is the root of the original, more complicated equation.

The procedures used in solving more complicated equations are used in a certain order. We will review this "order" in the following frames.

Answer to 181: (a) and (d)

183. When the equation contains a subtraction symbol or symbols, the first step is "converting all subtractions to additions." These conversions must be made before:

(1) combining terms,
or (2) using the addition axiom.

In which equations below would "converting all subtractions to additions" be the first step? ______________

(a) $7x + 9 = 37$ (c) $45 = 10 - 7m$

(b) $3y + 7 - 8y = 15 + 5y - 12$ (d) $9 + 2P = 5P + 2$

---

184. After converting all subtractions to additions, we "combine like terms" on each side of the equation. By "combining like terms," we mean:

(1) combining two number-terms,
or (2) combining two letter-terms.

The following equation is not ready for "combining like terms." Why not? ______________

$$10 + 3t - 20 = 2t + 40 - 9t$$

Both (b) and (c)

---

185. Of course, there is no need to "combine like terms" if there are not two number-terms or two letter-terms on the same side of the equation.

In which of the following is "combining like terms" necessary? ________

(a) $7x + (-9) = 5x + 9$ (c) $35 + 3b = 2b + 20 + 6b$

(b) $10 + 3q + (-30) = 2q$ (d) $3t = 15 + (-2t)$

The subtractions are not converted to additions.

---

186. When like terms are combined, we sometimes obtain an instance of the basic equation immediately. In which of the following would this be the case? ______________

(a) $5x + 3x = 40$ (b) $7y + 2y + 19 = 100$ (c) $35 = 3h + (-8h)$

Only in (b) and (c)

---

187. Here are some equations which contain both a letter-term and a number-term on one side and a single term on the other side:

$$3x + 5 = 26$$
$$9b = 4b + 25$$

Since the letter-term and number-term on the same side cannot be combined, we cannot obtain an instance of the basic equation immediately. We must use the "addition axiom" to do so.

If an equation has a number-term on both sides, we use the addition axiom to eliminate the number-term on the two-term side.

For example: In $3x + 5 = 26$,

we eliminate the "5" from the left side.

(Continued on following page.)

(a) and (c)

187. (Continued)

If an equation has a letter-term on both sides, we use the addition axiom to eliminate the letter-term on the two-term side.

For example: In $9b = 4b + 25$,

we eliminate the "4b" from the right side.

What term would we eliminate by means of the addition axiom in each of these?

(a) $6h + (-30) = 12$ _______ (c) $75 = 10x + 25$ _______

(b) $25 + (-3y) = 2y$ _______ (d) $5m = 49 + (-2m)$ _______

---

188. If an equation contains a letter-term and a number-term on both sides, we must use the addition axiom twice. We use it to eliminate the letter-term from one side and the number-term from the other side. Of the two pairs of terms we could eliminate, we eliminate that pair which leads to an instance of the basic equation in which the letter-term has a positive coefficient.

For example: In $3w + 18 = 7w + 6$,

we could eliminate either

(1) "3w" and "6" and get $12 = 4w$

or (2) "7w" and "18" and get $-4w = -12$.

We eliminate "3w" and "6" because the basic equation then has a positive coefficient.

Which pair of terms would we eliminate in each of these to obtain a positive coefficient in the basic equation?

(a) $5d + 40 = 9d + 12$ _______ and _______

(b) $7R + (-30) = 45 + (-8R)$ _______ and _______

(c) $100 + (-9a) = 65 + (-4a)$ _______ and _______

a) (-30) c) 25

b) (-3y) d) (-2m)

---

189. The addition axiom is never used until:

(1) all subtractions are converted to additions,

and (2) all like terms are combined on each side.

Which of the following equations are ready for use of the addition axiom? __________

(a) $10S - 6 = 5S - 36$ (c) $2R + 9 = 17 + 5R + (-20)$

(b) $3g + 21 + 6g = 7 + 11g$ (d) $30 + (-4b) = 70 + (-12b)$

a) 5d and 12

b) (-8R) and (-30)

c) (-9a) and 65

Only (d)

190. We have discussed three steps:

(1) Converting subtraction to addition
(2) Combining like terms
(3) Addition axiom

It is important that you recognize what your first step should be when solving an equation. For each equation below, write the name of your first step in the space provided. If you write "addition axiom," specify the term you would eliminate.

(a) $7x + 3x = 40$ ______________________

(b) $35 - 7m = 70$ ______________________

(c) $50 = 35 + 5y$ ______________________

---

191. Write the name of the first step for each of these:

(a) $7m - 11m = 40$ ______________________

(b) $4x + 50 + 2x = 35 + 3x + 30$ ______________________

(c) $10v = 60 + (-5v)$ ______________________

(d) $4q + 9q = 29 - 7q - 43$ ______________________

Answers to 190:
a) Combining like terms
b) Converting subtraction to addition
c) Addition axiom Eliminate "35"

---

192. All of the following are instances of the basic equation:

$7x = 14$ $27 = 3m$ $-9y = 45$

Equations of this type are easiest to solve if:

(1) the letter-term is on the left side,
and (2) the coefficient of the letter is positive.

Given any instance of the basic equation, we can use the interchange principle or the oppositing principle to obtain an equation which is easiest to solve.

(1) We can use the interchange principle to get the letter-term on the left.

$$27 = 3m$$

is equivalent to:

$$3m = 27$$

(2) We can use the oppositing principle to eliminate a negative coefficient.

$$-9y = 45$$

is equivalent to:

$$9y = -45$$

For each of the following, write the name of the principle you would use to obtain a basic equation which is easiest to solve.

(a) $-10d = 50$ ______________ principle

(b) $48 = 6y$ ______________ principle

(c) $-7d = -42$ ______________ principle

Answers to 191:
a) Converting subtraction to addition
b) Combining like terms
c) Addition axiom Eliminate "-5v"
d) Converting subtraction to addition

193. With the following basic equation, we need both the interchange principle and the oppositing principle to obtain an equation which is easiest to solve.

$$40 = -8x$$

is equivalent to:

$$-8x = 40$$

is equivalent to:

$$8x = -40$$

Which of the following are equations which require the use of both principles? ________

(a) $72 = 9y$ (b) $56 = -7d$ (c) $-5R = -45$

a) Oppositing principle
b) Interchange principle
c) Oppositing principle

---

194. It is important that you be able to identify the process used in each step of a solution of an equation. Here is an example:

$$7x - 15 = 50$$

$$7x + (-15) = 50$$

Identify by name what we did to the top equation in order to obtain the bottom one. ________________________

Only (b)

---

195. Identify by name the process used in each of the following steps:

(a) $7x + 3x = 40$
$10x = 40$ ________________

(b) $50 = 10y$
$10y = 50$ ________________

(c) $-6d = 54$ ________
$6d = -54$ ________________

Converting subtraction to addition.

---

196. Even though any use of the addition axiom includes:

(1) adding the same quantity to both sides,
and (2) then combining like terms,

we will consider this whole process to be one step. Here is an example:

$$5m + 20 = 50$$

$$\left.\begin{aligned} 5m + (20) + (-20) &= 50 + (-20) \\ 5m &= 30 \end{aligned}\right\} \quad \text{Addition axiom } (-20)$$

The brace "}" at the right shows that we consider both "adding -20" and "combining like terms" to be one step. Notice that we wrote (-20) after the words "addition axiom." We always identify in this way the term added to both sides when using the addition axiom.

(Continued on following page.)

a) Combining like terms
b) Interchange principle
c) Oppositing principle

196. (Continued)

Properly identify the use of the addition axiom for each of these:

(a) $9p = 5p + 28$

$9p + (-5p) = 5p + (-5p) + 28$

$4p = 28$ ____________________

(b) $50 = 8g + (-22)$

$50 + (+22) = 8g + (-22) + (+22)$

$72 = 8g$ ____________________

---

197. In the blank to the right, identify each of the following steps by name:

(a) $-49 = 7b$
$7b = -49$ ____________________

(b) $25 - 10x = 75 - 15x$
$25 + (-10x) = 75 + (-15x)$ ____________________

(c) $7h + (-20) + (-5h) = 40 + (-6h) + (-30)$
$2h + (-20) = 10 + (-6h)$ ____________________

a) Addition axiom (-5p)

b) Addition axiom (+22)

---

198. Identify each of the following steps:

(a) $3q + 19 = 6q + 7$

$3q + (-3q) + 19 = 6q + (-3q) + 7$

$19 = 3q + 7$ ____________________

(b) $-4t = -28$

$4t = 28$ ____________________

(c) $6d + (-10) = 9d + (-40)$

$6d + (-10) + (+10) = 9d + (-40) + (+10)$

$6d = 9d + (-30)$ ____________________

a) Interchange principle

b) Converting subtraction to addition

c) Combining like terms

---

199. When we have obtained an instance of the basic equation which is easiest to solve, we can write the root of the original equation. We call this step "solving the basic equation." Here is an example:

$7p = 42$

The root is 6. Solving the basic equation

Identify each of these steps:

(a) $70 = 10f$
$10f = 70$ ____________________

(b) $8R = 56$
The root is 7. ____________________

(c) $-9m = 54$
$9m = -54$ ____________________

a) Addition axiom (-3q)

b) Oppositing principle

c) Addition axiom (+10)

200. We have reviewed the following steps which occur in the solution of equations. The steps are listed in the order in which they are used. All of the steps are not necessary in the solution of any given equation. We only need those steps which are necessary to obtain an instance of the basic equation.

Converting subtraction to addition (if necessary)
Combining like terms (if necessary)
Addition axiom (if necessary)
Basic equation:
  Interchange principle (if necessary)
  Oppositing principle (if necessary)
  Solving the basic equation (always necessary)

a) Interchange principle
b) Solving the basic equation
c) Oppositing principle

2-17 IDENTIFYING ALL STEPS IN SOLVING EQUATIONS

In this section, we will examine complete solutions of equations. Two types of problems will be involved:

(1) All the steps will be shown. You must identify the process used in each step.
(2) Only the original equation will be given. You must generate each step and identify the process used.

---

201. In the following problem, we have shown all of the steps. Each step is numbered. The process used for each step is identified in the blank to the right.

$$\boxed{3x - 12 = 21}$$

Step 1: $3x + (-12) = 21$ — Converting subtraction to addition

Step 2: $3x + (-12) + (+12) = 21 + (+12)$, $3x = 33$ — Addition axiom (+12)

Step 3: The root is +11. — Solving the basic equation.

Identify the process used in each step in solving this equation:

$$\boxed{10y - 14y = 20}$$

Step 1: $10y + (-14y) = 20$ ________________

Step 2: $-4y = 20$ ________________

Step 3: $4y = -20$ ________________

Step 4: The root is −5. ________________

Step 1: Converting subtraction to addition
Step 2: Combining like terms
Step 3: Oppositing principle
Step 4: Solving the basic equation

202. Identify the process used in each step:

$$\boxed{5d + 40 = 7d + 32}$$

Step 1: $5d + (-5d) + 40 = 7d + (-5d) + 32$; $40 = 2d + 32$ ____________________

Step 2: $40 + (-32) = 2d + 32 + (-32)$; $8 = 2d$ ____________________

Step 3: $2d = 8$ ____________________

Step 4: The root is +4. ____________________

Answer to Frame 202:

Step 1: Addition axiom (-5d)
Step 2: Addition axiom (-32)
Step 3: Interchange principle
Step 4: Solving the basic equation

---

203. Identify the process used in each step:

$$\boxed{34 = 3p - 8 - 10p}$$

Step 1: $34 = 3p + (-8) + (-10p)$ ____________________

Step 2: $34 = (-7p) + (-8)$ ____________________

Step 3: $34 + (+8) = (-7p) + (-8) + (+8)$; $42 = -7p$ ____________________

Step 4: $-7p = 42$ ____________________

Step 5: $7p = -42$ ____________________

Step 6: The root is -6. ____________________

Answer to Frame 203:

Step 1: Converting subtraction to addition
Step 2: Combining like terms
Step 3: Addition axiom (+8)
Step 4: Interchange principle
Step 5: Oppositing principle
Step 6: Solving the basic equation

---

204. You will need three steps to solve the equation below. Show each step and identify the process in the space to the right.

$$\boxed{7t - 8 = 13}$$

Step 1: ____________________

Step 2: ____________________

Step 3: ____________________

Answer to Frame 204:

Step 1: $7t + (-8) = 13$ — Converting subtraction to addition

Step 2: $7t + (-8) + (+8) = 13 + (+8)$; $7t = 21$ — Addition axiom (+8)

Step 3: The root is +3. — Solving the basic equation

---

205. In these frames, we expect you to use the interchange principle or oppositing principle or both until you obtain a basic equation which is easiest to solve.

You need three steps for this one. Show each step and identify the process.

$$\boxed{40 + (-10) = 9m + (-3m)}$$

Step 1: ____________________

Step 2: ____________________

Step 3: ____________________

Answer to Frame 205:

Step 1: $30 = 6m$ — Combining like terms

Step 2: $6m = 30$ — Interchange principle

Step 3: The root is +5. — Solving the basic equation

---

206. You need four steps to solve this one. Show each step and identify the process.

$$\boxed{7d - 12d = 10 - 50}$$

Step 1: ____________________

Step 2: ____________________

Step 3: ____________________

Step 4: ____________________

Answer to Frame 206:

Step 1: $7d + (-12d) = 10 + (-50)$ — Converting subtraction to addition

Step 2: $-5d = -40$ — Combining like terms

Step 3: $5d = 40$ — Oppositing principle

Step 4: The root is +8. — Solving the basic equation

207. You will need <u>three</u> steps for this one:

$4x = 36 + 7x$

Step 1: ____________________

Step 2: ____________________

Step 3: ____________________

Answer to Frame 207:

Step 1: $4x + (-7x) = 36 + 7x + (-7x)$; $-3x = 36$ — Addition axiom (-7x)

Step 2: $3x = -36$ — Oppositing principle

Step 3: The root is -12. — Solving the basic equation

---

208. Whenever the addition axiom must be used <u>twice</u>, <u>eliminate the letter-term first</u>. Be sure to eliminate the letter-term which will lead to a <u>positive</u> coefficient in the basic equation.

You will need <u>three</u> steps for this one:

$10b + 9 = 7b + 18$

Step 1: ____________________

Step 2: ____________________

Step 3: ____________________

Answer to Frame 208:

Step 1: $10b + (-7b) + 9 = 7b + (-7b) + 18$; $3b + 9 = 18$ — Addition axiom (-7b)

Step 2: $3b + 9 + (-9) = 18 + (-9)$; $3b = 9$ — Addition axiom (-9)

Step 3: The root is +3. — Solving the basic equation

---

209. You will need <u>six</u> steps for this one:

$10 - 9x - 15 = 5x - 37 - 10x$

Step 1: ____________________

Step 2: ____________________

Step 3: ____________________

Step 4: ____________________

Step 5: ____________________

Step 6: ____________________

Answer to Frame 209:

Step 1:	$10 + (-9x) + (-15) = 5x + (-37) + (-10x)$	Converting subtraction to addition
Step 2:	$(-5) + (-9x) = (-5x) + (-37)$	Combining like terms
Step 3:	$(-5) + (-9x) + 9x = (-5x) + 9x + (-37)$ $(-5) = 4x + (-37)$	Addition axiom (9x)
Step 4:	$(-5) + 37 = 4x + (-37) + 37$ $32 = 4x$	Addition axiom (37)
Step 5:	$4x = 32$	Interchange principle
Step 6:	The root is +8.	Solving the basic equation

210. Each step in the solution of an equation has a definite purpose:

(1) "Converting subtraction to addition" is merely a preparatory step for either "combining like terms" or "using the addition axiom."
(2) "Combining like terms" and "using the addition axiom" reduce the number of terms in the equation.
(3) The "interchange principle" and the "oppositing principle" are used to write forms of the basic equation which are equivalent and easier to solve.

Each step in the solution leads to an equation which is equivalent to the preceding one and therefore equivalent to the original equation.

Since the basic equation is equivalent to all of the preceding equations in the steps of the solution, its root is the root of all of these preceding equations. Of course, we are mainly interested in the fact that its root is the root of the original equation, since that equation is the one we want to solve.

SELF-TEST 9 (Frames 181-210)

In Problems 1 to 8, state which of the principles at the right would be used as the first step in solving each equation. If the addition axiom is used as the first step, list the term which is to be added.

- Convert subtraction to addition.
- Combine like terms.
- Use addition axiom.
- Use interchange principle.
- Use oppositing principle.
- Solve basic equation.

1. $12 + 2d = 3d + 5d$ ____________
2. $24 + 3t = 5t$ ____________
3. $7p = 28$ ____________
4. $-4k = -36$ ____________
5. $11 - 2w = 5$ ____________
6. $5x + 8 = 2x + 17$ ____________
7. $-18 = 3b$ ____________
8. $8p + 3 = 9 + 3p + 4$ ____________

ANSWERS:

1. Combine like terms.
2. Use addition axiom. Add $(-3t)$.
3. Solve basic equation.
4. Use oppositing principle.
5. Convert subtraction to addition.
6. Use addition axiom. Add $(-2x)$.
7. Use interchange principle.
8. Combine like terms.

Chapter 3 MORE NON-FRACTIONAL EQUATIONS

The equations which are covered in this chapter are more difficult than those covered in the last chapter. Students typically have more trouble with them. However, we feel that they are easy if you use the principles and steps which we teach. These principles and steps build on those presented in the last chapter.

As equations become more complex, students have a tendency to begin to use shortcuts and to skip steps. For most students, shortcuts are fatal because they sacrifice accuracy for speed. We are more interested in accuracy than in speed. Therefore, we will insist that you use the principles which we teach.

3-1 THE IDENTITY PRINCIPLE OF MULTIPLICATION

In this section, we will introduce the "identity principle of multiplication." After showing what it means, we will show how it is used in the solution of equations.

---

1. Whenever a number is multiplied by +1 or 1, the product is identical to the original number. Examples: $(1)(8) = 8$ $(1)(-5) = -5$ "Multiplying by +1" is called THE IDENTITY PRINCIPLE OF MULTIPLICATION because the product is identical to the original number. We can write this principle with geometrical figures to show that it is true for any number. We get: $(1)(\square) = \square$ $(1)(17) = 17$ is an instance of the ______________ principle of multiplication.	
2. Since either geometrical figures or letters can be used to represent missing numbers, we can write the identity principle of multiplication in various ways: $(1)(\square) = \square$ $(1)(x) = x$ $(1)(n) = n$ (Any letter or geometrical figure could be used.) $(1)(10) = 10$ is an instance of this principle. On the left, "1" and "10" are factors. On the right, "10" is the product. Can any number ("n") be factored into the two factors "1" and "n"? _____	identity

3. The identity principle of multiplication can sometimes be helpful when combining terms.

Just as 3x means: "3 times x" or (3)(x),
1x means: "1 times x" or (1)(x).

Therefore: Since (1)(x) = x,
1 x = x

and using the interchange principle:

$x = 1x$

In many equations, a letter like "x" will appear without a numerical coefficient. Though it is not explicitly written, the numerical coefficient is really "1", since $x = 1x$.

Do these: (a) $7x + x = 7x +$ ____ = ____

(b) $4p + p = 4p +$ ____ = ____

(c) $(-5t) + t = (-5t) +$ ____ = ____

Answer: Yes

---

4. Solve each equation: (a) $5x + x = 42$ (b) $y + 10y = -33$

The root is ____. The root is ____.

Answers: a) $+ 1x = 8x$ b) $+ 1p = 5p$ c) $+ 1t = -4t$

---

5. Solve these: (a) $48 = m + (-7m)$ (b) $39 = 4d + 9 + d$

The root is ____. The root is ____.

Answers: a) 7 b) -3

---

6. In this expression, "x" follows a subtraction symbol:

$3x - x$

Its numerical coefficient is still "1". Therefore:

$3x - x = 3x - 1x$

Insert the numerical coefficient of the second term in these:

(a) $10y - y = 10y -$ ____ (b) $7q - q = 7q -$ ____

Answers: a) -8 b) 6

---

7. "1x" and "-1x" are a pair of opposites since their sum is "0".

Using this fact, we can convert the following subtraction to addition:

$3x - x = 3x - 1x$
$= 3x + (-1x)$
$= 2x$

Complete this one: $10t - t = 10t - 1t$
$= 10t + (\quad)$
= ____

Answers: a) $10y - 1y$ b) $7q - 1q$

8. Solve these: (a) $9m - m = -32$ (b) $45 = 6d - d$

The root is ____. The root is ____.

$= 10t + \underline{(-1t)}$

$= \underline{9t}$

---

9. Solve these: (a) $4x + 28 - x = 49$ (b) $27 = m - 4m$

The root is ____. The root is ____.

a) -4 b) 9

---

10. Here are two instances of the basic equation:

$3x = 27$ $4t = -24$

Any equation with a letter-term on one side and a number-term on the other is an instance of the basic equation. This is true even when the numerical coefficient of the letter is "1". Therefore, the following equations are also instances of the basic equation:

$1x = 9$ $1p = -7$

(a) The root of "$1x = 9$" is ____.

(b) The root of "$1p = -7$" is ____.

a) 7 b) -9

---

11. Here is a very simple equation. As is frequently the case, the numerical coefficient of the letter is not explicitly written because it is "1".

$x = 6$

By writing the numerical coefficient explicitly,

$1x = 6$

we can see that the equation above is really an instance of the basic equation.

What is the root of this basic equation? ______

a) 9, since $(1)(9) = 9$

b) -7, since $(1)(-7) = -7$

---

12. Since the following equations contain a letter-term on one side and a number-term on the other, they are instances of the basic equation <u>even though their numerical coefficients are not explicitly written.</u>

$m = 10$ $q = -6$

(a) The root of "$m = 10$" is ______.

(b) The root of "$q = -6$" is ______.

6, since $(1)(6) = 6$

---

a) 10, since $10 = 10$ or $1(10) = 10$

b) -6, since $-6 = -6$ or $(1)(-6) = -6$

13. When the numerical coefficient in a basic equation is "1" (no matter whether it is explicitly written or not), we call it a "root equation." We call it a "root equation" because the number-term is the root of the equation.

"p = 8" is a root equation because "8" is the root of the equation.

"2p = 8" is not a root equation because "8" is not the root of the equation.

Which of the following are root equations? ____________

(a) $t = 10$ (b) $3q = 33$ (c) $d = -19$

Both (a) and (c)

14. When solving a more complicated equation, sometimes we arrive at an instance of the basic equation which is a root equation. Here is an example:

$$\begin{aligned} 8x - 7x &= 10 \\ 8x + (-7x) &= 10 \\ 1x &= 10 \end{aligned}$$

or

$$x = 10$$

In such cases, we do not have to "solve the basic equation" because its root is obvious. In the case above, the root is obviously "10".

Solve: (a) $y + 6 = 15$ (b) $2b - b = -7$

The root is _______. The root is _______.

---

a) 9 b) -7

15. In the last frame, we found that the root of the following equation is "-7":

$$2b - b = -7$$

Let's check this root:

$$\begin{aligned} 2(-7) - (-7) &= -7 \\ (-14) - (-7) &= -7 \\ (-14) + (+7) &= -7 \\ -7 &= -7 \end{aligned}$$

Solve this one and check the root in the original equation:

(a) $2h - h = -5$ (b) Check:

The root is _______.

---

a) The root is -5.

b) Check:

$$\begin{aligned} 2(-5) - (-5) &= -5 \\ (-10) - (-5) &= -5 \\ (-10) + (+5) &= -5 \\ -5 &= -5 \end{aligned}$$

16. The root of the following equation is 5:

$$3x = 15$$

Instead of saying: "The root is 5," mathematicians frequently say: "x = 5".

That is, they write a root equation instead of using words.

The root of the following equation is "-10":

$$7t - 10t = 30$$

Instead of writing: "The root is -10," we can write ________.

Answer: t = -10

17. From now on, we will usually follow the convention of writing a root equation instead of the words "The root is ______."

Solve these: (a) $d - 3d = 14$ (b) $7h - h = 54$

d = ______ h = ______

Answer: a) -7 b) 9

3-2 THE OPPOSITE PRINCIPLE OF MULTIPLICATION

In this section, we will introduce the "opposite principle of multiplication." We will then show how it can be used to solve some equations.

---

18. Whenever we multiply any number by "-1", the product is the opposite of the original number.

Examples: $(-1)(+5) = -5$ $(-1)(-9) = +9$

"Multiplying by -1" is called THE OPPOSITE PRINCIPLE OF MULTIPLICATION because the product is the opposite of the original number. We can write this principle with geometrical figures or letters to show that it is true for any number. We get:

$$(-1)(\square) = -\square$$

$$(-1)(x) = -x$$

$$(-1)(n) = -n$$

$(-1)(8) = -8$ is an instance of the __________ principle of multiplication.

Answer: opposite

19. The opposite principle of multiplication is sometimes helpful when combining terms.

Just as -3x means: -3 times x or (-3)(x),
-1x means: -1 times x or (-1)(x).

Therefore: Since (-1)(x) = -x,
-1x = -x.

and using the interchange principle

-x = -1x

In many expressions, a term like "-x" will appear. Though it is not explicitly written, the numerical coefficient of the term is really "-1", since -x = -1x.

Therefore: 5x + (-x) = 5x + (-1x) = 4x

(a) 4b + (-b) = 4b + (___) = _____

(b) (-3g) + (-g) = (-3g) + (___) = _____

---

20. Solve: (a) 9x + (-x) = 24 (b) (-q) + (-5q) = 30

x = _______ q = _______

Answers to 19: a) + (-1b) = 3b b) + (-1g) = -4g

---

21. (a) In the equation "7F + F = 32", what can you substitute for "F"? _____

(b) In the equation "4s + (-s) = 24", what can you substitute for "-s"? _____

Answers to 20: a) 3 b) -5

---

22. (a) (1)(15) = 15 is an instance of the _______________ principle of multiplication.

(b) (-1)(15) = -15 is an instance of the _______________ principle of multiplication.

Answers to 21: a) 1F b) -1s

---

23. Here is a tricky equation: -x = 15

By substituting "-1x" for "-x", we get the following instance of the basic equation:

-1x = 15

or

(-1)(x) = 15

(a) The number-term in the equation is _______.

(b) The root of the equation is _______.

(c) Is the number-term the root of the equation? _______

Answers to 22: a) identity b) opposite

24. The equation "-y = -9" means "(-1)(y) = -9".

(a) The number-term in the equation is _______.

(b) The root of the equation is _______.

(c) Is the number-term the root of the equation? _______

a) 15

b) -15, since: (-1)(-15) = 15

c) No

---

25. Solve these:

(a) -m = 5 The root is _______.

(b) -x = 20 The root is _______.

(c) -p = -11 The root is _______.

a) -9

b) +9, since: (-1)(+9) = -9

c) No

---

26. Here are the equations we solved in the last frame. Their roots are given on the right.

-m = 5 The root is -5.

-x = 20 The root is -20.

-p = -11 The root is +11.

Though these equations are also instances of the basic equation, they are not root equations because the number-term in each is not the root of the equation.

In fact, the root of each equation is the ______________ of its number-term.

a) -5

b) -20

c) +11

---

27. The point we are trying to make is this:

> An equation like "-y = 14" is not a root equation, since "14" is not its root.
>
> Therefore, "-y = 14" is not solved. You must still find its root.

Solve: (a) 14 = -t t = _______ (b) -7 = -R R = _______

opposite

---

28. We can solve equations like "-y = 14" by means of the oppositing principle. First, however, we must introduce the opposites of terms like "-x", "-p", and so on.

+1x and -1x are a pair of opposites since their sum is "0".

Since +1x = x
and -1x = -x,

x and -x are also a pair of opposites.

Therefore: (a) The opposite of x is _____.

(b) The opposite of -x is _____.

a) -14, since: 14 = (-1)(-14)

b) +7, since: -7 = (-1)(+7)

---

a) -x

b) x

29. Write the opposite of each of the following: (a) 1m ______ (b) -1t ______	
30. Write the opposite of each of the following: (a) F ______ (b) -k ______	a) -1m b) +1t
31. -x = 17 We have solved equations like the one above by substituting "(-1)(x)" for "-x", as follows: (-1)(x) = 17 The root is -17. or x = -17. We can also solve it by using the <u>oppositing principle for equations</u> (taking the opposite of both sides). -x = 17 Applying the oppositing principle, we get the <u>root equation</u>: x = -17 "-x = 17" and "x = -17" are equivalent equations because the root of each is ______.	a) -F b) k
32. -m = -30 (a) Using the oppositing principle, you get the root equation: ______ = ______ (b) The root of the original equation must be ______.	-17
33. Solve: (a) 7x - 8x = 12 (b) -4 = 5y - 6y x = ______ y = ______	a) m = 30 b) 30
34. Solve: (a) 10 - d = 17 (b) 45 = 6p + 40 - 7p d = ______ p = ______	a) -12 b) +4
	a) -7 b) -5

3-3 EQUATIONS IN WHICH ONE SIDE IS "0"

In some equations, "0" is the only term on one side. In this section, we will show that the same procedures are used to solve such equations. "0" is handled just like any other number.

---

35. In this equation, "0" is the only term on the right side:

$$x + 2 = 0$$

Obviously, the root of this equation is "-2".

Let's solve the same equation by means of the addition axiom. We will add "-2" to both sides.

$$x + \underbrace{2 + (-2)}_{0} = \underbrace{0 + (-2)}_{-2}$$

$$x + 0 = -2$$

$$x = -2$$

Solve the following equations by means of the addition axiom:

$$x - 3 = 0$$

x = _______

---

36. Solve: (a) $3x + 21 = 0$ (c) $4F + 20 + F = 0$

x = _______ F = _______

(b) $0 = 4y - 12 - y$ (d) $0 = 4t + 27 - 5t$

y = _______ t = _______

Solution:

$$x + (-3) = 0$$

$$x + \underbrace{(-3) + 3}_{0} = \underbrace{0 + 3}_{3}$$

$$x + 0 = 3$$

$$x = +3$$

---

37. Here is an instance of the basic equation:

$$5x = 0$$

or

$$(5)(x) = 0$$

In this case, "0" is the <u>product</u>; "5" and "x" are <u>factors</u>.

(a) In order to get "0" as a product, one of the factors must be _______.

(b) Therefore, the root of this equation must be _______.

a) $x = -7$

b) $y = 4$

c) $F = -4$

d) $t = 27$
(From: $-27 = -t$)

---

a) 0

b) 0

38. Solve: (a) $0 = 10d$ (b) $-3t = 0$ $d =$ ______ $t =$ ______	
39. Solve: (a) $4m + 15 = 15$ (b) $20 = 35 - 9R - 15$ $m =$ ______ $R =$ ______	a) $d = 0$ b) $t = 0$
	a) $m = 0$ b) $R = 0$

SELF-TEST 1 (Frames 1-39)

1. $(1)(\square) = \square$ is called the __________ principle of multiplication.

Combine each of the following into a single term:

2. $16x + x =$ ______
3. $5t - t =$ ______
4. $r - 2r =$ ______

Find the root of each of the following equations:

5. $e - 5e = 20$ The root is ______.
6. $8 = 7h - h + 14$ The root is ______.
7. $9 + r = 3 - r$ The root is ______.

8. Which of the following are "root" equations? __________

(a) $-t = 2$ (c) $y = 7$ (e) $5 = -s$

(b) $t = -2$ (d) $-y = -7$ (f) $-5 = s$

9. $(-1)(\square) = -\square$ is called the __________ principle of multiplication.

10. Solve for d: $-d = 18$ $d =$ ______

11. Solve for w: $-10 = -w$ $w =$ ______

12. Solve for b: $8b + 2 - 9b = 1$ $b =$ ______

13. Solve for m: $3m + 4 - m = 0$ $m =$ ______

14. Solve for v: $v - 7 - 2v = 2v - 7$ $v =$ ______

ANSWERS:

1. identity
2. $17x$
3. $4t$
4. $-r$
5. -5
6. -1
7. -3
8. (b), (c), (f)
9. opposite
10. $d = -18$
11. $w = 10$
12. $b = 1$
13. $m = -2$
14. $v = 0$

3-4 EVALUATIONS AND CHECKING EQUATIONS

In this section, we will review the evaluation of expressions in which one of the terms contains a grouping. Having done so, we will then be able to check roots in equations which contain terms of that type.

---

40. Here is an evaluation of a two-term expression in which each term is a multiplication:

$$\underbrace{5(10)}_{\downarrow} + \underbrace{7(3)}_{\downarrow}$$
$$50 + 21 = 71$$

Notice the procedure: We perform the operation within each term before combining the two terms. That is, we reduce each term to a single number before combining the two terms.

Complete this one: $\underbrace{(-4)(7)}_{\downarrow} + \underbrace{(-5)(-2)}_{\downarrow}$

$(\quad) + (\quad) =$ ______

Answer to 40: (-28) + (+10) = -18

41. As we saw earlier, any multiplication is only one term, even if one of the factors is a grouping. Each of the following multiplications, for example, is only one term:

3(7 + 9) (-5)(6 - 3) (-4)[(-3) + (-2)]

Draw a box around each term in these expressions:

(a) 15 + 3(7 + 6) - 5(4)

(b) 27 - 2[(-3) + 4] - 10

(c) (-5)(7 - 4) - 6(-3) - (-9)(-8)

Answer to 41:

a) $\boxed{15} + \boxed{3(7+6)} - \boxed{5(4)}$

b) $\boxed{27} - \boxed{2[(-3)+4]} - \boxed{10}$

c) $\boxed{(-5)(7-4)} - \boxed{6(-3)} - \boxed{(-9)(-8)}$

42. Grouping symbols signify that the operation within the grouping should be performed first.

Therefore, when a term is a multiplication in which one of the factors is a grouping, we can reduce it to a single number most easily by:

(1) performing the operation within the grouping first,
and (2) then performing the multiplication.

Watch how we reduce this term to a single number:

$$5\underbrace{[(-4) + (-2)]}_{\downarrow}$$
$$5 \quad (-6) = -30$$

Complete this one: $(-3)\underbrace{[7 + (-2)]}_{\downarrow}$

$(\quad)(\quad) =$ ______

Answer to 42: (-3)(5) = -15

43. A grouping contains either an addition or subtraction. If it contains a subtraction, the subtraction should be converted to addition first. Here is an example:

$$8(7 - 12) = 8[7 + (-12)]$$
$$= 8 \quad (-5) \qquad = -40$$

Show all the steps for this one:

$$(-5)[(-4) - (-10)] = (\quad)[(\quad) + (\quad)]$$
$$= (\quad) \quad (\quad) \qquad = \_\_\_\_\_\_$$

---

44. The following expression contains two terms. When evaluating it, notice that we reduce each term to a single number and then combine the two number-terms.

$$(5)(-2) + 3(6 + 4)$$
$$+ 3(10)$$
$$(-10) + 30 \qquad = 20$$

Evaluate the following two-term expression. Use the blanks below to write each term as a single number.

$$(-4)(-5) + (-10)(7 - 13)$$

______ + ______ = ______

$(-5)[(-4) + (+10)]$

$(-5) \quad (+6) \qquad = -30$

---

45. When two terms are connected by a subtraction symbol, the subtraction is not converted to addition until each term is reduced to a single number.

$$(3)(5) - (-3)[(-2) + (-5)]$$
$$- (-3) \quad (-7)$$
$$(15) - \quad (+21) \qquad = 15 + (-21) = -6$$

Evaluate the following expression. Use the blanks below to write each term as a single number and convert to addition.

$$3(-4) - 4(5 - 2)$$

____ - ______ = ______ + ______ = ______

$(+20) + (+60) = 80$

or

$20 + 60 = 80$

---

46. Evaluate this three-term expression. Use the blanks below to write each term as a single number.

$$4(-2) - (4)(5 + 3) - (-40)$$

____ - ______ - ____ = ______ + ______ + ______ = ______

$(-12) - 12 = (-12) + (-12)$

$= -24$

$(-8) - (32) - (-40) =$

$(-8) + (-32) + (+40) = 0$

47. Here is an equation: $5(-4) + 3 = 25 - 2(3 + 1)$

To check whether the above equation is true, we must evaluate each side of the equation to see whether or not the two sides are equal. Here is the evaluation:

$$5(-4) + 3 = 25 - 2(3 + 1)$$

$$2\ (4)$$

$$(-20) + 3 = 25 - \ 8$$

$$-17 = 17$$

Is the equation true or false? __________

---

48. Evaluate each side of this equation. Then state whether it is true or false. __________

$$(-9) - (-2)(5 - 8) = 5[(-4) + 9] - (-6)(-5) - 10$$

False, since: $-17 \neq 17$

---

49. There is a letter within the grouping in the following term:

$$4(x + 1)$$

What is the value of the term: (a) If $x = 1$? _______

(b) If $x = 2$? _______

(c) If $x = 9$? _______

It is true, since:

$$(-9) - (+6) = 25 - (+30) - 10$$
$$(-9) + (-6) = 25 + (-30) + (-10)$$
$$-15 = -15$$

---

50. What is the value of the term below if: (a) $s = -1$? _______

$$2(s + 7)$$

(b) $s = -2$? _______

(c) $s = -9$? _______

a) +8

b) +12

c) +40

---

51. What is the value of the term below if: (a) $y = -1$? _______

$$(-10)[y + (-3)]$$

(b) $y = -3$? _______

(c) $y = -7$? _______

a) +12

b) +10

c) −4

---

52. $$2(x + 3) = 18$$

Let's see if the equation above is true if we plug in "6" for "x". We do so by evaluating the left side.

$$2(6 + 3) = 18$$

$$2(9) = 18$$

$$18 = 18$$

Since the equation is true when we plug in a "6" for "x", we say that "6" is the __________ of this equation.

a) +40

b) +60

c) +100

---

root

53. Evaluate the right side to answer the questions below: $-84 = 7[D + (-9)]$ (a) Is "-2" the root of this equation? _______ (b) Is "-4" the root of this equation? _______	
54. Evaluate the expression below: $15 - 3(x + 4)$ (a) If $x = -1$ _______ (b) If $x = -5$ _______	a) No, since: $-84 \neq -77$ b) No, since: $-84 \neq -91$
55. Evaluate the expression below: $10(R + 7) - 20$ (a) If $R = -5$ _______ (b) If $R = -9$ _______	a) +6 b) +18
56. $10 + 7(x + 3) = 33$ (a) Plug in "+2" for "x" and evaluate the <u>left side</u> of this equation. _____ (b) Is "+2" the root of this equation? _______	a) 0 b) -40
57. $30 - 2(x + 4) = 47$ (a) Plug in "-6" for "x" and evaluate the <u>left side</u> of this equation. _____ (b) Is "-6" the root of this equation? _______	a) 45 b) No, since: $45 \neq 33$
58. $19 = 17 + 2[t + (-7)]$ (a) Plug in "+8" for "t" and evaluate the <u>right</u> side of this equation. _____ (b) Is "+8" the root of this equation? _______	a) 34 b) No, since: $34 \neq 47$
59. $(-40) = 15 - 4[K + (-10)]$ (a) Plug in "17" for "K" and evaluate the <u>right side</u> of this equation. _____ (b) Is "17" the root of this equation? _______	a) 19 b) Yes, since: $19 = 19$
60. What is the value of the term below: $(-10)(D - 3)$ (a) If $D = 9$? _______ (b) If $D = 1$? _______	a) -13 b) No, since: $(-40) \neq (-13)$
61. Evaluate the term below: $7(x - 9)$ (a) If $x = -3$ _______ (b) If $x = -1$ _______	a) $(-10)(6) = \underline{-60}$ b) $(-10)(-2) = \underline{+20}$
	a) $7(-12) = \underline{-84}$ b) $7(-10) = \underline{-70}$

62. $(-3)(4 - S)$

If $S = -2$, we evaluate the expression above in the following way:

$$(-3)[4 - (-2)] = (-3)[4 + (+2)]$$
$$= (-3)\quad(6)\quad = -18$$

Evaluate the expression above: (a) If $S = +4$ _______

(b) If $S = -4$ _______

---

63. $2(x - 3) = -16$

To determine whether "-7" is the root of the equation above, we plug in "-7" for "x" and evaluate the left side.

(a) If we plug in "-7" for "x", the numerical value of the left side is _______.

(b) Is "-7" the root of this equation? _______

a) $(-3)(0) = \underline{0}$

b) $(-3)(8) = \underline{-24}$

---

64. $-45 = 5(b - 7)$

(a) What is the value of the right side if we plug in "-1" for "b"? _______

(b) Is "-1" the root of the equation? _______

(c) What is the value of the right side if we plug in "-2" for "b"? _______

(d) Is "-2" the root of the equation? _______

a) $2(-10) = \underline{-20}$

b) No, since: $-20 \neq -16$

---

65. $70 - 6(x - 4) = 30$

(a) Plug in "+9" for "x" and evaluate the <u>left side</u> of the equation. You get: _______

(b) Is "+9" the root of this equation? _______

a) -40

b) No, since: $-45 \neq -40$

c) -45

d) Yes, since: $-45 = -45$

---

66. $10 = 5(t - 7) + 60$

(a) Plug in "-3" for "t" and evaluate the <u>right side</u> of this equation. You get: _______

(b) Is "-3" the root of this equation? _______

a) 40

b) No, since: $40 \neq 30$

---

67. $3(d - 5) = 40 - 4(d + 5)$

Let's check to see whether "3" is the root of the equation above.

(a) If you plug in "3" for "d", the value of the <u>left side</u> is _______.

(b) If you plug in "3" for "d", the value of the <u>right side</u> is _______.

(c) Is "3" the root of the equation? _______

a) $(-50) + 60 = +10$

b) Yes, since: $10 = 10$

68. $14y - 5(y - 7) = 50 + 4(y + 5)$ Let's check to see whether "7" is the root of the equation above. (a) If you plug in "7" for "y", the value of the left side is _______. (b) If you plug in "7" for "y", the value of the right side is ______. (c) Is "7" the root? _______	a) -6 b) 8 c) No, since: $-6 \neq 8$

a) 98 b) 98 c) Yes, since: 98 = 98

SELF-TEST 2 (Frames 40-68)

How many terms are there in each expression?

1. $7(-2) - 5(8 - 6)$ _______
2. $6 - 3 + (-5)(9) - 4(6 - 3)$ _______

Evaluate the following expressions:

3. $17 + 8(1 - 3) =$ _______
4. $5[(-2) - (-6)] - 5 =$ _______
5. $12 - 7(-1) - 2(9 - 12) =$ _______
6. $(-4) - (-3)[(-1) - 2] + 6 - 5(4) =$ _______

Given this expression: $(-3)(6 - 2h)$ Find the value of this expression if:

7. $h = 0$ _______
8. $h = 3$ _______
9. $h = 5$ _______
10. $h = -1$ _______
11. $h = -3$ _______

Given this equation: $x - 3(x - 5) = 8 - (x - 2)$

12. If "-2" is plugged in for x, what is the value of the left side? _______
13. If "-2" is plugged in for x, what is the value of the right side? _______
14. Is "-2" the root of the equation? _______
15. If "5" is plugged in for x, the value of the left side is _______, and the value of the right side is _______.
16. Is "5" the root of the equation? _______

ANSWERS:

1. Two terms
2. Four terms
3. 1
4. 15
5. 25
6. -27
7. -18
8. 0
9. 12
10. -24
11. -36
12. 19
13. 12
14. No
15. 5; 5
16. Yes

3-5 MULTIPLYING BY THE DISTRIBUTIVE PRINCIPLE

In the equation below, there is only one term on the left side. One of the two factors is a grouping. A letter appears within the grouping.

$$7(x + 11) = 105$$

In this section, we will show the need for "multiplying by the distributive principle" in order to simplify any term which contains a grouping which includes a letter. In the next section, we will solve equations which contain terms of that type.

---

69. Up to this point, we have dealt with only two types of terms when solving equations.

(1) Number-terms: like 7 or -9
(2) Letter-terms: like 3x, -5d, y, or -m

We learned how to simplify equations by:

(1) combining terms of both types,
and (2) using the addition axiom to eliminate terms of both types.

Here are three terms which contain a grouping:

$3(4 + 2)$
$x(7 + 5)$
$3(x + 8)$

Terms of this type are more complicated than a simple number-term or a simple letter-term. When faced with a more complicated term of this type while solving an equation, we try to reduce it to the simpler types of terms which we already know how to handle:

If the grouping does not contain a letter, we can easily reduce the more complicated term to a simple term because we can combine the numbers within the grouping.

$3(4 + 2)$ reduces to a simple number-term:

$$3(4 + 2) = 3(6) = \underline{18}$$

$x(7 + 5)$ reduces to a simple letter-term:

$$x(7 + 5) = x(12) \text{ or } \underline{12x}$$

But if the grouping contains a letter, we cannot perform the addition within the grouping since we cannot combine a letter-term and a number-term.

Therefore, $3(x + 8)$ cannot be reduced to a simpler term in the same way that $3(4 + 2)$ and $x(7 + 5)$ were.

We need a different technique to simplify a term like $3(x + 8)$.

---

70. The special technique used to simplify a term like $3(x + 2)$ is the DISTRIBUTIVE PRINCIPLE FOR MULTIPLICATION OVER ADDITION. Earlier, we saw that this principle can be stated generally as:

$$\triangle(\bigcirc + \square) = \triangle(\bigcirc) + \triangle(\square)$$

At that time, it was introduced as an alternate technique for evaluating an expression like $3(4 + 2)$.

$$3(4 + 2) = 3(4) + 3(2)$$
$$= 12 + 6 = 18$$

Complete these:

(a) $8(7 + 6) = 8(\quad) + 8(\quad)$

(b) $(\quad)(5 + 4) = 3(5) + 3(4)$

Go to next frame.

71. This principle is called the distributive principle for multiplication over addition because the multiplication by the first factor is distributed to each term in the grouping.

$$5(7+3) = 5(7) + 5(3)$$

How many terms are there: (a) On the left side? ______

(b) On the right side? ______

a) $8(7) + 8(6)$

b) $3(5+4)$

---

72. When a letter is involved in the grouping, the distributive principle works the same way.

$$2(x+3) = 2(x) + 2(3)$$

Complete these:

(a) $7(x+10) = 7x + 7(\quad)$

(b) $10(x+2) = 10x + (\quad)(2)$

(c) $-3(x+4) = (-3)(\quad) + (-3)(4)$

(d) $-15(x+19) = (\quad)(x) + (-15)(19)$

a) One: $5(7+3)$

b) Two: $5(7)$ and $5(3)$

---

73. Complete:

(a) $(\quad)(x+2) = 7x + 7(2)$

(b) $(-6)(\_\_\_ + 5) = (-6)t + (-6)(5)$

(c) $(11)(R + \_\_\_) = 11R + 11(2)$

a) 10 c) x

b) 10 d) -15

---

74. When one of the factors in a multiplication is a grouping which contains a letter, we must use the distributive principle to perform the multiplication. We call this process "multiplying by the distributive principle." Here is an example:

$$5(x+3) = 5(x) + 5(3)$$

(a) The product in this multiplication is ______________.

(b) How many terms are there in the product? ______

a) 7

b) t

c) 2

---

75. After using the distributive principle to perform the multiplication, we simplify each term in the product:

$$5(x+3) = 5(x) + 5(3)$$
$$= 5x + 15$$

By simplifying the terms in the product, we obtain two simple terms of the type we know how to handle:

5x is a simple letter-term.
15 is a simple number-term.

Simplify the terms in each product:

(a) $3(7+y) = 3(7) + 3(y) = \_\_\_\_\_\_ + \_\_\_\_\_\_$

(b) $4(t+8) = 4(t) + 4(8) = \_\_\_\_\_\_ + \_\_\_\_\_\_$

(c) $5(d+2) = 5(d) + 5(2) = \_\_\_\_\_\_ + \_\_\_\_\_\_$

a) $5(x) + 5(3)$

b) Two

76. In one step, multiply by the distributive principle and write each term of the product in its simplest form:

(a) $5(x + 7) =$ ____ + ____ (c) $3(p + 3) =$ ____ + ____

(b) $6(2 + m) =$ ____ + ____ (d) $8(6 + R) =$ ____ + ____

a) $21 + 3y$

b) $4t + 32$

c) $5d + 10$

---

77. Multiply by the distributive principle and simplify the product:

(a) $9(y + 4) =$ ____ + ____ (b) $4(3 + t) =$ ____ + ____

a) $5x + 35$ c) $3p + 9$

b) $12 + 6m$ d) $48 + 8R$

---

78. A COMMON ERROR

There is one common error which occurs frequently when multiplying by the distributive principle. Here is an example:

$$2(x + 7) = 2x + 7$$

The error is this: The second term ("7") within the grouping was not multiplied by "2".

$2(x + 7)$ really equals ____________.

a) $9y + 36$ b) $12 + 4t$

---

79. In which of the following does this common error occur? ____________

(a) $10(d + 3) = 10d + 3$ (c) $3(7 + g) = 21 + g$

(b) $11(y + 2) = 11y + 22$ (d) $8(4 + A) = 32 + 8A$

$2x + 14$

---

80. For each of the following:

(1) Write the product you would get if you made the common error.

(2) Then write the correct product.

	Common-Error Product	Correct Product
$5(x + 9)$	$5x + 9$	$5x + 45$
(a) $8(y + 3)$	________	________
(b) $7(9 + p)$	________	________
(c) $4(t + 7)$	________	________

In both (a) and (c)

---

a) $8y + 3$ $8y + 24$

b) $63 + p$ $63 + 7p$

c) $4t + 7$ $4t + 28$

SELF-TEST 3 (Frames 69-80)

Using the distributive principle, complete the following:

1. $9(2 + 5) = (\quad)(\quad) + (\quad)(\quad)$ 2. $x(6 + 3) = (\quad)(\quad) + (\quad)(\quad)$

3. □(△ + ○) = ( )( ) + ( )( )

Multiply, and write each product in simplest form:

4. $2(p + 3) =$ ________ 6. $w(4 + 8) =$ ________

5. $5(7 + b) =$ ________ 7. $3(5 + 2) =$ ________

8. One of the following problems contains an error. Which problem is it? ________

(a) $12(k + 1) = 12k + 12$ (c) $h(4 + 2) = 4h + 2h$

(b) $3(s + 8) = 3s + 8$ (d) $7(6 + r) = 7r + 42$

ANSWERS:

1. $(9)(2) + (9)(5)$
2. $(x)(6) + (x)(3)$
3. (□)(△) + (□)(○)
4. $2p + 6$
5. $35 + 5b$
6. $12w$
7. 21
8. (b) contains the error. The correct answer is: $3s + 24$

3-6 USING THE DISTRIBUTIVE PRINCIPLE TO SOLVE EQUATIONS

Many equations contain a term with a grouping which contains a letter. In this section, we will solve equations of that type.

81. To simplify the following terms, we must multiply by the distributive principle.

$3(x + 7)$ $2(10 + b)$

Therefore, we will call terms of this type "instances of the distributive principle."

In the following equation, the left side is an instance of the distributive principle.

$$2(x + 3) = 6$$

Draw a circle around each "instance of the distributive principle" in the following equations:

(a) $45 = 7(x + 9)$ (c) $7h + 5 = 14 + 8(h + 4)$

(b) $11 + 3(t + 5) = 95$ (d) $8m + 5(m + 9) = 100$

You should have a circle drawn around:

a) $7(x + 9)$ c) $8(h + 4)$

b) $3(t + 5)$ d) $5(m + 9)$

82. When an equation contains an "instance of the distributive principle," <u>the first step is always to simplify that term by multiplying by the distributive principle.</u> $2(x + 3) = 16$ Multiplying by the distributive principle, we get: $2x + 6 = 16$ Now we have an equation which contains only a simple letter-term and two number-terms. We know how to solve this last equation. The root of the last equation is +5. By checking, show that +5 is the root of the <u>original</u> equation:	
83. When we multiply by the distributive principle to simplify an instance of the distributive principle, <u>the new equation is equivalent to the original one.</u> Here is another example: $14 + 4(t + 5) = 42$ Multiplying $4(t + 5)$ by the distributive principle, we get: $14 + 4t + 20 = 42$ (a) The root of this last equation is _______. (b) Therefore, the root of the original equation must be _______.	<u>Check:</u> $2(x + 3) = 16$ $2(5 + 3) = 16$ $2(8) = 16$ $16 = 16$
84. Find the roots of each of these: (a) $9(y + 5) = 135$ (b) $30 = 5(b + 9)$ y = _______ b = _______	a) 2 b) 2
85. Solve this one: $67 = 59 + 2(p + 7)$ p = _______	a) $y = 10$ b) $b = -3$
	$p = -3$

86. In the last frame, we found that the root of the following equation is "-3":

$$67 = 59 + 2(p + 7)$$

Let's check this root:

$$67 = 59 + 2[(-3) + 7]$$
$$67 = 59 + 2(4)$$
$$67 = 59 + 8$$
$$67 = 67$$

Solve the following equation and check the root yourself:

$75 + 3(x + 2) = 99$ Check:

$x =$ _______

If the root "checks," your solution is correct.

---

87. Solve this one: $9(x + 7) = 8x$

$x =$ _______

$x = -63$

(From: $63 = -x$)

---

88. Solve this one: $10d + 4(d + 10) = 7d + 12$

$d =$ _______

$d = -4$

(From: $7d = -28$)

---

89. Sometimes the letter within the grouping has a numerical coefficient:

Example: $5(2x + 4)$

Then multiplying by the distributive principle looks like this:

$$5(2x + 4) = 5(2x) + 5(4)$$

Let's examine "$5(2x)$", the first term in the product:

Since "$2x$" can be written "$(2)(x)$",

instead of $5(2x)$,

we could write $5(2)(x)$.

How many factors are there in the multiplication $5(2)(x)$? _______

Three

90. Though we distribute the multiplication when a grouping is involved

$$5(2 + x) = 5(2) + 5(x)$$

we do NOT distribute the multiplication in a three-factor term when none of the factors is a grouping.

$$5(2x) = 5(2)(x)$$
$$= 10(x) = 10x$$

Simplify each of the following by performing the multiplication:

(a) $(6)(2)(x) =$ ______ (c) $(7)(4)(x) =$ ______

(b) $4(5x) =$ ______ (d) $10(5x) =$ ______

---

91. Simplify these instances of the distributive principle:

(a) $3(2x + 4) = 3(2x) + 3(4) =$ ____ + ____

(b) $7(5x + 9) = 7(5x) + 7(9) =$ ____ + ____

(c) $10(3x + 5) =$ ____ + ____

(d) $9(4x + 8) =$ ____ + ____

(e) $5(7x + 6) =$ ____ + ____

Answers to 90: a) 12x c) 28x b) 20x d) 50x

---

92. Solve this equation: $3(2x + 8) = 0$

x = ______

Answers to 91: a) $6x + 12$ b) $35x + 63$ c) $30x + 50$ d) $36x + 72$ e) $35x + 30$

---

93. Solve this equation: $42 = 75 + 3(4c + 1)$

c = ______

Answer to 92: $x = -4$ (From: $6x = -24$)

---

94. Solve this one: $5x + 6(4x + 5) = 34x$

x = ______

Answer to 93: $c = -3$

Answer to 94: $x = 6$

3-7 USING THE DISTRIBUTIVE PRINCIPLE WHEN THE GROUPING CONTAINS A SUBTRACTION

In the following equation, the term on the left contains a grouping in which there is a subtraction:

$$7(x - 5) = 49$$

In this section, we will: (1) simplify terms of this type,
and (2) solve equations in which terms of this type occur.

---

95. The distributive principle for multiplication over addition can only be used when an addition appears within the grouping. When a subtraction appears within the grouping, the term is not an "instance of the distributive principle for multiplication over addition."

Therefore, the following term is not an "instance of the distributive principle over addition."

$$3(x - 7)$$

However, it can easily be converted into an instance of the distributive principle by converting the subtraction to addition.

$$3(x - 7) = 3[x + (-7)]$$

Having made this conversion, we can simplify the term by multiplying by the distributive principle:

$$3(x - 7) = 3[x + (-7)] = 3x + (-21)$$

Complete these:
(a) $5(y - 6) = 5[y + (-6)] =$ ______ + ______
(b) $10(3 - t) = 10[3 + (-t)] =$ ______ + ______
(c) $4(2m - 5) =$ ________ = ______ + ______
(d) $3(8 - 4d) =$ ________ = ______ + ______

---

96. When a term appears in an equation with a subtraction within the grouping, we must convert the subtraction to addition before multiplying by the distributive principle. Here is an example:

$$3(x - 2) = 15$$
$$3[x + (-2)] = 15$$
$$3x + (-6) = 15$$

(a) Solve the last equation.
Its root is ______.

(b) Show that this root satisfies the original equation.

Answers to 95:
a) $5y + (-30)$
b) $30 + (-10t)$
c) $4[2m + (-5)]$
$= 8m + (-20)$
d) $3[8 + (-4d)]$
$= 24 + (-12d)$

Answers to 96:
a) 7
b) $3(x - 2) = 15$
$3(7 - 2) = 15$
$3(5) = 15$
$15 = 15$

97. $6(2y - 5) + 7 = 25$

(a) Rewrite the equation, converting $6(2y - 5)$ into an instance of the distributive principle:

$$\_\_\_\_\_\_ + 7 = 25$$

(b) Now multiply by the distributive principle:

$$\_\_\_ + \_\_\_ + 7 = 25$$

(c) Complete the solution:

$y = \_\_\_\_$

a) $6[2y + (-5)]$

b) $\underline{12y} + \underline{(-30)}$

c) $y = 4$

---

98. Solve: (a) $15 = 5(F - 3)$ (b) $19 = 2(4p - 8) + 59$

$F = \_\_\_\_$ $p = \_\_\_\_$

a) $F = 6$ b) $p = -3$

---

99. Solve this one: $6(2z - 6) + 6z = 0$

$z = \_\_\_\_$

$z = 2$

---

100. When an equation contains an instance of the distributive principle over addition, we simplify that term before converting any subtraction to addition.

Here is an example: $5(x + 3) - 20 = 25 - 5x$

Multiplying by the distributive principle, we get:

$$5x + 15 - 20 = 25 - 5x$$

Now, converting the subtraction to addition, we get:

$$5x + 15 + (-20) = 25 + (-5x)$$

Complete the solution:

$x = \_\_\_\_$

$x = 3$

101. When an equation contains a term with a <u>subtraction in the grouping</u>, we simplify that term <u>before converting any other subtraction to addition.</u> In this case, "multiplying by the distributive principle" includes converting the subtraction within the grouping to addition first.

<u>Example</u>: $4(3d - 9) - 12 = 12 - 8d$

Multiplying by the distributive principle, we get:

$$4[3d + (-9)] - 12 = 12 - 8d$$

$$12d + (-36) - 12 = 12 - 8d$$

Converting the other subtraction to addition, we get:

$$12d + (-36) + (-12) = 12 + (-8d)$$

Complete the solution:

d = ______

---

d = 3

102. Solve this one: $4x + 2(4x - 9) = 10x - 2$

x = ______

---

x = 8

<u>SELF-TEST 4</u> (<u>Frames 81-102</u>)

Solve the following equations:

1. $2 + 3(2x + 1) = 23$

x = ______

2. $2(3r + 1) - r = 2r - 4$

r = ______

3. $38 + 5(4t - 3) = 3$

t = ______

4. $6(3p - 5) + 8 = 4(5p - 3) - p$

p = ______

<u>ANSWERS:</u> 1. $x = 3$ 2. $r = -2$ 3. $t = -1$ 4. $p = -10$

3-8 FORMAL STRATEGIES FOR SOLVING EQUATIONS

In the last chapter, we introduced some frames in which you formally wrote each step in solving an equation and identified the process used in each step. The only new process we have introduced in this chapter is "multiplying by the distributive principle." In the next section, we will review these "formal strategies" and incorporate this one new process into them.

---

103. Whenever a term contains a grouping as one factor, we "multiply by the distributive principle" first. By doing so, we replace the complicated term with a simple letter-term and a simple number-term which we can easily handle. We will identify this step with the words "multiplying (distributive principle)."

In this one, the first step is "multiplying (distributive principle)." Identify the rest of the steps:

$$\boxed{5(2x + 9) = 85}$$

Step 1: $10x + 45 = 85$ — Multiplying (distributive principle)

Step 2: $10x + (45) + (-45) = 85 + (-45)$

$10x = 40$ — ______________

Step 3: $x = 4$ — ______________

Answer to Frame 103: Step 2: Addition axiom (-45)
Step 3: Solving the basic equation

---

104. When a subtraction appears within the grouping, we must convert this subtraction to addition before multiplying by the distributive principle. (We do so because the distributive principle for multiplication over addition requires that the grouping contain an addition.)

Though there are two steps in this process, we will treat them as one process and call it "multiplying (distributive principle)." Complete this one:

$$\boxed{7(x - 6) = 14}$$

Step 1: $7[x + (-6)] = 14$
$7x + (-42) = 14$ } Multiplying (distributive principle)

Step 2: $7x + (-42) + (42) = 14 + 42$
$7x = 56$ } ______________

Step 3: $x = 8$ — ______________

Answer to Frame 104: Step 2: Addition axiom (42)
Step 3: Solving the basic equation

105. In the last chapter, we formally identified both the interchange principle and the oppositing principle as steps. To simplify matters, we will no longer do so. Both of these principles are used to make the solution of basic equations easier. From now on, whether these two principles are used or not, we will only identify "solving the basic equation." Here are some examples:

(1) Though you might use the interchange principle in this one, we will simply write:

$$45 = 9x$$

$$x = 5 \quad \text{Solving the basic equation}$$

(2) Though you might use the oppositing principle in this one, we will simply write:

$$-6y = 42$$

$$y = -7 \quad \text{Solving the basic equation}$$

Identify the steps in this solution. (Remember, we always combine like terms before using the addition axiom.)

$$\boxed{67 = 15 + 4(3x + 4)}$$

Step 1: $67 = 15 + 12x + 16$ ______________________

Step 2: $67 = 12x + 31$ ______________________

Step 3: $67 + (-31) = 12x + 31 + (-31)$ } ______________________
$36 = 12x$

Step 4: $x = 3$ ______________________

Answer to Frame 105:
Step 1: Multiplying (distributive principle)
Step 2: Combining like terms
Step 3: Addition axiom (-31)
Step 4: Solving the basic equation

106. When an "x" or "-x" has appeared in an equation, we have used the principles:

$1(\square) = \square$ to substitute "1x" for "x",

and

$-1(\square) = -\square$ to substitute "-1x" for "-x".

However, since this substitution is rather simple, we will not identify it as a separate step.

Identify the steps in this one:

$$\boxed{8(x - 2) = 6x}$$

Step 1: $8[x + (-2)] = 6x$ }
$8x + (-16) = 6x$ ______________________

Step 2: $8x + (-8x) + (-16) = 6x + (-8x)$ }
$-16 = -2x$ ______________________

Step 3: $x = 8$ ______________________

Answer to Frame 106: Step 1: Multiplying (distributive principle)
Step 2: Addition axiom $(-8x)$
Step 3: Solving the basic equation

---

107. Except for a subtraction within a grouping, we do not convert other subtractions to additions before multiplying by the distributive principle. We consider the conversion of subtraction to addition within a grouping to be part of "multiplying by the distributive principle."

Identify the steps in this one:

$$\boxed{3t = 2(4t + 3) - 21}$$

Step 1: $3t = 8t + 6 - 21$ __________

Step 2: $3t = 8t + 6 + (-21)$ __________

Step 3: $3t = 8t + (-15)$ __________

Step 4: $3t + (-8t) = 8t + (-8t) + (-15)$
$-5t = -15$ __________

Step 5: $t = +3$ __________

Answer to Frame 107: Step 1: Multiplying (distributive principle)
Step 2: Converting subtraction to addition
Step 3: Combining like terms
Step 4: Addition axiom $(-8t)$
Step 5: Solving the basic equation

---

108. Identify the steps in this one:

$$\boxed{2(5p - 6) - 13 = 7p - 1}$$

Step 1: $2[5p + (-6)] - 13 = 7p - 1$
$10p + (-12) - 13 = 7p - 1$ __________

Step 2: $10p + (-12) + (-13) = 7p + (-1)$ __________

Step 3: $10p + (-25) = 7p + (-1)$ __________

Step 4: $10p + (-7p) + (-25) = 7p + (-7p) + (-1)$
$3p + (-25) = (-1)$ __________

Step 5: $3p + (-25) + 25 = (-1) + 25$
$3p = 24$ __________

Step 6: $p = 8$ __________

Answer to Frame 108: Step 1: Multiplying (distributive principle)
Step 2: Converting subtraction to addition
Step 3: Combining like terms
Step 4: Addition axiom $(-7p)$
Step 5: Addition axiom (25)
Step 6: Solving the basic equation

109. Here are the five steps which we identify. Though all of them are not used in every solution, they are listed in the order in which they are used when needed.

(1) Multiplying (distributive principle)
(2) Converting subtraction to addition
(3) Combining like terms
(4) Addition axiom
(5) Solving the basic equation

In this problem, write each step and identify it in the space at the right. You will need four steps.

$$\boxed{17 = 2(1 - 4d) + 7}$$

Step 1: ______________________

Step 2: ______________________

Step 3: ______________________

Step 4: ______________________

Answer to Frame 109:

Step 1:	$17 = 2[1 + (-4d)] + 7$ $17 = 2 + (-8d) + 7$	Multiplying (distributive principle)
Step 2:	$17 = 9 + (-8d)$	Combining like terms
Step 3:	$17 + (-9) = 9 + (-9) + (-8d)$ $8 = -8d$	Addition axiom (-9)
Step 4:	$d = -1$	Solving the basic equation

---

110. You will need five steps for this one:

$$\boxed{9(x + 4) - 4 = x}$$

Step 1: ______________________

Step 2: ______________________

Step 3: ______________________

Step 4: ______________________

Step 5: ______________________

Answer to Frame 110:

Step 1:	$9x + 36 - 4 = 1x$	Multiplying (distributive principle)
Step 2:	$9x + 36 + (-4) = 1x$	Converting subtraction to addition
Step 3:	$9x + 32 = 1x$	Combining like terms
Step 4:	$9x + (-9x) + 32 = 1x + (-9x)$ $32 = -8x$	Addition axiom (-9x)
Step 5:	$x = -4$	Solving the basic equation

111. You will need six steps for this one:

$$\boxed{8 - 4h = 3(5h - 9) - 12h}$$

Step 1: ______________________

Step 2: ______________________

Step 3: ______________________

Step 4: ______________________

Step 5: ______________________

Step 6: ______________________

Answer to Frame 111:

Step 1:	$8 - 4h = 3[5h + (-9)] - 12h$ $8 - 4h = 15h + (-27) - 12h$	Multiplying (distributive principle)
Step 2:	$8 + (-4h) = 15h + (-27) + (-12h)$	Converting subtraction to addition
Step 3:	$8 + (-4h) = 3h + (-27)$	Combining like terms
Step 4:	$8 + (-4h) + 4h = 3h + 4h + (-27)$ $8 = 7h + (-27)$	Addition axiom (4h)
Step 5:	$8 + 27 = 7h + (-27) + 27$ $35 = 7h$	Addition axiom (27)
Step 6:	$h = 5$	Solving the basic equation

SELF-TEST 5 (Frames 103-111)

In Problems 1 to 8, state which of the principles listed at the right would be used as the next step in solving each equation. If the addition axiom is used as the next step, list the term which is to be added.

> Multiplying (distributive principle)
> Converting subtraction to addition
> Combining like terms
> Addition axiom
> Solving the basic equation

1. $3i + 8 = 5i$ ______________________
2. $r - 4r = 18$ ______________________
3. $-24 = -6d$ ______________________
4. $3(2w + 4) - w = 32$ ______________________
5. $h = 2h + 8 + 3h$ ______________________
6. $2t + 5(t - 3) = 6$ ______________________
7. $2v + 3 = 9 + 5v$ ______________________
8. $x - 1 = 3(x - 3)$ ______________________

ANSWERS:

1. Addition axiom (-3i)
2. Converting subtraction to addition
3. Solving the basic equation
4. Multiplying (distributive principle)
5. Combining like terms
6. Multiplying (distributive principle)
7. Addition axiom (-2v)
8. Multiplying (distributive principle)

3-9 GROUPINGS AS TERMS BY THEMSELVES

In this section, we will introduce a new type of term. This new type of term is a grouping which is not a factor in a multiplication. The grouping may contain either an addition or a subtraction. Here are some examples:

$$(7 + 3) \text{ or } (x + 2) \text{ or } (7 + x)$$

$$(5 - 9) \text{ or } (y - 4) \text{ or } (8 - y)$$

In this section, we will learn how to handle such terms in evaluations. In the next section, we will learn how to handle such terms when solving equations.

---

112. Up to this point, we have identified various types of terms in evaluations:

(1) Any single number like 7 or -10.

(2) Any multiplication of two or more factors:

(a) When the factors are single numbers

like 5(6) or 7(-8)(-2)

(b) When one factor is a grouping

like 5(3 + 7) or 5(3 - 7)

Draw a box around each term in these equations:

(a) $9 - 2(-6) = 7[8 + (-5)]$

(b) $10(5) - 3(4 - 6) = 14 + 6(5 + 2)$

Answers to 112:

a) $\boxed{9} - \boxed{2(-6)} = \boxed{7[8 + (-5)]}$

b) $\boxed{10(5)} - \boxed{3(4 - 6)} = \boxed{14} + \boxed{6(5 + 2)}$

113. When a grouping appears by itself (that is, not as a factor in a multiplication), the grouping is also a term.

For example: (1) In 10 - (7 + 5),

(7 + 5) is one term.

(2) In 15 + (8 - 6)

(8 - 6) is one term.

Draw a box around each term in these:

(a) $10 + (3 + 1) = 8(3) - (12 - 2)$

(b) $4(6 + 5) - (8 + 7) = 26 - (6 - 9)$

114. Mathematicians use grouping symbols to signify that a certain operation is to be performed first. This is also true when a grouping is a term by itself. In an evaluation, the grouping is reduced to a single number before the terms are combined. Here are two examples:

$$5 + \underbrace{(7 + 3)}_{\downarrow} \qquad 10 + \underbrace{(8 - 4)}_{\downarrow}$$

$$5 + \ 10 \ = 15 \qquad 10 + \ 4 \ = 14$$

When a grouping is preceded by a subtraction symbol, the grouping is reduced to a single number <u>before</u> the subtraction is converted to addition. Here is an example:

$$10 - \underbrace{(40 - 20)}_{\downarrow}$$

$$10 - \ 20 \ = 10 + (-20) = -10$$

Complete: $17 - \underbrace{(6 - 10)}_{\downarrow}$

____ - ______ = ____ + ____ = ____

a) [10] + [(3 + 1)]
= [8(3)] - [(12 - 2)]

b) [4(6 + 5)] - [(8 + 7)]
= [26] - [(6 - 9)]

115. In an evaluation, all terms are reduced to single numbers before the terms are combined.

Complete: (a) $\underbrace{7(-5)}_{\downarrow} + \underbrace{(10 - 20)}_{\downarrow}$

( ) + ( ) = ______

(b) $\underbrace{5(-9)}_{\downarrow} - \underbrace{(20 - 40)}_{\downarrow}$

( ) - ( ) = ______ + ______ = ______

17 - (-4) = 17 + (+4) = 21

116. If we plug in "10" for "x": (a) 17 + (x + 4) = ______

(b) 17 - (x + 4) = ______

a) (-35) + (-10) = -45

b) (-45) - (-20) = (-45) + (+20) = -25

117. If we plug in "5" for "m": (a) 18 + (m - 9) = ______

(b) 18 - (m - 9) = ______

a) 31

b) 3

118. 40 - (x + 9) = 17

(a) If we plug in "11" for "x", the <u>left side</u> of the equation equals _____.

(b) Is "11" the root of this equation? ______

a) 14

b) 22

119. 20 = 30 - (2y - 6)

(a) If we plug in "8" for "y", the <u>right side</u> of the equation equals _____.

(b) Is "8" the root of the equation? ______

a) 20

b) No, since $20 \neq 17$

120. 17 - (4d + 1) = 0

(a) If we plug in "-4" for "d", the <u>left side</u> of the equation equals _____.

(b) Is "-4" the root of the equation? ______

a) 20

b) Yes, since 20 = 20

121.

$$3p - (p - 2) = 40 - (4p + 8)$$

If we plug in "5" for "p":

(a) The left side of the equation equals _______.

(b) The right side of the equation equals _______.

(c) Is "5" the root of the equation? _______

Answers: a) 32 b) No, since $32 \neq 0$

---

122. There is an important difference between a grouping after an addition symbol and a grouping after a subtraction symbol.

WHEN A GROUPING FOLLOWS AN ADDITION SYMBOL, THE GROUPING SYMBOLS CAN BE DROPPED WITHOUT CHANGING THE VALUE OF THE EXPRESSION. Here is an example:

$$20 + (6 + 2)$$

Evaluating with the grouping symbols:

$$20 + \underbrace{(6 + 2)}$$
$$20 + \quad 8 \quad = 28$$

Evaluating without the grouping symbols:

$$\underbrace{20 + 6} + 2$$
$$26 \quad + 2 = 28$$

Did we get the same value when the grouping symbols were dropped? _____

Answers: a) 12 b) 12 c) Yes

---

123. Here is another expression in which the grouping is preceded by an addition symbol:

$$10 + (4 - 8)$$

Evaluating with the grouping symbol, we get:

$$10 + \underbrace{(4 - 8)}$$
$$10 + \quad (-4) \quad = 6$$

Evaluating without the grouping symbols, we get:

$$\underbrace{10 + 4} - 8$$
$$14 \quad - 8 = 6$$

Do we get the same value when the grouping symbols are dropped? _____

Answer: Yes

Answer: Yes

124. WHEN A GROUPING FOLLOWS A SUBTRACTION SYMBOL, THE GROUPING SYMBOLS CANNOT BE DROPPED WITHOUT CHANGING THE VALUE OF THE EXPRESSION. Here is an example:

$$20 - (6 + 2)$$

Evaluating with the grouping symbols:

$$20 - (6 + 2)$$

$$20 - 8 = 12$$

Evaluating without the grouping symbols:

$$20 - 6 + 2$$

$$14 + 2 = 16$$

Did we get the same value when we dropped the grouping symbols? _____

No

125. (a) Evaluate this expression with grouping symbols:

$$20 - (10 - 3)$$

$$(\ \ ) - (\ \ ) = \_\_\_\_\_\_$$

(b) Evaluate this expression without grouping symbols:

$$20 - 10 - 3$$

$$(\ \ ) - (\ \ ) = \_\_\_\_\_\_$$

(c) Are the two expressions equal? _______

---

a) 20 - 7 = 13

b) 10 - 3 = 7

c) No

126. (a) When a grouping appears after a "+", the grouping symbols __________ (can/cannot) be dropped.

(b) When a grouping appears after a "-", the grouping symbols __________ (can/cannot) be dropped.

---

a) can

b) cannot

127. The grouping symbols after a "+" can be dropped even if the grouping contains a letter.

That is: $10 + (x - 3) = 10 + x - 3$

If we plug in "4" for "x": (a) The left side equals _______.

(b) The right side equals _______.

---

a) 11

b) 11

128. The grouping symbols after a "-" cannot be dropped when the grouping contains a letter.

That is: $10 - (t - 2) \neq 10 - t - 2$

If we plug in "5" for "t": (a) The left side equals _______.

(b) The right side equals _______.

---

a) 7

b) 3

129. (a) Does $35 + (3x + 2) = 35 + 3x + 2$? ______ (b) Does $35 - (3x + 2) = 35 - 3x + 2$? ______	
130. Which of the following statements are <u>true</u>? ________ (a) $20 - (x - 2) = 20 - x - 2$ (b) $20 + (x - 2) = 20 + x - 2$	a) <u>Yes</u>, the grouping symbols can be dropped. b) <u>No</u>, the grouping symbols cannot be dropped.
131. Which of the following statements are <u>false</u>? ________ (a) $7x - (3x - 4) = 7x - 3x - 4$ (b) $7x + (3x - 4) = 7x + 3x - 4$	Only (b) is true.
	Only (a) is false.

SELF-TEST 6 (Frames 112-131)

1. Draw a box around each term: $5(-2) - 3 - (4 - 9)$
2. Evaluate the expression in Problem 1: ________

Given this equation: $\boxed{2 - (a - 3) = 9}$

3. If "-4" is plugged in for "a", the value of the left side is ______.
4. Is $a = -4$ the solution of the equation? ______

True or False? ________ 5. $7 - (w - 2) = 7 - w - 2$

________ 6. $7 + (w - 2) = 7 + w - 2$

________ 7. $(y - 3) - 5 = y - 3 - 5$

ANSWERS:

1. $\boxed{5(-2)} - \boxed{3} - \boxed{(4 - 9)}$
2. -8
3. 9
4. Yes
5. False
6. True
7. True

3-10 THE OPPOSITE OF AN ADDITION AND RELATED EQUATIONS

Groupings are easy to handle in evaluations because they can easily be eliminated by combining the two number-terms. However, when groupings appear in equations, they contain a letter-term and a number-term. Groupings of this type cannot be eliminated by combining terms since a letter-term and a number-term cannot be combined.

When a grouping follows an addition symbol in an equation, there is no problem because we CAN simply drop the grouping symbols. We then obtain an equation which we know how to solve.

Example: $20 + (3x + 2) = 3$

is equivalent to: $20 + 3x + 2 = 3$

However, when a grouping follows a subtraction symbol, we CANNOT simply drop the grouping symbols.

Example: $20 - (3x + 2) = 3$

is NOT equivalent to: $20 - 3x + 2 = 3$

In this section, we will introduce a method of solving equations in which a grouping is preceded by a subtraction symbol.

---

132. We have introduced the following types of terms in equations:

(1) Number-terms like 5 or −7

(2) Letter-terms like 3x or −5t

(3) Multiplications which contain a grouping as one factor like $3(x + 7)$ or $5(2p - 6)$

Draw a box around each term in these equations:

(a) $7x + 5(2x + 7) = 9x - 6$

(b) $10 - 4p = 4(3p - 9) - 15$

---

133. When a grouping appears by itself in an equation (that is, it is not a factor in a multiplication), it is also a term.

For example: In $10 - (x + 5) = 17$,

$(x + 5)$ is one term.

Draw a box around each term in these equations:

(a) $7b - (b + 9) = 17 + 2(b + 5)$

(b) $3(R - 6) - 9R = 40 - (3R - 2)$

Answers to 132:

a) $\boxed{7x} + \boxed{5(2x + 7)} = \boxed{9x} - \boxed{6}$

b) $\boxed{10} - \boxed{4p} = \boxed{4(3p - 9)} - \boxed{15}$

Answers to 133:

a) $\boxed{7b} - \boxed{(b + 9)} = \boxed{17} + \boxed{2(b + 5)}$

b) $\boxed{3(R - 6)} - \boxed{9R} = \boxed{40} - \boxed{(3R - 2)}$

134. When confronted with a more complex term, we try to reduce it to letter-terms or number-terms since we know

(1) how to combine terms of both types,

and (2) how to use the addition axiom to eliminate terms of both types.

We "multiply by the distributive principle" to reduce a term with a grouping as one factor to one letter-term and one number-term.

Example: $3(x + 7) = 10$

reduces to: $3x + 21 = 10$

In doing so, we increase the number of terms on the left side from one to ________.

Answer: two

135. A grouping by itself is also a complex term. If an equation contains a grouping by itself after a "+", we can drop the grouping symbols. By dropping the grouping symbols, we replace a complex term with one letter-term and one number-term.

Example: $7d + (5d - 3) = 33$

becomes: $7d + 5d - 3 = 33$

By doing so, we increase the number of terms on the left side from ________ to ________.

Answer: two to three

---

136. When a grouping comes after a "-", we cannot simply drop the grouping symbols to eliminate the complex term. We need another technique to replace the complex term with a letter-term and a number-term. The technique we use is this:

(1) We convert the subtraction of a grouping to the addition of a grouping.

(2) Then we can drop the grouping symbols since the grouping comes after a "+".

To convert the subtraction of a grouping to the addition of a grouping, we must "add the opposite of the grouping." Here is an example:

$$5 - (2x + 3) = 5 + [\text{the opposite of } (2x + 3)]$$

To complete this conversion, we must examine the "opposite" of the grouping (2x + 3). We will do so in the next few frames.

137. We know this fact: If opposites are added, their sum is "0".

Examples: $8 + (-8) = 0$ and $4x + (-4x) = 0$

The "opposite of a grouping" is also a grouping. We obtain it by replacing each term of the original grouping with its opposite. Here are two examples:

Example: The opposite of $(8 + 7)$ is $[(-8) + (-7)]$, since:

$$(8 + 7) + [(-8) + (-7)] = \underbrace{8 + (-8)}_{0} + \underbrace{7 + (-7)}_{0} = 0 + 0 = 0$$

Example: The opposite of $(3x + 2)$ is $[(-3x) + (-2)]$, since:

$$(3x + 2) + [(-3x) + (-2)] = \underbrace{3x + (-3x)}_{0} + \underbrace{2 + (-2)}_{0} = 0 + 0 = 0$$

---

138. Here is the general pattern for the opposite of a grouping:

The opposite of $(a + b)$ is $[(-a) + (-b)]$,

where "$-a$" is the opposite of "a"
and "$-b$" is the opposite of "b".

(Of course, "a" and "b" stand for any quantity.)

---

139. Write the opposite of each of the following groupings:

Grouping	Opposite
(a) $(x + 5)$	[_____ + _____]
(b) $(3x + 11)$	[_____ + _____]
(c) $[7 + (-4x)]$	[_____ + _____]
(d) $[5x + (-10)]$	[_____ + _____]

---

140. Write the opposite of each of the following groupings:

(a) $[(-2d) + (-9)]$ [_____ + _____]

(b) $[(-5) + (-8p)]$ [_____ + _____]

Answers to 139:
a) $[(-x) + (-5)]$
b) $[(-3x) + (-11)]$
c) $[(-7) + 4x]$
d) $[(-5x) + 10]$

---

141. We are now prepared to convert the subtraction of a grouping to the addition of a grouping.

$$5 - (3x + 2) = 5 + [\text{the opposite of } (3x + 2)]$$
$$= 5 + [(-3x) + (-2)]$$

Complete: (a) $5t - (t + 7) = 5t + [\text{the opposite of } (t + 7)]$
$= 5t +$ [_____ + _____]

(b) $4 - [2b + (-3)] = 4 + (\text{the opposite of } [2b + (-3)])$
$= 4 +$ [_____ + _____]

(c) $3y - [8 + (-5y)] = 3y + (\text{the opposite of } [8 + (-5y)])$
$= 3y +$ [_____ + _____]

Answers to 140:
a) $[2d + 9]$
b) $[5 + 8p]$

142. Complete these conversions to additions:

(a) $7 - (q + 4) = 7 + [\_\_\_\_ + \_\_\_\_]$

(b) $4x - [3x + (-9)] = 4x + [\_\_\_\_ + \_\_\_\_]$

(c) $5R - [(-10) + (-4R)] = 5R + [\_\_\_\_ + \_\_\_\_]$

a) $+ [(-t) + (-7)]$

b) $+ [(-2b) + 3]$

c) $+ [(-8) + 5y]$

---

143. After converting the subtraction of a grouping to the addition of a grouping, we can now drop the grouping symbols because the <u>grouping follows a "+"</u>.

$$35 - (y + 3) = 35 + [(-y) + (-3)]$$
$$= 35 + (-y) + (-3)$$

The last expression is the type we want because it does not contain a grouping. It contains only simple letter-terms and number-terms.

Complete these: (a) $21 - (2x + 7) = 21 + [(-2x) + (-7)]$

$= \_\_\_\_ + \_\_\_\_ + \_\_\_\_$

(b) $4t - [9t + (-5)] = 4t + [(-9t) + 5]$

$= \_\_\_\_ + \_\_\_\_ + \_\_\_\_$

a) $+ [(-q) + (-4)]$

b) $+ [(-3x) + 9]$

c) $+ [10 + 4R]$

---

144. You should be able to skip the step in which you write the grouping symbols around the opposite of the grouping.

Instead of: $7x - (x + 4) = 7x + [(-x) + (-4)]$
$= 7x + (-x) + (-4)$,

you should be able to write immediately:

$$7x - (x + 4) = 7x + (-x) + (-4)$$

Complete these: (a) $10 - (p + 6) = \_\_\_\_ + \_\_\_\_ + \_\_\_\_$

(b) $R - [3R + (-2)] = \_\_\_\_ + \_\_\_\_ + \_\_\_\_$

(c) $8 - [5 + (-4t)] = \_\_\_\_ + \_\_\_\_ + \_\_\_\_$

(d) $5q - [(-3q) + (-6)] = \_\_\_\_ + \_\_\_\_ + \_\_\_\_$

a) $21 + (-2x) + (-7)$

b) $4t + (-9t) + 5$

---

145. We are now prepared to solve the following equation:

$$20 - (3x + 2) = 3$$

Converting the subtraction of a grouping to the addition of a grouping, we get:

$$20 + [(-3x) + (-2)] = 3$$

Since the grouping now follows a "+", we can drop the grouping symbols and get:

$$20 + (-3x) + (-2) = 3$$

(a) Complete the solution of this last equation: $x = \_\_\_\_$

(b) Show that the root you obtained is also the root of the <u>original equation</u>:

a) $10 + (-p) + (-6)$

b) $R + (-3R) + 2$

c) $8 + (-5) + 4t$

d) $5q + 3q + 6$

146. When converting the subtraction of a grouping to the addition of a grouping in an equation, we can also skip the step in which we write grouping symbols after a "+".

<u>Example</u>: $26 - (5y + 6) = 10$

We can immediately write the conversion this way:

$$26 + (-5y) + (-6) = 10$$

Of course, we are adding the ______________ of each term in the grouping.

a) $x = 5$

b) Check:

$20 - (3x + 2) = 3$

$20 - [3(5) + 2] = 3$

$20 - [15 + 2] = 3$

$20 - 17 = 3$

$3 = 3$

---

147. Complete these:

(a) $5x - (x + 3) = 13$

$5x +$ ____ $+$ ____ $= 13$

(b) $10 = 35 - [6F + (-7)]$

$10 = 35 +$ ____ $+$ ____

(c) $25 - [8 + (-2d)] = 17$

$25 +$ ____ $+$ ____ $= 17$

(d) $9y = 6y - [(-4y) + (-5)]$

$9y = 6y +$ ____ $+$ ____

opposite

---

148. $7m - (m + 3) = 15$

(a) Converting the subtraction to addition, we get:

____ $+$ ____ $+$ ____ $=$ ____

(b) Solve the new equation:

$m =$ ______

(c) The root of the original equation is ______.

a) $+ (-x) + (-3)$

b) $+ (-6F) + 7$

c) $+ (-8) + 2d$

d) $+ 4y + 5$

---

149. $F = 35 - [6F + (-7)]$

(a) Convert the subtraction to an addition:

____ $=$ ____ $+$ ____ $+$ ____

(b) Solve the new equation:

$F =$ ______

(c) Show that this root satisfies the original equation:

a) $7m + (-m) + (-3) = 15$

b) $m = 3$

c) 3

150. The following equations contain the subtraction of a grouping on each side. You should be able to make both conversions to addition <u>in one step</u>. Do so:

(a) $10 - (p + 3) = 4p - [2p + (-7)]$

___ + ___ + ___ = ___ + ___ + ___

(b) $8x - [x + (-9)] = 14 - [(-x) + (-10)]$

___ + ___ + ___ = ___ + ___ + ___

a) $F = 35 + (-6F) + 7$

b) $F = 6$

c) Check:
$F = 35 - [6F + (-7)]$
$6 = 35 - [6(6) + (-7)]$
$6 = 35 - [36 + (-7)]$
$6 = 35 - 29$
$6 = 6$

---

151. The following equations contain a subtraction of a grouping and a subtraction of a simple term. You should be able to convert all subtractions to additions in one step. Do so:

(a) $5 - (t + 8) = 27 - 6t$

___ + ___ + ___ = ___ + ___

(b) $9d - [5 + (-3d)] - 17 = 0$

___ + ___ + ___ + ___ = 0

a) $10 + (-p) + (-3) = 4p + (-2p) + 7$

b) $8x + (-x) + 9 = 14 + x + 10$

---

152. $15 - (m + 7) = 2m - 13$

(a) Convert all subtractions to additions:

________ = ________

(b) Complete the solution:

m = _____

a) $5 + (-t) + (-8) = 27 + (-6t)$

b) $9d + (-5) + 3d + (-17) = 0$

---

153. $23 - [2x + (-7)] - 4x = 0$

(a) Convert all subtractions to additions:

________ = 0

(b) Complete the solution:

x = _____

a) $15 + (-m) + (-7) = 2m + (-13)$

b) $m = 7$

(From: $21 = 3m$)

---

a) $23 + (-2x) + 7 + (-4x) = 0$

b) $x = 5$

(From: $30 = 6x$)

154. In all of the equations we have solved, the grouping has contained an addition. In the following equation, the grouping contains a subtraction.

$$25 - (4x - 3) = 8$$

Our first step is to convert the subtraction within the grouping to addition:

$$25 - [4x + (-3)] = 8$$

Now we can proceed as usual:

$$25 + (-4x) + 3 = 8$$

(a) Complete the solution:

x = ______

(b) Show that this root satisfies the original equation:

---

155. We know how to find the opposite of a grouping which contains an addition. We do not know how to find the opposite of a grouping which contains a subtraction. Therefore, when a grouping after a "-" contains a subtraction, our first step is to convert the subtraction within the grouping to addition. Then we can proceed as usual.

Show the first step for each of these:

(a) $5p - (p - 2) = 30$ (b) $7d = 10 - (8 - 3d)$

______ ______

Answers to 154:

a) $x = 5$

b) Check:

$$25 - (4x - 3) = 8$$
$$25 - [4(5) - 3] = 8$$
$$25 - [20 - 3] = 8$$
$$25 - 17 = 8$$
$$8 = 8$$

---

156. $7m - (m - 4) = 15$

(a) Show the first step: ______

(b) Convert the subtraction of the grouping to addition:

______

Answers to 155:

In each case, we simply convert the subtraction within the grouping to addition:

a) $5p - [p + (-2)] = 30$

b) $7d = 10 - [8 + (-3d)]$

---

157. $0 = 12 - (5 - 3q)$

(a) Show the first step: ______

(b) Convert the subtraction of the grouping to addition:

______

Answers to 156:

a) $7m - [m + (-4)] = 15$

b) $7m + (-m) + 4 = 15$

Answers to 157:

a) $0 = 12 - [5 + (-3q)]$

b) $0 = 12 + (-5) + 3q$

158. This equation contains two groupings. Each follows a "-" and each contains a subtraction.

$$7m - (m - 4) = 15 - (7 - m)$$

(a) Show the first step for each grouping:

______________ = ______________

(b) Convert the subtraction of both groupings to addition:

______________ = ______________

---

159. This equation includes a subtraction of a grouping which contains a subtraction. It also includes a subtraction of a simple term.

$$2x - (3x - 9) = 14 - 6x$$

First we must convert the subtraction within the grouping to addition:

$$2x - [3x + (-9)] = 14 - 6x$$

Now we can convert all subtractions to additions in one step. Do so:

___ + ___ + ___ = ___ + ___

Answers to 158:

a) $7m - [m + (-4)]$
$= 15 - [7 + (-m)]$

b) $7m + (-m) + 4$
$= 15 + (-7) + m$

---

160. $5t - 7 = 15 - (8 - 4t) - 10t$

(a) We must convert the subtraction within the grouping to addition first. Do so:

__________ = __________________

(b) Now convert all subtractions to additions in one step:

___ + ___ = ___ + ___ + ___ + ___

Answer to 159:

$2x + (-3x) + 9$
$= 14 + (-6x)$

---

161. $5y - (y - 20) = 0$

(a) Show the first step: __________________ = 0

(b) Convert the subtraction of the grouping to addition:

___ + ___ + ___ = 0

(c) Complete the solution:

y = ______

(d) Show that this root satisfies the original equation:

Answers to 160:

a) $5t - 7$
$= 15 - [8 + (-4t)] - 10t$

b) $5t + (-7)$
$= 15 + (-8) + 4t + (-10t)$

162. Solve: $3(x + 3) = 5x - (x - 5)$

$x =$ _______

a) $5y - [y + (-20)] = 0$

b) $5y + (-y) + 20 = 0$

c) $y = -5$
(From $4y = -20$)

d) Check:

$5y - (y - 20) = 0$

$5(-5) - [(-5) - 20] = 0$

$(-25) - [(-5)+(-20)] = 0$

$(-25) - (-25) = 0$

$(-25) + (25) = 0$

$0 = 0$

---

163. Solve: $5q - (7 - 2q) = 14 - (3 - q)$

$q =$ _______

$x = 4$

---

164. Solve: $7t - (t + 5) = 10 - (4 - 5t)$

$t =$ _______

$q = 3$

$t = 11$

SELF-TEST 7 (Frames 132-164)

1. Draw a box around each term in the equation at the right: $x + 2(x - 12) = 15 - 8x - (x + 3)$

Complete the following:

2. $7 - (4x - 1) = 7 +$ the ____________ of $(4x - 1)$.
3. $14r - (2r + 5) = 14r +$ ______ + ______
4. $3 - (6 - e) = 3 +$ ______ + ______

Find the root of each equation:

5. $y - (3y + 2) = 8$ $y =$ ______
6. $15 - (h - 9) = 3h$ $h =$ ______

Solve each equation:

7. $1 - (t - 1) = t + (t - 1)$ $t =$ ______
8. $4 - 8x = 13 - (2x - 1) - x$ $x =$ ______

ANSWERS:

1. $\boxed{x} + \boxed{2(x - 12)} = \boxed{15} - \boxed{8x} - \boxed{(x + 3)}$
2. opposite
3. $14r + \underline{(-2r)} + \underline{(-5)}$
4. $\underline{3} + \underline{(-6)} + \underline{(+e)}$
5. $y = -5$
6. $h = 6$
7. $t = 1$
8. $x = -2$

3-11 SUBTRACTING INSTANCES OF THE DISTRIBUTIVE PRINCIPLE

We have already solved equations which contain instances of the distributive principle. These "instances" present no special problem when:

(1) They are the first term on either side of an equation:

Examples: $4(x + 3) = 16$

$5y = 7(y + 4) - 3y$

or (2) They follow an addition symbol:

Example: $10 + 6(x + 5) = 40$

The following equation is a new type. In it, the instance of the distributive principle follows a subtraction symbol:

$$10 - 3(x + 4) = 16$$

Extreme caution must be used in handling an "instance" when it follows a "-". In this section, we will learn how to solve equations of this type.

165. An instance of the distributive principle is <u>one</u> term. Since it is a complicated term, we always try to reduce it to a letter-term and a simple number-term.

In the following equation, an instance of the distributive principle is subtracted from 10:

$$10 - 3(x + 4) = 16$$

To show that we are subtracting one term from 10, we put brackets around $3(x + 4)$ and get:

$$10 - [3(x + 4)] = 16$$

We cannot convert the subtraction to addition as it stands because we do not know what the opposite of an instance of the distributive principle is. Therefore, we multiply by the distributive principle and obtain a simple grouping (that is, one which is not a factor in the distributive principle):

$$10 - [3x + 12] = 16$$

Now we can convert the subtraction to addition as usual:

$$10 + (-3x) + (-12) = 16$$

(a) Complete the solution:

x = ______

(b) Show that this root satisfies the original equation:

---

a) $x = -6$

(From $-3x = 18$)

b) Check:

$$10 - 3(x + 4) = 16$$
$$10 - 3[(-6) + 4] = 16$$
$$10 - 3(-2) = 16$$
$$10 - (-6) = 16$$
$$10 + 6 = 16$$
$$16 = 16$$

166. Here is another example:

$$40 - 2(3b + 2) = 6$$

When faced with an equation of this type, we immediately put brackets around the instance of the distributive principle:

$$40 - [2(3b + 2)] = 6$$

We use the brackets for two reasons:

(1) To remind us that we are subtracting one term from 40.
(2) To remind us that the "-" is the symbol for subtraction and not the sign of "2".

When multiplying by the distributive principle, we do not drop the grouping symbols since the grouping comes after a "-":

$$40 - [6b + 4] = 6$$

Now we can easily convert the subtraction to addition. Do so:

____ + ____ + ____ = ____

$40 + (-6b) + (-4) = 6$

167. Here is one more. We have put brackets around the instance of the distributive principle:

$$15 - [4(2t + 3)] = 27$$

(a) Multiply by the distributive principle:

_________________ = ____

(b) Now convert the subtraction to an addition:

____ + ____ + ____ = ____

---

168. **Bracketing an instance of the distributive principle when it follows a subtraction symbol is extremely useful. Using the brackets avoids many sign errors. We will insist that you use them.**

Put brackets around each instance of the distributive principle in this equation before you do anything else.

$$3 - 4(3 + x) = 5x - 6(2x + 7)$$

(a) Multiply by the distributive principle in each case. In the resulting equation, be sure to show the brackets.

_________________ = _________________

(b) Now convert both subtractions to additions:

____ + ____ + ____ = ____ + ____ + ____

Answers to 167:

a) $15 - [8t + 12] = 27$

b) $15 + (-8t) + (-12) = 27$

---

169. (a) Solve this equation: $0 = 7D - 9(D + 2)$

D = _______

(b) Show that the root satisfies the original equation:

Answers to 168:

a) $3 - [12 + 4x]$

$= 5x - [12x + 42]$

b) $3 + (-12) + (-4x)$

$= 5x + (-12x) + (-42)$

---

170. Solve; but be careful: $9S - 2(S + 6) = 4S - (S - 4)$

S = _______

Answers to 169:

a) $D = -9$

From $2D = -18$

or $-2D = 18$

b) Check:

$0 = 7D - 9(D + 2)$

$0 = 7(-9) - 9[(-9) + 2]$

$0 = (-63) - 9(-7)$

$0 = (-63) - (-63)$

$0 = (-63) + (63)$

$0 = 0$

Answer to 170: $S = 4$

171. The following term is not an instance of the distributive principle over addition because the grouping contains a subtraction:

$$7(x - 5)$$

However, since we can easily change it into an instance of the distributive principle over addition by converting the subtraction to addition:

$$7[x + (-5)]$$

from now on, we will call terms like the original one above "instances of the distributive principle."

When an instance of the distributive principle contains a subtraction within the grouping, we simply convert that subtraction to addition and proceed as usual.

The above procedure is applied to the following equation:

$$20 - 2(5x - 4) = 8$$

Bracketing the "instance," we get:

$$20 - \left[2(5x - 4)\right] = 8$$

Converting the subtraction within the grouping to addition, we get:

$$20 - \left[2[5x + (-4)]\right] = 8$$

Multiplying by the distributive principle, we get:

$$20 - \left[10x + (-8)\right] = 8$$

Converting the subtraction of a grouping to addition, we get:

$$20 + (-10x) + 8 = 8$$

In this frame, we began with this equation:

$$20 - 2(5x - 4) = 8$$

and obtained this equation as our last step:

$$20 + (-10x) + 8 = 8$$

(a) Complete the solution:

x = ______

(b) Show that this root satisfies the original equation:

$$20 - 2(5x - 4) = 8$$

---

a) $x = 2$

b) Check:

$$20 - 2(5x - 4) = 8$$
$$20 - 2[5(2) - 4] = 8$$
$$20 - 2[10 - 4] = 8$$
$$20 - 2(6) = 8$$
$$20 - 12 = 8$$
$$8 = 8$$

172. Here is another one. We have bracketed the instance of the distributive principle:

$$17 - [5(2k - 4)] = 7$$

(a) To multiply by the distributive principle, we must first convert the subtraction within the grouping to addition, and then perform the multiplication. Show both steps:

________________________ = ____

________________________ = ____

(b) Now convert the subtraction of the grouping to addition:

____ + ____ + ____ = ____

a) $17 - \left[5[2k + (-4)]\right] = 7$

$17 - [10k + (-20)] = 7$

b) $17 + (-10k) + 20 = 7$

---

173. Show all the steps needed to obtain an equation which does not contain a grouping:

$$35 = 47 - 6(4 - x)$$

$35 = 47 - [6(4 - x)]$

$35 = 47 - \left[6[4 + (-x)]\right]$

$35 = 47 - [24 + (-6x)]$

$35 = 47 + (-24) + 6x$

---

174. Show all the steps needed to obtain an equation which does not contain a grouping:

$$30 = 12p - 3(5p - 2)$$

$30 = 12p - [3(5p - 2)]$

$30 = 12p - \left[3[5p + (-2)]\right]$

$30 = 12p - [15p + (-6)]$

$30 = 12p + (-15p) + 6$

---

175. Solve this one: $19 - 3(5 - 4k) = 28$

k = ______

$k = 2$

---

176. Solve, but be careful: $7P - (P - 2) = 40P - 5(4P - 6)$

P = ______

177. Solving this equation is optional. However, <u>if you can find its root, you have an excellent grasp of all the principles we have taught.</u>

$$3(t - 5) - 10 - 4(t + 1) = t + 2(t + 7) - 5(5 - t)$$

t = ______

P = -2

t = -2

SELF-TEST 8 (Frames 165-177)

1. Solve for r: $8r - 3(r + 5) = 15$

   r = ______

2. Solve for w: $7w = 2 - 5(2 - w)$

   w = ______

3. Solve this equation:

   $$d - 2(d - 1) = 11 + 3(d + 1)$$

   d = ______

4. Find the root:

   $$5 - 3(1 - p) + 3p - 2(p - 1) = 0$$

   p = ______

ANSWERS: 1. r = 6 2. w = -4 3. d = -3 4. p = -1

3-12 A SUMMARY OF THE FORMAL STRATEGIES FOR SOLVING EQUATIONS

Although we have introduced other principles, the following five basic processes are the major steps used in solving the types of equations we have covered. They are listed in the order in which they are used when needed.

(1) Multiplying (distributive principle)
(2) Converting subtraction to addition
(3) Combining like terms
(4) Addition axiom
(5) Solving the basic equation

In this section, we will formally review the strategies used to solve equations.

---

178. Both of the following we now call instances of the distributive principle:

$$5(x + 7)$$

$$5(x - 7)$$

When an equation contains an instance of the distributive principle, we always "multiply by the distributive principle" first. (When the grouping-factor contains a subtraction, we need the added step of converting this subtraction to addition before multiplying.) Here are examples of the various types:

(1) When the "instance" is the first term on one side or comes after a "+", we can immediately obtain a simple letter-term and a simple number-term.

Example:
$$5(x - 7) = 45$$
$$5[x + (-7)] = 45$$
$$5x + (-35) = 45$$

Example:
$$15 = 7y + 3(y + 5)$$
$$15 = 7y + 3y + 15$$

(2) When the "instance" comes after a "-", we obtain a grouping which is not a factor in a multiplication.

Example:
$$20 - [4(p + 7)] = 8$$
$$20 - [4p + 28] = 8$$
$$20 + (-4p) + (-28) = 8$$

Example:
$$12 - [5(3d - 6)] = 7d$$
$$12 - \Big[5[3d + (-6)]\Big] = 7d$$
$$12 - [15d + (-30)] = 7d$$
$$12 + (-15d) + 30 = 7d$$

Note: After "multiplying by the distributive principle," we convert all the remaining subtractions to additions, as was done in the last two examples above.

---

179. When an isolated grouping contains a subtraction, the subtraction within the grouping must be converted to addition before the subtraction of the grouping is converted to addition. In the following two examples, the first subtraction to be converted is labeled with an arrow:

Example:
$$5 - (t \overset{\downarrow}{-} 9) = 17$$
$$5 - [t + (-9)] = 17$$
$$5 + (-t) + 9 = 17$$

Example:
$$10 = 3F - (2F \overset{\downarrow}{-} 6)$$
$$10 = 3F - [2F + (-6)]$$
$$10 = 3F + (-2F) + 6$$

180. After multiplying by the distributive principle and converting all subtractions to additions, we can proceed to combine terms, to use the addition axiom to obtain a basic equation, and to solve the basic equation.

Identify the steps used to solve the following equation:

$$\boxed{20 - (4P + 3) = 5}$$

Step 1: $20 + (-4P) + (-3) = 5$ ____________________

Step 2: $17 + (-4P) = 5$ ____________________

Step 3: $\left.\begin{array}{r} 17 + (-17) + (-4P) = 5 + (-17) \\ -4P = -12 \end{array}\right\}$ ____________________

Step 4: $P = 3$ ____________________

Answer to Frame 180:

Step 1: Converting subtraction to addition
Step 2: Combining like terms
Step 3: Addition axiom (-17)
Step 4: Solving the basic equation

---

181. Identify the steps in this solution:

$$\boxed{3R = 4 - (8 - R)}$$

Step 1: $\left.\begin{array}{r} 3R = 4 - [8 + (-R)] \\ 3R = 4 + (-8) + R \end{array}\right\}$ ____________________

Step 2: $3R = (-4) + R$ ____________________

Step 3: $\left.\begin{array}{r} 3R + (-R) = (-4) + R + (-R) \\ 2R = -4 \end{array}\right\}$ ____________________

Step 4: $R = -2$ ____________________

Answer to Frame 181:

Step 1: Converting subtraction to addition
Step 2: Combining like terms
Step 3: Addition axiom (-R)
Step 4: Solving the basic equation

182. Identify the steps in this one:

$$\boxed{8x - 5(x - 3) = 35 - 2x}$$

Step 1: $8x - \left[5[x + (-3)]\right] = 35 - 2x$
$8x - [5x + (-15)] = 35 - 2x$ ______________________

Step 2: $8x + (-5x) + 15 = 35 + (-2x)$ ______________________

Step 3: $3x + 15 = 35 + (-2x)$ ______________________

Step 4: $3x + 2x + 15 = 35 + (-2x) + 2x$
$5x + 15 = 35$ ______________________

Step 5: $5x + 15 + (-15) = 35 + (-15)$
$5x = 20$ ______________________

Step 6: $x = 4$ ______________________

Answer to Frame 182:
Step 1: Multiplying (distributive principle)
Step 2: Converting subtraction to addition
Step 3: Combining like terms
Step 4: Addition axiom (2x)
Step 5: Addition axiom (-15)
Step 6: Solving the basic equation

183. Solve this one. Show all the steps and identify them. You will need four steps:

$$\boxed{5m = 4m - (m + 12)}$$

Step 1: ______________________

Step 2: ______________________

Step 3: ______________________

Step 4: ______________________

Answer to Frame 183:

Step 1:	$5m = 4m + (-m) + (-12)$	Converting subtraction to addition
Step 2:	$5m = 3m + (-12)$	Combining like terms
Step 3:	$5m + (-3m) = 3m + (-3m) + (-12)$ $2m = -12$	Addition axiom (-3m)
Step 4:	$m = -6$	Solving the basic equation

184. You will need five steps for this one:

$$\boxed{26 = 10 - 4(x + 1)}$$

Step 1: ______________________

Step 2: ______________________

Step 3: ______________________

Step 4: ______________________

Step 5: ______________________

Answer to Frame 184:

Step 1:	$26 = 10 - [4x + 4]$	Multiplying (distributive principle)
Step 2:	$26 = 10 + (-4x) + (-4)$	Converting subtraction to addition
Step 3:	$26 = 6 + (-4x)$	Combining like terms
Step 4:	$26 + (-6) = 6 + (-6) + (-4x)$ $20 = -4x$	Addition axiom (-6)
Step 5:	$x = -5$	Solving the basic equation

SELF-TEST 9 (Frames 178-184)

1. In the equation at the right, which subtraction (A or B) should be converted to addition first? _______

$$6 - (2 - E) = 10$$

(A points to the first subtraction sign; B points to the second.)

2. The following principles or processes are used in solving equations: →

- Addition axiom
- Combining like terms
- Multiplying (distributive principle)
- Solving the basic equation
- Converting subtraction to addition

Identify each step in the solution below:

$$\boxed{5 - 2(x - 3) = 4(x + 2) - 3x}$$

Step 1: $5 - \left[2[x + (-3)]\right] = 4(x + 2) - 3x$
$5 - [2x + (-6)] = 4x + 8 - 3x$ ______________________

Step 2: $5 + (-2x) + 6 = 4x + 8 + (-3x)$ ______________________

Step 3: $11 + (-2x) = x + 8$ ______________________

Step 4: $11 + (-2x) + 2x = x + 8 + 2x$
$11 = 3x + 8$ ______________________

Step 5: $11 + (-8) = 3x + 8 + (-8)$
$3 = 3x$ ______________________

Step 6: $x = 1$ ______________________

ANSWERS:

1. B

2. Step 1: Multiplying (distributive principle)
Step 2: Converting subtraction to addition
Step 3: Combining like terms
Step 4: Addition axiom (+2x)
Step 5: Addition axiom (-8)
Step 6: Solving the basic equation

Chapter 4 MULTIPLICATION AND DIVISION OF FRACTIONS

A mastery of fractions is essential for anyone who hopes to develop skills in the type of algebra which is needed in basic science and technology. The "mastery" we are talking about is a mastery of both numerical fractions and fractions which contain letters. In this chapter, we will review the principles underlying the multiplication and division of fractions, factoring fractions, and reducing fractions to lower or lowest terms.

4-1 THE MEANING OF FRACTIONS

Fractions are frequently introduced as a way of stating a "part of a whole." For example, the meaning of $\frac{1}{8}$, $\frac{1}{4}$, or $\frac{1}{2}$ is frequently given in terms of slices of pie or parts of a circle. But, there is a much more useful and meaningful way of thinking about fractions. A fraction is merely another way of writing a division. In this section, we will show the relationship between a fraction and the ordinary way of writing a division.

---

1. A fraction is merely another way of writing a division:

   $\frac{6}{3}$ means $6 \div 3$ $\qquad$ $\frac{7}{8}$ means $7 \div 8$ $\qquad$ $\frac{5}{2}$ means ________

Answer: $5 \div 2$

2. Write each of the following divisions as a fraction:

   (a) $20 \div 10 =$ ______ $\qquad$ (c) $5 \div 9 =$ ______

   (b) $40 \div 27 =$ ______ $\qquad$ (d) $1 \div 4 =$ ______

Answers: a) $\frac{20}{10}$ $\quad$ b) $\frac{40}{27}$ $\quad$ c) $\frac{5}{9}$ $\quad$ d) $\frac{1}{4}$

3. Write each problem in ordinary division form using the "÷" symbol:

   (a) $\frac{35}{7} = 5$ ____________ $\qquad$ (b) $\frac{56}{8} = 7$ ____________

Answers: a) $35 \div 7 = 5$ $\quad$ b) $56 \div 8 = 7$

4. In the ordinary division form, there is a special name for each number.

   In $6 \div 3 = 2$, we call: "6" the dividend
"3" the divisor
"2" the quotient

   In $8 \div 2 = 4$: (a) The dividend is _____.

   (b) The divisor is _____.

   (c) The quotient is _____.

Answers: a) 8 $\quad$ b) 2 $\quad$ c) 4

5. In $28 \div 7 = 4$: (a) The quotient is _____.

   (b) The dividend is _____.

   (c) The divisor is _____.

6. Similarly, when we write a division as a fraction, there is a special name for each number.

In $\frac{6}{3} = 2$, we call: "6" the <u>numerator</u>
"3" the <u>denominator</u>
"2" the <u>quotient</u>

Whether it is written in the ordinary division form or as a fraction, the answer to a division is always called the ______________.

a) 4
b) 28
c) 7

---

7. In $\frac{5}{4}$: (a) _____ is the numerator.
(b) _____ is the denominator.

<u>quotient</u>

---

8. In $\frac{7}{6}$: (a) "7" is called the ______________.
(b) "6" is called the ______________.

a) 5
b) 4

---

9. (a) In the ordinary division form $8 \div 5$:
"8" is called the ______________.
"5" is called the ______________.

(b) In the fraction $\frac{8}{5}$:
"8" is called the ______________.
"5" is called the ______________.

(c) The <u>numerator</u> of a fraction is the same as the ______________ in ordinary division form.

(d) The <u>denominator</u> of a fraction is the same as the ______________ in ordinary division form.

a) numerator
b) denominator

---

a) dividend
divisor

b) numerator
denominator

c) dividend
d) divisor

> <u>IN ALGEBRA, WE ALWAYS WRITE ANY DIVISION AS A FRACTION.</u> Furthermore, most of the fractions you will meet in algebra make sense <u>only if you think of them as divisions</u>. They will not make sense if you think of them as "parts of a whole."

4-2 THE MULTIPLICATION OF TWO FRACTIONS

All of the basic operations which can be performed with whole numbers can also be performed with fractions. In this section, we will introduce the multiplication of two fractions. It is a fairly easy operation.

10. **DEFINITION OF THE MULTIPLICATION OF TWO FRACTIONS**

$$\left(\frac{\square}{\bigcirc}\right)\left(\frac{\triangle}{\text{⬡}}\right)=\frac{(\square)(\triangle)}{(\bigcirc)(\text{⬡})}$$

Notice that to get the product:

(1) The two numerators are multiplied.
(2) The two denominators are multiplied.

That is: $\left(\frac{2}{3}\right)\left(\frac{5}{7}\right)=\frac{(2)(5)}{(3)(7)}$

(a) $\left(\frac{3}{7}\right)\left(\frac{2}{5}\right)=\frac{(\quad)(\quad)}{(\quad)(\quad)}$ (b) $\left(\frac{6}{5}\right)\left(\frac{2}{11}\right)=\frac{(\quad)(\quad)}{(\quad)(\quad)}$

---

a) $\frac{(3)(2)}{(7)(5)}$ b) $\frac{(6)(2)}{(5)(11)}$

11. Let's examine the following multiplication of two fractions:

$$\left(\frac{2}{7}\right)\left(\frac{5}{3}\right)=\frac{(2)(5)}{(7)(3)}$$

In a fraction, the numerator and denominator are separated by a fraction line. There is one fraction line for each distinct fraction.

How many distinct fraction lines are there:

(a) On the left side above? __________

(b) On the right side above? __________

---

a) Two

b) One

12. Let's examine the same multiplication:

$$\left(\frac{2}{7}\right)\left(\frac{5}{3}\right)=\frac{(2)(5)}{(7)(3)}$$

On the left side:

Since there are two distinct fraction lines, there are two fractions. They are called factors. Each of the two factors has a single whole number as its numerator and as its denominator.

On the right side:

Since there is only one fraction line, there is only one fraction. It is called the product. The product has a multiplication of two factors in both its numerator and its denominator.

Identify each expression below as either "factors" or a "product:"

(a) $\frac{(4)(3)}{(7)(9)}$ __________ (b) $\left(\frac{5}{6}\right)\left(\frac{1}{8}\right)$ __________ (c) $\frac{(3)(7)}{(11)(5)}$ __________

---

a) Product

b) Factors

c) Product

13. A <u>letter</u> can also be the numerator or denominator of a fraction. The fraction still stands for a division.

$$\frac{x}{3} \text{ means } x \div 3$$

$$\frac{4}{d} \text{ means } 4 \div d$$

When a fraction contains letters, multiplication is performed the same way:

$$\left(\frac{2}{3}\right)\left(\frac{x}{7}\right) = \frac{(2)(x)}{(3)(7)}$$

Complete these:

(a) $\left(\frac{y}{4}\right)\left(\frac{3}{5}\right) = \frac{(\ )(\ )}{(\ )(\ )}$ (b) $\left(\frac{4}{7}\right)\left(\frac{11}{x}\right) = \frac{(\ )(\ )}{(\ )(\ )}$

---

a) $\frac{(y)(3)}{(4)(5)}$ b) $\frac{(4)(11)}{(7)(x)}$

14. A letter with a numerical coefficient can also be the numerator or denominator of a fraction.

$$\frac{2x}{3} \text{ means } 2x \div 3$$

$$\frac{5}{7R} \text{ means } 5 \div 7R$$

Multiplications are still performed the same way:

$$\left(\frac{5}{11}\right)\left(\frac{7t}{3}\right) = \frac{(5)(7t)}{(11)(3)}$$

Complete: (a) $\left(\frac{2b}{3}\right)\left(\frac{4}{7}\right) = \frac{(\ )(\ )}{(\ )(\ )}$ (b) $\left(\frac{5}{6}\right)\left(\frac{11}{3m}\right) = \frac{(\ )(\ )}{(\ )(\ )}$

---

a) $\frac{(2b)(4)}{(3)(7)}$ b) $\frac{(5)(11)}{(6)(3m)}$

15. The product of two fractions is always simplified as much as possible. We do so by performing the multiplication in the numerator and denominator.

Therefore: $\frac{(7)(3)}{(4)(11)}$ simplifies to $\frac{21}{44}$

$\frac{(x)(3)}{(2)(8)}$ simplifies to $\frac{3x}{16}$

$\frac{(5)(2y)}{(3)(11)}$ simplifies to $\frac{10y}{33}$

(Although "3x" and "10y" are still <u>two-factor</u> expressions, they cannot be simplified further. Notice that we write <u>the numerical coefficient in front of the letter.</u>)

Simplify each of these products:

(a) $\frac{(5)(9)}{(t)(8)} =$ ______ (b) $\frac{(7)(11)}{(4p)(5)} =$ ______

---

a) $\frac{45}{8t}$ b) $\frac{77}{20p}$

16. Ordinarily we write the product in its simplified form immediately. Do so with these:

(a) $\left(\frac{3}{8}\right)\left(\frac{7}{5}\right) =$ _____ (b) $\frac{(m)(5)}{(7)(8)} =$ _____ (c) $\frac{(3)(5)}{(7)(2R)} =$ _____

Answers: a) $\frac{21}{40}$ b) $\frac{5m}{56}$ c) $\frac{15}{14R}$

17. The most common error in multiplying fractions is adding the numerators or denominators instead of multiplying them.

Some of the following answers are wrong because the common error was committed. For those which are wrong, write the correct answer in the blank.

(a) $\left(\frac{1}{3}\right)\left(\frac{5}{6}\right) = \frac{6}{9}$ _____ (c) $\left(\frac{3}{7}\right)\left(\frac{2}{5}\right) = \frac{6}{12}$ _____

(b) $\left(\frac{1}{4}\right)\left(\frac{3}{7}\right) = \frac{4}{28}$ _____ (d) $\left(\frac{7}{8}\right)\left(\frac{3}{2}\right) = \frac{21}{16}$ _____

Answers: a) $\frac{5}{18}$ b) $\frac{3}{28}$ c) $\frac{6}{35}$ d) Answer is correct.

4-3 MULTIPLYING A FRACTION BY A NON-FRACTION

A fraction can also be multiplied by a non-fraction, such as 8, t, or 3x, since multiplication of this type can easily be converted into a multiplication of two fractions. We will show the method in this section.

---

18. The following principle can be used to convert a whole number to a fraction:

IF ANY QUANTITY IS DIVIDED BY +1, THE QUOTIENT IS THE ORIGINAL NUMBER.

That is: $\frac{8}{1} = 8$ (a) $\frac{15}{1} =$ _____ (b) $\frac{1976}{1} =$ _____

Answers: a) 15 b) 1976

19. In the box below, we have written this principle in symbol form:

$$\frac{\square}{1} = \square \quad \text{or} \quad \frac{n}{1} = n$$

We call it:

THE PRINCIPLE OF DIVIDING A QUANTITY BY +1

Which of the following is an instance of this principle? _____

(a) $\frac{8}{2} = 4$ (b) $\frac{17}{1} = 17$ (c) $\frac{10}{10} = 1$

Answer: Only (b)

20. Here is the same principle:

$$\frac{n}{1} = n$$

Using the interchange principle, we get:

$$n = \frac{n}{1}$$

The new form says this: <u>Any whole number can be written as a fraction in which the whole number is the numerator and "1" is the denominator.</u>

That is: 7 equals the fraction $\frac{7}{1}$.

Write each of these whole numbers as a fraction:

(a) 3 = ______ (b) 19 = ______

Answers: a) $\frac{3}{1}$ b) $\frac{19}{1}$

---

21. By <u>substituting a fraction for a whole number</u> in a multiplication, we can convert <u>the multiplication of a whole number and a fraction</u> into <u>the multiplication of two fractions.</u>

Here are two examples:

$$7\left(\frac{3}{8}\right) = \left(\frac{7}{1}\right)\left(\frac{3}{8}\right)$$

$$\frac{1}{6}(13) = \left(\frac{1}{6}\right)\left(\frac{13}{1}\right)$$

Convert each of these multiplications into a multiplication <u>of two fractions</u>:

(a) $8\left(\frac{3}{5}\right) = \left(\frac{\quad}{\quad}\right)\left(\frac{\quad}{\quad}\right)$ (b) $\left(\frac{5}{6}\right)(5) =$ ___ ___

Answers: a) $\left(\frac{8}{1}\right)\left(\frac{3}{5}\right)$ b) $\left(\frac{5}{6}\right)\left(\frac{5}{1}\right)$

---

22. Since a letter is simply a place where a number is plugged in, any letter (with or without a numerical coefficient) can also be written as a fraction. We use the same principle. For example:

$$t = \frac{t}{1} \qquad 2x = \frac{2x}{1}$$

Convert the following multiplications to a multiplication <u>of two fractions</u>:

(a) $\frac{7}{8}(x) = \left(\frac{\quad}{\quad}\right)\left(\frac{\quad}{\quad}\right)$ (c) $3y\left(\frac{1}{5}\right) = \left(\frac{\quad}{\quad}\right)\left(\frac{\quad}{\quad}\right)$

(b) $p\left(\frac{1}{4}\right) = \left(\frac{\quad}{\quad}\right)\left(\frac{\quad}{\quad}\right)$ (d) $2\left(\frac{4A}{3}\right) = \left(\frac{\quad}{\quad}\right)\left(\frac{\quad}{\quad}\right)$

Answers: a) $\left(\frac{7}{8}\right)\left(\frac{x}{1}\right)$ c) $\left(\frac{3y}{1}\right)\left(\frac{1}{5}\right)$ b) $\left(\frac{p}{1}\right)\left(\frac{1}{4}\right)$ d) $\left(\frac{2}{1}\right)\left(\frac{4A}{3}\right)$

23. Having converted the multiplication of a non-fraction and a fraction into the multiplication of two fractions, we can perform the multiplication in the usual way:

$$7\left(\frac{5}{8}\right) = \left(\frac{7}{1}\right)\left(\frac{5}{8}\right) = \frac{(7)(5)}{(1)(8)} = \frac{35}{8}$$

$$\left(\frac{1}{9}\right)(b) = \left(\frac{1}{9}\right)\left(\frac{b}{1}\right) = \frac{(1)(b)}{(9)(1)} = \frac{1b}{9} \text{ or } \frac{b}{9}$$

Convert each of the following to a multiplication of two fractions and write the simplified product:

(a) $5\left(\frac{1}{6}\right) = (\text{—})(\text{—}) = $ ____

(b) $\left(\frac{d}{4}\right)(9) = (\text{—})(\text{—}) = $ ____

(c) $(x)\left(\frac{1}{8}\right) = (\text{—})(\text{—}) = $ ____

(d) $\left(\frac{3}{5}\right)(6y) = (\text{—})(\text{—}) = $ ____

(e) $\left(\frac{1}{3m}\right)(7) = (\text{—})(\text{—}) = $ ____

Answers to 23:

a) $\left(\frac{5}{1}\right)\left(\frac{1}{6}\right) = \frac{5}{6}$

b) $\left(\frac{d}{4}\right)\left(\frac{9}{1}\right) = \frac{9d}{4}$

c) $\left(\frac{x}{1}\right)\left(\frac{1}{8}\right) = \frac{1x}{8} \text{ or } \frac{x}{8}$

d) $\left(\frac{3}{5}\right)\left(\frac{6y}{1}\right) = \frac{18y}{5}$

e) $\left(\frac{1}{3m}\right)\left(\frac{7}{1}\right) = \frac{7}{3m}$

---

24. In an earlier chapter, we introduced THE IDENTITY PRINCIPLE OF MULTIPLICATION:

$1(\square) = \square$	or	$1(n) = n$

That is, if any quantity is multiplied by "1", the product is the original quantity.

For example: $(1)(8) = 8$

$(1)(3t) = 3t$

This principle is also true for fractions. It is easy to show this fact because "1" can also be written as a fraction:

$$1 = \frac{1}{1}$$

Therefore, any multiplication of "1" and a fraction can also be converted into a multiplication of two fractions. Here are two examples:

$$1\left(\frac{2}{3}\right) = \left(\frac{1}{1}\right)\left(\frac{2}{3}\right) = \frac{2}{3} \qquad 1\left(\frac{p}{5}\right) = \left(\frac{1}{1}\right)\left(\frac{p}{5}\right) = \frac{p}{5}$$

Find each product:

(a) $1\left(\frac{7}{8}\right) = $ ____

(b) $1\left(\frac{x}{4}\right) = $ ____

(c) $1\left(\frac{3m}{5}\right) = $ ____

(d) $1\left(\frac{7}{4d}\right) = $ ____

Answers to 24:

a) $\frac{7}{8}$ b) $\frac{x}{4}$ c) $\frac{3m}{5}$ d) $\frac{7}{4d}$

SELF-TEST 1 (Frames 1-24)

1. $\frac{3}{8}$ means _____ divided by _____.

2. Write $32 \div 7$ in fraction form: _______

3. Write the fraction whose denominator is 4 and whose numerator is 15: _______

4. If a division problem is written as a fraction, the divisor is the _____________ (numerator/denominator).

Complete:

5. $\left(\frac{3}{7}\right)\left(\frac{2}{11}\right) = \frac{(\ \ )(\ \ )}{(\ \ )(\ \ )}$

6. $\left(\frac{5d}{2}\right)\left(\frac{7}{12}\right) = \frac{(\ \ )(\ \ )}{(\ \ )(\ \ )}$

Write each of the following as the multiplication of two fractions:

7. $\left(\frac{5}{8}\right)(3) = \left(\frac{\ }{\ }\right)\left(\frac{\ }{\ }\right)$

8. $(2t)\left(\frac{3}{7}\right) = \left(\frac{\ }{\ }\right)\left(\frac{\ }{\ }\right)$

Multiply and write your answer as a single fraction in simplest form:

9. $\left(\frac{2}{7}\right)\left(\frac{3}{5}\right) =$ _______

10. $\left(\frac{y}{3}\right)\left(\frac{2}{3}\right) =$ _______

11. $5\left(\frac{3}{8}\right) =$ _______

12. $\left(\frac{7}{10}\right)(3) =$ _______

13. $3\left(\frac{2w}{11}\right) =$ _______

14. $\left(\frac{9}{4}\right)(5R) =$ _______

ANSWERS:
1. 3 divided by 8
2. $\frac{32}{7}$
3. $\frac{15}{4}$
4. denominator
5. $\frac{(3)(2)}{(7)(11)}$
6. $\frac{(5d)(7)}{(2)(12)}$
7. $\left(\frac{5}{8}\right)\left(\frac{3}{1}\right)$
8. $\left(\frac{2t}{1}\right)\left(\frac{3}{7}\right)$
9. $\frac{6}{35}$
10. $\frac{2y}{9}$
11. $\frac{15}{8}$
12. $\frac{21}{10}$
13. $\frac{6w}{11}$
14. $\frac{45R}{4}$

4-4 FACTORING FRACTIONS: MEANING AND PROCEDURE

In this section, we will examine the meaning of factoring a fraction and the procedure for doing so. We will show: (1) that any fraction can be factored in some way, and (2) that the procedure for factoring is simply the reverse of the procedure for multiplying fractions.

25. Here is a multiplication of two fractions: $\left(\frac{3}{5}\right)\left(\frac{7}{2}\right) = \frac{21}{10}$

As in any multiplication:

(1) The answer $\frac{21}{10}$ is called the product.

(2) The two quantities which are multiplied, $\frac{3}{5}$ and $\frac{7}{2}$, are called factors.

Here is a multiplication of a non-fraction and a fraction:

$$(5)\left(\frac{1}{8}\right) = \frac{5}{8}$$

In this multiplication: (a) $\frac{5}{8}$ is called the ___________.

(b) 5 and $\frac{1}{8}$ are called ___________.

26. By using the interchange principle, we can write the product on the left and the factors on the right.

a) product
b) factors

Since $\left(\frac{3}{5}\right)\left(\frac{7}{2}\right) = \frac{21}{10}$

$$\frac{21}{10} = \left(\frac{3}{5}\right)\left(\frac{7}{2}\right)$$

We can read the last equation this way:

$\frac{21}{10}$ can be factored into $\frac{3}{5}$ and $\frac{7}{2}$.

Since $(5)\left(\frac{1}{8}\right) = \frac{5}{8}$,

$$\frac{5}{8} = (5)\left(\frac{1}{8}\right).$$

We can read the last equation this way:

$\frac{5}{8}$ can be ______________ into 5 and $\frac{1}{8}$.

factored

---

27. To factor a fraction, we merely reverse the procedure for multiplication. Here are two examples:

(1) The procedure for multiplying two fractions:

$$\left(\frac{7}{11}\right)\left(\frac{2}{3}\right) = \frac{(7)(2)}{(11)(3)} = \frac{14}{33}$$

The procedure for factoring $\frac{14}{33}$ into $\frac{7}{11}$ and $\frac{2}{3}$:

$$\frac{14}{33} = \frac{(7)(2)}{(11)(3)} = \left(\frac{7}{11}\right)\left(\frac{2}{3}\right)$$

(2) The procedure for multiplying a non-fraction and a fraction:

$$7\left(\frac{1}{5}\right) = \left(\frac{7}{1}\right)\left(\frac{1}{5}\right) = \frac{(7)(1)}{(1)(5)} = \frac{7}{5}$$

The procedure for factoring $\frac{7}{5}$ into 7 and $\frac{1}{5}$:

$$\frac{7}{5} = \frac{(7)(1)}{(1)(5)} = \left(\frac{7}{1}\right)\left(\frac{1}{5}\right) = 7\left(\frac{1}{5}\right)$$

Go to next frame.

28. When multiplying fractions, we always simplify the product. Any fraction is a simplified product for some multiplication.

$\frac{7}{11}$ is the simplified product from: $\frac{(7)(1)}{(1)(11)}$

$\frac{3x}{10}$ is the simplified product from: $\frac{(3)(x)}{(2)(5)}$

$\frac{5}{t}$ is the simplified product from: $\frac{(5)(1)}{(1)(t)}$

We will call the fractions on the left "simplified" products.

We will call the fractions on the right "unsimplified" products.

In the unsimplified products, how many factors are there:

(a) In each numerator? _____ (b) In each denominator? _____

---

a) Two b) Two

29. Here are some unsimplified products:

Example 1: $\frac{(3)(m)}{(5)(2)}$ Example 2: $\frac{(1)(3m)}{(5)(2)}$

In unsimplified products, there are two explicit factors in both the numerator and denominator. By "explicit," we mean that they are enclosed in parentheses.

Though $(3)(m) = (3m)$:

(a) In the first example, (3)(m) is treated as _______ factor(s).

(b) In the second example, (3m) is treated as _______ factor(s).

---

a) two

b) one

30. The first step in factoring a fraction is to convert from the simplified product back to the unsimplified product. To do so, both the numerator and denominator must be factored into two explicit factors. The numerator and denominator of simplified products are numbers (like 7 or 15) or letter-terms (like x or 3t). Let's examine the possibilities for factoring quantities of this type:

(1) Some quantities can only be factored by the principle $n = (1)(n)$.

For example: $7 = (1)(7)$
$x = (1)(x)$

(2) Some quantities can be factored both by this principle and in another way:

For example: $15 = (1)(15)$
$15 = (5)(3)$

$3t = (1)(3t)$
$3t = (3)(t)$

Which of the following quantities can be factored in another way besides the principle $n = (1)(n)$? _______________

(a) 5m (b) 5 (c) 11 (d) 6

31. An unsimplified product must have two explicit factors in both numerator and denominator.

Which of the following are unsimplified products? ________

(a) $\frac{(5)(3)}{(7)(11)}$ (b) $\frac{(5)}{(1)(8)}$ (c) $\frac{(1)(7)}{(3)(x)}$ (d) $\frac{(1)(5x)}{(3)}$

Answer: (a) and (d)

Since:

$5m = (5)(m)$

$6 = (3)(2)$

---

32. When converting back to an unsimplified product, you must write two factors in both the numerator and denominator:

(a) Instead of: $\frac{6}{7} = \frac{(3)(2)}{(7)}$,

you should get: $\frac{6}{7} = \frac{(\ )(\ )}{(\ )(\ )}$.

(b) Instead of: $\frac{p}{3} = \frac{(p)}{(1)(3)}$,

you should get: $\frac{p}{3} = \frac{(\ )(\ )}{(\ )(\ )}$.

Answer: Only (a) and (c)

---

33. After converting back to the unsimplified product, we can immediately write two factors:

$$\frac{14}{15} = \frac{(7)(2)}{(5)(3)} = \left(\frac{7}{5}\right)\left(\frac{2}{3}\right)$$

We have factored $\frac{14}{15}$ into ____ and ____.

Answer: a) $\frac{(3)(2)}{(1)(7)}$ b) $\frac{(p)(1)}{(1)(3)}$

---

34. When one of the factors is an instance of $\frac{n}{1}$, we always replace it with a non-fraction:

$$\frac{d}{5} = \frac{(d)(1)}{(1)(5)} = \left(\frac{d}{1}\right)\left(\frac{1}{5}\right) = d\left(\frac{1}{5}\right)$$

We have factored $\frac{d}{5}$ into ____ and ____.

Answer: $\frac{7}{5}$ and $\frac{2}{3}$

Answer: d and $\frac{1}{5}$

---

35. Whenever we convert from a simplified product back to the unsimplified product, we can write the factors in the numerator and denominator in four different orders:

$$\frac{3x}{14} = \frac{(3)(x)}{(7)(2)} \text{ or } \frac{(x)(3)}{(7)(2)} \text{ or } \frac{(x)(3)}{(2)(7)} \text{ or } \frac{(3)(x)}{(2)(7)}$$

Therefore, we can write factors in four different ways:

$$\frac{3x}{14} = \left(\frac{3}{7}\right)\left(\frac{x}{2}\right) \text{ or } \left(\frac{x}{7}\right)\left(\frac{3}{2}\right) \text{ or } \left(\frac{x}{2}\right)\left(\frac{3}{7}\right) \text{ or } \left(\frac{3}{2}\right)\left(\frac{x}{7}\right)$$

(Continued on following page.)

35. Continued

In practice, this flexibility means:

(1) After obtaining one pair of factors, we can obtain a different pair by interchanging their numerators:

$$\left(\frac{3}{7}\right)\left(\frac{x}{2}\right) \text{ or } \left(\frac{x}{7}\right)\left(\frac{3}{2}\right)$$

(2) Since multiplication is commutative, we can also commute the factors in each pair above:

$$\left(\frac{x}{2}\right)\left(\frac{3}{7}\right) \text{ or } \left(\frac{3}{2}\right)\left(\frac{x}{7}\right)$$

36.

$$\frac{21}{55} = \frac{(7)(3)}{(5)(11)}$$

(a) Write two different pairs of factors for $\frac{21}{55}$ by interchanging the two numerators:

$$\left(\frac{\quad}{\quad}\right)\left(\frac{\quad}{\quad}\right) \text{ and } \left(\frac{\quad}{\quad}\right)\left(\frac{\quad}{\quad}\right)$$

(b) Now commute the two factors in each pair:

$$\left(\frac{\quad}{\quad}\right)\left(\frac{\quad}{\quad}\right) \text{ and } \left(\frac{\quad}{\quad}\right)\left(\frac{\quad}{\quad}\right)$$

Go to next frame.

---

a) $\left(\frac{7}{5}\right)\left(\frac{3}{11}\right)$ and $\left(\frac{3}{5}\right)\left(\frac{7}{11}\right)$

b) $\left(\frac{3}{11}\right)\left(\frac{7}{5}\right)$ and $\left(\frac{7}{11}\right)\left(\frac{3}{5}\right)$

---

37. Let's examine the two different pairs of factors we get when both the numerator and denominator are factored by $n = (1)(n)$:

Pair 1: $\frac{7}{5} = \left(\frac{7}{1}\right)\left(\frac{1}{5}\right) = (7)\left(\frac{1}{5}\right)$

Interchanging the numerators, we get:

Pair 2: $\frac{7}{5} = \left(\frac{1}{1}\right)\left(\frac{7}{5}\right) = (1)\left(\frac{7}{5}\right)$

The second pair is an instance of $n = (1)(n)$:

$$\frac{7}{5} = (1)\left(\frac{7}{5}\right)$$

The first pair is not an instance of $n = (1)(n)$:

$$\frac{7}{5} = (7)\left(\frac{1}{5}\right)$$

Note: Any fraction can be factored by $n = (1)(n)$:

$$\frac{7}{8} = (1)\left(\frac{7}{8}\right) \qquad \frac{5}{x} = (1)\left(\frac{5}{x}\right) \qquad \frac{3t}{8} = (1)\left(\frac{3t}{8}\right)$$

Ordinarily, when we ask you to factor a fraction, we will exclude factoring it by $n = (1)(n)$.

38. Whenever we factor into a fraction and a non-fraction, the factors can be written in any order since multiplication is commutative:

$$\frac{x}{3} = (x)\left(\frac{1}{3}\right) \quad \text{or} \quad \frac{1}{3}(x)$$

$$\frac{2y}{5} = (y)\left(\frac{2}{5}\right) \quad \text{or} \quad \frac{2}{5}(y)$$

When the non-fractional factor is a letter, we usually write the fraction first. In this case, the fraction is the numerical coefficient of the letter.

Since $\frac{x}{3} = \frac{1}{3}(x)$, the numerical coefficient of "x" in $\frac{x}{3}$ is $\frac{1}{3}$.

Since $\frac{2y}{5} = \frac{2}{5}(y)$, the numerical coefficient of "y" in $\frac{2y}{5}$ is ____.

---

39. Write the numerical coefficient of "p" in each of the following expressions:

(a) $\frac{1}{7}(p)$ ____ (b) $\frac{p}{13}$ ____ (c) $\frac{4p}{17}$ ____ (d) $\frac{(p)(8)}{5}$ ____

Answer to 38: $\frac{2}{5}$

---

40. Whenever a fraction is factored, you should check your factoring. You do so by multiplying the two factors to see whether you obtain the original fraction.

Which of the following factorings are incorrect? ________

(a) $\frac{77}{40} = \left(\frac{7}{5}\right)\left(\frac{11}{8}\right)$ (c) $\frac{28}{5R} = \left(\frac{13}{R}\right)\left(\frac{2}{5}\right)$

(b) $\frac{56}{7d} = 8\left(\frac{9}{7d}\right)$ (d) $\frac{11t}{54} = \left(\frac{t}{6}\right)\left(\frac{11}{9}\right)$

Answers to 39: a) $\frac{1}{7}$ b) $\frac{1}{13}$ c) $\frac{4}{17}$ d) $\frac{8}{5}$

(b) and (c) are incorrect since: $8\left(\frac{9}{7d}\right) = \frac{72}{7d}$ $\left(\frac{13}{R}\right)\left(\frac{2}{5}\right) = \frac{26}{5R}$

4-5 FACTORING INTO A FRACTION AND A NON-FRACTION

In this section, we will discuss the cases in which one of the factors reduces to a non-fraction. We will show that any fraction can be factored into a fraction and a non-fraction in at least one way.

41. Whenever the number "1" is a factor in the denominator of an unsimplified product, one of the factors will be an instance of $\frac{n}{1}$ and therefore will reduce to a non-fraction.

$$\frac{(5)(1)}{(1)(11)} = \left(\frac{5}{1}\right)\left(\frac{1}{11}\right) = 5\left(\frac{1}{11}\right)$$

Therefore, whenever we factor the original denominator by $\underline{n = (1)(n)}$, one of the factors will reduce to a non-fraction.

Which of the following first steps will eventually lead to a non-fraction as a factor? ________

(a) $\frac{x}{6} = \frac{(x)(1)}{(3)(2)}$ (c) $\frac{5m}{14} = \frac{(5m)(1)}{(7)(2)}$

(b) $\frac{3x}{10} = \frac{(3)(x)}{(1)(10)}$ (d) $\frac{6}{5t} = \frac{(3)(2)}{(1)(5t)}$

Only (b) and (d)

42. Here are three complete factorings. In each case, we factored both the numerator and denominator by $\underline{n = (1)(n)}$:

$$\frac{5}{11} = \frac{(5)(1)}{(1)(11)} = \left(\frac{5}{1}\right)\left(\frac{1}{11}\right) = 5\left(\frac{1}{11}\right)$$

$$\frac{x}{7} = \frac{(x)(1)}{(1)(7)} = \left(\frac{x}{1}\right)\left(\frac{1}{7}\right) = x\left(\frac{1}{7}\right)$$

$$\frac{2t}{3} = \frac{(2t)(1)}{(1)(3)} = \left(\frac{2t}{1}\right)\left(\frac{1}{3}\right) = 2t\left(\frac{1}{3}\right)$$

Examine the two factors on the extreme right in each case:

The first factor is the numerator of the original fraction.

The second factor is a fraction whose numerator is "1" and whose denominator is the denominator of the original fraction.

The pattern is:
$\text{Fraction} = (\text{Numerator})\left(\frac{1}{\text{Denominator}}\right)$

Using this pattern, complete these:

(a) $\frac{11}{13} = (\quad)\left(\frac{}{}\right)$ (c) $\frac{7q}{19} = (\quad)\left(\frac{}{}\right)$

(b) $\frac{v}{9} = (\quad)\left(\frac{}{}\right)$ (d) $\frac{5}{2r} = (\quad)\left(\frac{}{}\right)$

---

a) $11\left(\frac{1}{13}\right)$ c) $7q\left(\frac{1}{19}\right)$

b) $v\left(\frac{1}{9}\right)$ d) $5\left(\frac{1}{2r}\right)$

43. Some numerators can only be factored by $n = (1)(n)$. However, when the numerator can be factored in some other way, we can factor into a fraction and a non-fraction in two other ways. Here is an example:

$$\frac{14}{5} = \frac{(7)(2)}{(1)(5)} = \left(\frac{7}{1}\right)\left(\frac{2}{5}\right) = 7\left(\frac{2}{5}\right)$$

$$\frac{14}{5} = \frac{(2)(7)}{(1)(5)} = \left(\frac{2}{1}\right)\left(\frac{7}{5}\right) = 2\left(\frac{7}{5}\right)$$

Examine the pair of factors on the extreme right:

The first factor is one of the factors of the original numerator.

The second factor is the other factor of the original numerator divided by the original denominator.

> Here is the pattern:
>
> $$\text{Fraction} = \frac{(\text{Factor 1})(\text{Factor 2})}{\text{Denominator}} = (\text{Factor 1})\left(\frac{\text{Factor 2}}{\text{Denominator}}\right)$$

Use this pattern for these:

(a) Since $\frac{15}{11} = \frac{(5)(3)}{11}$,

$$\frac{15}{11} = (\quad)\left(\frac{}{}\right) \text{ or } (\quad)\left(\frac{}{}\right)$$

(b) Since $\frac{21}{5} = \frac{(7)(3)}{5}$,

$$\frac{21}{5} = (\quad)\left(\frac{}{}\right) \text{ or } (\quad)\left(\frac{}{}\right)$$

Answers to 43:

a) $5\left(\frac{3}{11}\right)$ or $3\left(\frac{5}{11}\right)$

b) $7\left(\frac{3}{5}\right)$ or $3\left(\frac{7}{5}\right)$

44. The numerator of the following fraction can be factored without using $n = (1)(n)$. We can therefore use the new pattern:

$$\frac{5m}{13} = \frac{(5)(m)}{13} = 5\left(\frac{m}{13}\right) \text{ or } m\left(\frac{5}{13}\right)$$

Complete this one, using the new pattern:

$$\frac{11R}{3} = (\quad)\left(\frac{}{}\right) \text{ or } (\quad)\left(\frac{}{}\right)$$

Answer to 44: $11\left(\frac{R}{3}\right)$ or $R\left(\frac{11}{3}\right)$

45. When factoring into a fraction and a non-fraction:

(1) If the numerator can only be factored by $n = (1)(n)$, we can obtain only one pair of factors:

$$\frac{7}{8} = 7\left(\frac{1}{8}\right) \qquad \frac{x}{3} = x\left(\frac{1}{3}\right)$$

(2) If the numerator can be factored in one other way, we can obtain three pairs of factors:

$$\frac{6}{5} = 6\left(\frac{1}{5}\right) \text{ or } 3\left(\frac{2}{5}\right) \text{ or } 2\left(\frac{3}{5}\right)$$

$$\frac{5y}{17} = 5y\left(\frac{1}{17}\right) \text{ or } 5\left(\frac{y}{17}\right) \text{ or } y\left(\frac{5}{17}\right)$$

(Continued on following page.)

45. (Continued)

Factor each of these into a fraction and a non-fraction in as many ways as possible:

(a) $\frac{33}{7}$ ____________________

(b) $\frac{19}{x}$ ____________________

(c) $\frac{3x}{13}$ ____________________

---

a) $33\left(\frac{1}{7}\right)$ or $3\left(\frac{11}{7}\right)$ or $11\left(\frac{3}{7}\right)$ b) $19\left(\frac{1}{x}\right)$ c) $3x\left(\frac{1}{13}\right)$ or $3\left(\frac{x}{13}\right)$ or $x\left(\frac{3}{13}\right)$

4-6 FACTORING INTO TWO FRACTIONS

Any fraction can be factored into a fraction and a non-fraction. Only some fractions can be factored into two fractions without one reducing to a non-fraction. In this section, we will show when the latter type of factoring is possible.

---

46. Any factoring of a fraction begins by factoring into two fractions. However, if we factor the denominator by the principle $n = (1)(n)$, one of the fractions reduces to a non-fraction.

To factor into two fractions in which this type of reduction does not occur, we must be able to factor the denominator in some way other than $n = (1)(n)$.

Which of these fractions can be factored into two fractions without one reducing to a non-fraction? __________

(a) $\frac{7}{6}$ (b) $\frac{7}{11}$ (c) $\frac{7}{15}$

---

47. Here is a case in which the numerator can only be factored by $n = (1)(n)$. Watch the factoring into two fractions:

$$\frac{5}{6} = \frac{(5)(1)}{(3)(2)} = \left(\frac{5}{3}\right)\left(\frac{1}{2}\right)$$

Note: In the unsimplified product, there must be two factors in the numerator. Therefore, we factored "5" by $n = (1)(n)$.

In all of the following fractions, the numerator can only be factored by $n = (1)(n)$. Factor each one into two fractions:

(a) $\frac{7}{15} = \frac{(\quad)(\quad)}{(\quad)(\quad)} = \left(\frac{\quad}{\quad}\right)\left(\frac{\quad}{\quad}\right)$

(b) $\frac{y}{21} = \frac{(\quad)(\quad)}{(\quad)(\quad)} = \left(\frac{\quad}{\quad}\right)\left(\frac{\quad}{\quad}\right)$

(c) $\frac{5}{3t} = \frac{(\quad)(\quad)}{(\quad)(\quad)} = \left(\frac{\quad}{\quad}\right)\left(\frac{\quad}{\quad}\right)$

Answer to 46: (a) and (c)

Since $6 = (3)(2)$

and $15 = (5)(3)$

48. To factor into two fractions which do not reduce to a non-fraction, what number must we avoid as a factor in the denominator? _______

a) $\frac{(7)(1)}{(5)(3)} = \left(\frac{7}{5}\right)\left(\frac{1}{3}\right)$ or $\left(\frac{1}{5}\right)\left(\frac{7}{3}\right)$

b) $\frac{(y)(1)}{(7)(3)} = \left(\frac{y}{7}\right)\left(\frac{1}{3}\right)$ or $\left(\frac{1}{7}\right)\left(\frac{y}{3}\right)$

c) $\frac{(5)(1)}{(3)(t)} = \left(\frac{5}{3}\right)\left(\frac{1}{t}\right)$ or $\left(\frac{1}{3}\right)\left(\frac{5}{t}\right)$

---

49. The following numerator can be factored by $n = (1)(n)$ and in one other way. Therefore, we can factor into four different pairs of fractions.

$$\frac{21}{10} = \frac{(21)(1)}{(5)(2)} = \left(\frac{21}{5}\right)\left(\frac{1}{2}\right) \text{ or } \left(\frac{1}{5}\right)\left(\frac{21}{2}\right)$$

$$\frac{21}{10} = \frac{(7)(3)}{(5)(2)} = \left(\frac{7}{5}\right)\left(\frac{3}{2}\right) \text{ or } \left(\frac{3}{5}\right)\left(\frac{7}{2}\right)$$

Factor $\frac{5t}{6}$ into two fractions in two ways:

(a) $\frac{5t}{6} = \left(\frac{\quad}{\quad}\right)\left(\frac{\quad}{\quad}\right)$ or $\left(\frac{\quad}{\quad}\right)\left(\frac{\quad}{\quad}\right)$

(b) $\frac{5t}{6} = \left(\frac{\quad}{\quad}\right)\left(\frac{\quad}{\quad}\right)$ or $\left(\frac{\quad}{\quad}\right)\left(\frac{\quad}{\quad}\right)$

1

NOTE: When we use the words "different pairs" in the following frames, we will not count "commuting factors" as a different pair. By "commuting factors," we simply mean writing the same pair in a different order.

a) $\left(\frac{5t}{3}\right)\left(\frac{1}{2}\right)$ or $\left(\frac{1}{3}\right)\left(\frac{5t}{2}\right)$

b) $\left(\frac{5}{3}\right)\left(\frac{t}{2}\right)$ or $\left(\frac{t}{3}\right)\left(\frac{5}{2}\right)$

---

50. (a) Factor $\frac{x}{21}$ into a fraction and a non-fraction in as many ways as possible: _______________

(b) Factor $\frac{x}{21}$ into different pairs of fractions in as many ways as possible: _______________

---

51. (a) Factor $\frac{35}{11}$ into a fraction and a non-fraction in as many ways as possible: _______________

(b) Factor $\frac{35}{11}$ into different pairs of fractions in as many ways as possible: _______________

a) $x\left(\frac{1}{21}\right)$

b) $\left(\frac{x}{7}\right)\left(\frac{1}{3}\right)$ or $\left(\frac{1}{7}\right)\left(\frac{x}{3}\right)$

---

52. (a) Factor $\frac{7y}{15}$ into a fraction and a non-fraction in as many ways as possible: _______________

(b) Factor $\frac{7y}{15}$ into different pairs of fractions in as many ways as possible: _______________

a) $35\left(\frac{1}{11}\right)$, $7\left(\frac{5}{11}\right)$, $5\left(\frac{7}{11}\right)$

b) Not possible, since "11" can only be factored by $n = (1)(n)$.

53. (a) Factor $\frac{6}{5m}$ into a fraction and a non-fraction in as many ways as possible: ____________________

(b) Factor $\frac{6}{5m}$ into different pairs of fractions in as many ways as possible: ____________________

a) $7y\left(\frac{1}{15}\right)$, $7\left(\frac{y}{15}\right)$, $y\left(\frac{7}{15}\right)$

b) $\left(\frac{7y}{5}\right)\left(\frac{1}{3}\right)$, $\left(\frac{1}{5}\right)\left(\frac{7y}{3}\right)$, $\left(\frac{7}{5}\right)\left(\frac{y}{3}\right)$, $\left(\frac{y}{5}\right)\left(\frac{7}{3}\right)$

a) $6\left(\frac{1}{5m}\right)$, $3\left(\frac{2}{5m}\right)$, $2\left(\frac{3}{5m}\right)$ b) $\left(\frac{6}{5}\right)\left(\frac{1}{m}\right)$, $\left(\frac{1}{5}\right)\left(\frac{6}{m}\right)$, $\left(\frac{3}{5}\right)\left(\frac{2}{m}\right)$, $\left(\frac{2}{5}\right)\left(\frac{3}{m}\right)$

54. Up to this point, we have not factored a fraction whose numerator and denominator can both be factored in more than one way besides n = (1)(n). A fraction of this type can be factored in many ways. Here is an example:

$\frac{12}{24}$ can be factored into a fraction and a non-fraction in these ways:

$$12\left(\frac{1}{24}\right) \text{ or } 6\left(\frac{2}{24}\right) \text{ or } 2\left(\frac{6}{24}\right) \text{ or } 4\left(\frac{3}{24}\right) \text{ or } 3\left(\frac{4}{24}\right)$$

$\frac{12}{24}$ can be factored into various combinations of two fractions from these patterns:

$$\frac{(12)(1)}{(2)(12)} \text{ or } \frac{(12)(1)}{(3)(8)} \text{ or } \frac{(12)(1)}{(4)(6)}$$

$$\frac{(6)(2)}{(2)(12)} \text{ or } \frac{(6)(2)}{(3)(8)} \text{ or } \frac{(6)(2)}{(4)(6)}$$

$$\frac{(4)(3)}{(2)(12)} \text{ or } \frac{(4)(3)}{(3)(8)} \text{ or } \frac{(4)(3)}{(4)(6)}$$

There are all sorts of possible ways to factor fractions. We have given you the basic principles. You have to choose the "right" type of factoring for any given mathematical situation.

SELF-TEST 2 (Frames 25-54)

Factor each of the following into a non-fraction and a fraction:

1. $\frac{3}{5} = (\ )\left(\frac{\quad}{\quad}\right)$ 2. $\frac{a}{7} = (\ )\left(\frac{\quad}{\quad}\right)$ 3. $\frac{5}{t} = (\ )\left(\frac{\quad}{\quad}\right)$

Factor each of the following into a non-fraction and a fraction in three ways:

4. $\frac{6}{7} = (\ )\left(\frac{\quad}{\quad}\right) = (\ )\left(\frac{\quad}{\quad}\right) = (\ )\left(\frac{\quad}{\quad}\right)$ 5. $\frac{5h}{3} = (\ )\left(\frac{\quad}{\quad}\right) = (\ )\left(\frac{\quad}{\quad}\right) = (\ )\left(\frac{\quad}{\quad}\right)$

Write two different pairs of fractions for each of the following. Do not use a "1" in any numerator or denominator. Commuting a pair does not count as a different pair.

6. $\frac{14}{15} = \left(\frac{\quad}{\quad}\right)\left(\frac{\quad}{\quad}\right) = \left(\frac{\quad}{\quad}\right)\left(\frac{\quad}{\quad}\right)$ 7. $\frac{3w}{10} = \left(\frac{\quad}{\quad}\right)\left(\frac{\quad}{\quad}\right) = \left(\frac{\quad}{\quad}\right)\left(\frac{\quad}{\quad}\right)$

8. After you have factored a fraction, how should you check the correctness of your work?

____

9. Factor $\frac{14}{3p}$ into different pairs of fractions in as many ways as possible. Commuting a pair does not count as a different pair. A "1" can be used as a factor in the numerator.

$\frac{14}{3p} =$ ____________________

ANSWERS:

1. $3\left(\frac{1}{5}\right)$
2. $a\left(\frac{1}{7}\right)$
3. $5\left(\frac{1}{t}\right)$
4. $6\left(\frac{1}{7}\right)$, $3\left(\frac{2}{7}\right)$, $2\left(\frac{3}{7}\right)$
5. $5h\left(\frac{1}{3}\right)$, $5\left(\frac{h}{3}\right)$, $h\left(\frac{5}{3}\right)$
6. $\left(\frac{7}{5}\right)\left(\frac{2}{3}\right)$ and $\left(\frac{2}{5}\right)\left(\frac{7}{3}\right)$
7. $\left(\frac{3}{5}\right)\left(\frac{w}{2}\right)$ and $\left(\frac{w}{5}\right)\left(\frac{3}{2}\right)$
8. Multiply the factors. If correct, you will obtain the original fraction.
9. $\left(\frac{14}{3}\right)\left(\frac{1}{p}\right)$, $\left(\frac{1}{3}\right)\left(\frac{14}{p}\right)$, $\left(\frac{7}{3}\right)\left(\frac{2}{p}\right)$, $\left(\frac{2}{3}\right)\left(\frac{7}{p}\right)$

4-7 $\frac{n}{n}=1$ AND FAMILIES OF EQUIVALENT FRACTIONS

For any given fraction, there is an entire family of fractions which are equivalent to it. By "equivalent," we mean that they are equal. To obtain this family of fractions, we use the principle $\frac{n}{n}=1$. We will show the method for obtaining equivalent fractions in this section.

---

55. Here is an extremely important principle:

> IF ANY QUANTITY IS DIVIDED BY ITSELF, THE QUOTIENT IS +1.
>
> Here is the principle in symbol form:
>
> $$\frac{n}{n}=1 \qquad \text{or} \qquad \frac{\square}{\square}=1$$
>
> We call it:
>
> THE PRINCIPLE OF DIVIDING A QUANTITY BY ITSELF

Here are three instances of this principle:

$$\frac{8}{8}=1 \qquad \frac{37}{37}=1 \qquad \frac{150}{150}=1$$

Each is an instance of the principle of ______________ a quantity by itself.

---

56. We will show that any fraction can be written in many equivalent (equal) forms by means of the following two principles:

$$\boxed{n=(1)(n)} \qquad \text{and} \qquad \boxed{\frac{n}{n}=1}$$

If we multiply any fraction by "1", the product is the original fraction. That is:

$$(1)\left(\frac{4}{2}\right)=\frac{4}{2}$$

Since $\frac{2}{2}=1$, if we multiply any fraction by $\frac{2}{2}$, the product must equal the original fraction. That is:

$$\left(\frac{2}{2}\right)\left(\frac{4}{2}\right)=(1)\left(\frac{4}{2}\right)=\frac{4}{2}$$

(Continued on following page.)

Answer to 55: dividing

56. Continued

Since $\frac{3}{3} = 1$, if we multiply any fraction by $\frac{3}{3}$, the product must equal the original fraction. That is:

$$\left(\frac{3}{3}\right)\left(\frac{4}{2}\right) = (1)\left(\frac{4}{2}\right) = \frac{4}{2}$$

If we multiply $\frac{4}{2}$ by $\frac{7}{7}$, the product must equal $\frac{4}{2}$. Why?

____

Since:

$$\left(\frac{7}{7}\right)\left(\frac{4}{2}\right) = (1)\left(\frac{4}{2}\right) = \frac{4}{2}$$

---

57. Since $n = (1)(n)$: $\frac{4}{2} = (1)\left(\frac{4}{2}\right)$

Since $\frac{n}{n} = 1$, $\frac{2}{2} = 1$, and: $\frac{4}{2} = \left(\frac{2}{2}\right)\left(\frac{4}{2}\right)$

Performing the multiplication of fractions on the right, we get:

$$\frac{4}{2} = \frac{8}{4}$$

$\frac{4}{2}$ and $\frac{8}{4}$ are equivalent since they both equal <u>what</u> whole number? _____

2

---

58. $$\frac{4}{2} = (1)\left(\frac{4}{2}\right)$$

By replacing the "1" in the equation above with various instances of $\frac{n}{n}$ (such as $\frac{3}{3}$, $\frac{4}{4}$, $\frac{7}{7}$, etc.), we can find as many equivalent forms of $\frac{4}{2}$ as we want:

$$\frac{4}{2} = \left(\frac{3}{3}\right)\left(\frac{4}{2}\right) = \frac{12}{6}$$

$$\frac{4}{2} = \left(\frac{4}{4}\right)\left(\frac{4}{2}\right) = \frac{16}{8}$$

$$\frac{4}{2} = \left(\frac{7}{7}\right)\left(\frac{4}{2}\right) = \frac{28}{14}$$

$$\frac{4}{2} = \left(\frac{11}{11}\right)\left(\frac{4}{2}\right) = \frac{44}{22}$$

All of the fractions on the extreme right are equivalent to $\frac{4}{2}$ and to each other because they all equal what whole number? _______

2

59. Here is part of the family of fractions which are equivalent to $\frac{4}{2}$. We used the following two principles to generate the family:

$\boxed{n = (1)(n)}$ and $\boxed{\frac{n}{n} = 1}$

$$\frac{4}{2} = (1)\left(\frac{4}{2}\right) = \left(\frac{6}{6}\right)\left(\frac{4}{2}\right) = \frac{24}{12}$$
$$= \left(\frac{5}{5}\right)\left(\frac{4}{2}\right) = \frac{20}{10}$$
$$= \left(\frac{4}{4}\right)\left(\frac{4}{2}\right) = \frac{16}{8}$$
$$= \left(\frac{3}{3}\right)\left(\frac{4}{2}\right) = \frac{12}{6}$$
$$= \left(\frac{2}{2}\right)\left(\frac{4}{2}\right) = \frac{8}{4}$$
$$= \left(\frac{1}{1}\right)\left(\frac{4}{2}\right) = \frac{4}{2}$$

We could extend the family upwards as far as we wanted. What would the next member of the family be? ______

---

60. Let's generate a family of equivalent fractions beginning with $\frac{1}{2}$ (which equals 0.5):

$$\frac{1}{2} = (1)\left(\frac{1}{2}\right) = \left(\frac{6}{6}\right)\left(\frac{1}{2}\right) = \frac{6}{12}$$
$$= \left(\frac{5}{5}\right)\left(\frac{1}{2}\right) = \frac{5}{10}$$
$$= \left(\frac{4}{4}\right)\left(\frac{1}{2}\right) = \frac{4}{8}$$
$$= \left(\frac{3}{3}\right)\left(\frac{1}{2}\right) = \frac{3}{6}$$
$$= \left(\frac{2}{2}\right)\left(\frac{1}{2}\right) = \frac{2}{4}$$
$$= \left(\frac{1}{1}\right)\left(\frac{1}{2}\right) = \frac{1}{2}$$

All of the fractions in the family are equivalent because they all equal what decimal number? ______

$\frac{28}{14}$, since $\left(\frac{7}{7}\right)\left(\frac{4}{2}\right) = \frac{28}{14}$

---

61. It should be obvious to you that we can use the same principles to generate a family of equivalent fractions beginning with any fraction:

$$\frac{7}{4} = \frac{14}{8} = \frac{21}{12} = \frac{28}{16}, \text{ etc.}$$

$$\frac{5}{8} = \frac{10}{16} = \frac{15}{24} = \frac{20}{32}, \text{ etc.}$$

$$\frac{2}{3} = \_\_ = \_\_ = \_\_, \text{ etc.}$$

0.5

Answer to Frame 61:

$$\frac{2}{3} = \frac{4}{6} = \frac{6}{9} = \frac{8}{12}, \text{ etc.}$$

4-8 WRITING A FRACTION OR WHOLE NUMBER AS AN EQUIVALENT FRACTION WITH A SPECIFIC DENOMINATOR

At times it is necessary to replace a fraction or whole number with an equivalent fraction with a specific denominator. We can find this equivalent fraction by using the two principles: $n = (1)(n)$ and $\frac{n}{n} = 1$. We will show the method in this section.

62. In the problem below, we are asked to convert $\frac{5}{6}$ into a fraction whose denominator is "12".

$$\frac{5}{6} = \frac{(\ )}{12}$$

Naturally, we want the new fraction to be equivalent to $\frac{5}{6}$, and so the new fraction must be a member of the same family of fractions.

To make this conversion, we must multiply $\frac{5}{6}$ by the appropriate instance of $\frac{n}{n}$. In this case, we use $\frac{2}{2}$ and get:

$$\frac{5}{6}\left(\frac{2}{2}\right) = \frac{10}{12}$$

If we wanted to convert $\frac{5}{6}$ to a fraction whose denominator is "18":

$$\frac{5}{6} = \frac{(\ )}{18}$$

(a) What instance of $\frac{n}{n}$ would we use? ______

(b) Therefore, $\left(\frac{5}{6}\right)\left(\frac{}{}\right) = \frac{(\ )}{18}$

Answers to 62:

a) $\frac{3}{3}$

b) $\left(\frac{5}{6}\right)\left(\frac{3}{3}\right) = \frac{15}{18}$

63. When converting a fraction to an equivalent fraction with a specified denominator, the whole problem is determining what instance of $\frac{n}{n}$ to use. In this one, for example, we must convert to a fraction whose denominator is "32".

$$\frac{3}{4} = \frac{(\ )}{32}$$

To determine the correct instance of $\frac{n}{n}$, we can divide the denominator of the new fraction by the denominator of the original one.

Since $\frac{32}{4} = 8$, we must multiply 4 by 8 to get 32. Therefore, we use $\frac{8}{8}$.

Complete the problem: $\frac{3}{4} = \frac{3}{4}\left(\frac{8}{8}\right) = \frac{(\ )}{32}$

Answer to 63: $\frac{24}{32}$

64.

$$\frac{2}{7} = \frac{(\ \ )}{35}$$

Since $\frac{35}{7} = 5$, (a) what instance of $\frac{n}{n}$ do we use? ______

(b) Complete the problem: $\frac{2}{7}\left(\frac{\quad}{\quad}\right) = \frac{(\ \ )}{35}$

---

65. For each of these, insert the proper instance of $\frac{n}{n}$ and complete the problem:

(a) $\frac{5}{2}\left(\frac{\quad}{\quad}\right) = \frac{(\ \ )}{12}$ (c) $\frac{1}{5}\left(\frac{\quad}{\quad}\right) = \frac{(\ \ )}{45}$

(b) $\frac{2}{3}\left(\frac{\quad}{\quad}\right) = \frac{(\ \ )}{15}$ (d) $\frac{5}{8}\left(\frac{\quad}{\quad}\right) = \frac{(\ \ )}{56}$

a) $\frac{5}{5}$

b) $\left(\frac{2}{7}\right)\left(\frac{5}{5}\right) = \frac{10}{35}$

---

66. By means of the same principle, we can convert any whole number into a fraction with a specified denominator. For example:

$$4 = 4\left(\frac{2}{2}\right) = \frac{8}{2}$$

Note: In this case the denominator immediately tells us what instance of $\frac{n}{n}$ to use.

Insert the proper instance of $\frac{n}{n}$ and complete these:

(a) $7 = 7\left(\frac{\quad}{\quad}\right) = \frac{(\ \ )}{3}$ (c) $10 = 10\left(\frac{\quad}{\quad}\right) = \frac{(\ \ )}{5}$

(b) $9 = 9\left(\frac{\quad}{\quad}\right) = \frac{(\ \ )}{4}$ (d) $3 = 3\left(\frac{\quad}{\quad}\right) = \frac{(\ \ )}{8}$

a) $\frac{5}{2}\left(\frac{6}{6}\right) = \frac{30}{12}$

b) $\frac{2}{3}\left(\frac{5}{5}\right) = \frac{10}{15}$

c) $\frac{1}{5}\left(\frac{9}{9}\right) = \frac{9}{45}$

d) $\frac{5}{8}\left(\frac{7}{7}\right) = \frac{35}{56}$

---

67. Complete each of these:

(a) $\frac{7}{8} = \frac{(\ \ )}{24}$ (b) $5 = \frac{(\ \ )}{6}$ (c) $\frac{3}{2} = \frac{(\ \ )}{16}$ (d) $7 = \frac{(\ \ )}{9}$

a) $7\left(\frac{3}{3}\right) = \frac{21}{3}$ c) $10\left(\frac{5}{5}\right) = \frac{50}{5}$

b) $9\left(\frac{4}{4}\right) = \frac{36}{4}$ c) $3\left(\frac{8}{8}\right) = \frac{24}{8}$

---

68. Complete each of these:

(a) $11 = \frac{(\ \ )}{7}$ (b) $\frac{1}{4} = \frac{(\ \ )}{32}$ (c) $\frac{7}{16} = \frac{(\ \ )}{32}$ (d) $7 = \frac{(\ \ )}{20}$

a) $\frac{21}{24}$ b) $\frac{30}{6}$ c) $\frac{24}{16}$ d) $\frac{63}{9}$

---

a) $\frac{77}{7}$ b) $\frac{8}{32}$ c) $\frac{14}{32}$ d) $\frac{140}{20}$

SELF-TEST 3 (Frames 55-68)

Write the next member in each family of equivalent fractions:

1. $\frac{3}{7} = \frac{6}{14} = \frac{9}{21} = \frac{12}{28} =$ ____
2. $\frac{5}{2} = \frac{10}{4} = \frac{15}{6} = \frac{20}{8} =$ ____
3. $9 = \frac{9}{1} = \frac{18}{2} = \frac{27}{3} = \frac{36}{4} =$ ____

Complete:

4. $\frac{11}{3} = \frac{11}{3}\left(\frac{}{}\right) = \frac{(\;)}{15}$
5. $\frac{7}{9} = \frac{7}{9}\left(\frac{}{}\right) = \frac{(\;)}{54}$

Complete:

6. $6 = 6\left(\frac{}{}\right) = \frac{(\;)}{5}$
7. $11 = 11\left(\frac{}{}\right) = \frac{(\;)}{4}$

Complete:

8. $\frac{5}{7} = \frac{(\;)}{42}$
9. $\frac{2a}{3} = \frac{(\;)}{18}$
10. $4t = \frac{(\;)}{6}$

11. Using the principles $n(1) = n$ and $\frac{n}{n} = 1$,

prove that $\frac{3}{5} = \frac{18}{30}$. Show all steps: $\frac{3}{5} =$ ____________________

ANSWERS:

1. $\frac{15}{35}$
2. $\frac{25}{10}$
3. $\frac{45}{5}$
4. $\frac{11}{3}\left(\frac{5}{5}\right) = \frac{55}{15}$
5. $\frac{7}{9}\left(\frac{6}{6}\right) = \frac{42}{54}$
6. $6\left(\frac{5}{5}\right) = \frac{30}{5}$
7. $11\left(\frac{4}{4}\right) = \frac{44}{4}$
8. $\frac{30}{42}$
9. $\frac{12a}{18}$
10. $\frac{24t}{6}$
11. Proof:
$\frac{3}{5} = \frac{3}{5}(1) = \frac{3}{5}\left(\frac{6}{6}\right) = \frac{(3)(6)}{(5)(6)} = \frac{18}{30}$

4-9 REDUCING FRACTIONS TO LOWEST TERMS

The procedure for reducing fractions to lowest terms is simply the reverse of the procedure for generating a family of equivalent fractions. It uses the two principles: $\frac{n}{n} = 1$ and $n(1) = n$. In this section, we will define what is meant by "reducing to lowest terms" and then show the procedure.

69. Here is part of a family of equivalent fractions:

$$\frac{2}{3}\left(\frac{5}{5}\right) = \frac{10}{15}$$

$$\frac{2}{3}\left(\frac{4}{4}\right) = \frac{8}{12}$$

$$\frac{2}{3}\left(\frac{3}{3}\right) = \frac{6}{9}$$

$$\frac{2}{3}\left(\frac{2}{2}\right) = \frac{4}{6}$$

$$\frac{2}{3}\left(\frac{1}{1}\right) = \frac{2}{3}$$

(Continued on following page.)

69. (Continued)

Though all of the fractions on the right are equivalent, the ones on the top contain larger numbers in their numerators and denominators.

If two fractions are equivalent, the one containing the smaller numbers is said to be in "lower" terms.

Example: $\frac{4}{6}$ is in lower terms than $\frac{8}{12}$,

since 4 and 6 are smaller than 8 and 12.

In each pair below, encircle the one which is in lower terms:

(a) $\frac{2}{3}$ or $\frac{4}{6}$ (b) $\frac{10}{15}$ or $\frac{6}{9}$

---

70. "Reducing a fraction to lower terms" means "finding an equivalent fraction whose numerator and denominator are smaller." Here is the procedure:

Answers to 69: a) $\frac{2}{3}$ b) $\frac{6}{9}$

Step 1: We must factor the fraction into two fractions, one of which is an instance of $\frac{n}{n}$.

$$\frac{4}{8} = \frac{(4)(1)}{(4)(2)} = \left(\frac{4}{4}\right)\left(\frac{1}{2}\right)$$

Step 2: We substitute "1" for the instance of $\frac{n}{n}$.

$$\frac{4}{8} = \frac{(4)(1)}{(4)(2)} = \left(\frac{4}{4}\right)\left(\frac{1}{2}\right) = (1)\left(\frac{1}{2}\right)$$

Step 3: We drop the "1" since $(1)(n) = n$.

$$\frac{4}{8} = \frac{(4)(1)}{(4)(2)} = \left(\frac{4}{4}\right)\left(\frac{1}{2}\right) = (1)\left(\frac{1}{2}\right) = \frac{1}{2}$$

Complete this reduction to lower terms:

$$\frac{8}{12} = \frac{(2)(4)}{(2)(6)} = \left(\frac{2}{2}\right)\left(\frac{4}{6}\right) = (1)\left(\frac{4}{6}\right) = \underline{\qquad}$$

---

71. In the last frame, we reduced $\frac{8}{12}$ to $\frac{4}{6}$ which is "in lower terms."

Why do we say that it is "in lower terms?" ______________________

____

Answer to 70: $\frac{4}{6}$

---

72. Complete these reductions to lower terms:

(a) $\frac{3}{9} = \left(\frac{3}{3}\right)\left(\frac{1}{3}\right) = (\quad)\left(\frac{\quad}{\quad}\right) = \left(\frac{\quad}{\quad}\right)$

(b) $\frac{12}{16} = \left(\frac{2}{2}\right)\left(\frac{6}{8}\right) = (\quad)\left(\frac{\quad}{\quad}\right) = \left(\frac{\quad}{\quad}\right)$

Answer to 71: Because 4 and 6 are smaller than 8 and 12.

73. $\frac{20}{25}$ can be reduced to $\frac{4}{5}$. Show all the steps: $\frac{20}{25} = \left(\frac{\quad}{\quad}\right)\left(\frac{\quad}{\quad}\right) = (\quad)\left(\frac{\quad}{\quad}\right) =$ ______	a) $(1)\left(\frac{1}{3}\right) = \frac{1}{3}$ b) $(1)\left(\frac{6}{8}\right) = \frac{6}{8}$
74. $\frac{18}{24}$ can be reduced to $\frac{6}{8}$. Show all the steps: $\frac{18}{24} = \left(\frac{\quad}{\quad}\right)\left(\frac{\quad}{\quad}\right) = (\quad)\left(\frac{\quad}{\quad}\right) =$ ______	$\left(\frac{5}{5}\right)\left(\frac{4}{5}\right) = (1)\left(\frac{4}{5}\right) = \frac{4}{5}$
75. A fraction can be reduced to lower terms only if its numerator and denominator contain a common factor, because only then can we get an instance of $\frac{n}{n}$. (a) $\frac{10}{14} = \frac{(2)(5)}{(2)(7)}$ Can $\frac{10}{14}$ be reduced to lower terms? ______ (b) $\frac{6}{25} = \frac{(2)(3)}{(5)(5)}$ Can $\frac{6}{25}$ be reduced to lower terms? ______	$\left(\frac{3}{3}\right)\left(\frac{6}{8}\right) = (1)\left(\frac{6}{8}\right) = \frac{6}{8}$
76. For each of the following fractions, state the common factor in the numerator and denominator if there is one: (a) $\frac{14}{21}$ ______ (c) $\frac{7}{9}$ ______ (e) $\frac{15}{35}$ ______ (b) $\frac{4}{10}$ ______ (d) $\frac{14}{39}$ ______ (f) $\frac{9}{16}$ ______	a) Yes, since "2" is a common factor. b) No, since there is no common factor.
77. We reduce fractions to lower terms by means of two principles: $\boxed{\frac{n}{n} = 1}$ and $\boxed{n(1) = n}$ A fraction cannot be reduced to lower terms if its numerator and denominator do not contain a common factor. IF A FRACTION CANNOT BE REDUCED TO LOWER TERMS, WE SAY THAT THE FRACTION IS "IN LOWEST TERMS." Which of the following fractions are "in lowest terms?" ______ (a) $\frac{11}{13}$ (b) $\frac{14}{18}$ (c) $\frac{6}{8}$ (d) $\frac{5}{9}$ (e) $\frac{3}{12}$	a) 7 c) None e) 5 b) 2 d) None f) None
78. Here is part of a family of equivalent fractions: $\frac{3}{4} = \frac{6}{8} = \frac{9}{12} = \frac{12}{16} = \frac{15}{20}$ (a) Only one fraction is in lowest terms. Which one is it? ______ (b) If the other fractions in the family were reduced to lowest terms, they would all reduce to ______.	Only (a) and (d) are in lowest terms.
	a) $\frac{3}{4}$ b) $\frac{3}{4}$

79. The numerator and denominator of $\frac{16}{24}$ have more than one common factor. There are three of them: 2, 4, and 8. We can use any of them to reduce $\frac{16}{24}$ to lower terms.

$$\frac{16}{24} = \left(\frac{2}{2}\right)\left(\frac{8}{12}\right) = (1)\left(\frac{8}{12}\right) = \frac{8}{12}$$

$$\frac{16}{24} = \left(\frac{4}{4}\right)\left(\frac{4}{6}\right) = (1)\left(\frac{4}{6}\right) = \frac{4}{6}$$

$$\frac{16}{24} = \left(\frac{8}{8}\right)\left(\frac{2}{3}\right) = (1)\left(\frac{2}{3}\right) = \frac{2}{3}$$

All three fractions on the right are equivalent to $\frac{16}{24}$. Only one of them is in lowest terms. Which one? ______

---

80. When the numerator and denominator of a fraction have more than one common factor: | $\frac{2}{3}$

(1) We can reduce the fraction to lower terms by using any of the common factors to form $\frac{n}{n}$.

(2) We can reduce the fraction to lowest terms in one step only by using the largest common factor to form $\frac{n}{n}$.

What is the largest common factor for each of these?

(a) $\frac{33}{99}$ ______ (b) $\frac{18}{30}$ ______

---

81. It is not always easy to reduce a fraction to lowest terms in one step. When the numerator and denominator are large numbers, it is sometimes difficult to find the largest common factor. In such a case, use the largest one you can find. Then check to see whether the new fraction can be reduced even further. | a) 33 b) 6

$$\frac{33}{99} = \left(\frac{11}{11}\right)\left(\frac{3}{9}\right) = \frac{3}{9}$$

(a) $\frac{3}{9}$ can be further reduced to ______.

(b) Therefore, $\frac{33}{99}$ reduced to lowest terms is ______.

---

82. Reduce the following fractions to lowest terms: | a) $\frac{1}{3}$ b) $\frac{1}{3}$

(a) $\frac{2}{4}$ = ______ (c) $\frac{5}{15}$ = ______ (e) $\frac{14}{21}$ = ______

(b) $\frac{9}{6}$ = ______ (d) $\frac{12}{16}$ = ______ (f) $\frac{35}{15}$ = ______

---

a) $\frac{1}{2}$ b) $\frac{3}{2}$ c) $\frac{1}{3}$ d) $\frac{3}{4}$ e) $\frac{2}{3}$ f) $\frac{7}{3}$

83. If any of the following are not reduced to lowest terms, do so: (a) $\frac{3}{8}$ ______ (c) $\frac{7}{9}$ ______ (e) $\frac{4}{16}$ ______ (b) $\frac{10}{4}$ ______ (d) $\frac{11}{21}$ ______ (f) $\frac{4}{9}$ ______	
84. Reduce each fraction to lowest terms: (a) $\frac{24}{60}$ _____ (b) $\frac{18}{54}$ _____ (c) $\frac{27}{81}$ _____ (d) $\frac{98}{28}$ _____	All are in lowest terms except: b) $\frac{5}{2}$ and e) $\frac{1}{4}$
	a) $\frac{2}{5}$ b) $\frac{1}{3}$ c) $\frac{1}{3}$ d) $\frac{7}{2}$

SELF-TEST 4 (Frames 69-84)

1. In this family of equivalent fractions, which fraction is in lowest terms? ______ $\frac{5}{6} = \frac{10}{12} = \frac{15}{18} = \frac{20}{24} = \frac{25}{30}$

2. Which of the following fractions cannot be reduced to lower terms? __________

   (a) $\frac{9}{12}$ (b) $\frac{8}{25}$ (c) $\frac{16}{33}$ (d) $\frac{15}{27}$ (e) $\frac{21}{10}$

Examine this problem: $\frac{15}{27} = \frac{(3)(5)}{(3)(9)} = \left(\frac{3}{3}\right)\left(\frac{5}{9}\right) = (1)\left(\frac{5}{9}\right) = \frac{5}{9}$

3. The two principles used in reducing the fraction $\frac{15}{27}$ to lowest terms are: __________ and __________

Reduce each fraction to lowest terms:

4. $\frac{9}{15} =$ _____ 5. $\frac{30}{48} =$ _____ 6. $\frac{75}{90} =$ _____ 7. $\frac{12}{40} =$ _____ 8. $\frac{15}{36} =$ _____ 9. $\frac{20}{39} =$ _____

10. Using the principles $\frac{n}{n} = 1$ and $(1)(n) = n$, prove that $\frac{18}{42}$ can be reduced to $\frac{3}{7}$. Show all steps:

$\frac{18}{42} =$ ____________________

ANSWERS:
1. $\frac{5}{6}$
2. (b), (c), and (e)
3. $\frac{n}{n} = 1$ and $(1)(n) = n$
4. $\frac{3}{5}$
5. $\frac{5}{8}$
6. $\frac{5}{6}$
7. $\frac{3}{10}$
8. $\frac{5}{12}$
9. Cannot be reduced.
10. Proof: $\frac{18}{42} = \frac{(6)(3)}{(6)(7)} = \left(\frac{6}{6}\right)\left(\frac{3}{7}\right) = (1)\left(\frac{3}{7}\right) = \frac{3}{7}$

4-10 REDUCING PRODUCTS TO LOWEST TERMS

Any fractional product should be reduced to lowest terms. <u>In this course, if a product is not reduced to lowest terms, it will be considered wrong.</u>

When performing multiplications of fractions, frequently there are opportunities to simplify <u>before</u> multiplying. We will show some of these simplifications in this section.

---

85. Watch the difficulty we meet in this multiplication:

$$\left(\frac{9}{23}\right)\left(\frac{7}{9}\right) = \frac{(9)(7)}{(23)(9)} = \frac{63}{207}$$

$\frac{63}{207}$ must now be reduced to lowest terms.

By interchanging the original numerators, we can simplify <u>before</u> multiplying because we obtain an instance of $\frac{n}{n}$.

$$\left(\frac{9}{23}\right)\left(\frac{7}{9}\right) = \left(\frac{9}{9}\right)\left(\frac{7}{23}\right)$$

$$= (1)\left(\frac{7}{23}\right) = \frac{7}{23}$$

Notice that the product is already reduced to lowest terms. Obviously, this simplification before multiplying is easier.

Do the following in the same way: $\left(\frac{11}{13}\right)\left(\frac{27}{11}\right) =$ ______________

---

86. Watch how we can simplify this one by interchanging the numerators and reducing a fraction to lowest terms <u>before</u> multiplying:

$$\left(\frac{3}{11}\right)\left(\frac{7}{6}\right) = \left(\frac{3}{6}\right)\left(\frac{7}{11}\right)$$

$$\downarrow$$

$$= \left(\frac{1}{2}\right)\left(\frac{7}{11}\right) = \frac{7}{22}$$

By interchanging the numerators, we obtained $\frac{3}{6}$ which was reduced to $\frac{1}{2}$ <u>before</u> the actual multiplication was performed.

Using the same procedure, do this one: $\left(\frac{2}{15}\right)\left(\frac{10}{6}\right) =$ ______________

Answer to 85: $= \left(\frac{11}{11}\right)\left(\frac{27}{13}\right) = (1)\left(\frac{27}{13}\right) = \frac{27}{13}$

---

87. Do the following. Watch for opportunities to simplify <u>before</u> multiplying. Be sure that each product is reduced to lowest terms.

(a) $\left(\frac{7}{8}\right)\left(\frac{4}{3}\right) =$ ______________

(b) $\left(\frac{9}{7}\right)\left(\frac{14}{9}\right) =$ ______________

(c) $\left(\frac{35}{16}\right)\left(\frac{8}{7}\right) =$ ______________

(d) $\left(\frac{75}{100}\right)\left(\frac{100}{25}\right) =$ ______________

(e) $\left(\frac{33}{44}\right)\left(\frac{88}{99}\right) =$ ______________

Answer to 86: $= \left(\frac{2}{6}\right)\left(\frac{10}{15}\right) = \left(\frac{1}{3}\right)\left(\frac{2}{3}\right) = \frac{2}{9}$

88. The procedure for multiplying three fractions is similar to the procedure for multiplying two fractions.

For example: $\left(\frac{2}{3}\right)\left(\frac{3}{5}\right)\left(\frac{7}{2}\right) = \frac{(2)(3)(7)}{(3)(5)(2)} = \frac{42}{30}$

Reduce $\frac{42}{30}$ to lowest terms: ________

Answers: a) $\frac{7}{6}$ b) 2 c) $\frac{5}{2}$ d) 3 e) $\frac{2}{3}$

---

89. In the last frame, we saw that:

$$\left(\frac{2}{3}\right)\left(\frac{3}{5}\right)\left(\frac{7}{2}\right) = \frac{42}{30} = \frac{7}{5}$$

Again, there is an easier way. You can interchange numerators and get:

$$\left(\frac{2}{3}\right)\left(\frac{3}{5}\right)\left(\frac{7}{2}\right) = \left(\frac{3}{3}\right)\left(\frac{2}{2}\right)\left(\frac{7}{5}\right)$$

$$= (1)(1)\left(\frac{7}{5}\right) = \frac{7}{5}$$

Using the same procedure, do this one:

$$\left(\frac{4}{7}\right)\left(\frac{5}{4}\right)\left(\frac{7}{2}\right) = \_\_\_\_\_\_\_\_$$

Answer: $\frac{7}{5}$

---

90. Be sure each product is reduced to lowest terms:

(a) $\left(\frac{7}{8}\right)\left(\frac{4}{5}\right)\left(\frac{5}{16}\right) =$ ________ (b) $\left(\frac{7}{10}\right)\left(\frac{5}{9}\right)\left(\frac{6}{11}\right) =$ ________

Answer: $= \left(\frac{4}{4}\right)\left(\frac{7}{7}\right)\left(\frac{5}{2}\right) = (1)(1)\left(\frac{5}{2}\right) = \frac{5}{2}$

Answers: a) $\frac{7}{32}$ b) $\frac{7}{33}$

4-11 REDUCING FRACTIONS WITH LETTERS TO LOWEST TERMS

In this section, we will show that fractions with letters can also be reduced to lowest terms. The same principles are used for these reductions.

---

91. The principle $\frac{n}{n} = 1$ also holds for single letters since the same number must be plugged in for the same letter in any expression. Therefore:

$$\frac{x}{x} = 1 \qquad \frac{t}{t} = 1 \qquad \frac{p}{p} = 1$$

Using this principle, we can reduce the following fraction with letters to lowest terms.

$$\frac{3m}{5m} = \left(\frac{3}{5}\right)\left(\frac{m}{m}\right) = \left(\frac{3}{5}\right)(1) = \frac{3}{5}$$

Reduce these to lowest terms: (a) $\frac{7R}{3R} =$ ______ (b) $\frac{9b}{10b} =$ ______

92. Watch for cases in which the numerical fraction can also be reduced to lowest terms. For example:

$$\frac{4F}{10F} = \left(\frac{4}{10}\right)\left(\frac{F}{F}\right) = \left(\frac{2}{5}\right)(1) = \frac{2}{5}$$

Reduce each of these to lowest terms:

(a) $\frac{7s}{21s}$ = ______ (c) $\frac{14y}{7y}$ = ______

(b) $\frac{6t}{6t}$ = ______ (d) $\frac{6x}{15x}$ = ______

Answers: a) $\frac{7}{3}$ b) $\frac{9}{10}$

---

93. Occasionally a fraction contains a letter without a numerical coefficient as its numerator or denominator. For example, the first fraction below has "d" as its denominator; the second fraction below has "R" as its numerator. When reducing fractions of this type to lowest terms, we begin by substituting "1d" for "d" and "1R" for "R". Then we proceed in the usual way.

Examples: $\frac{4d}{d} = \frac{4d}{1d} = \left(\frac{4}{1}\right)\left(\frac{d}{d}\right) = 4(1) = 4$

$$\frac{R}{3R} = \frac{1R}{3R} = \left(\frac{1}{3}\right)\left(\frac{R}{R}\right) = \left(\frac{1}{3}\right)(1) = \frac{1}{3}$$

Reduce each of these to lowest terms:

(a) $\frac{8m}{m}$ = ______ (b) $\frac{t}{5t}$ = ______ (c) $\frac{p}{10p}$ = ______

Answers: a) $\frac{1}{3}$ b) 1 c) 2 d) $\frac{2}{5}$

---

94. Reduce each of these to lowest terms:

(a) $\frac{5x}{5x}$ = ______ (c) $\frac{9y}{y}$ = ______

(b) $\frac{18R}{30R}$ = ______ (d) $\frac{a}{3a}$ = ______

Answers: a) 8 b) $\frac{1}{5}$ c) $\frac{1}{10}$

Answers: a) 1 b) $\frac{3}{5}$ c) 9 d) $\frac{1}{3}$

4-12 MULTIPLICATIONS WITH NON-FRACTIONAL PRODUCTS

In multiplications involving fractions, it is possible to get a product which is a non-fraction. We will examine multiplications of this type in this section.

---

95. Here is a multiplication of a non-fraction and a fraction. The non-fraction and the denominator of the fraction are identical. Watch the steps:

$$6\left(\frac{7}{6}\right) = \frac{(6)(7)}{6} = \left(\frac{6}{6}\right)(7) = (1)(7) = 7$$

Complete: $x\left(\frac{3}{x}\right) = \frac{(x)(3)}{x} = \left(\frac{\quad}{\quad}\right)(\quad) = (\quad)(\quad) =$ ______

96. Here are the two problems from the last frame:

$$6\left(\frac{7}{6}\right) = 7 \qquad x\left(\frac{3}{x}\right) = 3$$

In each one, the non-fraction is identical to the denominator of the fraction. When this is the case:

(1) The product is a non-fraction.
(2) The product is identical to the numerator of the original fraction.

Multiply: (a) $8\left(\frac{5}{8}\right) =$ _____ (c) $t\left(\frac{10}{t}\right) =$ _____

(b) $7\left(\frac{3y}{7}\right) =$ _____ (d) $2x\left(\frac{5}{2x}\right) =$ _____

Answer: $\left(\frac{x}{x}\right)(3) = (1)(3) = 3$

---

97. Even when the non-fraction is not identical to the denominator of the fraction, the product will sometimes be a non-fraction.

Example: $6\left(\frac{x}{2}\right) = \frac{6x}{2} = \left(\frac{6}{2}\right)(x) = (3)(x) = 3x$

Complete: (a) $27\left(\frac{7}{9}\right) = \frac{(27)(7)}{9} = \left(\frac{27}{9}\right)(7) = (\ \ )(\ \ ) =$ _____

(b) $9\left(\frac{4}{3}\right) = \frac{(9)(4)}{3} = \left(\frac{\ }{\ }\right)(\ \ ) = (\ \ )(\ \ ) =$ _____

(c) $8\left(\frac{7x}{4}\right) = \frac{(8)(7x)}{4} = \left(\frac{\ }{\ }\right)(\ \ ) = (\ \ )(\ \ ) =$ _____

Answers: a) 5 c) 10 b) 3y d) 5

---

98. Perform each of these multiplications:

(a) $40\left(\frac{7}{8}\right) =$ _____ (b) $16\left(\frac{x}{4}\right) =$ _____ (c) $56\left(\frac{2t}{7}\right) =$ _____

Answers:
a) $(3)(7) = 21$
b) $\left(\frac{9}{3}\right)(4) = (3)(4) = 12$
c) $\left(\frac{8}{4}\right)(7x) = (2)(7x) = 14x$

---

99. In multiplications of this type, we attempt to isolate a fraction which reduces to a whole number. Watch the steps in this one:

$$4x\left(\frac{7}{4}\right) = \frac{(4x)(7)}{4} = \frac{(4)(7x)}{4} = \left(\frac{4}{4}\right)(7x) = (1)(7x) = 7x$$

Complete: (a) $8m\left(\frac{5}{4}\right) = \frac{(8m)(5)}{4} = \frac{8(5m)}{4} = \left(\frac{\ }{\ }\right)(\ \ ) = (\ \ )(\ \ ) =$ _____

(b) $5x\left(\frac{4}{x}\right) = \frac{(5x)(4)}{x} = \frac{20x}{x} = (\ \ )\left(\frac{\ }{\ }\right) = (\ \ )(\ \ ) =$ _____

(c) $10d\left(\frac{9}{5d}\right) = \frac{(10d)(9)}{5d} = \left(\frac{10d}{5d}\right)(9) = (\ \ )(\ \ ) =$ _____

Answers: a) $(5)(7) = 35$ b) $(4)(x) = 4x$ c) $(8)(2t) = 16t$

Answers:
a) $\left(\frac{8}{4}\right)(5m) = (2)(5m) = 10m$
b) $20\left(\frac{x}{x}\right) = 20(1) = 20$
c) $(2)(9) = 18$

100. Perform each of these multiplications:

(a) $4\left(\frac{7v}{4}\right) =$ ______ (c) $21\left(\frac{3m}{7}\right) =$ ______

(b) $3b\left(\frac{7}{b}\right) =$ ______ (d) $35\left(\frac{9v}{5}\right) =$ ______

---

101. Do these: (a) $6x\left(\frac{7}{6x}\right) =$ ______ (c) $9p\left(\frac{4}{3p}\right) =$ ______

(b) $4y\left(\frac{5}{y}\right) =$ ______ (d) $8r\left(\frac{3}{r}\right) =$ ______

Answers to frame 100: a) 7v c) 9m b) 21 d) 63v

---

102. When both factors are fractions, the product can also be a non-fraction when it is reduced to lowest terms. Do these:

(a) $\left(\frac{x}{2}\right)\left(\frac{2}{x}\right) =$ ______ (d) $\left(\frac{7}{d}\right)\left(\frac{2d}{7}\right) =$ ______

(b) $\left(\frac{9}{8}\right)\left(\frac{8}{9}\right) =$ ______ (e) $5S\left(\frac{1}{S}\right) =$ ______

(c) $\left(\frac{2R}{5}\right)\left(\frac{10}{R}\right) =$ ______ (f) $t\left(\frac{1}{10t}\right) =$ ______

Answers to frame 101: a) 7 c) 12 b) 20 d) 24

Answers to frame 102: a) 1 b) 1 c) 4 d) 2 e) 5 f) $\frac{1}{10}$

SELF-TEST 5 (Frames 85-102)

Reduce each fraction to lowest terms:

1. $\frac{18}{24} =$ ______ 2. $\frac{9w}{w} =$ ______ 3. $\frac{16h}{10h} =$ ______ 4. $\frac{t}{6t} =$ ______

Multiply. Be sure each product is in lowest terms:

5. $\left(\frac{2}{3}\right)\left(\frac{3}{4}\right) =$ ______
6. $\left(\frac{5}{8}\right)\left(\frac{2}{5}\right) =$ ______
7. $\left(\frac{3}{5}\right)\left(\frac{8}{3}\right)\left(\frac{5}{6}\right) =$ ______
8. $8\left(\frac{3}{8}\right) =$ ______
9. $3R\left(\frac{5}{R}\right) =$ ______
10. $4b\left(\frac{7}{12b}\right) =$ ______

ANSWERS:

1. $\frac{3}{4}$ 3. $\frac{8}{5}$ 5. $\frac{1}{2}$ 7. $\frac{4}{3}$ 9. 15

2. 9 4. $\frac{1}{6}$ 6. $\frac{1}{4}$ 8. 3 10. $\frac{7}{3}$

4-13 MULTIPLICATION OF SIGNED FRACTIONS

The rules for the "sign of the product" of signed fractions are the same as the rules for signed whole numbers. We will apply these rules to signed fractions in this section.

103. You know the rules for obtaining the sign of the product when multiplying two whole numbers.

(a) If both are positive, the product is ______________.

(b) If both are negative, the product is ______________.

(c) If one is positive and the other negative, the product is ______________.

Answers: a) positive b) positive c) negative

104. Do these: (a) $(7)(8) =$ ______ (c) $(4)(-3) =$ ______

(b) $(-6)(4) =$ ______ (d) $(-5)(-2) =$ ______

Answers: a) +56 c) -12 b) -24 d) +10

105. When multiplying fractions, the same rules for the "sign of the product" apply.

$$\left(\frac{1}{2}\right)\left(-\frac{3}{4}\right) = -\frac{3}{8}$$

$$\left(-\frac{5}{6}\right)\left(-\frac{7}{3}\right) = +\frac{35}{18}$$

Do these: (a) $\left(-\frac{1}{3}\right)\left(-\frac{4}{5}\right) =$ ______ (c) $\left(-\frac{5}{8}\right)\left(-\frac{7}{4}\right) =$ ______

(b) $\left(\frac{7}{8}\right)\left(-\frac{1}{3}\right) =$ ______ (d) $\left(\frac{3}{2}\right)\left(-\frac{5}{4}\right) =$ ______

Answers: a) $+\frac{4}{15}$ c) $+\frac{35}{32}$ b) $-\frac{7}{24}$ d) $-\frac{15}{8}$

106. Do the following. Be sure to <u>reduce to lowest terms</u>:

(a) $\left(\frac{4}{5}\right)\left(-\frac{7}{8}\right) =$ ______ (c) $\left(-\frac{21}{4}\right)\left(\frac{5}{7}\right) =$ ______

(b) $\left(-\frac{7}{6}\right)\left(-\frac{9}{5}\right) =$ ______ (d) $\left(-\frac{4}{3}\right)\left(-\frac{15}{2}\right) =$ ______

Answers: a) $-\frac{7}{10}$ c) $-\frac{15}{4}$ b) $+\frac{21}{10}$ d) $+10$

107. The same rules apply when the fractions contain letters. Do these:

(a) $\left(\frac{1}{7}\right)\left(-\frac{5x}{3}\right) =$ ______ (c) $\left(-\frac{3}{8}\right)\left(-\frac{1}{x}\right) =$ ______

(b) $\left(-\frac{7}{d}\right)\left(\frac{2}{9}\right) =$ ______ (d) $\left(-\frac{t}{6}\right)\left(-\frac{3}{2}\right) =$ ______

Answers: a) $-\frac{5x}{21}$ c) $+\frac{3}{8x}$ b) $-\frac{14}{9d}$ d) $+\frac{t}{4}$

108. Do these. Be sure to <u>reduce to lowest terms</u>:

(a) $\left(-\frac{5x}{7}\right)\left(\frac{3}{x}\right) =$ ______ (c) $\left(-\frac{3}{q}\right)\left(\frac{q}{3}\right) =$ ______

(b) $\left(-\frac{5}{2b}\right)\left(-\frac{4b}{15}\right) =$ ______ (d) $\left(-\frac{s}{3}\right)\left(-\frac{3}{4s}\right) =$ ______

109. We use the same rules when one factor is a non-fraction. Do these:

(a) $10\left(-\frac{1}{5}\right) =$ ______ (c) $(-8)\left(\frac{3w}{16}\right) =$ ______

(b) $\left(-\frac{1}{2t}\right)(-t) =$ ______ (d) $\left(-\frac{2}{3h}\right)(-h) =$ ______

Answers: a) $-\frac{15}{7}$ b) $+\frac{2}{3}$ c) -1 d) $+\frac{1}{4}$

---

110. Do these: (a) $(-6)\left(\frac{1}{6}\right) =$ ______ (c) $\left(-\frac{1}{11}\right)(-11) =$ ______

(b) $\left(-\frac{x}{4}\right)\left(\frac{4}{x}\right) =$ ______ (d) $\left(-\frac{5}{2y}\right)\left(-\frac{2y}{5}\right) =$ ______

Answers: a) -2 b) $+\frac{1}{2}$ c) $-\frac{3w}{2}$ d) $+\frac{2}{3}$

Answers: a) -1 b) -1 c) $+1$ d) $+1$

4-14 PAIRS OF RECIPROCALS

In this section, we will define the meaning of "reciprocals." Understanding the meaning of reciprocals is extremely important because reciprocals are very useful in mathematics. You will see that the concept of reciprocals is not difficult.

---

111. Definition: **TWO QUANTITIES ARE A PAIR OF RECIPROCALS IF THEIR PRODUCT IS +1.**

Since $\left(\frac{1}{7}\right)(7) = 1$, $\frac{1}{7}$ and 7 are a pair of ______________.

---

112. We say: "The reciprocal of $\frac{1}{7}$ is 7."
"The reciprocal of 7 is $\frac{1}{7}$."

Since $27\left(\frac{1}{27}\right) = 1$: (a) The reciprocal of 27 is ______.

(b) The reciprocal of $\frac{1}{27}$ is ______.

Answer: reciprocals

---

113. Write the reciprocals of the following:

(a) 12 ______ (b) 13 ______ (c) 42 ______ (d) 99 ______

Answers: a) $\frac{1}{27}$ b) 27

---

114. Write the reciprocals of the following:

(a) $\frac{1}{8}$ ______ (b) $\frac{1}{4}$ ______ (c) $\frac{1}{67}$ ______ (d) $\frac{1}{93}$ ______

Answers: a) $\frac{1}{12}$ b) $\frac{1}{13}$ c) $\frac{1}{42}$ d) $\frac{1}{99}$

---

115. Since $\left(\frac{7}{6}\right)\left(\frac{6}{7}\right) = 1$, $\frac{7}{6}$ and $\frac{6}{7}$ are a pair of ______________.

Answers: a) 8 b) 4 c) 67 d) 93

Answer: reciprocals

116. Since $\left(\frac{3}{4}\right)\left(\frac{4}{3}\right) = 1$: (a) The reciprocal of $\frac{3}{4}$ is ______.

(b) The reciprocal of $\frac{4}{3}$ is ______.

---

117. It is easy to see that:

The reciprocal of any fraction $\frac{\square}{\bigcirc}$ is $\frac{\bigcirc}{\square}$.

Write the reciprocals of the following:

(a) $\frac{2}{3}$ ______ (b) $\frac{9}{11}$ ______ (c) $\frac{44}{57}$ ______

a) $\frac{4}{3}$

b) $\frac{3}{4}$

---

118. There are two numbers which are their own reciprocals.

(a) Since $(+1)(+1) = 1$, the reciprocal of "+1" is ______.

(b) Since $(-1)(-1) = 1$, the reciprocal of "-1" is ______.

a) $\frac{3}{2}$ b) $\frac{11}{9}$ c) $\frac{57}{44}$

---

119. Let's see if "0" has a reciprocal.

Whenever "0" is a factor in any multiplication, the product is always "0". In each multiplication below, the product is "0":

$$(0)(-8) = 0 \qquad (+1)(0) = 0 \qquad (0)\left(\frac{3}{4}\right) = 0$$

$$(0)(0) = 0 \qquad (0)(-1) = 0 \qquad \left(-\frac{7}{6}\right)(0) = 0$$

(a) Can we ever get "+1" as a product when one of the "factors" is "0"? ______

(b) Therefore, does "0" have a reciprocal? ______

a) +1

b) -1

---

120. $(-1)(+1) = -1 \qquad (-1)(-1) = +1$

(a) Are "-1" and "+1" a pair of reciprocals? ______

(b) Are "-1" and "-1" a pair of reciprocals? ______

a) No. If "0" is one of the factors, the product must be "0". It cannot be "+1".

b) No. "0" has no reciprocal.

---

121. (a) The reciprocal of "+1" is ______.

(b) The reciprocal of "-1" is ______.

(c) The reciprocal of "0" is ______.

a) No. Their product is not "+1".

b) Yes. Their product is "+1".

---

a) +1

b) -1

c) "0" has no reciprocal.

122. Let's examine the reciprocals of <u>negative</u> <u>quantities</u>.

(a) $(-6)\left(\frac{1}{6}\right) = -1$ Are -6 and $\frac{1}{6}$ a pair of reciprocals? ______

(b) $(-6)\left(-\frac{1}{6}\right) = +1$ Are -6 and $-\frac{1}{6}$ a pair of reciprocals? ______

---

123. $\left(-\frac{1}{8}\right)(8) = -1$ $\left(-\frac{1}{8}\right)(-8) = +1$

(a) The reciprocal of $-\frac{1}{8}$ is ______.

(b) The reciprocal of -8 is ______.

a) No. Their product <u>is</u> <u>not</u> "+1".

b) Yes. Their product <u>is</u> "+1".

---

124. Complete these. In each case, you are inserting the reciprocal of the given factor:

(a) $(-10)(\quad) = +1$ (b) $\left(-\frac{1}{7}\right)(\quad) = +1$ (c) $(-47)(\quad) = +1$

a) -8

b) $-\frac{1}{8}$

---

125. It should be obvious that:

(a) The reciprocal of a <u>negative</u> quantity must be a ______________ (positive/negative) quantity.

(b) The reciprocal of a <u>positive</u> quantity must be a ______________ (positive/negative) quantity.

a) $-\frac{1}{10}$ b) -7 c) $-\frac{1}{47}$

---

126. Since $(23.7)\left(\frac{1}{23.7}\right) = \frac{23.7}{23.7} = 1$:

(a) The reciprocal of 23.7 is ______.

(b) The reciprocal of $\frac{1}{23.7}$ is ______.

a) negative

b) positive

---

127. Since $(0.303)\left(\frac{1}{0.303}\right) = \frac{0.303}{0.303} = 1$:

(a) The reciprocal of $\frac{1}{0.303}$ is ______.

(b) The reciprocal of 0.303 is ______.

a) $\frac{1}{23.7}$

b) 23.7

---

128. Write the reciprocals of the following:

(a) $\frac{1}{16.9}$ ______ (c) 14.8 ______

(b) 0.066 ______ (d) $\frac{1}{0.099}$ ______

a) 0.303

b) $\frac{1}{0.303}$

---

a) 16.9 c) $\frac{1}{14.8}$

b) $\frac{1}{0.066}$ d) 0.099

129. Write the reciprocal of:

(a) 14 ______ (e) $\frac{2}{5}$ ______

(b) +1 ______ (f) $-\frac{7}{3}$ ______

(c) −15 ______

(d) 0 ______ (g) −1 ______

---

130. Write the reciprocal of:

(a) $\frac{142}{697}$ ______ (d) $\frac{2}{9}$ ______

(b) $-\frac{1}{3}$ ______ (e) $-\frac{3}{10}$ ______

(c) $-\frac{7}{8}$ ______

Answers to 129:

a) $\frac{1}{14}$ e) $\frac{5}{2}$

b) +1 f) $-\frac{3}{7}$

c) $-\frac{1}{15}$ g) −1

d) "0" has no reciprocal.

---

131. The reciprocal of any negative quantity is always ______ (positive/negative).

Answers to 130:

a) $\frac{697}{142}$ d) $\frac{9}{2}$

b) −3 e) $-\frac{10}{3}$

c) $-\frac{8}{7}$

---

132. There is one number which has no reciprocal. What is it? ______

negative

---

133. Since $(b)\left(\frac{1}{b}\right) = \frac{b}{b} = 1$:

(a) The reciprocal of b is ______.

(b) The reciprocal of $\frac{1}{b}$ is ______.

0

---

134. Since $2x\left(\frac{1}{2x}\right) = \frac{2x}{2x} = 1$:

(a) The reciprocal of 2x is ______.

(b) The reciprocal of $\frac{1}{2x}$ is ______.

a) $\frac{1}{b}$

b) b

---

135. Write the reciprocal of each:

(a) m ______ (c) $\frac{1}{3q}$ ______

(b) $\frac{1}{t}$ ______ (d) 10y ______

a) $\frac{1}{2x}$

b) 2x

---

136. Write the reciprocal of each:

(a) −x ______ (c) $-\frac{1}{5d}$ ______

(b) $-\frac{1}{y}$ ______ (d) −7R ______

a) $\frac{1}{m}$ c) 3q

b) t d) $\frac{1}{10y}$

All are negative:

a) $-\frac{1}{x}$ c) −5d

b) −y d) $-\frac{1}{7R}$

137. Do not confuse "reciprocals" with "opposites." Two quantities are a pair of opposites if their sum is "0". Two quantities are a pair of reciprocals if their product is "+1". Therefore: (a) Opposites are defined by what operation, addition or multiplication? ________ (b) Reciprocals are defined by what operation, addition or multiplication? ________	
138. (a) The opposite of +2 is ______. (b) The reciprocal of +2 is ______.	a) Addition b) Multiplication
139. (a) The opposite of -5 is ______. (b) The reciprocal of -5 is ______.	a) -2 b) $\frac{1}{2}$
140. (a) The opposite of +1 is _____. (c) The opposite of -1 is _____. (b) The reciprocal of +1 is _____. (d) The reciprocal of -1 is _____.	a) +5 b) $-\frac{1}{5}$
	a) -1 c) +1 b) +1 d) -1

SELF-TEST 6 (Frames 103-140)

Multiply. Reduce each product to lowest terms:

1. $\left(-\frac{9}{4}\right)\left(-\frac{8}{3}\right) =$ ____
2. $\left(-\frac{3t}{8}\right)\left(\frac{2}{t}\right) =$ ____
3. $\left(-\frac{1}{p}\right)(-p) =$ ____
4. $4w\left(-\frac{3}{2w}\right) =$ ____

Write the reciprocal of each of the following:

5. $\frac{7}{12}$ ____
6. $\frac{1}{5}$ ____
7. $-\frac{r}{2}$ ____
8. 51.4 ____
9. -1 ____
10. 0 ____

11. The reciprocal of $-\frac{1}{4}$ is ____.
12. The opposite of $-\frac{1}{4}$ is ____.

ANSWERS:
1. +6
2. $-\frac{3}{4}$
3. +1
4. -6
5. $\frac{12}{7}$
6. 5
7. $-\frac{2}{r}$
8. $\frac{1}{51.4}$
9. -1
10. None
11. -4
12. $+\frac{1}{4}$

4-15 CONVERTING DIVISION TO MULTIPLICATION

In this section, we will show that <u>any division can be converted to a multiplication.</u> To make this conversion, we use the concept of <u>reciprocals.</u>

141. We showed earlier that a fraction is simply another way of writing a division. That is: $\frac{16}{8}$ means "16 divided by 8" (a) Write the fraction $\frac{9}{4}$ as a division: ________ (b) Write the division $7 \div 9$ as a fraction: ______	
142. Let's examine the following multiplication of a non-fraction and a fraction: $15\left(\frac{1}{3}\right) = \frac{15}{3}$ The <u>left side</u> indicates a <u>multiplication</u> of the two factors "15" and $\frac{"1"}{3}$. The <u>right side</u> indicates a <u>division</u> of "15" by "3". It should be obvious that the multiplication and division are equal since both equal ______.	a) $9 \div 4$ b) $\frac{7}{9}$
143. In the following equation, we are saying that the multiplication on the left equals the division on the right: $7\left(\frac{1}{8}\right) = \frac{7}{8}$ If we interchange the sides, we get: $\frac{7}{8} = 7\left(\frac{1}{8}\right)$ In the new equation, we are saying that the <u>division</u> on the left equals the ____________ on the right.	5
144. $\frac{15}{3} = 15\left(\frac{1}{3}\right)$ Let's compare the division $\frac{15}{3}$ and the multiplication $15\left(\frac{1}{3}\right)$. The denominator of $\frac{15}{3}$ is "3". The second factor of $15\left(\frac{1}{3}\right)$ is $\frac{"1"}{3}$. 3 and $\frac{1}{3}$ are a pair of ______________.	multiplication
	reciprocals

145. It should be obvious that any fraction can be converted to a multiplication. For example:

$$\frac{7}{8} = 7\left(\frac{1}{8}\right)$$

On the right side:

The first factor in the multiplication (the "7") is the numerator of the fraction.

The second factor in the multiplication (the $\frac{\text{"1"}}{8}$) is the reciprocal of the denominator of the fraction.

Convert each of these fractions to a multiplication:

(a) $\frac{8}{4} = (\ \ )$ ——   (b) $\frac{5}{7} = (\ \ )$ ——

Answers: a) $8\left(\frac{1}{4}\right)$ b) $5\left(\frac{1}{7}\right)$

146. Since any fraction stands for a division, we can convert any division to a multiplication. Here is the method:

> TO DIVIDE A FIRST QUANTITY BY A SECOND QUANTITY, MULTIPLY THE FIRST BY THE RECIPROCAL OF THE SECOND.
>
> Example: To divide "20" by "10", we write:
>
> $$\frac{20}{10} = 20(\text{the reciprocal of } 10) = 20\left(\frac{1}{10}\right)$$
>
> Example: To divide "5" by "16", we write:
>
> $$\frac{5}{16} = 5(\text{the reciprocal of } 16) = 5\left(\frac{1}{16}\right)$$

Complete these: (a) $\frac{12}{6} = 12(\text{the reciprocal of } \_\_\_\_)$

(b) $\frac{6}{13} = 6(\text{the reciprocal of } \_\_\_\_)$

Answers: a) 6   b) 13

147. Complete: (a) $\frac{7}{11} = 7(\text{the reciprocal of } 11) = 7(\ \ \ )$

(b) $\frac{14}{9} = 14(\text{the reciprocal of } 9) = 14(\ \ \ )$

Answers: a) $\frac{1}{11}$ b) $\frac{1}{9}$

148. Translate the following division problems into multiplications:

(a) $\frac{6}{17} =$ ______   (c) $\frac{17.8}{6.2} =$ ______

(b) $\frac{23}{9} =$ ______   (d) $\frac{2.91}{8.35} =$ ______

Answers: a) $6\left(\frac{1}{17}\right)$ c) $17.8\left(\frac{1}{6.2}\right)$

b) $23\left(\frac{1}{9}\right)$ d) $2.91\left(\frac{1}{8.35}\right)$

149. Any division can be written as a single fraction. If one of the quantities involved in the division is itself a fraction, there will be more than one fraction line. Here are some examples:

(1) "Divide $\frac{7}{8}$ by 5" can be written as: $\dfrac{\frac{7}{8}}{5}$ ←

(2) "Divide 3 by $\frac{7}{11}$" can be written as: $\dfrac{3}{\frac{7}{11}}$ ←

(3) "Divide $\frac{5}{9}$ by $\frac{4}{11}$" can be written as: $\dfrac{\frac{5}{9}}{\frac{4}{11}}$ ←

Note: In each case, one of the fraction lines is longer. See the arrows. This longer fraction line is called the "major" fraction line. It separates the numerator and denominator of the major division.

Complete these: (a) $\dfrac{\frac{1}{2}}{\frac{3}{4}}$ means "divide ____ by ____".

(b) $\dfrac{1}{\frac{2}{3}}$ means "divide ____ by ____".

(c) $\dfrac{\frac{5}{2}}{3}$ means "divide ____ by ____".

a) $\frac{1}{2}$ by $\frac{3}{4}$

b) 1 by $\frac{2}{3}$

c) $\frac{5}{2}$ by 3

---

150. Even when the numerator or denominator in a division is a fraction, we can convert the division to multiplication:

(a) $\dfrac{\frac{5}{9}}{\frac{4}{11}} = \frac{5}{9}(\text{the reciprocal of } \frac{4}{11}) = \frac{5}{9}\left(\frac{\quad}{\quad}\right)$

(b) $\dfrac{7}{\frac{8}{5}} = 7(\text{the reciprocal of } \frac{8}{5}) = 7\left(\frac{\quad}{\quad}\right)$

(c) $\dfrac{\frac{4}{5}}{11} = \frac{4}{5}(\text{the reciprocal of } 11) = \frac{4}{5}\left(\frac{\quad}{\quad}\right)$

a) $\frac{11}{4}$

b) $\frac{5}{8}$

c) $\frac{1}{11}$

---

151. Complete the following conversions from division to multiplication:

(a) $\dfrac{\frac{5}{7}}{\frac{4}{9}} = \left(\frac{5}{7}\right)\left(\frac{\quad}{\quad}\right)$ (b) $\dfrac{\frac{11}{17}}{\frac{47}{5}} = \left(\frac{11}{17}\right)\left(\frac{\quad}{\quad}\right)$

152. Complete the following conversions from division to multiplication:

(a) $\dfrac{\frac{2}{3}}{7} = \left(\dfrac{2}{3}\right)\left(\dfrac{\quad}{\quad}\right)$ (b) $\dfrac{7}{\frac{3}{11}} = (7)\left(\dfrac{\quad}{\quad}\right)$

a) $\left(\dfrac{5}{7}\right)\left(\dfrac{9}{4}\right)$ b) $\left(\dfrac{11}{17}\right)\left(\dfrac{5}{47}\right)$

---

153. Once we have converted the division to multiplication, we can easily find the final answer.

Complete: $\dfrac{\frac{5}{7}}{\frac{3}{2}} = \left(\dfrac{5}{7}\right)(\text{the reciprocal of } \dfrac{3}{2}) = \left(\dfrac{\quad}{\quad}\right)\left(\dfrac{\quad}{\quad}\right) = \underline{\qquad}$

a) $\left(\dfrac{2}{3}\right)\left(\dfrac{1}{7}\right)$ b) $(7)\left(\dfrac{11}{3}\right)$

---

154. Complete: $\dfrac{3}{\frac{8}{5}} = 3(\text{the reciprocal of } \dfrac{8}{5}) = (\quad)\left(\dfrac{\quad}{\quad}\right) = \underline{\qquad}$

$\left(\dfrac{5}{7}\right)\left(\dfrac{2}{3}\right) = \dfrac{10}{21}$

---

155. Complete: $\dfrac{\frac{5}{6}}{2} = \dfrac{5}{6}(\text{the reciprocal of } 2) = \left(\dfrac{\quad}{\quad}\right)\left(\dfrac{\quad}{\quad}\right) = \underline{\qquad}$

$3\left(\dfrac{5}{8}\right) = \dfrac{15}{8}$

---

156. Do each of the following division problems:

(a) $\dfrac{\frac{2x}{3}}{\frac{3}{2}} = \left(\dfrac{\quad}{\quad}\right)\left(\dfrac{\quad}{\quad}\right) = \underline{\qquad}$ (b) $\dfrac{m}{\frac{7}{9}} = (\quad)\left(\dfrac{\quad}{\quad}\right) = \underline{\qquad}$

$\left(\dfrac{5}{6}\right)\left(\dfrac{1}{2}\right) = \dfrac{5}{12}$

---

157. Complete each of these:

(a) $\dfrac{b}{\frac{1}{2}} = (\quad)(\quad) = \underline{\qquad}$ (b) $\dfrac{\frac{3}{4}}{5x} = (\quad)(\quad) = \underline{\qquad}$

a) $\left(\dfrac{2x}{3}\right)\left(\dfrac{2}{3}\right) = \dfrac{4x}{9}$

b) $(m)\left(\dfrac{9}{7}\right) = \dfrac{9m}{7}$

---

158. Do these. Be sure to reduce to lowest terms:

(a) $\dfrac{20p}{\frac{10}{7}} = \underline{\qquad\qquad}$ (c) $\dfrac{\frac{h}{4}}{\frac{h}{2}} = \underline{\qquad\qquad}$

(b) $\dfrac{\frac{9t}{5}}{\frac{3t}{10}} = \underline{\qquad\qquad}$ (d) $\dfrac{\frac{3w}{4}}{6w} = \underline{\qquad\qquad}$

a) $(b)\left(\dfrac{2}{1}\right) = 2b$

b) $\left(\dfrac{3}{4}\right)\left(\dfrac{1}{5x}\right) = \dfrac{3}{20x}$

---

a) 14p c) $\dfrac{1}{2}$

b) 6 d) $\dfrac{1}{8}$

4-16 DIVISIONS WHICH INVOLVE "0"

There are two types of divisions which involve "0":

(1) division of "0" by some other quantity, and

(2) division of some other quantity by "0".

We will examine both of these types of division in this section.

---

159. When a division of "0" by some other quantity is written as a fraction, the numerator of the fraction is "0". For example:

$$0 \div 3 = \frac{0}{3}$$

Write each of these divisions as a fraction:

(a) $0 \div 4 =$ _____ (b) $0 \div \frac{2}{3} =$ _____

Answers: a) $\frac{0}{4}$ b) $\frac{0}{\frac{2}{3}}$

160. When a division of some other quantity by "0" is written as a fraction, the denominator of the fraction is "0". For example:

$$7 \div 0 = \frac{7}{0}$$

Write each of these divisions as a fraction:

(a) $10 \div 0 =$ _____ (b) $\frac{5}{6} \div 0 =$ _____

Answers: a) $\frac{10}{0}$ b) $\frac{\frac{5}{6}}{0}$

161. Write each of the following fractions in ordinary division form:

(a) $\frac{8}{0} =$ __________ (b) $\frac{0}{6} =$ __________

Answers: a) $8 \div 0$ b) $0 \div 6$

162. Any division involving "0" can be converted to a multiplication in the usual way. That is, we simply multiply the numerator by the reciprocal of the denominator. For example:

$$\frac{0}{2} = 0(\text{the reciprocal of } 2)$$

$$\frac{9}{0} = 9(\text{the reciprocal of } 0)$$

Complete each of these:

(a) $\frac{0}{\frac{4}{5}} = 0(\text{the reciprocal of } \_\_\_\_)$ (b) $\frac{\frac{2}{3}}{0} = \frac{2}{3}(\text{the reciprocal of } \_\_\_\_)$

Answers: a) $\frac{4}{5}$ b) 0

163. If we convert the division of "0" by some other quantity to a multiplication, we obtain a multiplication in which "0" is one of the factors. For example:

$$\frac{0}{5} = 0\left(\frac{1}{5}\right) \qquad \frac{0}{\frac{3}{4}} = 0\left(\frac{4}{3}\right)$$

Since the product of "0" times any quantity is "0", the product in each multiplication above must be _____.

---

164. Complete these:

(a) $\frac{0}{9} = 0\left(\frac{\quad}{\quad}\right) =$ _____ (b) $\frac{0}{\frac{8}{7}} = 0\left(\frac{\quad}{\quad}\right) =$ _____

Answer: 0

---

165. It should be obvious that the value of any fraction whose numerator is "0" and whose denominator is some other quantity is "0". This is true even when the denominator contains a letter. For example:

$$\frac{0}{x} = 0\left(\frac{1}{x}\right)$$

$$\frac{0}{5x} = 0\left(\frac{1}{5x}\right)$$

The product of each multiplication above is _______.

Answers: a) $0\left(\frac{1}{9}\right) = 0$ b) $0\left(\frac{7}{8}\right) = 0$

---

166. If we convert the division of some other quantity by "0" to a multiplication, we obtain a multiplication in which one of the factors is "the reciprocal of 0". For example:

$$\frac{10}{0} = 10(\text{the reciprocal of } 0)$$

$$\frac{\frac{5}{9}}{0} = \frac{5}{9}(\text{the reciprocal of } 0)$$

What is the reciprocal of 0? That is, what number can we multiply times 0 to get a product of +1? _______

Answer: 0

---

167. When we have converted a division of some quantity by "0" to a multiplication, we cannot perform the multiplication because "0" has no reciprocal. Therefore:

> DIVISION BY "0" IS BOTH MEANINGLESS AND IMPOSSIBLE.

Which of the following divisions are impossible to do? _________

(a) $\frac{6}{0}$ (b) $\frac{0}{\frac{2}{3}}$ (c) $\frac{0}{5d}$ (d) $\frac{x}{0}$

Answer: There is none. "0" has no reciprocal since there is no number which we can multiply times 0 to get a product of +1.

Answer: Both (a) and (d)

168. Perform each possible division. If the division cannot be done, write "impossible" in the blank.

(a) $\frac{0}{t} =$ ______ (c) $\frac{3y}{0} =$ ______

(b) $\frac{9}{0} =$ ______ (d) $\frac{0}{\frac{5}{7}} =$ ______

---

169. If $t = 0$, $\frac{t}{8} =$ ______

a) 0
b) impossible
c) impossible
d) 0

---

170. If $m = 0$, $\frac{5}{m} =$ ______

0

---

171. If $a = 2$, $\frac{a - 2}{7} =$ ______

There is no answer since division by "0" is impossible.

---

172. If $b = 7$, $\frac{9}{b - 7} =$ ______

0

---

173. Do these: (a) $\frac{0}{324} =$ ______ (c) $\frac{29.8}{0} =$ ______

(b) $\frac{87}{0} =$ ______ (d) $\frac{0}{62.4} =$ ______

There is no answer since $7 - 7 = 0$ and division by 0 is impossible.

---

174. Do these: (a) $\frac{R}{0} =$ ______ (c) $\frac{0}{5t} =$ ______

(b) $\frac{0}{p} =$ ______ (d) $\frac{10y}{0} =$ ______

a) 0
b) No answer is possible.
c) No answer is possible.
d) 0

---

a) No answer is possible.
b) 0
c) 0
d) No answer is possible.

SELF-TEST 7 (Frames 141-174)

Convert each division into a multiplication:

1. $\frac{5}{9} =$ ______ 2. $\frac{16.9}{12.1} =$ ______ 3. $\frac{\frac{3}{7}}{\frac{5}{2}} =$ ______

Divide each. Report answers in lowest terms:

4. $\frac{\frac{8h}{9}}{\frac{h}{6}} =$ ______ 5. $\frac{\frac{4}{5}}{2w} =$ ______ 6. $\frac{4P}{\frac{P}{2}} =$ ______

Divide each:

7. $\frac{0}{\frac{1}{2}} =$ ______ 8. $\frac{19.2}{0} =$ ______ 9. $\frac{\frac{1}{2}}{0} =$ ______

ANSWERS:

1. $5\left(\frac{1}{9}\right)$
2. $16.9\left(\frac{1}{12.1}\right)$
3. $\frac{3}{7}\left(\frac{2}{5}\right)$
4. $\frac{16}{3}$
5. $\frac{2}{5w}$
6. 8
7. 0
8. No answer is possible.
9. No answer is possible.

4-17 DIVISION OF SIGNED QUANTITIES

When dividing signed quantities, the major problem is determining the sign of the quotient. Since any division can be converted to a multiplication and since we know how to determine the signs of products, it is easy to determine the sign of a quotient. We will examine the different cases in this section.

175. It is easy to determine the sign of a quotient by converting the division to multiplication.

Example: $\frac{-6}{3} = (-6)(\text{the reciprocal of } 3) = (-6)\left(\frac{1}{3}\right) =$ ______

176. Complete: $\frac{8}{-2} = 8(\text{the reciprocal of } -2) = 8\left(-\frac{1}{2}\right) =$ ______

Answer: -2, since: One factor is negative and the other is positive.

177. Complete: $\frac{-16}{-4} = (-16)(\text{the reciprocal of } -4) = (-16)\left(-\frac{1}{4}\right) =$ ______

Answer: -4, since: One factor is positive and the other is negative.

Answer: +4, since: Both factors are negative.

178. (a) $\frac{10}{2} = (10)(\ \ ) =$ ____ (c) $\frac{-10}{2} = (-10)(\ \ ) =$ ____

(b) $\frac{10}{-2} = (10)(\ \ ) =$ ____ (d) $\frac{-10}{-2} = (-10)(\ \ ) =$ ____

---

a) $(10)\left(\frac{1}{2}\right) = +5$

b) $(10)\left(-\frac{1}{2}\right) = -5$

c) $(-10)\left(\frac{1}{2}\right) = -5$

d) $(-10)\left(-\frac{1}{2}\right) = +5$

179. Do these:

(a) $\frac{9}{3} =$ ________ (c) $\frac{9}{-3} =$ ________

(b) $\frac{-9}{-3} =$ ________ (d) $\frac{-9}{3} =$ ________

---

a) +3 c) −3

b) +3 d) −3

180. Referring to the last frame, answer these:

(a) The sign of the quotient of two positive quantities is ________.

(b) The sign of the quotient of two negative quantities is ________.

(c) The sign of the quotient of one positive and one negative quantity is ________.

---

a) positive

b) positive

c) negative

181. Do these:

(a) $\frac{-12}{-3} =$ ____ (b) $\frac{-20}{4} =$ ____ (c) $\frac{14}{-7} =$ ____ (d) $\frac{-70}{-7} =$ ____

---

a) +4 c) −2

b) −5 d) +10

182. The same rules for the sign of the quotient apply when fractions are involved in the division.

Complete: (a) $\frac{-\frac{2}{3}}{-\frac{7}{11}} = \left(-\frac{2}{3}\right)\left(-\frac{11}{7}\right) =$ ____

(b) $\frac{1}{-\frac{1}{4}} = (1)(-4) =$ ____

(c) $\frac{-\frac{3}{5}}{4} = \left(-\frac{3}{5}\right)\left(\frac{1}{4}\right) =$ ____

---

a) $+\frac{22}{21}$

b) −4

c) $-\frac{3}{20}$

183. Do these: (a) $\frac{-5}{\frac{x}{7}} = (\ \ )(\ \ ) =$ ____

(b) $\frac{-\frac{m}{8}}{-3} = (\ \ )(\ \ ) =$ ____

(c) $\frac{-\frac{1}{5}}{\frac{1}{a}} = (\ \ )(\ \ ) =$ ____

184. All three of the following expressions equal what number? ______ $-\frac{27}{3}$ $\frac{-27}{3}$ $\frac{27}{-3}$	a) $(-5)\left(\frac{7}{x}\right) = -\frac{35}{x}$ b) $\left(-\frac{m}{8}\right)\left(-\frac{1}{3}\right) = +\frac{m}{24}$ c) $\left(-\frac{1}{5}\right)(a) = -\frac{a}{5}$
185. All three of the following expressions equal what number? ______ $\frac{-42}{6}$ $-\frac{42}{6}$ $\frac{42}{-6}$	-9
186. There are three equivalent ways of showing that a fraction is negative. (1) We can put a negative sign in front of the fraction: $-\frac{28}{7}$ (2) We can put a negative sign in the numerator alone: $\frac{-28}{7}$ (3) We can put a negative sign in the denominator alone: $\frac{28}{-7}$ Which one of the following four fractions is not equal to "-2"? ______ $-\frac{18}{9}$ $\frac{-18}{9}$ $\frac{-18}{-9}$ $\frac{18}{-9}$	-7
187. Which one of the following statements is false? ______ (a) $\frac{-7}{8} = \frac{7}{-8}$ (c) $\frac{4}{-3d} = -\frac{4}{3d}$ (b) $-\frac{x}{2} = \frac{-x}{2}$ (d) $\frac{-16}{t} = \frac{-16}{-t}$	$\frac{-18}{-9}$ equals +2.
188. Write $\frac{-7}{9}$ in two other equivalent ways: ______ and ______	(d) is false
189. Write $-\frac{3t}{2}$ in two other equivalent ways: ______ and ______	$\frac{7}{-9}$ and $-\frac{7}{9}$
190. Write $\frac{x}{-4}$ in two other equivalent ways: ______ and ______	$\frac{-3t}{2}$ and $\frac{3t}{-2}$
191. What is the overall sign of a fraction: (a) If the numerator is positive and the denominator is negative? ______ (b) If the numerator is negative and the denominator is positive? ______ (c) If both the numerator and denominator are negative? ______	$-\frac{x}{4}$ and $\frac{-x}{4}$
192. Complete: (a) $\frac{0}{-5} = 0\left(-\frac{1}{5}\right) =$ ______ (b) $\frac{0}{-\frac{7}{8}} = 0\left(-\frac{8}{7}\right) =$ ______	a) Negative b) Negative c) Positive

193. In each example below, a negative quantity is divided by "0". Each division has been converted to a multiplication.

$$\frac{-7}{0} = (-7)(\text{the reciprocal of } 0)$$

$$\frac{-\frac{3}{8}}{0} = \left(-\frac{3}{8}\right)(\text{the reciprocal of } 0)$$

Why can't we complete the multiplication on the right? ________________________

Answers to 192: a) 0 b) 0

---

194. Which of the following two divisions cannot be done? ________

(a) $\frac{-4}{0}$ (b) $\frac{0}{-4}$

Answer to 193: Because "0" has no reciprocal.

---

195. Do these:

(a) $\frac{0}{-x} =$ ______ (c) $\frac{-\frac{9}{8}}{0} =$ ______

(b) $\frac{-t}{0} =$ ______ (d) $\frac{0}{-\frac{1}{2}} =$ ______

Answer to 194: (a) Division by "0" is impossible since "0" has no reciprocal.

Answers to 195:
a) 0
b) No answer is possible.
c) No answer is possible.
d) 0

4-18 EXTENDING $\frac{\square}{\square} = 1$ AND $\frac{\square}{1} = \square$ TO FRACTIONS AND NEGATIVE QUANTITIES

In this section, we will show that the principle of dividing a quantity by itself and the principle of dividing a quantity by +1 are also true for fractions and negative quantities.

---

196. An instance of $\frac{\square}{\square}$ is $\frac{-3}{-3}$.

<u>Simplifying</u>: $\frac{-3}{-3} = (-3)(\text{the reciprocal of } -3) = (-3)\left(-\frac{1}{3}\right) =$ ______

---

197. Here are two more instances of $\frac{\square}{\square}$:

(a) $\frac{\frac{3}{2}}{\frac{3}{2}} = \left(\frac{3}{2}\right)(\text{the reciprocal of } \frac{3}{2}) = \left(\frac{3}{2}\right)\left(\frac{2}{3}\right) =$ ______

(b) $\frac{-\frac{5}{6}}{-\frac{5}{6}} = \left(-\frac{5}{6}\right)(\text{the reciprocal of } -\frac{5}{6}) = \left(-\frac{5}{6}\right)\left(-\frac{6}{5}\right) =$ ______

Answer to 196: +1

198. An instance of $\frac{\square}{1}$ is $\frac{-5}{1}$. Simplifying: $\frac{-5}{1}$ = (-5)(the reciprocal of 1) = (-5)(1) = ____	The answer is +1 for both.
199. Here are two more instances of $\frac{\square}{1}$: (a) $\frac{\frac{7}{8}}{1} = \frac{7}{8}$(the reciprocal of 1) $= \frac{7}{8}(1) =$ ____ (b) $\frac{-\frac{5}{9}}{1} = \left(-\frac{5}{9}\right)$(the reciprocal of 1) $= \left(-\frac{5}{9}\right)(1) =$ ____	-5
200. It should be obvious that the principles $\frac{\square}{\square} = 1$ and $\frac{\square}{1} = \square$ are true for: (1) Positive and negative quantities. (2) Non-fractions and fractions. Complete these: (a) $\frac{-12}{-12} =$ ____ (b) $\frac{\frac{3}{4}}{1} =$ ____ (c) $\frac{-10}{1} =$ ____ (d) $\frac{\frac{5}{2}}{\frac{5}{2}} =$ ____	a) $\frac{7}{8}$ b) $-\frac{5}{9}$
201. Complete these: (a) $\frac{-\frac{7}{8}}{-\frac{7}{8}} =$ ____ (c) $\frac{-\frac{9}{10}}{1} =$ ____ (b) $\frac{-t}{-t} =$ ____ (d) $\frac{-m}{1} =$ ____	a) +1 c) -10 b) $\frac{3}{4}$ d) +1
202. Complete these: (a) $\frac{\frac{a}{4}}{\frac{a}{4}} =$ ____ (c) $\frac{-\frac{2}{3x}}{-\frac{2}{3x}} =$ ____ (b) $\frac{\frac{4S}{3}}{1} =$ ____ (d) $\frac{-\frac{t}{14}}{1} =$ ____	a) +1 c) $-\frac{9}{10}$ b) +1 d) -m
	a) +1 c) +1 b) $\frac{4S}{3}$ d) $-\frac{t}{14}$

SELF-TEST 8 (Frames 175-202)

Do these divisions: 1. $\frac{-27}{3} =$ ____ 2. $\frac{56}{-8} =$ ____ 3. $\frac{-18}{-9} =$ ____ 4. $\frac{0}{-5} =$ ____

Divide: 5. $\frac{\frac{a}{2}}{-\frac{3a}{8}} =$ ____ 6. $\frac{-3}{-\frac{1}{r}} =$ ____ 7. $\frac{-\frac{3y}{4}}{2} =$ ____

8. Write $-\frac{5}{8}$ in two other equivalent ways: ____ and ____

9. Write $\frac{-w}{3}$ in two other equivalent ways: ____ and ____

Simplify: 10. $\frac{-\frac{h}{3}}{-\frac{h}{3}} =$ ____ 11. $\frac{\frac{5}{-2d}}{1} =$ ____ 12. $\frac{\frac{-1}{x}}{\frac{1}{-x}} =$ ____

ANSWERS:

1. -9
2. -7
3. 2
4. 0
5. $-\frac{4}{3}$
6. 3r
7. $-\frac{3y}{8}$
8. $\frac{-5}{8}$ and $\frac{5}{-8}$
9. $-\frac{w}{3}$ and $\frac{w}{-3}$
10. 1
11. $\frac{5}{-2d}$ or $-\frac{5}{2d}$ or $\frac{-5}{2d}$
12. 1

Chapter 5 ADDITION AND SUBTRACTION OF FRACTIONS

In this chapter, we will cover the following topics:

(1) the addition and subtraction of fractions
(2) the addition and subtraction of mixed numbers
(3) combined operations with fractions which are needed to check fractional equations and non-fractional equations with fractional roots.

The content of this chapter becomes progressively more difficult. On the basis of our past experience, we know that most of you, for example, have had little practice in problems involving combined operations with fractions. Therefore, be alert. Many of the skills taught in this chapter will be used frequently throughout the rest of the course.

5-1 THE NECESSARY CONDITION FOR ADDING TWO FRACTIONS

In this section, we will show that "factoring by the distributive principle" is the key step in adding two fractions. And since it is the key step, we will show that two fractions can be added only if they have identical denominators.

---

1. The key step in adding fractions is "factoring by the distributive principle." Therefore, we will begin by reviewing that type of factoring.

The following expressions are instances of a product in the distributive principle, since:

(1) each of the two terms is a multiplication,
and (2) there is a common factor in each term.

Examples: $3(4) + 5(4)$

$2x + 3x$

Any product of this type can be factored by the distributive principle.

$$3(4) + 5(4) = (3 + 5)(4)$$

Complete: $2x + 3x \ = (\quad + \quad)(\quad)$

---

2. The common factor of a product in the distributive principle can be a fraction. Here are two examples:

$$3\left(\frac{1}{8}\right) + 4\left(\frac{1}{8}\right)$$

$$5\left(\frac{1}{x}\right) + 2\left(\frac{1}{x}\right)$$

(Continued on following page.)

Answer to frame 1: $(2 + 3)(x)$

Frame	Answer
2. (Continued) We still factor in the same way: $3\left(\frac{1}{8}\right) + 4\left(\frac{1}{8}\right) = (3 + 4)\left(\frac{1}{8}\right)$ Complete: $5\left(\frac{1}{x}\right) + 2\left(\frac{1}{x}\right) = (\quad + \quad)(\quad)$	
3. We <u>cannot</u> factor by the distributive principle <u>unless</u> each term contains a common factor. In which of the following cases can we factor by the distributive principle? ________ (a) $7\left(\frac{1}{5}\right) + 6\left(\frac{1}{5}\right)$ (b) $8\left(\frac{1}{7}\right) + 3\left(\frac{1}{10}\right)$ (c) $3\left(\frac{1}{x}\right) + 2\left(\frac{1}{5}\right)$	$(5 + 2)\left(\frac{1}{x}\right)$
4. In an addition of two fractions, <u>each fraction is a division.</u> We can convert each of these divisions to a multiplication by "multiplying the numerator by the reciprocal of the denominator." <u>That is:</u> $\frac{3}{8} + \frac{4}{8} = 3\left(\frac{1}{8}\right) + 4\left(\frac{1}{8}\right)$ $\frac{7}{5} + \frac{6}{11} = 7\left(\frac{1}{5}\right) + 6\left(\frac{1}{11}\right)$ Complete: $\frac{5}{y} + \frac{6}{7} = (\quad)\left(\frac{}{}\right) + (\quad)\left(\frac{}{}\right)$	Only in (a). There is no common factor in the others.
5. When each of the two fractions has been converted to a multiplication, we can reduce the number of fractions from two to one <u>when the same fractional factor appears in each term.</u> The reduction is made by factoring by the distributive principle. Here is an example: $\frac{3}{5} + \frac{4}{5} = 3\left(\frac{1}{5}\right) + 4\left(\frac{1}{5}\right) = (3 + 4)\left(\frac{1}{5}\right)$ <u>Note</u>: There is only one fraction in the expression on the far right. Following the steps above, reduce the number of fractions from two to one in this addition problem: $\frac{7}{8} + \frac{6}{8} = (\quad)\left(\frac{}{}\right) + (\quad)\left(\frac{}{}\right) = (\quad + \quad)\left(\frac{}{}\right)$	$5\left(\frac{1}{y}\right) + 6\left(\frac{1}{7}\right)$
6. After converting each fraction to a multiplication, we can reduce the number of fractions from two to one <u>only if the fractional factors are identical in each term.</u> Here are two cases in which we cannot reduce the number of fractions from two to one: <u>Example:</u> $\frac{3}{7} + \frac{5}{8} = 3\left(\frac{1}{7}\right) + 5\left(\frac{1}{8}\right)$ <u>Note</u>: We cannot factor by the distributive principle. <u>Example:</u> $\frac{3}{5} + \frac{3}{4} = 3\left(\frac{1}{5}\right) + 3\left(\frac{1}{4}\right) = 3\left(\frac{1}{5} + \frac{1}{4}\right)$ <u>Note</u>: We can factor by the distributive principle. However, there are still <u>two</u> fractions in the expression on the far right. (Continued on following page.)	$7\left(\frac{1}{8}\right) + 6\left(\frac{1}{8}\right) = (7 + 6)\left(\frac{1}{8}\right)$

6. (Continued)

In which of the following cases can we reduce the number of fractions from two to one by factoring by the distributive principle? __________

(a) $\frac{4}{9}+\frac{5}{2}=4\left(\frac{1}{9}\right)+5\left(\frac{1}{2}\right)$

(b) $\frac{3}{11}+\frac{4}{11}=3\left(\frac{1}{11}\right)+4\left(\frac{1}{11}\right)$

(c) $\frac{5}{7}+\frac{5}{8}=5\left(\frac{1}{7}\right)+5\left(\frac{1}{8}\right)$

---

7. The number of fractions is reduced from two to one by factoring by the distributive principle. This reduction occurs only when the fractional factors are identical. It should be obvious that the fractional factors will be identical only in those cases in which the denominators of the original fractions are identical.

In the following example, since the denominators of the original fractions are identical, the fractional factors on the right are identical:

$$\frac{3}{7}+\frac{2}{7}=3\left(\frac{1}{7}\right)+2\left(\frac{1}{7}\right)$$

In the following example, since the denominators of the original fraction are not identical, the fractional factors on the right are not identical:

$$\frac{5}{6}+\frac{2}{7}=5\left(\frac{1}{6}\right)+2\left(\frac{1}{7}\right)$$

In which case below will we get identical fractional factors when we convert the fractions to multiplications? __________

(a) $\frac{7}{10}+\frac{4}{9}$ (b) $\frac{5}{7}+\frac{8}{7}$

Only in (b)

---

8. Addition of fractions is an operation in which two fractions are combined into one. We have not yet shown all of the steps in an addition of two fractions. However, it should already be obvious to you that two fractions can be added or combined into one only if they have identical denominators.

Which one of the following pairs of fractions can be added as they stand? __________

(a) $\frac{3}{5}+\frac{8}{5}$ (b) $\frac{2}{5}+\frac{4}{7}$

Only in (b), since the denominators in (a) are not identical.

---

9. In order to show all steps in the addition of two fractions, we must extend three principles to include groupings.

The first principle to be extended is the principle of dividing a quantity by +1. We earlier used this principle to replace a non-fraction by a fraction, as follows:

$$\boxed{\square=\frac{\square}{1}}$$

For example: $7=\frac{7}{1}$ $5t=\frac{5t}{1}$

(Continued on following page.)

Only the pair in (a)

9. (Continued)

This principle also holds when the non-fraction is a grouping:

Examples: $(3+2) = \dfrac{(3+2)}{1}$ (since both equal 5)

$(4+6) = \dfrac{(4+6)}{1}$ (since both equal 10)

Complete these: (a) $(7+8) = \dfrac{\boxed{}}{1}$ (b) $(3x+7) = \dfrac{\boxed{}}{1}$

(a) $\dfrac{\boxed{7+8}}{1}$

(b) $\dfrac{\boxed{3x+7}}{1}$

10. We must also extend <u>the principle for multiplication of two fractions</u> to include a grouping. We earlier used this definition with fractions which do not contain a grouping, as follows:

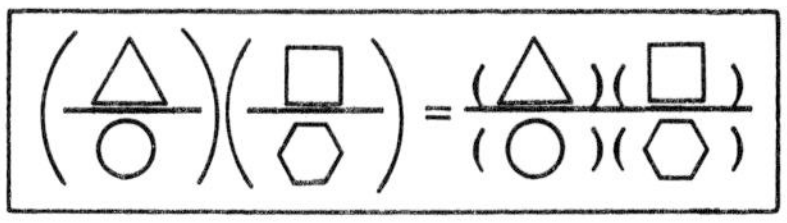

For example: $\left(\dfrac{7}{8}\right)\left(\dfrac{3}{5}\right) = \dfrac{(7)(3)}{(8)(5)}$

This principle can also be used when the numerator of a fraction is a grouping:

Example: $\left(\dfrac{5+6}{1}\right)\left(\dfrac{1}{8}\right) = \dfrac{(5+6)(1)}{(1)(8)}$

Complete this one: $\left(\dfrac{3+4}{1}\right)\left(\dfrac{1}{7}\right) = \dfrac{(\quad+\quad)(\quad)}{(\quad)(\quad)}$

$\dfrac{(3+4)(1)}{(1)(7)}$

11. We must also extend <u>the identity principle of multiplication</u> to include a grouping. We earlier used this principle with non-groupings, as follows:

$\boxed{\square(1) = \square}$ For example: $(12)(1) = 12$ $(5x)(1) = 5x$

This principle also holds when a grouping is written in the box:

Example: $(3+4)(1) = 3+4$

Complete: (a) $(7+9)(1) =$ _______ (b) $(t+6)(1) =$ _______

---

a) $7+9$ b) $t+6$

12. Now we can show all of the steps in the addition of two fractions. Each step is described in words on the right:

Step 1:	$\dfrac{3}{7}+\dfrac{2}{7} = 3\left(\dfrac{1}{7}\right)+2\left(\dfrac{1}{7}\right)$	Converting each fraction to a multiplication
Step 2:	$= (3+2)\left(\dfrac{1}{7}\right)$	Factoring by the distributive principle
Step 3:	$= \left(\dfrac{3+2}{1}\right)\left(\dfrac{1}{7}\right)$	Substituting $\dfrac{3+2}{1}$ for $3+2$
Step 4:	$= \dfrac{(3+2)(1)}{(1)(7)}$	Multiplying two fractions
Step 5:	$= \dfrac{3+2}{7}$	Substituting $3+2$ for $(3+2)(1)$ and substituting 7 for $(1)(7)$
Step 6:	$= \dfrac{5}{7}$	Adding the two numbers in the numerator

(Continued on following page.)

12. (Continued)

Ordinarily, we write the steps horizontally (in a line). We have done so for the problem below. Complete the steps:

$$\frac{5}{3}+\frac{2}{3}=5\left(\frac{1}{3}\right)+2\left(\frac{1}{3}\right)=(5+2)\left(\frac{1}{3}\right)=\left(\frac{5+2}{1}\right)\left(\frac{1}{3}\right)$$

$$=\frac{(\quad)(1)}{(1)(\quad)}=\frac{(\quad)+(\quad)}{(\quad)}=\frac{(\quad)}{(\quad)}$$

Answer: $\frac{(5+2)(1)}{(1)(3)}=\frac{5+2}{3}=\frac{7}{3}$

13. In the last frame, we described the formal steps used to add two fractions. The key step is Step 2 (factoring by the distributive principle) since two fractions can be reduced to one only by means of that step. And this reduction from two fractions to one fraction is possible <u>only if the original fractions have identical denominators.</u> Therefore, you must remember this fact:

TWO FRACTIONS CAN BE ADDED AS THEY STAND ONLY IF THEY HAVE IDENTICAL DENOMINATORS

Which of the following pairs of fractions can be added as they stand? ________

(a) $\frac{7}{9}+\frac{1}{9}$ (b) $\frac{3}{8}+\frac{3}{4}$ (c) $\frac{1}{6}+\frac{1}{5}$ (d) $\frac{3}{10}+\frac{2}{10}$

Answer: Only (a) and (d)

---

14. Identical denominators are sometimes called "like" denominators. For example, in the following additions, the denominators in each pair are "like":

$$\frac{7}{10}+\frac{1}{10} \qquad \frac{3}{x}+\frac{5}{x} \qquad \frac{12}{5d}+\frac{21}{5d}$$

In our course, however, we will generally avoid the use of the word "like" and will use the word "identical" instead. The word "identical" is more explicit and meaningful.

5-2 THE PATTERN FOR ADDING TWO FRACTIONS

In the last section, we described the formal procedure for adding two fractions. Since this formal procedure contains many steps, it is long and clumsy. There is a shorter method we can use. It is called the "pattern for adding two fractions." In this section, we will discuss the "pattern for adding two fractions" and show that it is based on the long formal procedure.

15. Using the long formal procedure, two additions of fractions are shown below:

$$\boxed{\frac{3}{5}+\frac{2}{5}} = 3\left(\frac{1}{5}\right) + 2\left(\frac{1}{5}\right) = (3+2)\left(\frac{1}{5}\right) = \left(\frac{3+2}{1}\right)\left(\frac{1}{5}\right) = \frac{(3+2)(1)}{(1)(5)} = \boxed{\frac{3+2}{5}}$$

$$\boxed{\frac{7}{4}+\frac{6}{4}} = 7\left(\frac{1}{4}\right) + 6\left(\frac{1}{4}\right) = (7+6)\left(\frac{1}{4}\right) = \left(\frac{7+6}{1}\right)\left(\frac{1}{4}\right) = \frac{(7+6)(1)}{(1)(4)} = \boxed{\frac{7+6}{4}}$$

Examine the boxed expressions in each problem above. You can see that there is a pattern in adding two fractions with identical denominators. The pattern is this:

$$\boxed{\frac{\triangle}{\bigcirc} + \frac{\square}{\bigcirc} = \frac{\triangle + \square}{\bigcirc}}$$

That is: Two fractions with identical denominators can be added by simply adding their numerators.

Using the pattern above, complete the following: (a) $\frac{3}{9}+\frac{5}{9} = \frac{(\ \ )+(\ \ )}{(\ \ )}$ (b) $\frac{5}{3}+\frac{7}{3} = \frac{(\ \ )+(\ \ )}{(\ \ )}$

---

a) $\frac{3+5}{9}$ b) $\frac{5+7}{3}$

16. Using the pattern for the addition of two fractions is a shortcut, since it enables us to skip many steps in the formal procedure. Here is the pattern again:

$$\frac{\triangle}{\bigcirc} + \frac{\square}{\bigcirc} = \frac{\triangle + \square}{\bigcirc}$$

Notice these points:

(1) The two fractions on the left are combined into one fraction on the right.

(2) The single fraction on the right has an addition as its numerator. In this addition, the numerators of the original two fractions are added.

(3) The single fraction on the right has the same denominator as each of the two fractions on the left.

Here is an example of this pattern: $\frac{5}{13}+\frac{2}{13} = \frac{5+2}{13}$

How many fractions are there: (a) on the left? _______

(b) on the right? _______

---

a) Two

b) One

17. Using the pattern for adding fractions with identical denominators, complete these:

(a) $\frac{3}{8}+\frac{1}{8} = \frac{(\ \ )+(\ \ )}{(\ \ )}$ (b) $\frac{4}{7}+\frac{5}{7} = \frac{(\ \ )+(\ \ )}{(\ \ )}$

---

a) $\frac{3+1}{8}$ b) $\frac{4+5}{7}$

18. The pattern for adding two fractions can be used only if the original fractions have identical denominators.

Which of the following pairs of fractions can be combined into one by means of the pattern? ________

(a) $\frac{5}{9}+\frac{5}{7}$ (b) $\frac{3}{5}+\frac{8}{5}$ (c) $\frac{3}{4}+\frac{3}{2}$

Only the pair in (b)

19. This pattern can also be used with fractions whose denominators contain a letter, provided that the fractions have identical denominators.

Using the pattern, complete these:

(a) $\frac{5}{y}+\frac{1}{y}=\frac{(\ )+(\ )}{(\ )}$ (b) $\frac{6}{5t}+\frac{8}{5t}=\frac{(\ )+(\ )}{(\ )}$

---

a) $\frac{5+1}{y}$ b) $\frac{6+8}{5t}$

20. Which of the following pairs of fractions can be combined into one by means of the pattern? ________

(a) $\frac{2}{y}+\frac{5}{7}$ (b) $\frac{7}{3m}+\frac{5}{2m}$ (c) $\frac{10}{7x}+\frac{9}{7x}$

Only the pair in (c)

21. Complete the following, using the pattern for addition:

(a) $\frac{5}{D}+\frac{4}{D}=\frac{(\ )+(\ )}{(\ )}$ (c) $\frac{5t}{7}+\frac{3t}{7}=\frac{(\ )+(\ )}{(\ )}$

(b) $\frac{x}{8}+\frac{7}{8}=\frac{(\ )+(\ )}{(\ )}$ (d) $\frac{3}{4b}+\frac{8}{4b}=\frac{(\ )+(\ )}{(\ )}$

---

a) $\frac{5+4}{D}$ c) $\frac{5t+3t}{7}$

b) $\frac{x+7}{8}$ d) $\frac{3+8}{4b}$

22. When using the pattern for adding fractions, the sum is always a single fraction whose numerator is an addition.

Sometimes we can simplify the sum by performing the addition in its numerator.

Examples: $\frac{3}{7}+\frac{2}{7}=\frac{3+2}{7}=\frac{5}{7}$

$$\frac{5t}{3}+\frac{6t}{3}=\frac{5t+6t}{3}=\frac{11t}{3}$$

Sometimes we cannot simplify the sum by performing the addition in its numerator.

Example: $\frac{2x}{5}+\frac{3}{5}=\frac{2x+3}{5}$

Since we cannot combine "2x" and "3", we cannot simplify the sum any further.

Add the following. Simplify the sum whenever possible:

(a) $\frac{5}{2p}+\frac{12}{2p}=$ ________ (c) $\frac{2t}{6}+\frac{1}{6}=$ ________

(b) $\frac{3x}{7}+\frac{2x}{7}=$ ________ (d) $\frac{6}{5}+\frac{m}{5}=$ ________

23. When the sum of two fractions can be simplified by performing the addition in the numerator, <u>be sure to reduce the simplified sum to lowest terms whenever possible</u>. <u>Example:</u> $\frac{7}{8} + \frac{5}{8} = \frac{7+5}{8} = \frac{12}{8} = \frac{3}{2}$ Perform the following additions. Reduce to lowest terms whenever possible: (a) $\frac{5}{4} + \frac{9}{4} =$ ________ (c) $\frac{5w}{8} + \frac{w}{8} =$ ________ (b) $\frac{7}{2x} + \frac{3}{2x} =$ ________ (d) $\frac{3r}{5r} + \frac{2r}{5r} =$ ________	a) $\frac{17}{2p}$ c) $\frac{2t+1}{6}$ b) $\frac{5x}{7}$ d) $\frac{6+m}{5}$
	a) $\frac{7}{2}$ c) $\frac{3w}{4}$ b) $\frac{5}{x}$ d) 1

5-3 THE CONCEPT OF "MULTIPLES"

To use the pattern for adding fractions, the <u>denominator</u> of the fractions <u>must be identical</u> or "like." We must also be able to add fractions where denominators are <u>non-identical</u> or "unlike."

We will first discuss the addition of two fractions, having non-identical denominators, in which the larger denominator is a multiple of the smaller denominator. Before doing so, however, it is necessary that the concept of "multiples" be clearly understood. That concept will now be discussed.

24. Below, we have multiplied the number "4" by a series of whole numbers beginning with the number "2". All of the products on the right are called "multiples of 4": $2(4) = 8$ $3(4) = 12$ $4(4) = 16$ $5(4) = 20$ $6(4) = 24$ Whenever we multiply "4" by a <u>whole number</u> larger than 1, the product is called a "multiple" of 4. Therefore: Since $7(4) = 28$, 28 is called a ____________ of 4.	
25. Whenever we multiply "6" by a <u>whole number</u> larger than 1, the product is a "multiple" of 6. (a) Since $2(6) = 12$, 12 is a ____________ of 6. (b) Since $9(6) = 54$, 54 is a ____________ of 6.	multiple
	a) multiple b) multiple

26. Whenever we multiply any whole number by another whole number larger than 1, the product is a multiple of the original whole number.

If we multiply 6 by 2 and get 12,
we say that 12 is a multiple of 6.

If we multiply 8 by 7 and get 56,
we say that 56 is a multiple of _____.

Answer: 8

27. The smallest multiple of a number is obtained by multiplying the number by 2. Therefore:

(a) The smallest multiple of 7 is _____.

(b) The smallest multiple of 11 is _____.

Answer: a) 14; b) 22

28. It should be obvious that any multiple of a number must be larger than that number. For example:

7 cannot be a multiple of 9, since 7 is smaller than 9.

Why can't 4 be a multiple of 6? ______________________

Answer: Because 4 is smaller than 6.

29. A multiple of 5 is produced by multiplying 5 by some whole number. Therefore, if a multiple of 5 is divided by 5, the quotient must be a whole number. For example:

$$\text{Since } 3(5) = 15, \quad \frac{15}{5} = 3$$

$$\text{Since } 7(5) = 35, \quad \frac{35}{5} = 7$$

Note: In each case, the whole number quotient on the extreme right is identical to the whole number we multiplied by on the extreme left.

27 is a multiple of 9 since $3(9) = 27$. If we divide 27 by 9, we must get the whole number _____.

Answer: 3

30. We know that a multiple of 7 must be larger than 7. We can test whether a number larger than 7 is a multiple of 7 by dividing the number by 7. The original number is a multiple of 7 only if the quotient is a whole number.

(a) Since $\frac{14}{7}$ equals the whole number 2, we know that 14 __________ (is/is not) a multiple of 7.

(b) Since $\frac{25}{7}$ does not equal a whole number, we know that 25 __________ (is/is not) a multiple of 7.

Answer: a) is; b) is not

31. If we divide any multiple of 6 by 6, the quotient must be a whole number.

(a) Since $\frac{30}{6}$ equals the whole number, we know that 30 ________ (is/is not) a multiple of 6.

(b) Since $\frac{28}{6}$ does not equal a whole number, we know that 28 ________ (is/is not) a multiple of 6.

---

32. If we divide any multiple of a number by that number, the quotient must be a whole number.

(a) Is 48 a multiple of 8? ______

(b) Is 39 a multiple of 5? ______

a) is

b) is not

---

33. Why can't 10 be a multiple of 30?

____

a) Yes, since $\frac{48}{8} = 6$

b) No, since $\frac{39}{5}$ does not equal a whole number.

---

34. (a) Is 30 a multiple of 10? ______

(b) Is 40 a multiple of 5? ______

(c) Is 9 a multiple of 18? ______

Because 10 is smaller than 30. Also, $\frac{10}{30}$ is not a whole number.

---

35. If we multiply a letter by any whole number larger than 1, the product is called a multiple of the letter. For example:

$$2(x) = 2x$$
$$3(x) = 3x$$
$$4(x) = 4x$$

All of the products on the right are called "multiples of x."

Since $7(t) = 7t$, we say that 7t is a ____________ of t.

a) Yes, since $\frac{30}{10} = 3$

b) Yes, since $\frac{40}{5} = 8$

c) No, since 9 is smaller than 18. Also, $\frac{9}{18}$ is not a whole number.

---

36. If we multiply a letter-term by any whole number larger than 1, the product is called a multiple of the letter-term. For example:

$$2(3y) = 6y$$
$$3(3y) = 9y$$
$$4(3y) = 12y$$

All of the products on the right are called "multiples of 3y."

Since $8(4p) = 32p$, we say that 32p is a ____________ of 4p.

multiple

---

multiple

37. To decide whether a letter-term is a multiple of "x", we divide the letter-term by "x". The letter-term is a multiple of "x" <u>only if the quotient is a whole number.</u> For example:

Since $\frac{5x}{x} = 5$, we know that 5x is a multiple of x.

<u>Note:</u> $\frac{5x}{x} = 5\left(\frac{x}{x}\right) = 5(1) = 5$

(a) Since $\frac{8x}{x} = 8$, we know that ____ is a multiple of ____.

(b) Since $\frac{10d}{d} = 10$, we know that ____ is a multiple of ____.

---

38. If we multiply a whole number larger than 1 by a letter, the product is called a multiple of the whole number. For example:

$$x(5) = 5x$$
$$t(5) = 5t$$
$$y(5) = 5y$$

All of the products on the right are called "multiples of 5."

Since m(7) = 7m, we say that 7m is a ______________ of 7.

Answers to 37:

a) <u>8x</u> is a multiple of <u>x</u>.

b) <u>10d</u> is a multiple of <u>d</u>.

---

39. To decide whether a letter-term is a multiple of "4m", we divide the letter-term by "4m". The letter-term is a multiple of "4m" <u>only if the quotient is a whole number.</u> For example:

Since $\frac{12m}{4m} = 3$, we know that 12m is a multiple of 4m.

<u>Note:</u> $\frac{12m}{4m} = \left(\frac{12}{4}\right)\left(\frac{m}{m}\right) = (3)(1) = 3$

(a) Since $\frac{10m}{5m} = 2$, we know that ____ is a multiple of ____.

(b) Since $\frac{28k}{7k} = 4$, we know that ____ is a multiple of ____.

Answer to 38: multiple

---

40. (a) Is 3R a multiple of R? _______

(b) Is 18t a multiple of 2t? _______

(c) Is 27x a multiple of 8x? _______

Answers to 39:

a) <u>10m</u> is a multiple of <u>5m</u>.

b) <u>28k</u> is a multiple of <u>7k</u>.

---

41. (a) Is 10x a multiple of x? _______

(b) Is 7p a multiple of 2p? _______

(c) Is 20y a multiple of 4y? _______

(d) Is 7w a multiple of w? _______

Answers to 40:

a) Yes, since $\frac{3R}{R} = 3$

b) Yes, since $\frac{18t}{2t} = 9$

c) No. The quotient is not a whole number.

42. Here is a trickier type of multiple. If we multiply a whole number by a letter or a letter-term, the product is called a multiple of the whole number. For example:

$$x(5) = 5x$$
$$3x(5) = 15x$$
$$4t(5) = 20t$$

All of the letter-terms on the right are called "multiples of 5."

(a) Since $m(7) = 7m$, we say that 7m is a _____________ of 7.

(b) Since $5d(8) = 40d$, we say that 40d is a multiple of ____.

a) Yes, since $\frac{10x}{x} = 10$

b) No. The quotient is not a whole number.

c) Yes, since $\frac{20y}{4y} = 5$

d) Yes, since $\frac{7w}{w} = 7$

---

43. We test whether a first quantity is a multiple of a second quantity by dividing the first by the second. Up to this point, the test has been whether the quotient is a whole number or not. In order to test whether a letter-term is a multiple of a whole number, we must expand our definition of the result of the test, as follows:

If a first quantity is divided by a second quantity and the quotient is a non-fraction, the first quantity is a multiple of the second.

To test whether a letter-term is a multiple of 7, we divide the letter-term by 7. The letter term is a multiple of 7 only if the quotient is a non-fraction. Here are two examples:

Since $\frac{7t}{7} = t$, we say that 7t is a multiple of 7 because "t" is a non-fraction.

Since $\frac{42x}{7} = 6x$, we say that 42x is a multiple of 7 because "6x" is a non-fraction.

(a) Is 48y a multiple of 6? _______

(b) Is 33t a multiple of 7? _______

a) multiple

b) 8

---

44. (a) Is 5d a multiple of 5? _______

(b) Is 24x a multiple of 9? _______

(c) Is 25m a multiple of 5? _______

a) Yes, since $\frac{48y}{6} = 8y$

b) No, since the quotient is $\frac{33t}{7}$, a fraction.

---

45. Is 27 a multiple of: (a) 3? _______ (b) 7? _______ (c) 2? _______ (d) 9? _______

a) Yes. The quotient is d, a non-fraction.

b) No. The quotient is a fraction.

c) Yes. The quotient is 5m, a non-fraction.

---

46. Is 40p a multiple of: (a) p? _______ (b) 7p? _______ (c) 5? _______ (d) 8p? _______

a) Yes c) No

b) No d) Yes

47. Is 10q a multiple of: (a) 2q? _______ (c) 3q? _______ (b) q? _______ (d) 10? _______	a) Yes c) Yes b) No d) Yes
	a) Yes c) No b) Yes d) Yes

5-4 ADDING TWO FRACTIONS IN WHICH THE LARGER DENOMINATOR IS A MULTIPLE OF THE SMALLER DENOMINATOR

In the last section, we discussed the concept of "multiples." That concept was introduced so that we could discuss the addition of fractions in which the larger denominator is a multiple of the smaller denominator. We will discuss that type of addition in this section.

48. When you are adding fractions with non-identical or "unlike" denominators, always check to see whether the larger denominator is a multiple of the smaller one: (a) In $\frac{3}{4}+\frac{7}{32}$, is 32 a multiple of 4? _______ (b) In $\frac{4x}{5}+\frac{6x}{7}$, is 7 a multiple of 5? _______	
49. (a) In $\frac{7}{2y}+\frac{11}{7y}$, is 7y a multiple of 2y? _______ (b) In $\frac{5}{d}+\frac{6}{3d}$, is 3d a multiple of d? _______	a) Yes b) No
50. (a) In $\frac{3}{7}+\frac{5}{7q}$, is 7q a multiple of 7? _______ (b) In $\frac{10}{9p}+\frac{11}{27p}$, is 27p a multiple of 9p? _______	a) No b) Yes
	a) Yes b) Yes

51. Since the denominators in the addition below are non-identical, we cannot add them as they stand:

$$\frac{3}{8} + \frac{5}{16}$$

In checking the denominators, we see that 16 is a multiple of 8. Therefore, we can easily find a fraction equivalent to $\frac{3}{8}$ whose denominator is 16. To do so, we:

(1) Divide 16 by 8 and get 2.

(2) Then multiply $\frac{3}{8}$ by $\frac{2}{2}$: $\left(\frac{3}{8}\right)\left(\frac{2}{2}\right) = \frac{6}{16}$

<u>Note</u>: The quotient 2 told us what instance of $\frac{n}{n}$ to use.

By substituting $\frac{6}{16}$ for $\frac{3}{8}$, we obtain an equivalent addition in which the denominators are identical, as follows:

$$\frac{3}{8} + \frac{5}{16} = \frac{6}{16} + \frac{5}{16} = \left(\frac{\quad}{\quad}\right)$$

Answer: $\frac{11}{16}$

52. In the addition below, 35b is a multiple of 7:

$$\frac{2}{7} + \frac{3}{35b}$$

We can find a fraction equivalent to $\frac{2}{7}$ whose denominator is 35b. To do so:

(1) We divide 35b by 7 and get 5b.

(2) We multiply $\frac{2}{7}$ by $\frac{5b}{5b}$: $\left(\frac{2}{7}\right)\left(\frac{5b}{5b}\right) = \frac{10b}{35b}$

<u>Note</u>: The quotient 5b tells us what instance of $\frac{n}{n}$ to use.

Therefore: $\frac{2}{7} + \frac{3}{35b} = \left(\frac{\quad}{\quad}\right) + \frac{3}{35b} =$ ________

Answer: $\frac{10b}{35b} + \frac{3}{35b} = \frac{10b + 3}{35b}$

53. To convert a fraction to an equivalent fraction whose denominator is a multiple of the original denominator, we multiply by an instance of $\frac{n}{n}$. The proper instance of $\frac{n}{n}$ is determined by <u>dividing the larger denominator by the smaller and using the quotient to set up</u> $\frac{n}{n}$. For example:

To add $\frac{3}{4}$ and $\frac{7}{8}$, we must convert $\frac{3}{4}$ to an equivalent fraction whose denominator is 8.

Since $\frac{8}{4} = 2$, we get: $\frac{3}{4}\left(\frac{2}{2}\right) = \frac{6}{8}$

To add $\frac{2}{3}$ and $\frac{1}{3x}$, we must convert $\frac{2}{3}$ to an equivalent fraction whose denominator is 3x.

Since $\frac{3x}{3} = x$, we get: $\frac{2}{3}\left(\frac{x}{x}\right) = \frac{2x}{3x}$

(Continued on following page.)

53. (Continued)

To add $\frac{1}{2}$ and $\frac{5}{12y}$, we must convert $\frac{1}{2}$ to an equivalent fraction whose denominator is 12y.

Since $\frac{12y}{2} = 6y$, we get: $\frac{1}{2}\left(\frac{\quad}{\quad}\right) = \frac{(\ \ )}{12y}$

$\frac{1}{2}\left(\frac{6y}{6y}\right) = \frac{6y}{12y}$

54. Complete each of these:

(a) $\frac{1}{7}\left(\frac{\quad}{\quad}\right) = \frac{(\ \ )}{14}$ (c) $\frac{3m}{5}\left(\frac{\quad}{\quad}\right) = \frac{(\ \ )}{10}$

(b) $\frac{3}{5}\left(\frac{\quad}{\quad}\right) = \frac{(\ \ )}{15}$ (d) $\frac{10}{x}\left(\frac{\quad}{\quad}\right) = \frac{(\ \ )}{7x}$

---

a) $\frac{1}{7}\left(\frac{2}{2}\right) = \frac{2}{14}$ c) $\frac{3m}{5}\left(\frac{2}{2}\right) = \frac{6m}{10}$

b) $\frac{3}{5}\left(\frac{3}{3}\right) = \frac{9}{15}$ d) $\frac{10}{x}\left(\frac{7}{7}\right) = \frac{70}{7x}$

55. Complete each of these:

(a) $\frac{1}{3d}\left(\frac{\quad}{\quad}\right) = \frac{(\ \ )}{12d}$ (c) $\frac{5}{3}\left(\frac{\quad}{\quad}\right) = \frac{(\ \ )}{15b}$

(b) $\frac{2}{7}\left(\frac{\quad}{\quad}\right) = \frac{(\ \ )}{7d}$ (d) $\frac{2}{y}\left(\frac{\quad}{\quad}\right) = \frac{(\ \ )}{3y}$

---

a) $\frac{1}{3d}\left(\frac{4}{4}\right) = \frac{4}{12d}$

b) $\frac{2}{7}\left(\frac{d}{d}\right) = \frac{2d}{7d}$

c) $\frac{5}{3}\left(\frac{5b}{5b}\right) = \frac{25b}{15b}$

d) $\frac{2}{y}\left(\frac{3}{3}\right) = \frac{6}{3y}$

56. You should be able to do these mentally, without writing the instance of $\frac{n}{n}$:

(a) $\frac{1}{6} = \frac{(\ \ )}{18}$ (c) $\frac{3}{8} = \frac{(\ \ )}{8t}$

(b) $\frac{2}{5x} = \frac{(\ \ )}{10x}$ (d) $\frac{8}{7} = \frac{(\ \ )}{14m}$

---

a) $\frac{3}{18}$ c) $\frac{3t}{8t}$

b) $\frac{4}{10x}$ d) $\frac{16m}{14m}$

57. Complete: (a) $\frac{5}{27} + \frac{4}{9} = \frac{5}{27} + \frac{(\ \ )}{27}$ (c) $\frac{5}{16} + \frac{1}{4} = \frac{5}{16} + \frac{(\ \ )}{16}$

(b) $\frac{1}{2} + \frac{3}{16} = \frac{(\ \ )}{16} + \frac{3}{16}$ (d) $\frac{3}{4} + \frac{3}{32} = \frac{(\ \ )}{32} + \frac{3}{32}$

---

a) $\frac{(12)}{27}$ c) $\frac{(4)}{16}$

b) $\frac{(8)}{16}$ d) $\frac{(24)}{32}$

58. Substitute the required equivalent fraction in each problem below:

(a) $\frac{7}{8} + \frac{3m}{2} = \frac{7}{8} + \left(\frac{\quad}{\quad}\right)$ (c) $\frac{1}{p} + \frac{8}{7p} = \left(\frac{\quad}{\quad}\right) + \frac{8}{7p}$

(b) $\frac{7}{5x} + \frac{1}{5} = \frac{7}{5x} + \left(\frac{\quad}{\quad}\right)$ (d) $\frac{3}{2d} + \frac{11}{16d} = \left(\frac{\quad}{\quad}\right) + \frac{11}{16d}$

---

a) $\frac{12m}{8}$ c) $\frac{7}{7p}$

b) $\frac{x}{5x}$ d) $\frac{24}{16d}$

59. Do these additions:

(a) $\frac{1}{2} + \frac{3}{8} =$ ____________ (b) $\frac{t}{4} + \frac{t}{32} =$ ____________

---

a) $\frac{7}{8}$ b) $\frac{9t}{32}$

60. Do these: (a) $\frac{10}{3g} + \frac{1}{g} =$ ____________ (b) $\frac{5}{4} + \frac{7}{4p} =$ ____________

61. Do these additions. Be sure each answer is in lowest terms:

(a) $\frac{2}{3} + \frac{2}{15} =$ ______ (c) $\frac{9}{24y} + \frac{7}{8y} =$ ______

(b) $\frac{x}{4} + \frac{5x}{12} =$ ______ (d) $\frac{1}{a} + \frac{12}{2a} =$ ______

a) $\frac{13}{3g}$ b) $\frac{5p + 7}{4p}$

a) $\frac{4}{5}$ b) $\frac{2x}{3}$ c) $\frac{5}{4y}$ d) $\frac{7}{a}$

SELF-TEST 1 (Frames 1-61)

Factor: 1. $w\left(\frac{1}{3}\right) + 2\left(\frac{1}{3}\right) =$ ______ 2. $5\left(\frac{1}{t}\right) + a\left(\frac{1}{t}\right) =$ ______

Add the following. Write each sum in simplest form:

3. $\frac{2}{9} + \frac{5}{9} =$ ______ 4. $\frac{5r}{4} + \frac{r}{4} =$ ______ 5. $\frac{1}{6b} + \frac{11}{6b} =$ ______ 6. $\frac{2}{d} + \frac{7d}{d} =$ ______

7. Which of these are multiples of 3h? ______ (a) 12h (b) 12 (c) 5h (d) h (e) 45h

Complete the following: 8. $\frac{5}{t} = \frac{(\ )}{4t}$ 9. $\frac{2x}{3} = \frac{(\ )}{12}$

Add the following. Write each sum in simplest form.

10. $\frac{2}{3} + \frac{4}{9} =$ ______ 11. $\frac{x}{6} + \frac{3x}{2} =$ ______ 12. $\frac{5}{6w} + \frac{1}{2w} =$ ______ 13. $\frac{7}{3} + \frac{5}{12a} =$ ______

ANSWERS:

1. $(w + 2)\left(\frac{1}{3}\right)$
2. $(5 + a)\left(\frac{1}{t}\right)$
3. $\frac{7}{9}$
4. $\frac{3r}{2}$
5. $\frac{2}{b}$
6. $\frac{2 + 7d}{d}$
7. (a), (e)
8. 20
9. 8x
10. $\frac{10}{9}$
11. $\frac{5x}{3}$
12. $\frac{4}{3w}$
13. $\frac{28a + 5}{12a}$

5-5 ADDING TWO FRACTIONS IN WHICH NEITHER DENOMINATOR IS A MULTIPLE OF THE OTHER

In this section, we will show the method for adding fractions with non-identical or "unlike" denominators, when neither denominator is a multiple of the other. As in any addition of fractions, we must eventually obtain identical denominators. We will begin by showing a two-step process of obtaining identical denominators. Then we will show a method by which both of these steps can be performed at the same time.

62. In the following addition, 7 is not a multiple of 5:

$$\frac{3}{7}+\frac{2}{5}$$

However, we can convert the addition into one in which one denominator is a multiple of the other. It is easy to find a fraction equivalent to $\frac{3}{7}$ whose denominator is a multiple of 5. We do so by multiplying $\frac{3}{7}$ by $\frac{5}{5}$, as follows: $\boxed{\frac{3}{7}=\frac{3}{7}\left(\frac{5}{5}\right)=\frac{15}{35}}$

Substituting $\frac{15}{35}$ for $\frac{3}{7}$ in the addition, we get:

$$\frac{3}{7}+\frac{2}{5}=\frac{15}{35}+\frac{2}{5}$$

In the new addition on the right, one denominator is a multiple of the other, since _____ is a multiple of _____.

Answer: 35 is a multiple of 5.

63. Here is the same addition: $\frac{3}{7}+\frac{2}{5}$

There is a second way in which we can convert it into an addition in which one denominator is a multiple of the other. We can find a fraction equivalent to $\frac{2}{5}$ whose denominator is a multiple of 7. To do so, we multiply $\frac{2}{5}$ by $\frac{7}{7}$, as follows: $\frac{2}{5}=\frac{2}{5}\left(\frac{7}{7}\right)=\frac{14}{35}$

Substituting $\frac{14}{35}$ for $\frac{2}{5}$ in the addition, we get:

$$\frac{3}{7}+\frac{2}{5}=\frac{3}{7}+\frac{14}{35}$$

In the new addition on the right, one denominator is a multiple of the other, since _____ is a multiple of _____.

Answer: 35 is a multiple of 7.

64. When confronted with an addition in which one denominator is not a multiple of the other, the first step is converting the addition to an equivalent addition in which one denominator is a multiple of the other. To do so, we substitute an equivalent fraction for either of two fractions. To make sure that the new fraction has a denominator which is a multiple of the other denominator, we multiply by an instance of $\frac{n}{n}$, in which "n" is the denominator of the other fraction.

Here is an example:

$$\frac{5}{3} + \frac{1}{4}$$

We can convert the addition above into an equivalent addition in which one denominator is a multiple of the other in either of two ways:

(1) We can substitute an equivalent fraction for $\frac{5}{3}$. To make sure that this equivalent fraction has a denominator which is a multiple of 4, we multiply $\frac{5}{3}$ by $\frac{4}{4}$ and get $\frac{20}{12}$. Substituting, we get:

$$\frac{5}{3} + \frac{1}{4} = \frac{20}{12} + \frac{1}{4}$$

(2) We can substitute an equivalent fraction for $\frac{1}{4}$. To make sure that this equivalent fraction has a denominator which is a multiple of 3, we multiply $\frac{1}{4}$ by $\frac{3}{3}$ and get $\frac{3}{12}$. Substituting, we get:

$$\frac{5}{3} + \frac{1}{4} = \frac{5}{3} + \frac{3}{12}$$

(a) After substituting in (1) above, the new denominators are 12 and 4. Is 12 a multiple of 4? _______

(b) After substituting in (2) above, the new denominators are 3 and 12. Is 12 a multiple of 3? _______

---

65. Here is another addition: $\frac{3}{2} + \frac{4}{5}$

(a) To find a fraction equivalent to $\frac{3}{2}$ whose denominator is a multiple of 5, we should multiply $\frac{3}{2}$ by $\left(\frac{\ }{\ }\right)$.

(b) Do so, and substitute this new fraction for $\frac{3}{2}$:

$$\frac{3}{2} + \frac{4}{5} = \boxed{\frac{\ }{\ }} + \frac{4}{5}$$

(c) To find a fraction equivalent to $\frac{4}{5}$ whose denominator is a multiple of 2, we should multiply $\frac{4}{5}$ by $\left(\frac{\ }{\ }\right)$.

(d) Do so, and substitute this new fraction for $\frac{4}{5}$:

$$\frac{3}{2} + \frac{4}{5} = \frac{3}{2} + \boxed{\frac{\ }{\ }}$$

a) Yes

b) Yes

66. Here is another addition in which neither denominator is a multiple of the other:

$$\frac{5}{x} + \frac{4}{7}$$

Even though one denominator is a letter, we can use the same procedure to convert the addition into an equivalent addition in which one denominator is a multiple of the other. There are two possibilities:

(1) We can find a fraction equivalent to $\frac{5}{x}$ whose denominator is a multiple of 7. To do so, we multiply $\frac{5}{x}$ by $\frac{7}{7}$ and get $\frac{35}{7x}$. Substituting, we get:

$$\boxed{\frac{5}{x}} + \frac{4}{7} = \boxed{\frac{35}{7x}} + \frac{4}{7}$$

(2) We can find a fraction equivalent to $\frac{4}{7}$ whose denominator is a multiply of x. To do so, we multiply $\frac{4}{7}$ by $\frac{x}{x}$ and get $\frac{4x}{7x}$. Substituting, we get:

$$\frac{5}{x} + \boxed{\frac{4}{7}} = \frac{5}{x} + \boxed{\frac{4x}{7x}}$$

(a) After substituting in (1) above, the new denominators are 7x and 7. Which one is a multiple of the other? ______________

(b) After substituting in (2) above, the new denominators are x and 7x. Which one is a multiple of the other? ______________

a) $\frac{5}{5}$

b) $\boxed{\frac{15}{10}} + \frac{4}{5}$

c) $\frac{2}{2}$

d) $\frac{3}{2} + \boxed{\frac{8}{10}}$

---

67. Here is another addition in which one denominator is not a multiple of the other:

$$\frac{3}{2x} + \frac{5}{7}$$

We can obtain an addition in which one denominator is a multiple of the other in either of two ways:

(1) Multiplying $\frac{3}{2x}$ by $\frac{7}{7}$, we get $\frac{21}{14x}$. Substituting, we get:

$$\boxed{\frac{3}{2x}} + \frac{5}{7} = \boxed{\frac{21}{14x}} + \frac{5}{7}$$

(2) Multiplying $\frac{5}{7}$ by $\frac{2x}{2x}$, we get $\frac{10x}{14x}$. Substituting, we get:

$$\frac{3}{2x} + \boxed{\frac{5}{7}} = \frac{3}{2x} + \boxed{\frac{10x}{14x}}$$

(a) After substituting in (1) above, which denominator is a multiple of the other? ______________

(b) After substituting in (2) above, which denominator is a multiple of the other? ______________

a) 7x is a multiple of 7.

b) 7x is a multiple of x.

68. By making the two possible substitutions in each case below, write two equivalent additions in which one denominator is a multiple of the other:

(a) $\frac{2}{5} + \frac{3}{8} = \boxed{\frac{}{}} + \frac{3}{8}$ or $\frac{2}{5} + \boxed{\frac{}{}}$

(b) $\frac{6}{7} + \frac{5}{3} = \boxed{\frac{}{}} + \frac{5}{3}$ or $\frac{6}{7} + \boxed{\frac{}{}}$

a) 14x is a multiple of 7.

b) 14x is a multiple of 2x.

---

69. Write two equivalent additions in which one denominator is a multiple of the other:

(a) $\frac{4}{5} + \frac{6}{y} = \boxed{\frac{}{}} + \frac{6}{y}$ or $\frac{4}{5} + \boxed{\frac{}{}}$

(b) $\frac{1}{3t} + \frac{7}{4} = \boxed{\frac{}{}} + \frac{7}{4}$ or $\frac{1}{3t} + \boxed{\frac{}{}}$

a) $\boxed{\frac{16}{40}} + \frac{3}{8}$ or $\frac{2}{5} + \boxed{\frac{15}{40}}$

b) $\boxed{\frac{18}{21}} + \frac{5}{3}$ or $\frac{6}{7} + \boxed{\frac{35}{21}}$

Answer to Frame 69: a) $\boxed{\frac{4y}{5y}} + \frac{6}{y}$ or $\frac{4}{5} + \boxed{\frac{30}{5y}}$ b) $\boxed{\frac{4}{12t}} + \frac{7}{4}$ or $\frac{1}{3t} + \boxed{\frac{21t}{12t}}$

---

70. In the following addition, neither denominator is a multiple of the other:

$$\frac{2}{7} + \frac{3}{5}$$

To perform this addition, we can use two steps to obtain an equivalent addition with identical denominators:

Step 1: Obtaining an equivalent addition in which one denominator is a multiple of the other:

To do so, we substitute $\frac{10}{35}$ for $\frac{2}{7}$, since $\frac{2}{7}\left(\frac{5}{5}\right) = \frac{10}{35}$. We get:

$$\frac{2}{7} + \frac{3}{5} = \frac{10}{35} + \frac{3}{5}$$

Step 2: Obtaining an equivalent addition with identical denominators:

To do so, we substitute $\frac{21}{35}$ for $\frac{3}{5}$, since $\frac{3}{5}\left(\frac{7}{7}\right) = \frac{21}{35}$. We get:

$$\frac{10}{35} + \frac{3}{5} = \frac{10}{35} + \frac{21}{35}$$

Complete the addition. The sum is _______.

Answer to Frame 70: $\frac{31}{35}$

71. Here is the same addition: $\frac{2}{7} + \frac{3}{5}$

In the last frame, we substituted $\frac{10}{35}$ for $\frac{2}{7}$ as the first step (Step 1), to get an addition in which one denominator is a multiple of the other. In this frame, we will substitute an equivalent fraction for $\frac{3}{5}$ as the first step in doing the addition. <u>We want to show that we get the same answer both ways.</u>

<u>Step 1</u>: <u>Obtaining a new addition in which one denominator is a multiple of the other:</u>

This time we will substitute $\frac{21}{35}$ for $\frac{3}{5}$, since $\frac{3}{5} = \frac{3}{5}\left(\frac{7}{7}\right) = \frac{21}{35}$. We get:

$$\frac{2}{7} + \frac{3}{5} = \frac{2}{7} + \frac{21}{35}$$

<u>Step 2</u>: <u>Obtaining an equivalent addition with identical denominators:</u>

We substitute $\frac{10}{35}$ for $\frac{2}{7}$, since $\frac{2}{7} = \frac{2}{7}\left(\frac{5}{5}\right) = \frac{10}{35}$. We get:

$$\frac{2}{7} + \frac{21}{35} = \frac{10}{35} + \frac{21}{35}$$

(a) Complete the addition. The sum is ______.

(b) Is this the same sum we got in the last frame? ______

<u>Answer to Frame 71</u>: a) $\frac{31}{35}$ b) Yes

72. The point of the last frame is to show this fact: <u>When substituting for one fraction to obtain an addition in which one denominator is a multiple of the other, we have two choices. Since both choices lead to the same answer, you can use either one.</u>

Here is another addition in which neither denominator is a multiple of the other:

$$\frac{7}{4} + \frac{3}{t}$$

We can use two steps to perform the addition:

<u>Step 1</u>: <u>Obtaining an equivalent addition in which one denominator is a multiple of the other:</u>

Let's substitute $\frac{12}{4t}$ for $\frac{3}{t}$, since $\frac{3}{t} = \frac{3}{t}\left(\frac{4}{4}\right) = \frac{12}{4t}$. We get:

$$\frac{7}{4} + \frac{3}{t} = \frac{7}{4} + \frac{12}{4t}$$

<u>Step 2</u>: <u>Obtaining an equivalent addition with identical denominators:</u>

We substitute $\frac{7t}{4t}$ for $\frac{7}{4}$, since $\frac{7}{4} = \frac{7}{4}\left(\frac{t}{t}\right) = \frac{7t}{4t}$. We get:

$$\frac{7}{4} + \frac{12}{4t} = \frac{7t}{4t} + \frac{12}{4t}$$

Complete the addition. The sum is ______.

<u>Answer to Frame 72</u>: $\frac{7t + 12}{4t}$

73. Let's do one more in two steps: $\frac{3}{4x} + \frac{2}{5}$

<u>Step 1</u>: <u>Obtaining an equivalent addition in which one denominator is a multiple of the other.</u>

Let's substitute $\frac{8x}{20x}$ for $\frac{2}{5}$, since $\frac{2}{5} = \frac{2}{5}\left(\frac{4x}{4x}\right) = \frac{8x}{20x}$. We get:

$$\frac{3}{4x} + \frac{2}{5} = \frac{3}{4x} + \frac{8x}{20x}$$

<u>Step 2</u>: <u>Obtaining an equivalent addition with identical denominators.</u>

We substitute $\frac{15}{20x}$ for $\frac{3}{4x}$, since $\frac{3}{4x}\left(\frac{5}{5}\right) = \frac{15}{20x}$. We get:

$$\frac{3}{4x} + \frac{8x}{20x} = \frac{15}{20x} + \frac{8x}{20x}$$

Complete the addition. The sum is ____________.

Answer: $\frac{15 + 8x}{20x}$

---

74. In the last few frames, we obtained an equivalent addition with identical denominators in two steps. In the next few frames, we will show that these two steps can be done at the same time.

Let's examine the essential part of each step in the addition we did in Frame 71:

$$\frac{2}{7} + \frac{3}{5}$$

<u>In Step 1</u>, we multiplied $\frac{3}{5}$ by $\frac{7}{7}$.

<u>In Step 2</u>, we multiplied $\frac{2}{7}$ by $\frac{5}{5}$.

<u>Therefore</u>: $\frac{2}{7} + \frac{3}{5} = \frac{10}{35} + \frac{21}{35}$

In each step, the instance of $\frac{n}{n}$ used to obtain an equivalent fraction was determined by examining the ______________(numerator/denominator) of the other fraction.

Answer: denominator

---

75. When neither denominator is a multiple of the other, we must replace both fractions with equivalent fractions whose denominators are <u>identical</u>. These equivalent fractions are obtained by multiplying each fraction by $\frac{n}{n}$, where "n" is the same as the denominator of the other original fraction.

Let's examine the essential part of each step in the addition we did in Frame 72:

$$\frac{7}{4} + \frac{3}{t}$$

<u>In Step 1</u>, we multiplied $\frac{3}{t}$ by $\frac{4}{4}$. <u>In Step 2</u>, we multiplied $\frac{7}{4}$ by $\frac{t}{t}$.

<u>Therefore</u>: $\frac{7}{4} + \frac{3}{t} = \frac{7t}{4t} + \frac{12}{4t}$

In each step, the instance of $\frac{n}{n}$ was determined by examining the ______________ of the other original fraction.

76. Since we know what instance of $\frac{n}{n}$ to use with each fraction, we can get an equivalent addition of fractions with identical denominators <u>in one step</u>. Here is an example:

$$\frac{1}{2}+\frac{3}{7}$$

Since the denominators are "2" and "7", the instances of $\frac{n}{n}$ to use are $\frac{7}{7}$ and $\frac{2}{2}$. Therefore:

$$\frac{1}{2}+\frac{3}{7}=\frac{1}{2}\left(\frac{7}{7}\right)+\frac{3}{7}\left(\frac{2}{2}\right)=\frac{7}{14}+\frac{6}{14}$$

Complete the addition. The sum is ______.

Answer: denominator

---

77. Here is another example of obtaining identical denominators <u>in one step</u>:

$$\frac{5}{d}+\frac{2}{3}$$

Since the denominators are "d" and "3", the instances of $\frac{n}{n}$ to use are $\frac{3}{3}$ and $\frac{d}{d}$. Therefore:

$$\frac{5}{d}+\frac{2}{3}=\frac{5}{d}\left(\frac{3}{3}\right)+\frac{2}{3}\left(\frac{d}{d}\right)=\frac{15}{3d}+\frac{2d}{3d}$$

Complete the addition. The sum is ______.

Answer: $\frac{13}{14}$

---

78. Here is another example: $\frac{1}{2}+\frac{8}{5m}$

Since the denominators are "2" and "5m", the instances of $\frac{n}{n}$ to use are $\frac{5m}{5m}$ and $\frac{2}{2}$. Therefore:

$$\frac{1}{2}+\frac{8}{5m}=\frac{1}{2}\left(\frac{5m}{5m}\right)+\frac{8}{5m}\left(\frac{2}{2}\right)=\frac{5m}{10m}+\frac{16}{10m}$$

Complete the addition. The sum is ______.

Answer: $\frac{15+2d}{3d}$

---

79. Complete: $\frac{4}{7}+\frac{3}{5}=\frac{4}{7}\left(\frac{5}{5}\right)+\frac{3}{5}\left(\frac{7}{7}\right)=\left(\frac{}{}\right)+\left(\frac{}{}\right)=\left(\frac{}{}\right)$

Answer: $\frac{5m+16}{10m}$

---

80. Complete this addition:

$$\frac{1}{x}+\frac{2}{7}=\frac{1}{x}\left(\frac{}{}\right)+\frac{2}{7}\left(\frac{}{}\right)=\left(\frac{}{}\right)+\left(\frac{}{}\right)=\frac{(\quad)+(\quad)}{(\quad)}$$

Answer: $\frac{20}{35}+\frac{21}{35}=\frac{41}{35}$

---

81. Complete this one:

$$\frac{1}{5}+\frac{7}{6x}=\frac{1}{5}\left(\frac{}{}\right)+\frac{7}{6x}\left(\frac{}{}\right)=\left(\frac{}{}\right)+\left(\frac{}{}\right)=\frac{(\quad)+(\quad)}{(\quad)}$$

Answer: $\frac{1}{x}\left(\frac{7}{7}\right)+\frac{2}{7}\left(\frac{x}{x}\right)=\frac{7}{7x}+\frac{2x}{7x}=\frac{7+2x}{7x}$

Answer: $\frac{1}{5}\left(\frac{6x}{6x}\right)+\frac{7}{6x}\left(\frac{5}{5}\right)=\frac{6x}{30x}+\frac{35}{30x}=\frac{6x+35}{30x}$

82. Do these: (a) $\frac{1}{2} + \frac{1}{7} =$ ________

(b) $\frac{2x}{3} + \frac{x}{2} =$ ________

Answers: a) $\frac{9}{14}$ b) $\frac{7x}{6}$

83. Do these: (a) $\frac{10}{p} + \frac{7}{5} =$ ________

(b) $\frac{5}{8} + \frac{1}{3q} =$ ________

Answers: a) $\frac{50 + 7p}{5p}$ b) $\frac{15q + 8}{24q}$

84. Always check first to see whether one denominator is a multiple of the other. If it is, only one fraction has to be replaced.

Do these: (a) $\frac{1}{7} + \frac{1}{p} =$ ________

(b) $\frac{1}{8x} + \frac{1}{8} =$ ________

Answers: a) $\frac{p + 7}{7p}$ b) $\frac{1 + x}{8x}$ $\left(\text{or } \frac{x + 1}{8x}\right)$

85. Always make sure that the sum is reduced to lowest terms.

Do these: (a) $\frac{3}{4} + \frac{1}{6} =$ ________

(b) $\frac{5}{8} + \frac{1}{12} =$ ________

Answers: a) $\frac{11}{12}$ b) $\frac{17}{24}$

5-6 ADDING A FRACTION AND A NON-FRACTION

Since any non-fraction can easily be converted into a fraction, it is easy to add a non-fraction and a fraction. We will show the method in this section.

---

86. Here is an addition of a non-fraction and a fraction:

$$3 + \frac{1}{8}$$

To perform the addition, we must convert the 3 to a fraction with an identical denominator. It is easy to convert 3 to an equivalent fraction whose denominator is 8. We simply multiply 3 by $\frac{8}{8}$: $3\left(\frac{8}{8}\right) = \frac{24}{8}$

Therefore: $3 + \frac{1}{8} = \frac{24}{8} + \frac{1}{8} =$ ______

Answer: $\frac{25}{8}$

87. To convert each of the following non-fractions to an equivalent fraction whose denominator is "5", we multiply each by $\frac{5}{5}$. Do so:

(a) $7 = \frac{(\quad)}{5}$ (b) $t = \frac{(\quad)}{5}$ (c) $2m = \frac{(\quad)}{5}$

---

88. Complete each of these conversions of a non-fraction to a fraction:

(a) $10 = \frac{(\quad)}{3}$ (b) $x = \frac{(\quad)}{7}$ (c) $5m = \frac{(\quad)}{4}$ (d) $4 = \frac{(\quad)}{x}$

a) $\frac{35}{5}$ b) $\frac{5t}{5}$ c) $\frac{10m}{5}$

---

89. To add a fraction and a non-fraction, we simply convert the non-fraction with an identical denominator. Study these examples:

$\boxed{5} + \frac{2}{3} = \frac{\boxed{15}}{3} + \frac{2}{3}$ $\boxed{7} + \frac{2}{x} = \frac{\boxed{7x}}{x} + \frac{2}{x}$ $\boxed{5d} + \frac{3}{4} = \frac{\boxed{20d}}{4} + \frac{3}{4}$

Complete: (a) $10 + \frac{3}{7} = \left(\frac{\quad}{\quad}\right) + \frac{3}{7}$ (c) $\frac{5}{R} + 3 = \frac{5}{R} + \left(\frac{\quad}{\quad}\right)$

(b) $y + \frac{1}{7} = \left(\frac{\quad}{\quad}\right) + \frac{1}{7}$ (d) $\frac{2m}{7} + 3m = \frac{2m}{7} + \left(\frac{\quad}{\quad}\right)$

a) $\frac{30}{3}$

b) $\frac{7x}{7}$

c) $\frac{20m}{4}$

d) $\frac{4x}{x}$

---

90. Here's a complete solution:

$$\frac{5}{2} + 6 = \frac{5}{2} + \frac{12}{2} = \frac{17}{2}$$

Complete these: (a) $3 + \frac{2}{5} =$ ____ + ____ = ____

(b) $5 + \frac{3}{t} =$ ____ + ____ = ____

a) $\frac{70}{7}$ c) $\frac{3R}{R}$

b) $\frac{7y}{7}$ d) $\frac{21m}{7}$

---

91. Do these:

(a) $4 + \frac{1}{8} =$ ________ (c) $\frac{t}{7} + 6 =$ ________

(b) $2b + \frac{3b}{5} =$ ________ (d) $1 + \frac{1}{R} =$ ________

a) $\frac{15}{5} + \frac{2}{5} = \frac{17}{5}$

b) $\frac{5t}{t} + \frac{3}{t} = \frac{5t + 3}{t}$

---

a) $\frac{33}{8}$ c) $\frac{t + 42}{7}$

b) $\frac{13b}{5}$ d) $\frac{R + 1}{R}$

SELF-TEST 2 (Frames 62-91)

Add the following. Write each answer in simplest form.

1. $\frac{3}{4} + \frac{5}{6} =$ ______ 3. $\frac{5}{3t} + \frac{1}{4} =$ ______ 5. $2 + \frac{3r}{4} =$ ______

2. $\frac{4}{5} + \frac{2}{d} =$ ______ 4. $5 + \frac{3}{8} =$ ______ 6. $\frac{1}{2} + y =$ ______

ANSWERS: 1. $\frac{19}{12}$ 2. $\frac{4d + 10}{5d}$ 3. $\frac{20 + 3t}{12t}$ 4. $\frac{43}{8}$ 5. $\frac{8 + 3r}{4}$ 6. $\frac{1 + 2y}{2}$

5-7 CONVERTING MIXED NUMBERS TO FRACTIONS AND FRACTIONS TO MIXED NUMBERS

In this section, we will show what we mean by a mixed number. Then, we will convert mixed numbers to fractions and fractions to mixed numbers.

92. Instead of writing $2 + \frac{1}{8}$, mathematicians frequently write $2\frac{1}{8}$. They skip the "+" and write the fraction next to the whole number.

$2\frac{1}{8}$ is called a mixed number because it includes both a whole number and a fraction.

You must remember that a mixed number stands for an addition.

That is: $3\frac{5}{7}$ means $3 + \frac{5}{7}$ Also: $4\frac{5}{6}$ means ___ + ___

Answer: $4 + \frac{5}{6}$

93. A mixed number is a shorthand way of writing an addition. When writing a mixed number, we write the whole number and fraction next to each other with no other symbols. If parentheses are written around the fraction, the expression is not a mixed number since the parentheses indicate a multiplication and not an addition.

Though $2\frac{3}{5}$ means $2 + \frac{3}{5}$, $2\left(\frac{3}{5}\right)$ means 2 times $\frac{3}{5}$.

Which of the following are mixed numbers? ______

(a) $3\left(\frac{7}{11}\right)$ (b) $5\left(\frac{4}{9}\right)$ (c) $7\frac{1}{3}$

Answer: Only (c)

94. Mixed numbers are a shorthand way of writing an addition which contains a <u>whole number and a numerical fraction.</u> There is no comparable shorthand when a letter is involved in either the non-fraction or the fraction. The longer addition form must be retained.

$$x + \frac{1}{3} \text{ cannot be written } x\frac{1}{3}.$$

$$5 + \frac{y}{7} \text{ cannot be written } 5\frac{y}{7}.$$

$$3 + \frac{4}{m} \text{ cannot be written } 3\frac{4}{m}.$$

In fact, there are no expressions like $x\frac{1}{3}$, $5\frac{y}{7}$, $3\frac{4}{m}$ in algebra.

Which of the following statements are true? ______

(a) $3t + \frac{1}{8} = 3t\frac{1}{8}$ (b) $7 + \frac{4}{5} = 7\frac{4}{5}$ (c) $5 + \frac{q}{7} = 5\frac{q}{7}$

---

95. It is easy to convert a mixed number into a single fraction. Watch the steps below:

$$2\frac{1}{8} = 2 + \frac{1}{8} = \frac{16}{8} + \frac{1}{8} = \frac{17}{8}$$

Complete this one: $3\frac{4}{7} = 3 + \frac{4}{7} = \underline{\qquad} + \frac{4}{7} = \underline{\qquad}$

Only (b)

---

96. Complete the following:

(a) $5\frac{3}{4} = 5 + \frac{3}{4} = \underline{\qquad} + \frac{3}{4} = \underline{\qquad}$

(b) $1\frac{7}{8} = 1 + \frac{7}{8} = \underline{\qquad} + \frac{7}{8} = \underline{\qquad}$

$\frac{21}{7} + \frac{4}{7} = \frac{25}{7}$

---

97. The following example shows all steps involved in converting a mixed number to a fraction:

$$3\frac{2}{5} = 3 + \frac{2}{5} = 3\left(\frac{5}{5}\right) + \frac{2}{5} = \frac{15}{5} + \frac{2}{5} = \frac{17}{5}$$

There is a shorter method we can use for this conversion. The steps in this shorter method are shown below.

$$3\frac{2}{5} = \boxed{\frac{5(3) + 2}{5}} = \frac{15 + 2}{5} = \frac{17}{5}$$

Let's examine the complicated fraction in the box above:

(1) The denominator is "5".
(2) In the numerator, we multiplied the whole number "3" by "5", the denominator of the fraction. Then we added "2", the numerator of the fraction, to this product.

Using the same steps, complete this one:

$$4\frac{1}{3} = \frac{(\quad)(\quad) + (\quad)}{3} = \frac{(\quad) + (\quad)}{3} = \underline{\qquad}$$

a) $\frac{20}{4} + \frac{3}{4} = \frac{23}{4}$

b) $\frac{8}{8} + \frac{7}{8} = \frac{15}{8}$

98. Complete: (a) $2\frac{3}{7} = \frac{(\quad)(\quad) + (\quad)}{7} = \frac{(\quad) + (\quad)}{7} = $ ______

(b) $1\frac{5}{8} = \frac{(\quad)(\quad) + (\quad)}{8} = \frac{(\quad) + (\quad)}{8} = $ ______

$\frac{3(4) + 1}{3} = \frac{12 + 1}{3} = \frac{13}{3}$

---

99. You should be able to do some of the shortcut mentally. To convert the mixed numbers below to fractions, do the following:

(1) Multiply the whole number by the denominator, and write that product in the first parentheses.

(2) Then write the numerator of the fraction in the second parentheses and simplify.

Complete: (a) $3\frac{4}{5} = \frac{(\quad) + (\quad)}{5} = $ ______

(b) $1\frac{2}{9} = \frac{(\quad) + (\quad)}{9} = $ ______

a) $\frac{7(2) + 3}{7} = \frac{14 + 3}{7} = \frac{17}{7}$

b) $\frac{8(1) + 5}{8} = \frac{8 + 5}{8} = \frac{13}{8}$

---

100. Convert each to a fraction:

(a) $5\frac{1}{3} = $ ______ (b) $1\frac{4}{7} = $ ______

a) $\frac{15 + 4}{5} = \frac{19}{5}$

b) $\frac{9 + 2}{9} = \frac{11}{9}$

---

101. Whenever we convert a mixed number to a single fraction, we obtain a fraction whose numerator is larger than its denominator.

Here are two definitions:

> A fraction is called a "proper" fraction if its numerator is smaller than its denominator.
>
> A fraction is called an "improper" fraction if its numerator is larger than its denominator.

Which of the following are "improper" fractions? ______

(a) $\frac{7}{8}$ (b) $\frac{31}{5}$ (c) $\frac{9}{11}$

a) $\frac{16}{3}$ b) $\frac{11}{7}$

---

102. All mixed numbers are greater than 1. Therefore:

Proper fractions (like $\frac{2}{3}, \frac{1}{4}$) cannot be converted to mixed numbers since they are all less than 1.

Improper fractions (like $\frac{9}{8}, \frac{25}{3}$) can be converted to mixed numbers since they are all greater than 1.

Here is an example of a conversion from an improper fraction to a mixed number:

$$\frac{9}{4} = \frac{8 + 1}{4} = \frac{8}{4} + \frac{1}{4} = 2 + \frac{1}{4} = 2\frac{1}{4}$$

(Continued on following page.)

Only (b)

102. (Continued)

Complete each of these conversions:

(a) $\frac{11}{3} = \frac{9+2}{3} = \frac{9}{3} + \frac{2}{3} =$ ____ $+ \frac{2}{3} =$ ____

(b) $\frac{41}{8} = \frac{40+1}{8} = \frac{40}{8} + \frac{1}{8} =$ ____ $+ \frac{1}{8} =$ ____

(c) $\frac{35}{16} = \frac{32+3}{16} = \frac{32}{16} + \frac{3}{16} =$ ____ $+ \frac{3}{16} =$ ____

---

a) $3 + \frac{2}{3} = 3\frac{2}{3}$

b) $5 + \frac{1}{8} = 5\frac{1}{8}$

c) $2 + \frac{3}{16} = 2\frac{3}{16}$

103. When converting an improper fraction to a mixed number, we break up the original fraction into two fractions. For example:

$$\frac{8}{3} = \frac{6}{3} + \frac{2}{3}$$

Notice these points about the two new numerators:

(1) The <u>second</u> numerator "2" must be less than the denominator "3".
(2) The first numerator "6" is a multiple of 3. It is the <u>largest multiple of 3</u> which is less than the original numerator "8".

Using the principles above, complete these:

(a) $\frac{17}{5} = \frac{(\ \ )}{5} + \frac{(\ \ )}{5}$ (c) $\frac{7}{4} = \frac{(\ \ )}{4} + \frac{(\ \ )}{4}$

(b) $\frac{13}{6} = \frac{(\ \ )}{6} + \frac{(\ \ )}{6}$ (d) $\frac{3}{2} = \frac{(\ \ )}{2} + \frac{(\ \ )}{2}$

---

a) $\frac{15}{5} + \frac{2}{5}$ c) $\frac{4}{4} + \frac{3}{4}$

b) $\frac{12}{6} + \frac{1}{6}$ d) $\frac{2}{2} + \frac{1}{2}$

104. Complete each conversion to a mixed number:

(a) $\frac{9}{2} = \frac{(\ \ )}{2} + \frac{(\ \ )}{2} =$ ____ (b) $\frac{10}{3} = \frac{(\ \ )}{3} + \frac{(\ \ )}{3} =$ ____

---

a) $\frac{8}{2} + \frac{1}{2} = 4\frac{1}{2}$

b) $\frac{9}{3} + \frac{1}{3} = 3\frac{1}{3}$

105. There is a shortcut we can use to convert an improper fraction to a mixed number. This shortcut gives us the whole-number part and the fraction-part of the mixed number immediately.

Let's use $\frac{25}{4}$ as an example:

We divide 25 by 4 and get:

$4\overline{)25}$ quotient 6 (with a remainder of 1)

The quotient "6" is the <u>whole-number part</u> of the mixed number. The remainder "1" is the <u>numerator of the fraction-part</u> of the mixed number.

Therefore: $\frac{25}{4} = 6\frac{1}{4}$

Let's convert $\frac{35}{8}$ to a mixed number by the shortcut:

Since: $8\overline{)35}$ quotient 4 (with a remainder of 3)

$\frac{35}{8} =$ ____

106. Use the shortcut to do these: (a) $\frac{22}{5}$ = ______ (c) $\frac{17}{9}$ = ______ (b) $\frac{13}{2}$ = ______ (d) $\frac{29}{8}$ = ______	$4\frac{3}{8}$
107. When writing a mixed number, the "fraction-part" should be reduced to lowest terms. That is: Instead of $3\frac{6}{8}$, we write $3\frac{3}{4}$. Therefore, when converting improper fractions to mixed numbers, always make sure that the "fraction-part" is reduced to lowest terms. Example: $\frac{52}{8} = 6\frac{4}{8} = 6\frac{1}{2}$ Convert each improper fraction to a mixed number, in lowest terms: (a) $\frac{36}{16} = 2\frac{4}{16}$ = ______ (b) $\frac{14}{4}$ = ______ (c) $\frac{48}{32}$ = ______	a) $4\frac{2}{5}$ c) $1\frac{8}{9}$ b) $6\frac{1}{2}$ d) $3\frac{5}{8}$
	a) $2\frac{1}{4}$ b) $3\frac{1}{2}$ c) $1\frac{1}{2}$

SELF-TEST 3 (Frames 92-107)

Write each of the following as a single fraction:

1. $2\frac{5}{6}$ = ______ 2. $4\left(\frac{5}{8}\right)$ = ______ 3. $w + \frac{2}{3}$ = ______

4. Which of the following are improper fractions? ______

(a) $\frac{1}{16}$ (b) $\frac{5}{4}$ (c) $\frac{8}{8}$ (d) $\frac{3}{10}$ (e) $\frac{10}{3}$

Write each of the following as a mixed number:

5. $\frac{11}{4}$ = ______ 6. $\frac{25}{3}$ = ______ 7. $\frac{30}{12}$ = ______

ANSWERS: 1. $\frac{17}{6}$ 2. $\frac{5}{2}$ 3. $\frac{3w+2}{3}$ 4. (b), (e) 5. $2\frac{3}{4}$ 6. $8\frac{1}{3}$ 7. $2\frac{1}{2}$

5-8 ADDING MIXED NUMBERS

There are two types of additions involving mixed numbers:

(1) Adding a whole number and a mixed number.
(2) Adding two mixed numbers.

We will show the method for each in this section.

---

108. It is easy to add a whole number and a mixed number since the mixed number can be written as an addition.

$$4 + 2\frac{3}{5} = \underbrace{4 + 2}_{\downarrow} + \frac{3}{5}$$

$$= \quad 6 \quad + \frac{3}{5} = 6\frac{3}{5}$$

Write $6\frac{3}{5}$ as an improper fraction: ______

$\frac{33}{5}$

109. When adding a whole number and a mixed number, the sum can be written as either a mixed number or an improper fraction. For each of the following, write the sum in both ways:

(a) $5 + 3\frac{3}{8} =$ ______ or ______

(b) $1\frac{7}{16} + 2 =$ ______ or ______

---

a) $8\frac{3}{8}$ or $\frac{67}{8}$

b) $3\frac{7}{16}$ or $\frac{55}{16}$

110. There are two methods for adding two mixed numbers. We will show both methods. As you will see later on, the first method is very tedious with some numbers.

The first method of adding mixed numbers is to convert both to improper fractions and then add.

Example: $2\frac{1}{5} + 1\frac{3}{5} = \frac{11}{5} + \frac{8}{5} = \frac{19}{5}$

Complete this one: $2\frac{3}{7} + 3\frac{1}{7} =$ ____ + ____ = ____

$\frac{17}{7} + \frac{22}{7} = \frac{39}{7}$

111. When two mixed numbers are added by this method, the sum is an improper fraction. This improper fraction should then be converted back to a mixed number.

For each of these, write the sum as an improper fraction and then as a mixed number:

(a) $1\frac{2}{5} + 3\frac{1}{5} =$ ______ or ______

(b) $5\frac{1}{7} + 1\frac{4}{7} =$ ______ or ______

112. When the sum of two mixed numbers is converted back to a mixed number, be sure that the "fraction-part" is reduced to lowest terms. Write each of the following sums as a mixed number:

(a) $1\frac{3}{16} + 2\frac{5}{16} =$ ________ (b) $2\frac{7}{8} + 3\frac{3}{8} =$ ________

a) $\frac{23}{5}$ or $4\frac{3}{5}$

b) $\frac{47}{7}$ or $6\frac{5}{7}$

---

113. When the mixed numbers in the following example are converted to improper fractions, the denominators are not identical. We must obtain identical denominators before adding the fractions.

Example: $3\frac{1}{2} + 1\frac{3}{4} = \frac{7}{2} + \frac{7}{4}$

$= \frac{14}{4} + \frac{7}{4} = \frac{21}{4}$ or $5\frac{1}{4}$

Do these: (a) $2\frac{1}{8} + 1\frac{3}{16} =$ ________

(b) $5\frac{1}{2} + 3\frac{2}{3} =$ ________

a) $3\frac{1}{2}$ b) $6\frac{1}{4}$

---

a) $\frac{53}{16}$ or $3\frac{5}{16}$ b) $\frac{55}{6}$ or $9\frac{1}{6}$

---

114. We have shown one method for adding mixed numbers. This method involves two types of conversions:

(1) Converting each mixed number to an improper fraction before adding them.
(2) Converting the sum from an improper fraction back to a mixed number.

This method becomes tedious when the mixed numbers are large:

Example: $85\frac{1}{32} + 99\frac{7}{32} = \frac{2,721}{32} + \frac{3,175}{32}$

$= \frac{5,896}{32} = 184\frac{8}{32} = 184\frac{1}{4}$

(As you can see, the chance for error is very great when making the conversions above.)

There is a second method for adding mixed numbers. This second method is much more efficient since it avoids conversions and immediately gives the sum in mixed-number form. It is based on the fact that any mixed number can be written as an addition of a whole number and a fraction. Here is an example:

$5\frac{3}{8} + 6\frac{2}{8} = 5 + \frac{3}{8} + 6 + \frac{2}{8}$

Step 1: $= 5 + 6 + \frac{3}{8} + \frac{2}{8}$

Step 2: $= 11 + \frac{5}{8}$

Step 3: $= 11\frac{5}{8}$

In Step 1, we rearranged the order of addition.
In Step 2, we added the whole numbers and the fractions separately.
In Step 3, we wrote the addition as a mixed number.

(Continued on following page.)

114. (Continued)

Complete this addition: $5\frac{7}{16} + 4\frac{4}{16} = 5 + \frac{7}{16} + 4 + \frac{4}{16} = 5 + 4 + \frac{7}{16} + \frac{4}{16} =$ ____ + ____ = ____

---

115. The second method is based on adding the "whole-number parts" and the "fraction-parts" separately. If you use this second method, the following are easy.

(a) $70\frac{1}{4} + 30\frac{2}{4} =$ ____________

(b) $85\frac{1}{16} + 10\frac{4}{16} =$ ____________

(c) $100\frac{7}{32} + 200\frac{4}{32} =$ ____________

Answer: $9 + \frac{11}{16} = 9\frac{11}{16}$

---

116. Be sure to reduce the "fraction-part" of the sum to lowest terms:

(a) $87\frac{3}{8} + 13\frac{3}{8} =$ ____________

(b) $35\frac{7}{16} + 15\frac{3}{16} =$ ____________

(c) $27\frac{11}{32} + 13\frac{5}{32} =$ ____________

Answers: a) $100\frac{3}{4}$ b) $95\frac{5}{16}$ c) $300\frac{11}{32}$

---

117. When adding mixed numbers by the second method, the "fraction-part" of the sum can be an improper fraction.

$$2\frac{3}{4} + 1\frac{2}{4} = 3\frac{5}{4}$$

The sum should not be left in this form. Since $\frac{5}{4} = 1\frac{1}{4}$:

$$3\frac{5}{4} = 3 + \frac{5}{4} = 3 + 1\frac{1}{4} = 4\frac{1}{4}$$

Complete this one: $5\frac{7}{8} + 6\frac{4}{8} = 11 + \frac{11}{8} = 11 +$ ____ = ____

Answers: a) $100\frac{3}{4}$ b) $50\frac{5}{8}$ c) $40\frac{1}{2}$

---

118. Complete these. Be sure all final fractions are reduced to lowest terms.

(a) $40\frac{2}{4} + 90\frac{3}{4} =$ ____________

(b) $14\frac{7}{8} + 10\frac{1}{8} =$ ____________

(c) $25\frac{7}{16} + 10\frac{11}{16} =$ ____________

(d) $90\frac{29}{32} + 50\frac{9}{32} =$ ____________

Answer: $11 + \underline{1\frac{3}{8}} = \underline{12\frac{3}{8}}$

---

119. Here is an addition in which the "fraction-parts" of the mixed numbers do not have "like" denominators. Of course, when adding the "fraction-parts" separately, you must get "like" denominators first. Do this one:

$3\frac{1}{8} + 2\frac{3}{16} =$ ____________

Answers: a) $131\frac{1}{4}$ b) 25 c) $36\frac{1}{8}$ d) $141\frac{3}{16}$

Answer: $5\frac{5}{16}$

120. Do these: (a) $25\frac{1}{2} + 25\frac{1}{4} =$ __________

(b) $27\frac{11}{16} + 3\frac{3}{8} =$ __________

---

121. Add: $44\frac{1}{8} + 22\frac{1}{3} =$ __________

a) $50\frac{3}{4}$

b) $31\frac{1}{16}$ (Not $30\frac{17}{16}$)

$66\frac{11}{24}$

SELF-TEST 4 (Frames 108-121)

Add the following. Write each sum as a mixed number.

1. $15 + 9\frac{3}{4} =$ __________
2. $2\frac{3}{8} + 1\frac{17}{32} =$ __________
3. $86\frac{1}{2} + 104\frac{7}{8} =$ __________
4. $20\frac{13}{16} + 31\frac{15}{32} =$ __________

ANSWERS: 1. $24\frac{3}{4}$ 2. $3\frac{29}{32}$ 3. $191\frac{3}{8}$ 4. $52\frac{9}{32}$

5-9 ADDITIONS INVOLVING SIGNED FRACTIONS

We have used the pattern at the right to add fractions:

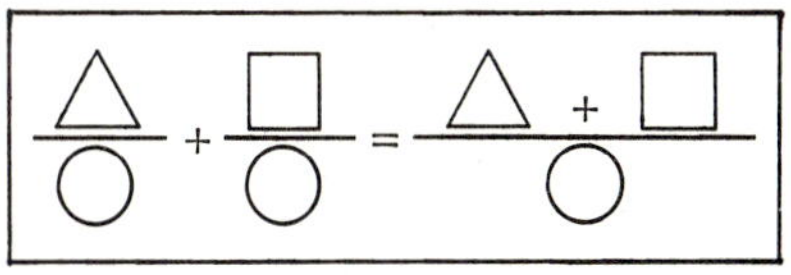

In this section, we will use the same pattern to add signed fractions.

---

122. Here is an example of using the pattern to add signed fractions with identical denominators:

$$\frac{(-3)}{7} + \frac{5}{7} = \frac{(-3) + 5}{7} = \frac{2}{7}$$

Complete these additions: (a) $\frac{(-2)}{5} + \frac{(-4)}{5} = \frac{(-2) + (-4)}{5} =$ ______

(b) $\frac{10}{4} + \frac{(-7)}{4} = \frac{10 + (-7)}{4} =$ ______

123. Do these additions:

(a) $\frac{x}{8} + \frac{(-7)}{8} =$ ____ (b) $\frac{7}{2x} + \frac{(-3)}{2x} =$ ____ (c) $\frac{(-3x)}{5} + \frac{(-4x)}{5} =$ ____

a) $\frac{-6}{5}$

b) $\frac{3}{4}$

---

124. To use the pattern: $\frac{\triangle}{\bigcirc} + \frac{\square}{\bigcirc} = \frac{\triangle + \square}{\bigcirc}$

<u>all negative signs must appear in the numerators.</u>

Here is a case in which the negative sign appears in front of the fraction:

$$\frac{7}{8} + \left(-\frac{2}{8}\right)$$

Since $-\frac{2}{8}$ is equivalent to $\frac{(-2)}{8}$, we can replace $-\frac{2}{8}$ with $\frac{(-2)}{8}$.

Then the negative sign is <u>in the numerator.</u> We get:

$$\frac{7}{8} + \left(-\frac{2}{8}\right) = \frac{7}{8} + \frac{(-2)}{8} = \_\_\_\_$$

a) $\frac{x + (-7)}{8}$

b) $\frac{2}{x}$ (Did you reduce?)

c) $\frac{-7x}{5}$

---

125. Complete these: (a) $\frac{4}{5} + \left(-\frac{3}{5}\right) = \frac{(\ \ )}{5} + \frac{(\ \ )}{5} = \frac{(\ \ )}{5}$

(b) $\left(-\frac{7}{16}\right) + \left(-\frac{2}{16}\right) = \frac{(\ \ )}{16} + \frac{(\ \ )}{16} = \frac{(\ \ )}{16}$

$\frac{5}{8}$

---

126. Add the following:

(a) $\frac{13}{7} + \left(-\frac{9}{7}\right) =$ ____ (b) $\frac{7}{16} + \left(-\frac{10}{16}\right) =$ ____ (c) $\frac{t}{32} + \left(-\frac{10}{32}\right) =$ ____

a) $\frac{(4)}{5} + \frac{(-3)}{5} = \frac{1}{5}$

b) $\frac{(-7)}{16} + \frac{(-2)}{16} = \frac{-9}{16}$

---

127. Add the following. Be sure to reduce the sum to lowest terms whenever possible.

(a) $\frac{3}{8} + \left(-\frac{7}{8}\right) =$ ____ (b) $\frac{5}{3x} + \left(-\frac{11}{3x}\right) =$ ____ (c) $\frac{11y}{21} + \left(-\frac{4y}{21}\right) =$ ____

a) $\frac{4}{7}$ b) $\frac{-3}{16}$ c) $\frac{t + (-10)}{32}$

---

128. Here is a case in which one denominator is a multiple of the other:

$$\left(-\frac{1}{4}\right) + \frac{3}{8}$$

We proceed as usual, substituting $-\frac{2}{8}$ for $-\frac{1}{4}$:

$$\left(-\frac{2}{8}\right) + \frac{3}{8} = \_\_\_\_$$

a) $\frac{-1}{2}$ or $-\frac{1}{2}$

b) $\frac{-2}{x}$ or $-\frac{2}{x}$

c) $\frac{y}{3}$

$\frac{1}{8}$

129. Do these:

(a) $\frac{3}{16} + \left(-\frac{1}{2}\right) =$ ______________

(b) $\frac{5}{x} + \left(-\frac{3}{4x}\right) =$ ______________

(c) $\left(-\frac{t}{16}\right) + \left(-\frac{1}{8}\right) =$ ______________

---

130. Here is a case in which one denominator is not a multiple of the other. We proceed as usual:

$$\frac{3}{5} + \left(-\frac{2}{7}\right) = \frac{3}{5}\left(\frac{7}{7}\right) + \left(-\frac{2}{7}\right)\left(\frac{5}{5}\right)$$

$$= \frac{21}{35} + \left(-\frac{10}{35}\right) = \frac{11}{35}$$

Complete these:

(a) $\left(-\frac{5}{3}\right) + \left(-\frac{1}{2}\right) = \left(\text{—}\right) + \left(\text{—}\right) =$ ______________

(b) $\frac{x}{5} + \left(-\frac{1}{3}\right) = \left(\text{—}\right) + \left(\text{—}\right) =$ ______________

(c) $\frac{5}{4} + \left(-\frac{2}{m}\right) = \left(\text{—}\right) + \left(\text{—}\right) =$ ______________

Answers to 129:

a) $\frac{-5}{16}$ or $-\frac{5}{16}$

b) $\frac{17}{4x}$

c) $\frac{(-t) + (-2)}{16}$

---

131. Here is an addition of a whole number and a fraction. We proceed as usual:

$$4 + \left(-\frac{3}{5}\right) = \frac{20}{5} + \left(-\frac{3}{5}\right) = \frac{17}{5}$$

Complete these:

(a) $7 + \left(-\frac{2}{3}\right) = \left(\text{—}\right) + \left(\text{—}\right) =$ ______________

(b) $(-2) + \frac{7}{8} = \left(\text{—}\right) + \left(\text{—}\right) =$ ______________

(c) $\left(-\frac{1}{4}\right) + (-1) = \left(\text{—}\right) + \left(\text{—}\right) =$ ______________

Answers to 130:

a) $\left(-\frac{10}{6}\right) + \left(-\frac{3}{6}\right) = \frac{-13}{6}$

b) $\frac{3x}{15} + \left(-\frac{5}{15}\right) = \frac{3x + (-5)}{15}$

b) $\frac{5m}{4m} + \left(-\frac{8}{4m}\right) = \frac{5m + (-8)}{4m}$

---

132. Do these additions:

(a) $t + \left(-\frac{1}{3}\right) =$ ______________

(b) $1 + \left(-\frac{5}{2R}\right) =$ ______________

Answers to 131:

a) $\frac{21}{3} + \left(-\frac{2}{3}\right) = \frac{19}{3}$

b) $\left(-\frac{16}{8}\right) + \frac{7}{8} = \frac{-9}{8}$ or $-\frac{9}{8}$

c) $\left(-\frac{1}{4}\right) + \left(-\frac{4}{4}\right) = \frac{-5}{4}$ or $-\frac{5}{4}$

Answers to 132:

a) $\frac{3t + (-1)}{3}$

b) $\frac{2R + (-5)}{2R}$

5-10 ADDING "SIGNED" MIXED NUMBERS

In this section, we will show the meaning of a negative mixed number. Then, we will show the method for additions which contain negative mixed numbers.

---

133. Just as $3\frac{1}{2}$ is shorthand for the addition: $3 + \frac{1}{2}$,

$-3\frac{1}{2}$ is shorthand for the addition: $(-3) + (-\frac{1}{2})$.

(Notice that both the whole number and the fraction in the addition are negative.)

Which one of the following statements is true? _____

(a) $-7\frac{1}{4}$ means $(-7) + (-\frac{1}{4})$ (b) $-7\frac{1}{4}$ means $(-7) + \frac{1}{4}$

---

134. $-5\frac{1}{8}$ is shorthand for which one of the following? _____

(a) $(-5) + \frac{1}{8}$ (b) $(-5) + (-\frac{1}{8})$

Answer to 133: (a)

Both the whole number and the fraction are negative.

---

135. An addition can be immediately converted to a mixed number only if both parts are positive or both parts are negative.

$$7 + \frac{1}{4} = 7\frac{1}{4}$$

$$(-7) + (-\frac{1}{4}) = -7\frac{1}{4}$$

$-7\frac{1}{4}$ is not shorthand for $(-7) + \frac{1}{4}$. Why not?

____

Answer to 134: (b)

Both parts are negative.

---

136. Any negative mixed number can be converted to an improper fraction in the regular way. The improper fraction, of course, is negative. Complete these conversions:

Example: $-2\frac{1}{4} = -\frac{9}{4}$

(a) $-1\frac{3}{8} =$ _____ (b) $-2\frac{3}{16} =$ _____

Answer to 135: Because the fraction "$\frac{1}{4}$" is not negative.

---

137. Any negative improper fraction can be converted to a negative mixed number. Complete these conversions:

Example: $-\frac{13}{8} = -1\frac{5}{8}$

(a) $-\frac{5}{4} =$ _____ (b) $-\frac{7}{2} =$ _____

Answer to 136: a) $-\frac{11}{8}$ b) $-\frac{35}{16}$

Answer to 137: a) $-1\frac{1}{4}$ b) $-3\frac{1}{2}$

138. The following addition is easy if you know what the mixed numbers mean. Notice that we rearrange the order of the parts, and then add the "whole-number parts" and the "fraction-parts":

$$5\frac{4}{8} + (-1\frac{1}{8}) = 5 + \frac{4}{8} + (-1) + (-\frac{1}{8})$$

$$= 5 + (-1) + \frac{4}{8} + (-\frac{1}{8})$$

$$= 4 + \frac{3}{8} = 4\frac{3}{8}$$

Complete: (a) $7\frac{7}{8} + (-4\frac{2}{8}) = 7 + \frac{7}{8} + (-4) + (\quad)$

(b) $3\frac{3}{4} + (-1\frac{2}{4}) = 3 + \frac{3}{4} + (\quad) + (\quad)$

---

139. Add: (a) $37\frac{7}{16} + (-27\frac{1}{16}) =$ ______________

(b) $150\frac{13}{32} + (-100\frac{5}{32}) =$ ______________

a) $(-\frac{2}{8})$

b) $(-1) + (-\frac{2}{4})$

---

140. Watch what happens in this addition:

$$40\frac{5}{16} + (-31\frac{8}{16}) = 40 + \frac{5}{16} + (-31) + (-\frac{8}{16})$$

$$= 40 + (-31) + \frac{5}{16} + (-\frac{8}{16})$$

$$= 9 + (-\frac{3}{16})$$

Note: $9 + (-\frac{3}{16})$ is not equal to $9\frac{3}{16}$. Why not?

____

a) $10\frac{3}{8}$

b) $50\frac{1}{4}$

Did you reduce to lowest terms?

---

141. The addition of a positive whole number and a fraction can be converted immediately to a mixed number only if the fraction is also positive.

To convert $9 + (-\frac{3}{16})$ to a mixed number, we must get rid of the negative fraction. To do so we:

(1) Substitute "8 + 1" for "9" $9 + (-\frac{3}{16}) = 8 + 1 + (-\frac{3}{16})$

(2) Then add the "1" to $(-\frac{3}{16})$ $9 + (-\frac{3}{16}) = 8 + 1 + (-\frac{3}{16})$

$$= 8 + \frac{16}{16} + (-\frac{3}{16})$$

$$= 8 + \frac{13}{16}$$

Since both "8" and "$\frac{13}{16}$" are positive, we can immediately convert the last addition to a mixed number. The mixed number is ______.

Because:

$$9\frac{3}{16} = 9 + (+\frac{3}{16})$$

142. To convert any addition of a positive whole number and a negative proper fraction to a mixed number, we must take "1" away from the whole number and add it to the fraction. For example:

(a) $15 + (-\frac{7}{8}) = 14 + 1 + (-\frac{7}{8})$

$= 14 + \frac{8}{8} + (-\frac{7}{8}) =$ ______

(b) $24 + (-\frac{5}{32}) = 23 + 1 + (-\frac{5}{32})$

$= 23 + \frac{32}{32} + (-\frac{5}{32}) =$ ______

$8\frac{13}{16}$

---

143. Convert each of these to a mixed number.

(a) $9 + (-\frac{3}{4}) =$ ______

(b) $17 + (-\frac{1}{8}) =$ ______

a) $14\frac{1}{8}$

b) $23\frac{27}{32}$

---

144. Write each as a mixed number:

(a) $7 + \frac{1}{4} =$ ______

(b) $(-7) + (-\frac{1}{4}) =$ ______

(c) $7 + (-\frac{1}{4}) =$ ______

a) $8\frac{1}{4}$

b) $16\frac{7}{8}$

---

145. Do these:

(a) $17\frac{1}{8} + (-12\frac{7}{8}) =$ ______

(b) $25\frac{9}{16} + (-24\frac{10}{16}) =$ ______

a) $7\frac{1}{4}$

b) $-7\frac{1}{4}$

c) $6\frac{3}{4}$

---

146. Here are some additions in which the denominators are <u>not</u> identical. You can use the same method.

(a) $6\frac{3}{4} + (-4\frac{1}{2}) =$ ______

(b) $8\frac{7}{16} + (-5\frac{7}{8}) =$ ______

a) $4\frac{1}{4}$

b) $\frac{15}{16}$ From: $1 + (-\frac{1}{16})$

---

147. Do these:

(a) $17\frac{5}{6} + (-15\frac{2}{5}) =$ ______

(b) $20\frac{1}{3} + (-15\frac{3}{4}) =$ ______

a) $2\frac{1}{4}$

b) $2\frac{9}{16}$

---

a) $2\frac{13}{30}$ b) $4\frac{7}{12}$

SELF-TEST 5 (Frames 122-147)

Add the following:

1. $\frac{3}{10} + \left(-\frac{4}{5}\right) =$ ______

3. $4 + \left(-\frac{3}{4}\right) =$ ______

2. $\frac{e}{3} + \left(-\frac{3e}{2}\right) =$ ______

4. $1 + \left(-\frac{2}{h}\right) =$ ______

5. Complete: $-8\frac{3}{4} = -8 + \left(\quad\right)$

6. Convert $-6\frac{2}{3}$ to an improper fraction.

7. Convert $-\frac{19}{8}$ to a mixed number.

Add the following:

8. $12 + \left(-\frac{5}{6}\right) =$ ______

9. $(-8) + \left(-\frac{4}{9}\right) =$ ______

10. $53\frac{2}{5} + \left(-18\frac{1}{2}\right) =$ ______

ANSWERS:
1. $-\frac{1}{2}$
2. $-\frac{7e}{6}$
3. $\frac{13}{4}$ or $3\frac{1}{4}$
4. $\frac{h + (-2)}{h}$
5. $\left(-\frac{3}{4}\right)$
6. $-\frac{20}{3}$
7. $-2\frac{3}{8}$
8. $\frac{67}{6}$ or $11\frac{1}{6}$
9. $-\frac{76}{9}$ or $-8\frac{4}{9}$
10. $34\frac{9}{10}$

5-11 SUBTRACTION OF FRACTIONS

Any subtraction can be performed by converting it to an addition. This is true even when the subtraction involves a fraction. We will show the method in this section.

148. To convert the subtraction of a fraction to an addition, we "add the opposite of the fraction." That is:

$$\frac{2}{8} - \frac{7}{8} = \frac{2}{8} + (\text{the opposite of } \frac{7}{8})$$

To make this conversion, we must examine the opposites of fractions. Here is an example of a pair of opposites, since their sum is zero:

$$\frac{7}{8} + \left(-\frac{7}{8}\right) = \frac{7}{8} + \frac{(-7)}{8} = \frac{0}{8} = 0$$

Since $-\frac{7}{8} = \frac{-7}{8}$, the "opposite of $\frac{7}{8}$" can be written in two ways:

(1) $-\frac{7}{8}$, with the negative sign in front of the fraction.

(2) $\frac{-7}{8}$, with the negative sign in the numerator.

Write the opposite of $\frac{5}{3}$ in two ways: ______ and ______

149. Since there are two different ways of writing the opposite of a positive fraction, there are two different ways of defining a pair of opposites for fractions:

(1) Two fractions are a pair of opposites if they have identical numerators and denominators, and one of the fractions is positive and the other is negative.

Examples: $\frac{3}{4}$ and $-\frac{3}{4}$ $-\frac{5}{h}$ and $\frac{5}{h}$

(2) Two fractions are a pair of opposites if they have identical denominators, and their numerators are a pair of opposites.

Examples: $\frac{3}{4}$ and $\frac{-3}{4}$ $\frac{-5}{h}$ and $\frac{5}{h}$

Write the opposite of: (a) $\frac{-8}{x}$ ______ (b) $-\frac{t}{5}$ ______

Answer: $-\frac{5}{3}$ and $\frac{-5}{3}$

---

150. Write the opposite of: (a) $\frac{t}{9}$ ____ (b) $-\frac{6}{5}$ ____ (c) $\frac{-3b}{16}$ ____

Answer: a) $\frac{8}{x}$ b) $\frac{t}{5}$

---

151. (a) When are two numbers a pair of opposites? ______________________

(b) When are two numbers a pair of reciprocals? ______________________

Answer: a) $-\frac{t}{9}$ or $\frac{-t}{9}$ b) $\frac{6}{5}$ c) $\frac{3b}{16}$

---

152. (a) The opposite of $\frac{7}{8}$ is ____ . (c) The opposite of $-\frac{2}{3}$ is ____ .

(b) The reciprocal of $\frac{7}{8}$ is ____ . (d) The reciprocal of $-\frac{2}{3}$ is ____ .

Answer: a) When their sum is 0. b) When their product is +1.

---

153. Complete: (a) $\frac{7}{x}$ is the ______________ of $\frac{x}{7}$.

(b) $-\frac{3m}{2}$ is the ______________ of $\frac{3m}{2}$.

(c) $-\frac{y}{4}$ is the ______________ of $-\frac{4}{y}$.

Answer: a) $-\frac{7}{8}$ b) $\frac{8}{7}$ c) $\frac{2}{3}$ d) $-\frac{3}{2}$

---

154. In a subtraction of two fractions, we can convert the subtraction to addition by "adding the opposite" of the second fraction:

$$\frac{2}{8} - \frac{7}{8} = \frac{2}{8} + (\text{the opposite of } \frac{7}{8}) = \frac{2}{8} + \frac{(-7)}{8}$$

Complete: $\frac{5}{7} - \left(-\frac{3}{7}\right) = \frac{5}{7} + (\text{the opposite of } -\frac{3}{7}) = \frac{5}{7} +$ ______

Answer: a) reciprocal b) opposite c) reciprocal

Answer: $+\frac{3}{7}$

155. Complete these conversions to addition. Write negative signs in the numerator so that the pattern for adding can be immediately used:

(a) $\frac{7}{6} - \frac{11}{6} = \frac{7}{6} + \_\_\_$ (b) $\frac{7}{3} - \left(-\frac{5}{3}\right) = \frac{7}{3} + \_\_\_$ (c) $\left(-\frac{2}{9}\right) - \frac{7}{9} = \frac{(-2)}{9} + \_\_\_$

---

156. Complete these conversions to additions:

(a) $\frac{3d}{5} - \frac{7d}{5} = \frac{3d}{5} + \_\_\_$ (c) $\left(-\frac{5}{t}\right) - \left(-\frac{7}{t}\right) = \frac{-5}{t} + \_\_\_$

(b) $\frac{7b}{10} - \left(-\frac{4b}{10}\right) = \frac{7b}{10} + \_\_\_$ (d) $\left(-\frac{6}{5p}\right) - \left(-\frac{1}{5p}\right) = \frac{-6}{5p} + \_\_\_$

Answers to 155: a) $+\frac{(-11)}{6}$ b) $+\frac{5}{3}$ c) $+\frac{(-7)}{9}$

---

157. For each of the following, convert to addition and find the sum:

(a) $\frac{5}{9} - \frac{10}{9} = \_\_\_ + \_\_\_ = \_\_\_$

(b) $\frac{13}{16} - \left(-\frac{4}{16}\right) = \_\_\_ + \_\_\_ = \_\_\_$

(c) $\left(-\frac{2}{7}\right) - \frac{3}{7} = \_\_\_ + \_\_\_ = \_\_\_$

Answers to 156: a) $+\frac{(-7d)}{5}$ b) $+\frac{4b}{10}$ c) $+\frac{7}{t}$ d) $+\frac{1}{5p}$

---

158. Complete: (a) $\frac{8}{d} - \frac{11}{d} = \_\_\_ + \_\_\_ = \_\_\_$

(b) $\frac{10t}{7} - \left(-\frac{5t}{7}\right) = \_\_\_ + \_\_\_ = \_\_\_$

(c) $\left(-\frac{3g}{11}\right) - \frac{8}{11} = \_\_\_ + \_\_\_ = \_\_\_$

Answers to 157: a) $\frac{5}{9} + \frac{(-10)}{9} = \frac{-5}{9}$ or $-\frac{5}{9}$ b) $\frac{13}{16} + \frac{4}{16} = \frac{17}{16}$ c) $\frac{(-2)}{7} + \frac{(-3)}{7} = \frac{-5}{7}$ or $-\frac{5}{7}$

---

159. Do these subtractions:

(a) $\frac{x}{3} - \frac{7}{3} = \_\_\_$ (c) $\frac{3}{r} - \frac{1}{r} = \_\_\_$

(b) $\frac{y}{7} - \left(-\frac{5}{7}\right) = \_\_\_$ (d) $\frac{10}{3a} - \left(-\frac{4}{3a}\right) = \_\_\_$

Answers to 158: a) $\frac{8}{d} + \frac{(-11)}{d} = \frac{-3}{d}$ or $-\frac{3}{d}$ b) $\frac{10t}{7} + \frac{5t}{7} = \frac{15t}{7}$ c) $\frac{(-3g)}{11} + \frac{(-8)}{11} = \frac{(-3g)+(-8)}{11}$

---

160. Complete these. Be sure to reduce to lowest terms whenever possible.

(a) $\frac{7}{16} - \frac{1}{16} = \_\_\_$ (c) $\frac{5q}{14} - \frac{12q}{14} = \_\_\_$

(b) $\frac{11}{8} - \frac{19}{8} = \_\_\_$ (d) $\frac{V}{6} - \frac{5}{6} = \_\_\_$

Answers to 159: a) $\frac{x+(-7)}{3}$ b) $\frac{y+5}{7}$ c) $\frac{2}{r}$ d) $\frac{14}{3a}$

Answers to 160: a) $\frac{3}{8}$ b) -1 c) $\frac{-1q}{2}$ or $\frac{-q}{2}$ or $-\frac{q}{2}$ d) $\frac{V+(-5)}{6}$

161. When a subtraction has been converted to addition, we cannot add unless we have identical denominators. Here is a subtraction with non-identical denominators, with one denominator a multiple of the other. Notice that we convert to addition before obtaining identical denominators:

Example: $\frac{5}{2} - \frac{3}{4} = \frac{5}{2} + \frac{(-3)}{4} = \frac{10}{4} + \frac{(-3)}{4} = \frac{7}{4}$

Complete: $\frac{1}{8} - \frac{1}{4} = \frac{1}{8} + \frac{(-1)}{4} =$ ______ + ______ = ______

Answer: $\frac{1}{8} + \frac{(-2)}{8} = \frac{-1}{8}$ or $-\frac{1}{8}$

162. Do these:

(a) $\frac{x}{2} - \frac{7}{8} =$ ______

(b) $\frac{5}{8} - \left(-\frac{3}{16}\right) =$ ______

(c) $\left(-\frac{7}{2}\right) - \frac{5}{8} =$ ______

Answer: a) $\frac{4x + (-7)}{8}$ b) $\frac{13}{16}$ c) $\frac{-33}{8}$ or $-\frac{33}{8}$

163. Do these:

(a) $\frac{5m}{16} - \frac{m}{2} =$ ______

(b) $\frac{4}{b} - \frac{5}{2b} =$ ______

Answer: a) $\frac{-3m}{16}$ or $-\frac{3m}{16}$ b) $\frac{3}{2b}$

164. In this example, neither denominator is a multiple of the other. After converting to addition, we find identical denominators in the usual way:

$$\frac{3}{5} - \frac{1}{3} = \frac{3}{5} + \frac{(-1)}{3} = \frac{3}{5}\left(\frac{3}{3}\right) + \frac{(-1)}{3}\left(\frac{5}{5}\right) = \frac{9}{15} + \frac{(-5)}{15} = \frac{4}{5}$$

Subtract:

(a) $\frac{1}{3} - \frac{1}{2} =$ ______

(b) $\left(-\frac{1}{7}\right) - \left(-\frac{1}{2}\right) =$ ______

Answer: a) $\frac{(-1)}{6}$ or $-\frac{1}{6}$ b) $\frac{5}{14}$

165. Subtract:

(a) $\frac{x}{3} - \frac{5}{7} =$ ______

(b) $\frac{9}{V} - \left(-\frac{3}{4}\right) =$ ______

Answer: a) $\frac{7x + (-15)}{21}$ b) $\frac{36 + 3V}{4V}$

166. The following subtraction includes a non-fraction. After converting to addition, we replace the non-fraction with a fraction, as shown:

Example: $4 - \frac{6}{5} = 4 + \frac{(-6)}{5} = \frac{20}{5} + \frac{(-6)}{5} = \frac{14}{5}$

Complete:

(a) $\frac{9}{2} - 1 = \frac{9}{2} + (-1) = (\quad) + (\quad) =$ ______

(b) $7 - \left(-\frac{8}{3}\right) =$ ______

167. Subtract: (a) $\left(-\frac{3}{4}\right) - 1 =$ ________ (b) $(-2) - \left(-\frac{5}{7}\right) =$ ________	a) $\frac{9}{2} + \frac{(-2)}{2} = \frac{7}{2}$ b) $\frac{29}{3}$
168. Do these: (a) $t - \frac{2}{3} =$ ________ (b) $5f - \frac{13}{7} =$ ________	a) $-\frac{7}{4}$ b) $-\frac{9}{7}$
	a) $\frac{3t + (-2)}{3}$ b) $\frac{35f + (-13)}{7}$

SELF-TEST 6 (Frames 148-168)

1. Which of the following are the opposite of $\frac{5}{8}$? (a) $\frac{-5}{-8}$ (b) $\frac{8}{5}$ (c) $-\frac{5}{8}$ (d) $-\frac{8}{5}$ (e) $\frac{-5}{8}$

Convert to addition: 2. $\frac{t}{3} - \frac{4t}{3} = \frac{t}{3} +$ ______ 3. $\left(-\frac{4}{x}\right) - \left(-\frac{5}{2x}\right) = \left(\frac{-4}{x}\right) +$ ______

Do the following subtraction problems:

4. $\frac{4}{9} - \frac{7}{9} =$ ________
5. $\frac{3}{2a} - \frac{1}{a} =$ ________
6. $\left(-\frac{5}{7}\right) - 2 =$ ________
7. $x - \left(-\frac{1}{5}\right) =$ ________
8. $1 - \frac{w}{3} =$ ________
9. $\left(-\frac{2}{h}\right) - \left(-\frac{2}{3h}\right) =$ ________

ANSWERS: 1. (c), (e) 2. $\frac{(-4t)}{3}$ 3. $\frac{5}{2x}$ 4. $-\frac{1}{3}$ 5. $\frac{1}{2a}$ 6. $-\frac{19}{7}$ or $-2\frac{5}{7}$ 7. $\frac{5x+1}{5}$ 8. $\frac{3 + (-w)}{3}$ 9 $\frac{-4}{3h}$ or $-\frac{4}{3h}$

5-12 SUBTRACTING MIXED NUMBERS

Just as we have converted any subtraction to addition, we will convert the subtraction of a mixed number to addition. Before doing so, however, we must discuss the opposite of a mixed number.

169. It is easy to show that two mixed numbers are a pair of opposites if the only difference between them is their signs. For example:

$3\frac{1}{2}$ and $-3\frac{1}{2}$ are a pair of opposites, since:

$$3\frac{1}{2} + (-3\frac{1}{2}) = 3 + \frac{1}{2} + (-3) + (-\frac{1}{2})$$

$$= 3 + (-3) + \frac{1}{2} + (-\frac{1}{2})$$

$$= 0 + 0 = 0$$

Write the opposite of: (a) $7\frac{3}{8}$ ______ (b) $-10\frac{7}{16}$ ______

---

a) $-7\frac{3}{8}$ b) $+10\frac{7}{16}$

170. Complete the following conversions to addition:

(a) $13\frac{7}{8} - 4\frac{5}{8} = 13\frac{7}{8} + (\text{the opposite of } 4\frac{5}{8}) = 13\frac{7}{8} + (\quad)$

(b) $37\frac{3}{4} - 31\frac{1}{4} = 37\frac{3}{4} + (\text{the opposite of } 31\frac{1}{4}) = 37\frac{3}{4} + (\quad)$

---

a) $-4\frac{5}{8}$

b) $-31\frac{1}{4}$

171. Having converted to addition, we proceed in the usual way. Notice that both parts of the second mixed number are negative.

$$39\frac{7}{16} - 28\frac{1}{16} = 39\frac{7}{16} + (-28\frac{1}{16}) = 39 + \frac{7}{16} + (-28) + (-\frac{1}{16})$$

$$= 39 + (-28) + \frac{7}{16} + (-\frac{1}{16}) = \_\_\_\_\_\_$$

$11\frac{3}{8}$ (from $11\frac{6}{16}$)

172. Do these: (a) $111\frac{15}{32} - 111\frac{7}{32} =$ ______

(b) $99\frac{7}{8} - 79\frac{5}{8} =$ ______

---

a) $\frac{1}{4}$

b) $20\frac{1}{4}$

173. Complete: $44\frac{3}{8} - 33\frac{6}{8} = 44\frac{3}{8} + (-33\frac{6}{8}) = 44 + (-33) + \frac{3}{8} + (-\frac{6}{8})$

$$= 11 + (-\frac{3}{8})$$

$$= \_\_\_\_\_\_$$

$10\frac{5}{8}$

174. Complete: $26\frac{3}{8} - 25\frac{7}{8} = 26\frac{3}{8} + (-25\frac{7}{8}) = \underbrace{26 + (-25)} + \underbrace{\frac{3}{8} + (-\frac{7}{8})}$ $= \quad 1 \quad + \quad (-\frac{4}{8})$ $=$ __________	
175. Do these: (a) $144\frac{1}{8} - 143\frac{7}{8} =$ __________ (b) $196\frac{5}{16} - 190\frac{11}{16} =$ __________ (c) $120\frac{1}{4} - 90\frac{3}{4} =$ __________	$\frac{1}{2}$ (Did you reduce to lowest terms?)
176. Complete this one in which the denominators are non-identical: $50\frac{3}{4} - 45\frac{5}{8} = 50\frac{3}{4} + (-45\frac{5}{8})$ $= 50 + (-45) + \frac{3}{4} + (-\frac{5}{8})$ $=$ __________	a) $\frac{1}{4}$ b) $5\frac{5}{8}$ c) $29\frac{1}{2}$
177. Do these: (a) $44\frac{1}{4} - 33\frac{1}{8} =$ __________ (b) $35\frac{5}{16} - 30\frac{7}{8} =$ __________	$5\frac{1}{8}$
178. Complete: (a) $17\frac{2}{5} - 13\frac{1}{4} =$ __________ (b) $25\frac{2}{5} - 20\frac{2}{3} =$ __________	a) $11\frac{1}{8}$ b) $4\frac{7}{16}$
	a) $4\frac{3}{20}$ b) $4\frac{11}{15}$

5-13 MULTIPLICATIONS AND DIVISIONS CONTAINING MIXED NUMBERS

In this section, we will briefly show the method for performing multiplications and divisions which contain mixed numbers. In each case, we will convert the mixed numbers to an improper fraction before performing the operation.

179. Here is a multiplication of a whole number and a mixed number: $3 \times 2\frac{1}{2}$

By converting the mixed number to an improper fraction, we obtain a multiplication of a whole number and a fraction: $3 \times \frac{5}{2} = \frac{15}{2}$

Now convert the product back to a mixed number: $\frac{15}{2} =$ ________

Answer: $7\frac{1}{2}$

180. Do these multiplications. Write each product as a mixed number:

(a) $10 \times 3\frac{3}{4} =$ ________

(b) $1\frac{5}{8} \times 2 =$ ________

Answers: a) $\frac{150}{4} = 37\frac{1}{2}$ b) $\frac{26}{8} = 3\frac{1}{4}$

181. If both factors are mixed numbers, each must be converted to an improper fraction before multiplying:

Example: $1\frac{3}{8} \times 2\frac{1}{4} = \frac{11}{8} \times \frac{9}{4} = \frac{99}{32} = 3\frac{3}{32}$

Do these multiplications of two mixed numbers:

(a) $3\frac{1}{2} \times 1\frac{3}{4} =$ ________ (b) $2\frac{1}{8} \times 4\frac{1}{2} =$ ________

Answers: a) $\frac{49}{8} = 6\frac{1}{8}$ b) $\frac{153}{16} = 9\frac{9}{16}$

182. Like any division, a division which contains a mixed number can be written as a fraction. That is:

$$3\frac{1}{2} \text{ divided by } 4 = \frac{3\frac{1}{2}}{4}$$

(a) 15 divided by $1\frac{7}{8} =$ ________ (b) $2\frac{1}{4}$ divided by $3\frac{1}{16} =$ ________

Answers: a) $\frac{15}{1\frac{7}{8}}$ b) $\frac{2\frac{1}{4}}{3\frac{1}{16}}$

183. To perform a division which contains a mixed number, we convert the mixed number to an improper fraction first. Then we convert the division to a multiplication:

Example: $\frac{3\frac{1}{2}}{4} = \frac{\frac{7}{2}}{4}$

$$= \frac{7}{2} \times \text{the reciprocal of } 4 = \frac{7}{2} \times \frac{1}{4} = \frac{7}{8}$$

Note: The answer cannot be converted back to a mixed number since the numerator is smaller than the denominator.

Do these divisions. If possible, write your answer as a mixed number:

(a) $\frac{10\frac{7}{8}}{2} =$ ________ (b) $\frac{50}{7\frac{1}{2}} =$ ________

184. In a division, if both the numerator and denominator are mixed numbers, <u>both</u> must be converted to improper fractions first:

<u>Example:</u> $\frac{5\frac{1}{8}}{4\frac{1}{2}} = \frac{\frac{41}{8}}{\frac{9}{2}} = \frac{41}{8} \times \frac{2}{9} = \frac{82}{72} = \frac{41}{36}$

Convert the answer to a mixed number: $\frac{41}{36}$ ____________

a) $\frac{87}{16} = 5\frac{7}{16}$

b) $\frac{100}{15} = 6\frac{2}{3}$

---

185. Do these. If possible write the answer as a mixed number:

(a) $\frac{10\frac{1}{2}}{2\frac{1}{4}} =$ ____________ (b) $\frac{1\frac{3}{4}}{3\frac{1}{8}} =$ ____________

$1\frac{5}{36}$

---

a) $4\frac{2}{3}$ (from $\frac{84}{18} = \frac{14}{3}$) b) $\frac{14}{25}$ (from $\frac{56}{100}$)

<u>SELF-TEST 7 (Frames 169-185)</u>

1. $82\frac{5}{16} - 59\frac{13}{16} =$ ____________
2. $118\frac{5}{8} - 117\frac{3}{4} =$ ____________
3. $5 \times 2\frac{7}{10} =$ ____________
4. $\frac{3\frac{3}{4}}{2\frac{5}{8}} =$ ____________
5. $\frac{2\frac{2}{3}}{12} =$ ____________

<u>ANSWERS:</u> 1. $22\frac{1}{2}$ 2. $\frac{7}{8}$ 3. $\frac{27}{2}$ or $13\frac{1}{2}$ 4. $\frac{10}{7}$ or $1\frac{3}{7}$ 5. $\frac{2}{9}$

5-14 IDENTIFYING TERMS IN COMPLEX EXPRESSIONS

In this section, you will learn how to evaluate complex expressions which contain fractions. Here are some examples:

$$3\left(\frac{1}{4}\right) + 5 \qquad 5\left(\frac{3}{8} - 7\right) - 2\left(\frac{3}{8}\right) \qquad \frac{3\left(\frac{1}{2}\right) + 5}{\frac{1}{2}} - \frac{2}{5}$$

A complex expression usually consists of a string of two or more terms connected by "+" and "-" signs. To evaluate a complex expression, we reduce each term to a single whole number or fraction <u>before</u> combining terms. Therefore, it is important that you be able to identify the terms. In this section, we will identify the types of terms which occur in these complex expressions.

186. Any whole number or simple fraction is a term. For example, there are three terms in the expression below:

$\frac{1}{8} + 5 - \frac{7}{8}$ The three terms are $\frac{1}{8}$, 5, and $\frac{7}{8}$.

Draw a box around each term in this expression: $7 - \frac{1}{5} + 9$

---

187. Any multiplication is a term, even if one of the factors is a fraction.

Just as 5(6) is one term, $2\left(\frac{5}{6}\right)$ is also one term.

Note: $2\left(\frac{5}{6}\right)$ is a multiplication, not a mixed number.

There are two terms in the following expression: $3\left(\frac{7}{8}\right) + 9$

The two terms are _______ and _______.

Answer: $\boxed{7} - \boxed{\frac{1}{5}} + \boxed{9}$

---

188. Draw a box around each term in this expression: $\frac{5}{7} - 3 + 2\left(\frac{1}{6}\right)$

Answer: $3\left(\frac{7}{8}\right)$ and 9

---

189. Any grouping is a term, even if the grouping contains a fraction. Therefore, there are two terms in the following expression:

$3\left(\frac{1}{8}\right) - \left(2 - \frac{1}{8}\right)$ The two terms are $3\left(\frac{1}{8}\right)$ and $\left(2 - \frac{1}{8}\right)$.

Draw a box around each term in these expressions:

(a) $\frac{7}{6} - \left(\frac{1}{6} + 4\right) - 9$ (b) $10 - 3\left(\frac{1}{8}\right) - \left[\left(-\frac{5}{8}\right) - 2\right]$

Answer: $\boxed{\frac{5}{7}} - \boxed{3} + \boxed{2\left(\frac{1}{6}\right)}$

---

190. Any instance of the distributive principle is one term, even if the grouping contains a fraction. For example:

$5\left(\frac{3}{4} + 1\right)$ is one term. $7\left[3 - \left(-\frac{9}{8}\right)\right]$ is one term.

Draw a box around each term in these expressions:

(a) $10 - 3\left(\frac{5}{6} - 4\right)$ (b) $2\left[7 + \left(-\frac{1}{8}\right)\right] - \left(\frac{1}{8} - 1\right)$

Answers:

a) $\boxed{\frac{7}{6}} - \boxed{\left(\frac{1}{6} + 4\right)} - \boxed{9}$

b) $\boxed{10} - \boxed{3\left(\frac{1}{8}\right)} - \boxed{\left[\left(-\frac{5}{8}\right) - 2\right]}$

---

191. Any single fraction is one term, no matter how complex its numerator and denominator. All of the following fractions, for example, are one term:

$$\frac{\frac{1}{5}}{6} \qquad \frac{\frac{3}{8} + 5}{7} \qquad \frac{2\left(\frac{1}{9}\right) - 5}{\frac{1}{9}} \qquad \frac{3\left(1 - \frac{2}{5}\right)}{4\left(\frac{2}{5}\right) - 9}$$

A single fraction is identified by the fact that it contains one major fraction line. In the examples above, notice that simple fractions can be part of the numerator or denominator of a single fraction.

(Continued on following page.)

Answers:

a) $\boxed{10} - \boxed{3\left(\frac{5}{6} - 4\right)}$

b) $\boxed{2\left[7 + \left(-\frac{1}{8}\right)\right]} - \boxed{\left(\frac{1}{8} - 1\right)}$

191. (Continued)

Draw a box around each term in these expressions:

(a) $7\left(\frac{2}{9}\right) - \frac{\frac{2}{9}}{6}$ (b) $\frac{\frac{4}{7}+6}{9} - 5\left(\frac{4}{7}-3\right)$

---

192. Draw a box around each term in these expressions:

(a) $\frac{9}{\frac{1}{5}} - \frac{\frac{1}{5}+7}{8}$ (b) $\frac{5\left(\frac{3}{8}\right)+1}{2\left(\frac{3}{8}\right)-9} - 9$

a) $\boxed{7\left(\frac{2}{9}\right)} - \boxed{\frac{\frac{2}{9}}{6}}$

b) $\boxed{\frac{\frac{4}{7}+6}{9}} - \boxed{5\left(\frac{4}{7}-3\right)}$

---

193. Draw a box around each term on both sides of the following sentences:

(a) $\frac{3\left(-\frac{7}{6}\right)-5}{\left(-\frac{7}{6}\right)+5} - \frac{1}{6} = 5 - \frac{1}{\left(-\frac{7}{6}\right)+5}$

(b) $3\left(\frac{1}{8}\right) - \frac{\frac{1}{8}}{7} = 8 + \frac{7\left[\left(-\frac{1}{8}\right)-1\right]}{\frac{1}{8}}$

a) $\boxed{\frac{9}{\frac{1}{5}}} - \boxed{\frac{\frac{1}{5}+7}{8}}$

b) $\boxed{\frac{5\left(\frac{3}{8}\right)+1}{2\left(\frac{3}{8}\right)-9}} - \boxed{9}$

Answers to Frame 193: a) $\boxed{\frac{3\left(-\frac{7}{6}\right)-5}{\left(-\frac{7}{6}\right)+5}} - \boxed{\frac{1}{6}} = \boxed{5} - \boxed{\frac{1}{\left(-\frac{7}{6}\right)+5}}$ b) $\boxed{3\left(\frac{1}{8}\right)} - \boxed{\frac{\frac{1}{8}}{7}} = \boxed{8} + \boxed{\frac{7\left[\left(-\frac{1}{8}\right)-1\right]}{\frac{1}{8}}}$

5-15 EVALUATING COMPLEX EXPRESSIONS WHICH CONTAIN FRACTIONS

To evaluate a complex expression, we reduce each term to a whole number or simple fraction before combining terms. In the process of doing so, we use some combination of adding, subtracting, multiplying and dividing fractions. We will show the method in this section.

---

194. The first step in evaluating a complex expression is to reduce each term to a whole number or a simple fraction. We will begin by showing the types of reductions which occur.

In each of the following terms, a multiplication is required:

$$3\left(\frac{5}{8}\right) \qquad 5\left(-\frac{4}{3}\right)$$

Each term can be reduced to a simple fraction by performing the multiplication.

Just as $3\left(\frac{5}{8}\right) = \frac{15}{8}$, $5\left(-\frac{4}{3}\right) =$ ______

195. By reducing a term to a simple fraction, we mean reducing it to a proper or improper fraction. Mixed numbers are not used in this context because they are too clumsy.

Any grouping is a term. It can be reduced to a whole number or simple fraction by performing the addition or subtraction within the grouping. Reduce each of the following:

(a) $\left(3 - \frac{9}{8}\right) =$ __________ (c) $\left[\left(-\frac{8}{3}\right) - 2\right] =$ __________

(b) $\left[1 - \left(-\frac{2}{5}\right)\right] =$ __________

$-\frac{20}{3}$

---

196. If a term is an instance of the distributive principle, we simplify the grouping before performing the multiplication:

Example: $3\left(\frac{4}{5} + 1\right)$

$3\left(\frac{9}{5}\right) = \frac{27}{5}$

Complete this one: $2\left[\left(-\frac{7}{4}\right) + 1\right] =$

$2\left(\quad\right)$ __________

a) $\frac{15}{8}$ b) $\frac{7}{5}$ c) $-\frac{14}{3}$

---

197. Reduce each of these to a whole number or simple fraction:

(a) $5\left(\frac{6}{5} - 1\right) =$ __________ (b) $2\left[\left(-\frac{7}{3}\right) - 2\right] =$ __________

$2\left(-\frac{3}{4}\right) = -\frac{6}{4} = -\frac{3}{2}$

---

198. To evaluate the following expression, we reduce each term to a whole number or simple fraction first. Then we can proceed as usual:

$3\left(\frac{5}{8}\right) - 1$

$\frac{15}{8} - 1 =$ __________

a) 1 (from $\frac{5}{5}$)

b) $-\frac{26}{3}$

---

199. Complete this simplification: $3\left(\frac{5}{4}\right) - \left(1 - \frac{5}{4}\right)$

$\left(\quad\right) - \left(\quad\right) =$ __________

$\frac{7}{8}$

---

200. Complete this one: $5\left(\frac{1}{4} + 3\right) - 3\left(\frac{1}{4}\right)$

$5\left(\frac{13}{4}\right)$

$\left(\quad\right) - \left(\quad\right) =$ __________

$\frac{15}{4} - \left(-\frac{1}{4}\right) = \frac{15}{4} + \frac{1}{4}$

$= \frac{16}{4} = 4$

$\frac{65}{4} - \frac{3}{4} = \frac{62}{4} = \frac{31}{2}$

201. Complete this one: $\frac{5}{8} + 2\left(1 - \frac{11}{8}\right)$

$$(\quad) + (\quad) = \_\_\_\_\_\_$$

202. Let's check to see whether the following equation is true:

$$3 - 5\left(\frac{4}{7}\right) = 2\left(\frac{4}{7} - 1\right) + 1$$

(a) Evaluate the left side: ____ (b) Evaluate the right side: ____

(c) Is the equation true or false? ________

Answer to 201: $\frac{5}{8} + \left(-\frac{6}{8}\right) = -\frac{1}{8}$

203. Evaluate each side to see whether this equation is true:

$$3\left(-\frac{1}{2}\right) - \left[\left(-\frac{1}{2}\right) - 2\right] = \left(-\frac{1}{2}\right) + 5\left[1 - \left(-\frac{1}{2}\right)\right]$$

(a) The left side equals ____. (b) The right side equals ____.

(c) Is the equation true or false? ________

Answer to 202: a) $\frac{1}{7}$ b) $\frac{1}{7}$

c) True, since: $\frac{1}{7} = \frac{1}{7}$

204. In the following term, the numerator of the fraction is itself a fraction. We can reduce this complex fraction to a simple fraction by performing the division. To do so, we convert to multiplication first:

$$\frac{\frac{1}{2}}{6} = \frac{1}{2}(\text{the reciprocal of } 6) = \frac{1}{2}\left(\frac{1}{6}\right) = \frac{1}{12}$$

Complete: $\frac{2}{-\frac{3}{4}} = 2(\text{the reciprocal of } -\frac{3}{4}) = 2(\quad) = \_\_\_\_\_$

Answer to 203: a) +1 b) +7

c) False, since: $1 \neq 7$

205. In this term, a multiplication is called for in the numerator. We must perform this multiplication before reducing to a simple fraction:

$$\frac{3\left(\frac{2}{5}\right)}{7} = \frac{\frac{6}{5}}{7} = \frac{6}{5}\left(\frac{1}{7}\right) = \frac{6}{35}$$

Reduce each of these to a simple fraction:

(a) $\frac{5\left(-\frac{1}{3}\right)}{2} = \_\_\_\_\_\_$ (b) $\frac{1}{3\left(\frac{7}{8}\right)} = \_\_\_\_\_\_$

Answer to 204: $2\left(-\frac{4}{3}\right) = -\frac{8}{3}$

206. Evaluate this expression: $\dfrac{\frac{1}{3}}{5} - 7\left(\frac{1}{3}\right)$

$(\quad) - (\quad) =$ ______

Answer to previous frame: a) $-\frac{5}{6}$ b) $\frac{8}{21}$

---

207. Evaluate this expression: $\dfrac{\frac{3}{4}}{3} - \dfrac{2\left(\frac{4}{3}\right)}{3}$

$(\quad) - (\quad) =$ ______

Answer to 206: $\frac{1}{15} - \frac{7}{3} = -\frac{34}{15}$

---

208. <u>We always simplify the numerator and denominator of a complex fraction first.</u> Then, if the numerator or denominator is still a fraction, we use one more step to complete the simplification of the term.

In the following example, when the numerator and denominator are simplified, the term is simplified:

$$\frac{3(-4) + 5}{(-4) + 7} = \frac{-7}{3} \text{ or } -\frac{7}{3}$$

In the following example, when the numerator and denominator are simplified, we use one more step to simplify the term:

$$\frac{3\left(\frac{1}{4} + 1\right)}{\frac{1}{4} - 1} = \frac{\frac{15}{4}}{-\frac{3}{4}} = \frac{15}{4}\left(-\frac{4}{3}\right) = -\frac{15}{3} = -5$$

Simplify each of these complex fractions:

(a) $\dfrac{2\left(-\frac{3}{5}\right) + 1}{3} =$ ______ (b) $\dfrac{-\frac{3}{2}}{4\left[\left(-\frac{3}{2}\right) - 3\right]} =$ ______

Answer to 207: $\frac{9}{4} - \frac{8}{9} = \frac{49}{36}$

---

209. Simplify each of these terms:

(a) $\dfrac{\left(-\frac{6}{5}\right) + 2}{3\left(-\frac{6}{5}\right) + 1} =$ ______

(b) $\dfrac{1 - 2\left(\frac{5}{2}\right)}{3\left(2 - \frac{5}{2}\right)} =$ ______

Answer to 208: a) $\dfrac{-\frac{1}{5}}{3} = -\frac{1}{15}$ b) $\dfrac{-\frac{3}{2}}{-18} = +\frac{1}{12}$

210. Evaluate this expression: $\dfrac{5\left(\frac{1}{2}\right)}{3} - \dfrac{3\left(\frac{1}{2}\right) - 5}{\frac{1}{2} + 2}$

$(\quad) - (\quad) =$ ________

a) $-\frac{4}{13}$ (from $\dfrac{\frac{4}{5}}{-\frac{13}{5}}$)

b) $+\frac{8}{3}$ (from $\dfrac{-4}{-\frac{3}{2}}$)

---

211. Evaluate this expression: $\dfrac{5\left[\left(-\frac{3}{4}\right) + 1\right]}{\left(-\frac{3}{4}\right)} + 3\left(-\frac{3}{4}\right)$

$(\quad) + (\quad) =$ ________

$\frac{5}{6} - \left(-\frac{7}{5}\right) = \frac{67}{30}$

---

212. Let's check to see whether the following equation is true:

$$\frac{1}{5} - \frac{3\left(\frac{1}{5}\right) - 1}{5} = \frac{\frac{1}{5}}{2} - \frac{4\left(\frac{1}{5}\right) - 5}{2}$$

(a) The left side equals ____. (b) The right side equals ____.

(c) Is the equation true? ________

$\left(-\frac{5}{3}\right) + \left(-\frac{9}{4}\right) = -\frac{47}{12}$

---

213. Let's check to see whether the following equation is true:

$$3\left(\frac{4}{3}\right) + \frac{1}{2\left(\frac{4}{3}\right) - 1} = 1 - \frac{3\left(1 - \frac{4}{3}\right)}{2}$$

(a) The left side equals ____. (b) The right side equals ____.

(c) Is the equation true? ________

a) $+\frac{7}{25}$

b) $+\frac{11}{5}$

c) No.

---

a) $\frac{23}{5}$ (or $\frac{46}{10}$) b) $\frac{3}{2}$ (or $\frac{15}{10}$) c) No, since: $\frac{23}{5} \neq \frac{3}{2}$

SELF-TEST 8 (Frames 186-213)

Draw a box around each term in these expressions:

1. $7 + \frac{2}{3} - 3\left(\frac{5}{3} + 6\right)$

2. $\frac{3 - \frac{1}{2}}{4} + 5\left(\frac{3}{4}\right)$

Find the numerical value of each:

3. $2\left(3 - \frac{2}{3}\right) =$ ______

4. $\frac{\frac{3}{5}}{2} - 3\left(\frac{5}{2}\right) =$ ______

Evaluate each of the following expressions:

5. $1 - \frac{2 - \frac{9}{4}}{\frac{9}{4}} =$ ______

6. $\frac{\frac{4}{5} - 2}{3} - \frac{3\left(\frac{4}{5}\right)}{2} =$ ______

7. Refer to the equation at the right.

$$2\left(\frac{3}{11}\right) - \frac{\frac{3}{11} - 1}{3} = 1 - \frac{\frac{3}{11} + 1}{6}$$

   (a) The numerical value of the left side is ______.

   (b) The numerical value of the right side is ______.

   (c) Is the equation true? ______

ANSWERS:

1. $\boxed{7} + \boxed{\frac{2}{3}} - \boxed{3\left(\frac{5}{3} + 6\right)}$

2. $\boxed{\frac{3 - \frac{1}{2}}{4}} + \boxed{5\left(\frac{3}{4}\right)}$

3. $\frac{14}{3}$

4. $-\frac{36}{5}$

5. $\frac{10}{9}$

6. $-\frac{8}{5}$

7. (a) $\frac{26}{33}$
   (b) $\frac{26}{33}$
   (c) Yes.

5-16 CHECKING FRACTIONAL ROOTS IN NON-FRACTIONAL EQUATIONS

In all the solutions of equations we did in earlier chapters, the roots were always whole numbers. In the next chapter, we will solve equations which have fractional roots. Before doing so, we will show how to check fractional roots in non-fractional equations.

214. Before checking fractional roots, we will review the types of terms which occur in non-fractional equations. They are:

(1) Any whole number: like 8 or -3.

(2) Any letter with or without a numerical coefficient: like x or p, 3t or 5m.

(3) Any grouping: like (x + 2) or (3d - 5).

(4) Any instance of the distributive principle: like 4(y + 7) or 7(2S - 1).

Draw a box around each term in each of the following equations:

(a) $10 - 3m = m - (2 - m)$

(b) $1 = 3x - 3(x - 5)$

(c) $10(1 - y) - (y - 1) = y$

Answers to 214:

a) $\boxed{10} - \boxed{3m} = \boxed{m} - \boxed{(2 - m)}$

b) $\boxed{1} = \boxed{3x} - \boxed{3(x - 5)}$

c) $\boxed{10(1 - y)} - \boxed{(y - 1)} = \boxed{y}$

215. To check a fractional root in an equation, we:

(1) Plug in the root for the letter.
(2) Reduce each term to a whole number or simple fraction.
(3) Then combine numerical terms on each side to see whether the numerical value of the left side is equal to the numerical value of the right side.

Example: Let's check to see whether $\frac{8}{3}$ is the root of this equation:

$$1 = 2Q - (7 - Q)$$

(1) Plugging in $\frac{8}{3}$ for Q, we get: $1 = 2\left(\frac{8}{3}\right) - \left(7 - \frac{8}{3}\right)$

(2) Reducing each term to a whole number or simple fraction, we get: $1 = \frac{16}{3} - \left(\frac{13}{3}\right)$

(3) Combining terms on each side, we get: $1 = 1$

(a) Does $\frac{8}{3}$ satisfy this equation? _______

(b) Is $\frac{8}{3}$ the root of this equation? _______

Answers to 215:

a) Yes.

b) Yes.

216. Let's check to see whether $\frac{9}{8}$ is the root of this equation:

$$3W = 6 + 5(1 - W)$$

(1) Plugging in $\frac{9}{8}$ for W, we get: $3\left(\frac{9}{8}\right) = 6 + 5\left(1 - \frac{9}{8}\right)$

(2) Reducing each term, we get: $\frac{27}{8} = 6 + \left(-\frac{5}{8}\right)$

(3) Combining terms, we get: $\frac{27}{8} = \frac{43}{8}$

Is $\frac{9}{8}$ the root of the equation? _______

217. Plug in $\left(-\frac{3}{2}\right)$ for S in the equation below:

$$6S = 7 - (19 + 2S)$$

(a) The left side equals _____. (b) The right side equals _____.

(c) Is $-\frac{3}{2}$ the root of the equation? _______

No. It does not satisfy the equation, since the left side equals $\frac{27}{8}$ and the right side equals $\frac{43}{8}$.

---

218. Plug in $\frac{1}{2}$ for y in the following equation:

$$4y + 9 = 9 - 2y$$

(a) The left side equals _____. (b) The right side equals _____.

(c) Is $\frac{1}{2}$ the root of the equation? _______

a) -9

b) -9 (from 7 - 16)

c) Yes.

---

219. Is $-\frac{3}{5}$ the root of the equation at the right? _______

$$6F - (4 + F) = -7$$

a) 11

b) 8

c) No, since: $11 \neq 8$

---

220. Is $\frac{2}{5}$ the root of this equation? _______

$$2W + 3(2 - 4W) = 3$$

Yes, since the right side is -7, and the left side reduces to -7:

$$\left(-\frac{18}{5}\right) - \left(\frac{17}{5}\right) = -\frac{35}{5} = -7$$

No, since the left side reduces to 2, and $2 \neq 3$.

5-17 CHECKING ROOTS IN FRACTIONAL EQUATIONS

In the next chapter, we will solve equations like the following:

$$\frac{4x}{5} - 3 = \frac{x}{2} \qquad \frac{t+3}{5t} = \frac{3}{10t} + 4$$

Equations of this type are called "fractional" equations. They can have either whole numbers or fractions as their roots. In this section, we will check both types of roots in fractional equations.

221. Before checking roots in fractional equations, we must be able to identify the terms in them.

As we saw earlier, any fraction is one term, no matter how complex its numerator or denominator. This is also true when letters appear in the fraction. All of the following fractions, for example, are <u>one term</u>:

$$\frac{3}{8} \qquad \frac{2x}{7} \qquad \frac{5}{3t} \qquad \frac{3m-1}{9} \qquad \frac{5(y+1)}{y-1}$$

Draw a box around each term on both sides of the following equations:

(a) $\frac{1}{3} - \frac{2}{R} = \frac{5}{R} - \frac{5}{6}$ (b) $\frac{2V}{3} - 5 = \frac{V}{2}$

---

222. Draw a box around each term in these equations:

(a) $\frac{2}{P} = \frac{2}{3(P-7)}$ (b) $\frac{T-1}{T} = \frac{1}{5}$

Answer (221):
a) $\boxed{\frac{1}{3}} - \boxed{\frac{2}{R}} = \boxed{\frac{5}{R}} - \boxed{\frac{5}{6}}$
b) $\boxed{\frac{2V}{3}} - \boxed{5} = \boxed{\frac{V}{2}}$

---

223. Draw a box around each term in these equations:

(a) $1 + \frac{R-2}{4(R-3)} = \frac{3}{4}$ (b) $\frac{7}{8} = 5 - \frac{2(x-7)}{8x}$

Answer (222):
a) $\boxed{\frac{2}{P}} = \boxed{\frac{2}{3(P-7)}}$
b) $\boxed{\frac{T-1}{T}} = \boxed{\frac{1}{5}}$

---

224. To check a root in a fractional equation, we use the same procedure that is used for any equation.

Let's check to see whether "5" is the root of this equation: $\frac{45}{3t} + 2 = 5$

Plugging in 5 for t, we get: $\frac{45}{3(5)} + 2 = 5$

Reducing the fractional term, we get: $3 + 2 = 5$

Is "5" the root? _______

Answer (223):
a) $\boxed{1} + \boxed{\frac{R-2}{4(R-3)}} = \boxed{\frac{3}{4}}$
b) $\boxed{\frac{7}{8}} = \boxed{5} - \boxed{\frac{2(x-7)}{8x}}$

---

225. Let's check to see whether $\frac{25}{6}$ is the root of this equation: $2 = \frac{6L}{5} - 3$

Plugging in $\frac{25}{6}$ for L, we get: $2 = \frac{6\left(\frac{25}{6}\right)}{5} - 3$

Reducing the fractional term, we get: $2 = 5 - 3$

Is $\frac{25}{6}$ the root? _______

Answer (224): Yes, since both sides equal "5".

---

226. Is $\frac{24}{7}$ the root of this equation? _______ $\frac{7Q}{3} = 8$

Answer (225): Yes, since both sides equal "2".

---

227. Is -5 the root of this one? _______ $7 = \frac{x}{3} + 9$

Answer (226): Yes.

228. Does $f = 12$ satisfy this equation? ______	$\frac{3f}{9(f - 4)} = \frac{1}{2}$	No, because $7 \neq \frac{22}{3}$.
229. Does $Q = -1$ satisfy this equation? ______	$\frac{Q}{4} - 1 = \frac{3Q}{5} + \frac{2Q}{3}$	Yes, since both sides equal $\frac{1}{2}$.
230. Is $t = \frac{7}{12}$ the root of this equation? ______	$\frac{t}{2} - \frac{1}{3} = \frac{5t - 3}{2}$	No, since: $-\frac{5}{4} \neq -\frac{19}{15}$
		Yes, since both sides equal $-\frac{1}{24}$.

SELF-TEST 9 (Frames 214-230)

1. (a) If $x = \frac{5}{2}$, what is the numerical value of the left side of the equation shown at the right? ______

   $2x + 4(3 - x) = 7$

   (b) Is $x = \frac{5}{2}$ a root of the equation? ______

2. (a) If $t = \frac{1}{2}$, what is the numerical value of the right side of the equation? ______

   $\frac{3}{2} = 1 - \frac{t + 1}{2t}$

   (b) Is $t = \frac{1}{2}$ a root of the equation? ______

3. (a) If $h = -\frac{3}{5}$, what is the numerical value of the left side of the equation? ______

   $1 - \frac{h}{3} = 2 + \frac{h - 1}{2}$

   (b) If $h = -\frac{3}{5}$, what is the numerical value of the right side of the equation? ______

   (c) Is $h = -\frac{3}{5}$ a root of the equation? ______

ANSWERS: 1. (a) 7 (b) Yes 2. (a) $-\frac{1}{2}$ (b) No 3. (a) $\frac{6}{5}$ (b) $\frac{6}{5}$ (c) Yes

Chapter 6 FRACTIONAL EQUATIONS

In the non-fractional equations solved in Chapters 2 and 3, all of the roots were whole numbers. Many non-fractional equations of the same types, however, have fractions as their roots. In the first part of this chapter, equations of the latter type will be solved.

The major part of this chapter will discuss methods for solving various types of fractional equations. A fractional equation is an equation containing fractions in which a letter appears in either the numerator or denominator. Examples are:

$$\frac{15}{2x} = 4 \qquad 2h = 3 - \frac{4h}{5} \qquad \frac{7}{y} = \frac{10}{y - 2} \qquad \frac{t}{8} - \frac{t}{2} = \frac{1}{4} - t$$

You will see that the roots of fractional equations can also be either whole numbers or fractions.

6-1 THE MULTIPLICATION AXIOM FOR EQUATIONS

To solve a complex equation, we apply axioms and principles to obtain simpler but equivalent equations. The last step is always "solving an instance of the basic equation."

Up to this point in the course, we have solved these basic equations by trial-and-error. Using trial-and-error with equations like the following has been possible because the roots have always been whole numbers:

$$2x = 16 \qquad 5r = -20$$

But many instances of the basic equation do not have whole numbers for their roots. Here are some examples:

$$7x = 9 \qquad 4w = -13$$

Since it is difficult to solve equations of this type by trial-and-error, we will introduce another method. This method involves the use of a new principle called the "multiplication axiom for equations."

1. We want to show that we obtain an equivalent equation if we multiply both sides by the same quantity. We will use the following equation as an example:

$$2x = 10$$

The root of this equation is "5".

(a) If we multiply both sides of the original equation by "2", we get the new equation:

$$2(2x) = 2(10)$$

or $4x = 20$

What is the root of this new equation? _______

(b) If we multiply both sides of the original equation by "5", we get the new equation:

$$5(2x) = 5(10)$$

or $10x = 50$

What is the root of this new equation? _______

(Continued on following page.)

1. (Continued) $2x = 10$

(c) If we multiply both sides of the original equation by "-3", we get the new equation:

$$-3(2x) = -3(10)$$
$$-6x = -30$$

What is the root of this new equation? _______

---

2. In the last frame, the original equation was: $2x = 10$

By multiplying both sides by "2", "5", and "-3", respectively, we obtained the new equations shown at the right:

$$4x = 20$$
$$10x = 50$$
$$-6x = -30$$

Since these three new equations still have "5" as their root, we say they are _______________ to the original equation.

Answers: a) 5 b) 5 c) 5

---

3. Let's try the same procedure with a more complex equation:

$$x + 2 = 6 \quad \text{(The root is "4".)}$$

(a) Let's multiply both sides of the original equation by "3". Notice that we multiply the <u>entire left side</u> by "3". We get:

$$3(x + 2) = 3(6)$$
$$\text{or} \quad 3x + 6 = 18$$

What is the root of this new equation? _______

(b) Let's multiply both sides of the original equation by "10". We get:

$$10(x + 2) = 10(6)$$
$$\text{or} \quad 10x + 20 = 60$$

The root of this new equation is _______.

(c) Let's multiply both sides of the original equation by "-2". We get:

$$(-2)(x + 2) = (-2)(6)$$
$$\text{or} \quad (-2x) + (-4) = -12$$

Is "4" the root of this equation? _______

Answer: equivalent

Answers:

a) 4

b) 4

c) Yes, since:

$(-8) + (-4) = -12$

4. In Frames 1 to 3, both sides of an equation were multiplied by the same quantity. This multiplication process is an important and useful principle in solving equations. This principle is called the <u>MULTIPLICATION AXIOM FOR EQUATIONS</u>:

IF YOU MULTIPLY BOTH SIDES OF AN EQUATION BY THE SAME QUANTITY, THE NEW EQUATION IS EQUIVALENT BECAUSE IT HAS THE <u>SAME ROOT</u>.

Any equation having this form: 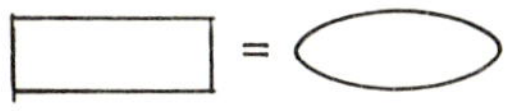

has the same root as one having this form: 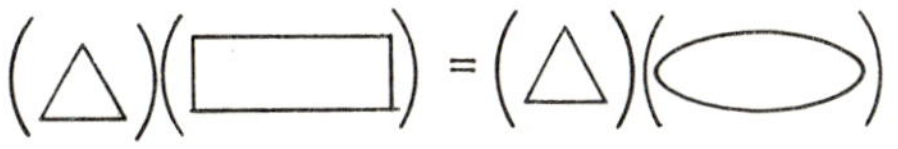

(The two triangles show symbolically that both sides of the equation are multiplied by the same quantity.)

6-2 USING THE MULTIPLICATION AXIOM TO SOLVE BASIC EQUATIONS

5. As we saw earlier, some basic equations are called <u>root equations</u> because their roots appear in the equation. This is true only when the coefficient of the letter is "1". Here are two examples:

$1x = 5$ and $1m = -7$

Usually a "1" coefficient is not explicitly written. Instead we write:

$x = 5$ and $m = -7$

The following basic equation is not a root equation because the coefficient of the letter is not "1": $3x = 21$ We can solve this equation by trial-and-error and obtain the root equation: $x = 7$

However, we can also use the multiplication axiom to solve $3x = 21$ and obtain the root equation. To do so, we multiply both sides by "$\frac{1}{3}$". We get:

$$3x = 21$$

$$\frac{1}{3}(3x) = \frac{1}{3}(21)$$

$$\frac{1}{3}(3)(x) = \frac{1}{3}(21)$$

$$(1)(x) = \frac{21}{3}$$

$$x = 7$$

To obtain "1x" or "x", we eliminated the factor "3" by multiplying both sides by "$\frac{1}{3}$", the ______________ of "3".

---

reciprocal

6. Another basic equation is shown at the right: $7t = 56$

Solving it means finding its root equation. To do so, we must eliminate the "7" which is the coefficient of "t". We can eliminate the "7" by multiplying both sides by "$\frac{1}{7}$", the reciprocal of 7. Do so, showing all steps:

---

7. To solve a basic equation, we must eliminate the coefficient of the letter. We can do so by using the multiplication axiom. We multiply both sides by the <u>reciprocal of the coefficient.</u>

<u>Problem:</u> $-4M = 36$

(a) The coefficient of M is _______.

(b) The reciprocal of this coefficient is _______.

(c) Solve by means of the multiplication axiom: M = _______

<u>Solution:</u>

$$7t = 56$$
$$\frac{1}{7}(7t) = \frac{1}{7}(56)$$
$$\frac{1}{7}(7)(t) = \frac{56}{7}$$
$$(1)(t) = 8$$
$$t = 8$$

---

8. When the root of a basic equation is a whole number, we can solve it by trial-and-error. We do not need the multiplication axiom in such cases.

However, roots are usually not whole numbers, not even for basic equations. When the root is a fraction, the multiplication axiom should be used. Here is an example:

$$7x = 8$$

We can solve this one by multiplying both sides by $\frac{1}{7}$.

$$\frac{1}{7}(7x) = \frac{1}{7}(8)$$
$$(1)(x) = \frac{8}{7}$$
$$x = \frac{8}{7}$$

<u>Check:</u>

Check to see whether $\frac{8}{7}$ satisfies the original equation.

a) -4

b) $-\frac{1}{4}$

c) <u>Solution:</u> $-4M = 36$

$$\left(-\frac{1}{4}\right)(-4M) = \left(-\frac{1}{4}\right)36$$
$$(1)(M) = -\frac{36}{4}$$
$$M = -9$$

---

9. Use the multiplication axiom to solve each of these equations:

(a) $5x = 4$ (b) $52 = 47R$ (c) $9t = -5$

x = _____ R = _____ t = _____

It does, since:

$$7x = 8$$
$$7\left(\frac{8}{7}\right) = 8$$
$$8 = 8$$

---

a) $x = \frac{4}{5}$

b) $R = \frac{52}{47}$

c) $t = -\frac{5}{9}$

10. Solve each. Be sure your root is reduced to lowest terms:

(a) $-36 = 49S$ (b) $9m = 24$ (c) $-11 = 99t$

S = ______ m = ______ t = ______

---

11. Earlier we introduced the "oppositing principle for equations." This principle states that we get an equivalent equation if we replace each side by its opposite.

For example: $-5x = 15$ and $5x = -15$ are equivalent equations since the root of each is "-3".

Though applying the oppositing principle is not necessary, it is useful for some students when the coefficient of the letter is negative.

Here is an example which can be solved as it stands.

$-5x = 30$ The root is -6.

(a) Apply the oppositing axiom to the equation above: ______ = ______

(b) The root of this new equation is also ______.

Answers to 10:

a) $S = -\frac{36}{49}$

b) $m = \frac{8}{3}$

c) $t = -\frac{1}{9}$

---

12. In the last frame, we saw that we could solve "$-5x = 30$" as it stands or after applying the oppositing principle. These two procedures are possible for any basic equation, even when the root is a fraction.

Refer to the equation at the right. We will solve it first as it stands.

$$-3y = 8$$

$$\left(-\frac{1}{3}\right)(-3y) = -\frac{1}{3}(8)$$

$$(+1)y = -\frac{8}{3}$$

$$y = -\frac{8}{3}$$

(a) Apply the oppositing principle to "$-3y = 8$": ______ = ______

(b) Now solve by means of the multiplication axiom:

y = ______

(c) Do you obtain the same root using this method? ______

Answers to 11:

a) $5x = -30$

b) -6

13. When the coefficient of the letter in a basic equation is negative, you can use the oppositing principle or not. Use the method in which you make the least mistakes.

Solve: (a) $-10t = 90$ (b) $-23m = 17$

$t =$ ______ $m =$ ______

a) $3y = -8$

b) $\frac{1}{3}(3y) = \frac{1}{3}(-8)$

$1\ y = -\frac{8}{3}$

$y = -\frac{8}{3}$

c) Yes.

---

14. Solve: (a) $-v = 13$ (b) $-7S = 2$ (c) $-8 = 5x$

$v =$ ______ $S =$ ______ $x =$ ______

a) $t = -9$

b) $m = -\frac{17}{23}$

---

15. Solve. Be sure to reduce the root to lowest terms:

(a) $-10t = 25$ (b) $36 = -48R$

$t =$ ______ $R =$ ______

a) $v = -13$

b) $S = -\frac{2}{7}$

c) $x = -\frac{8}{5}$

---

a) $t = -\frac{5}{2}$

b) $R = -\frac{3}{4}$

6-3 BASIC EQUATIONS WITH "0" OR "1" AS ONE SIDE

Students are sometimes confused when one side of a basic equation is "0" or "1". Such equations are easy to solve by means of the multiplication axiom. We will solve equations of this type in this section.

---

16. Solve by means of the multiplication axiom: $3x = 1$

$x =$ ______

---

17. Solve this equation: $-4d = 1$

$d =$ ______

$x = \frac{1}{3}$, since:

$\frac{1}{3}(3x) = \frac{1}{3}(1)$

$1x = \frac{1}{3}$ $x = \frac{1}{3}$

18. Solve this one: $1 = -7m$ $m =$ ______	$d = -\frac{1}{4}$, since: $\left(-\frac{1}{4}\right)(-4d) = -\frac{1}{4}(1)$ $(1)\ d = -\frac{1}{4}$ $d = -\frac{1}{4}$
19. (a) Solve by trial-and-error: $2y = 0$ "y" must be ______. (b) Solve by means of the multiplication axiom: $2y = 0$ $y =$ ______	$m = -\frac{1}{7}$
20. (a) Solve by trial-and-error: $-4F = 0$ $F =$ ______ (b) Solve by means of the multiplication axiom: $-4F = 0$ $F =$ ______	a) 0 b) $\frac{1}{2}(2y) = \frac{1}{2}(0)$ $(1)(y) = 0$ $y = 0$
21. Solve each of these: (a) $0 = -5t$ (b) $1 = -5t$ $t =$ ______ $t =$ ______	a) $F = 0$ b) $\left(-\frac{1}{4}\right)(-4F) = \left(-\frac{1}{4}\right)(0)$ $(1)(F) = 0$ $F = 0$
	a) $t = 0$ b) $t = -\frac{1}{5}$

SELF-TEST 1 (Frames 1-21)

Use the multiplication axiom to solve the following equations:

1. $8P = 13$ $P =$ ______	3. $-6t = -22$ $t =$ ______	5. $1 = -12h$ $h =$ ______
2. $12 = -21w$ $w =$ ______	4. $-5x = 0$ $x =$ ______	6. $0 = 29a$ $a =$ ______

ANSWERS: 1. $P = \frac{13}{8}$ 2. $w = -\frac{4}{7}$ 3. $t = \frac{11}{3}$ 4. $x = 0$ 5. $h = -\frac{1}{12}$ 6. $a = 0$

6-4 A REVIEW OF SOLVING NON-FRACTIONAL EQUATIONS

When solving a more complicated equation with a fractional root, we use the methods learned earlier to simplify the equation until we obtain an instance of the basic equation. The multiplication axiom is only used to solve the basic equation. In this section, we will review the solution of the types of equations introduced earlier.

---

22. Students sometimes confuse the use of the addition axiom and the multiplication axiom.

The addition axiom is used to eliminate a term from one side of an equation.

In $5x + 3 = 17$, "3" is a term on the left side. We can eliminate it by adding its opposite to both sides:

$$5x + 3 + (-3) = 17 + (-3)$$

$$5x = 14$$

The multiplication axiom is used to eliminate a factor in a term.

In $3x = 7$, "3" is a factor of a term. We can eliminate this factor by multiplying both sides by its reciprocal:

$$\frac{1}{3}(3x) = \frac{1}{3}(7)$$

$$(1)(x) = \frac{7}{3}$$

$$x = \frac{7}{3}$$

Would we use the addition axiom or multiplication axiom to eliminate the "7" from the right side in:

(a) $9 = 5t + 7$? ________ (b) $13 = 7t$? ________

---

23. In this frame and in those immediately following, you will use the multiplication axiom only for the final step of solving the basic equation.

Solve: (a) $7t + 8 = 9$ (b) $3R + 8 = 8$

$t =$ ______ $R =$ ______

Answers to frame 22:

a) Addition axiom, since "7" is a term.

b) Multiplication axiom, since "7" is a factor of a term.

---

24. Solve the following equations:

(a) $5 = 4 - 11p$ (b) $7 = 7 - 2q$ (c) $4y + 9 = 9 - 2y$

$p =$ ______ $q =$ ______ $y =$ ______

Answers to frame 23:

a) $t = \frac{1}{7}$ b) $R = 0$

25. Solve: (a) $7n + 6 = -21$ (b) $4 + 5(3 - x) = -8$ n = ______ x = ______	a) $p = -\frac{1}{11}$ b) $q = 0$ c) $y = 0$
26. Solve: (a) $15R - 18R + 16 = 17$ (b) $2(3 + y) + y = -7$ R = ______ y = ______	a) $n = -\frac{27}{7}$ b) $x = \frac{27}{5}$
27. Solve: (a) $7 = 6(5 - 2m) + 9m$ (b) $15 - 9V = 8$ m = ______ V = ______	a) $R = -\frac{1}{3}$ b) $y = -\frac{13}{3}$
28. Solve: (a) $15x - 7x = 9$ (b) $2W + 3(7 - 4W) = -5$ x = ______ W = ______	a) $m = \frac{23}{3}$ b) $V = \frac{7}{9}$
29. Solve the following equations: (a) $4R + 5 = 6$ (b) $2 = 6b - 9b$ (c) $4(d + 1) + 3 = 2$ R = ______ b = ______ d = ______	a) $x = \frac{9}{8}$ b) $W = \frac{13}{5}$
	a) $R = \frac{1}{4}$ b) $b = -\frac{2}{3}$ c) $d = -\frac{5}{4}$

30. The most difficult equations are those in which a grouping appears after a subtraction symbol. For example:

$$7 - (8 + t) = 5$$

To solve these equations, we eliminate the grouping by converting the subtraction to addition. We do so by adding the opposite of the grouping.

If the grouping contains an addition, <u>we obtain its opposite by taking the opposite of each term in the addition.</u> Therefore:

The opposite of $(8 + t)$ is $[(-8) + (-t)]$

The opposite of $(2y + 6)$ is $[(\quad) + (\quad)]$

$[(-2y) + (-6)]$

31. If the grouping contains a subtraction, we must convert the subtraction to addition before writing its opposite.

Since $(x - 7) = [x + (-7)]$,

the opposite of $(x - 7)$ is $[(-x) + 7]$.

Since $(11 - 5p) = [11 + (-5p)]$,

the opposite of $11 - 5p$ is $[(\quad) + (\quad)]$.

$[(-11) + 5p]$

32. Complete: (a) $10 - (R + 6) = 10 +$ <u>the opposite of (R + 6)</u>

$= 10 +$ ______ $+$ ______

(b) $3x - (5 - 4x) = 3x - [5 + (-4x)]$

$= 3x +$ <u>the opposite of [5 + (-4x)]</u>

$= 3x +$ ______ $+$ ______

---

a) $10 + \underline{(-R)} + \underline{(-6)}$

b) $3x + \underline{(-5)} + \underline{4x}$

33. Complete this solution:

$$3 - (3 + y) = 15$$

$$3 - (3 + y) = 15$$

$$3 + (-3) + (-y) = 15$$

y = ______

$y = -15$

34. Complete this solution:

$$3m - (5 - 2m) = -5$$

$$3m - (5 - 2m) = -5$$

$$3m + (-5) + 2m = -5$$

m = ______

$m = 0$

35. Solve: (a) $6F - (4 + F) = -7$ (b) $-5 = 3t - (5 - 2t)$

F = ______ t = ______

---

a) $F = -\frac{3}{5}$ b) $t = 0$

36. In the following equation, the left side involves the subtraction of a grouping which is an instance of the distributive principle.

$$5 - 3(3 + 2S) = -11$$

Students often make "sign" errors in handling such a subtraction. To prevent errors, always put brackets around the instance of the distributive principle first, as follows:

$$5 - [3(3 + 2S)] = -11$$

Then, multiply the instance of the distributive principle in the brackets:

$$5 - [9 + 6S] = -11$$

Then, perform the subtraction by "adding the opposite" of the grouping:

$$5 + (-9) + (-6S) = -11$$

Complete the solution:

S = ______

$S = \frac{7}{6}$

37. Solve this equation. As your first step, put brackets around the instance of the distributive principle:

$$1 = 4 - 3(2h + 1)$$

h = ______

$h = 0$

38. To prevent "sign" errors in the following equation, first insert brackets around the instance of the distributive principle:

$$2x - 3(2 - x) = 8$$

$$2x - [3(2 - x)] = 8$$

Then convert the subtraction within the grouping to addition:

$$2x - \left[3[2 + (-x)]\right] = 8$$

Then multiply the instance of the distributive principle in the large brackets:

$$2x - [6 + (-3x)] = 8$$

Then perform the subtraction by "adding the opposite" of the grouping:

$$2x + (-6) + (3x) = 8$$

Complete the solution:

x = ______

39. Solve this equation. As your first step, <u>put brackets around the instance of the distributive principle</u>:

$$17 = 11 - 5(2M - 3)$$

M = ______

$x = \frac{14}{5}$

$M = \frac{9}{10}$

SELF-TEST 2 (Frames 22-39)

Solve the following equations:

1. $9 + 3y = 10 + 5y$

   y = ______

2. $12 = 3(d + 4)$

   d = ______

3. $i + 3(4 + i) = 1$

   i = ______

Solve these equations. As your first step, put brackets around the grouping:

4. $3 = b - (4b - 7)$

   b = ______

5. $5 + 3(e - 4) = e$

   e = ______

6. $5 + 10r = r - 4(1 - r)$

   r = ______

ANSWERS: 1. $y = -\frac{1}{2}$ 2. $d = 0$ 3. $i = -\frac{11}{4}$ 4. $b = \frac{4}{3}$ 5. $e = \frac{7}{2}$ 6. $r = -\frac{9}{5}$

6-5 FRACTIONAL EQUATIONS WHICH CONTAIN A SINGLE FRACTION

Here are some fractional equations which contain a single fraction:

$$\frac{x}{2} = 6 \qquad \frac{45}{3t} + 2 = 5 \qquad \frac{d}{3} - 2 = 9$$

We will solve equations of this type in this section.

---

40. When fractional equations are simple and have whole number roots, they can be solved by trial-and-error. Solve the following by trial-and-error:

(a) $\frac{x}{2} = 6$ x = ______ (c) $4 = \frac{R}{6}$ R = ______

(b) $\frac{27}{d} = 9$ d = ______ (d) $8 = \frac{40}{m}$ m = ______

---

41. However, when fractional equations are more complicated or have fractional roots, it is too difficult to solve them by trial-and-error. We need another method. This method involves <u>using the multiplication axiom to obtain an equivalent but non-fractional equation</u> which is easier to solve. Here is an example:

$$\frac{3x}{2} = 6$$

We can obtain a non-fractional equation by multiplying both sides by "2", the denominator of the fraction. We get:

$$2\left(\frac{3x}{2}\right) = 2(6)$$

$$\left(\frac{2}{2}\right)(3x) = 12$$

$$(1)(3x) = 12$$

$$3x = 12$$

(a) The root of the <u>non-fractional</u> equation is ______.

(b) Show that this root satisfies the original fractional equation: $\frac{3x}{2} = 6$

Answers to 40:

a) x = 12 c) R = 24

b) d = 3 d) m = 5

Answers to 41:

a) 4

b) <u>Check:</u>

$$\frac{3x}{2} = 6$$

$$\frac{3(4)}{2} = 6$$

$$\frac{12}{2} = 6$$

$$6 = 6$$

42. Let's try another: $\frac{2x}{3} = 4$

We can obtain a non-fractional equation by multiplying both sides by "3". We get:

$$3\left(\frac{2x}{3}\right) = 3(4)$$

$$\left(\frac{3}{3}\right)(2x) = 12$$

$$(1)(2x) = 12$$

$$2x = 12$$

(a) The root of this last equation is _____.

(b) Show that this root satisfies the original equation: $\frac{2x}{3} = 4$

---

43. Let's try one with a letter in the denominator:

$$\frac{15}{x} = 3$$

We will multiply both sides by "x", the denominator of the fraction:

$$x\left(\frac{15}{x}\right) = x(3)$$

$$\left(\frac{x}{x}\right)(15) = 3x$$

$$(1)(15) = 3x$$

$$15 = 3x$$

(a) The root of $15 = 3x$ is _____.

(b) Show that this root satisfies the original equation: $\frac{15}{x} = 3$

Answer to 42:

a) 6

b) Check:

$$\frac{2x}{3} = 4$$

$$\frac{2(6)}{3} = 4$$

$$\frac{12}{3} = 4$$

$$4 = 4$$

---

44. Let's try one more: $\frac{36}{3t} = 2$

We will multiply both sides by "3t":

$$3t\left(\frac{36}{3t}\right) = 3t(2)$$

$$\left(\frac{3t}{3t}\right)(36) = 6t$$

$$(1)(36) = 6t$$

$$36 = 6t$$

(a) The root of $36 = 6t$ is _____.

(b) Show that this root satisfies the original equation: $\frac{36}{3t} = 2$

Answer to 43:

a) 5

b) Check:

$$\frac{15}{x} = 3$$

$$\frac{15}{5} = 3$$

$$3 = 3$$

45. By using the multiplication axiom, we can simplify any fractional equation and obtain an equivalent non-fractional equation which is easier to solve. We will call this use of the multiplication axiom:

CLEARING THE FRACTION

When "clearing the fraction" in a fractional equation, we encounter multiplications like this one. Watch the steps:

$$3\left(\frac{2}{3}\right) = \frac{(3)(2)}{3} = \left(\frac{3}{3}\right)(2) = (1)(2) = 2$$

Write the final product for each of these multiplications:

(a) $4\left(\frac{7x}{4}\right) =$ ____ (b) $m\left(\frac{8}{m}\right) =$ ____ (c) $5y\left(\frac{9}{5y}\right) =$ ____

a) 6

b) Check:

$$\frac{36}{3t} = 2$$

$$\frac{36}{3(6)} = 2$$

$$\frac{36}{18} = 2$$

$$2 = 2$$

---

46. To "clear the fraction" in each of the equations below, we multiply each side by the denominator of the fraction. Do so, and write the root of the original equation:

(a) $\frac{54}{3x} = 9$ (b) $14 = \frac{7t}{5}$

x = ______ t = ______

a) 7x b) 8 c) 9

---

47. Solve each: (a) $\frac{3R}{7} = 4$ (b) $3 = \frac{14}{5S}$

R = ______ S = ______

a) $x = 2$
(From $54 = 27x$)

b) $t = 10$
(From $70 = 7t$)

---

48. In the last frame, we decided that $S = \frac{14}{15}$ is the root of this equation: $3 = \frac{14}{5S}$

Show that this value satisfies the equation:

a) $\frac{28}{3}$ or $9\frac{1}{3}$
(From $3R = 28$)

b) $\frac{14}{15}$
(From $15S = 14$)

Check:

$$3 = \frac{14}{5S}$$

$$3 = \frac{14}{5\left(\frac{14}{15}\right)}$$

$$3 = \frac{14}{\frac{14}{3}}$$

$$3 = 14\left(\frac{3}{14}\right)$$

$$3 = 3$$

49. In the following equation, there are two terms on the left side:

$$\frac{x}{3} + 2 = 6$$

To clear the fraction, we multiply both sides by "3".

When using the multiplication axiom, the whole left side of the equation must be multiplied by "3". To remember this fact, we draw large parentheses around the left side, like this:

$$3\left(\frac{x}{3} + 2\right) = 3(6)$$

Notice that $3\left(\frac{x}{3} + 2\right)$ is an instance of the distributive principle.

Simplifying the left side, we get:

$$3\left(\frac{x}{3}\right) + 3(2) = 3(6)$$

$$x + 6 = 18$$

(a) The root of $x + 6 = 18$ is _______.

(b) The root of $\frac{x}{3} + 2 = 6$ must be _______.

---

a) 12

b) 12

50. Whenever one side of a fractional equation contains two terms, we encounter an instance of the distributive principle when clearing the fraction. Here is another example:

$$7 = \frac{7y}{14} + 5$$

To clear the fraction, we multiply both sides of the equation by "14". Notice how we enclose the right side in parentheses and obtain an instance of the distributive principle:

$$14(7) = 14\left(\frac{7y}{14} + 5\right)$$

$$14(7) = 14\left(\frac{7y}{14}\right) + 14(5)$$

$$98 = 7y + 70$$

(a) The root of $98 = 7y + 70$ is _______.

(b) The root of $7 = \frac{7y}{14} + 5$ is _______

---

a) 4

b) 4

51.

$$\frac{45}{3t} + 2 = 5$$

To clear the fraction, we multiply both sides by "3t".

$$3t\left(\frac{45}{3t} + 2\right) = 3t(5)$$

There are two pitfalls to avoid in clearing the fraction:

(1) Forgetting to multiply both sides by "3t".

(2) Forgetting to multiply both terms $\left(\frac{45}{3t} \text{ and } 2\right)$ by "3t" in the instance of the distributive principle.

Clear the fraction above. What equation do you get?

____

52. In the last frame, the fractional equation was: $\frac{45}{3t} + 2 = 5$

When the fraction was cleared, we got: $45 + 6t = 15t$

The root of this non-fractional equation is "5". Plug this root into the <u>original</u> fractional equation. Does it satisfy the original equation? ______

Answer: $3t\left(\frac{45}{3t}\right) + 3t(2) = 3t(5)$

$45 + 6t = 15t$

---

53. To solve a fractional equation, the first step is "clearing the fraction" to obtain a non-fractional equation. Let's clear the fraction in this one:

$$18 = \frac{5V}{3} + 3$$

Complete: (a) $(\quad)(18) = (\quad)\left(\frac{5V}{3} + 3\right)$

(b) ______ $= (\quad)\left(\frac{5V}{3}\right) + (\quad)(3)$

(c) ______ = ______ + ______

Answer: Yes, since:

$\frac{45}{3(5)} + 2 = 5$

$\frac{45}{15} + 2 = 5$

$3 + 2 = 5$

---

54. Let's clear the fraction in this one:

$$\frac{8}{7x} + 9 = 10$$

Complete: (a) $(\quad)\left(\frac{8}{7x} + 9\right) = (\quad)(10)$

(b) $(\quad)\left(\frac{8}{7x}\right) + (\quad)(9) =$ ______

(c) ______ + ______ = ______

Answer:

a) $3(18) = 3\left(\frac{5V}{3} + 3\right)$

b) $54 = 3\left(\frac{5V}{3}\right) + 3(3)$

c) $54 = 5V + 9$

---

55. (a) Clear the fraction in the equation at the right. The non-fractional equation obtained is: ______ $\qquad \frac{4p}{7} + 5 = 2p$

(b) The root of this non-fractional equation is ______.

(c) The root of the original fractional equation must be ______.

Answer:

a) $7x\left(\frac{8}{7x} + 9\right) = 7x(10)$

b) $7x\left(\frac{8}{7x}\right) + 7x(9) = 70x$

c) $8 + 63x = 70x$

Answer:

a) $4p + 35 = 14p$

b) $\frac{7}{2}$ (from $\frac{35}{10}$)

c) $\frac{7}{2}$

SELF-TEST 3 (Frames 40-55)

Solve these equations. In each problem, write down the actual multiplication used to clear the fraction.

1. $\frac{5R}{9} = 15$ R = ______	2. $6 = \frac{9}{5w}$ w = ______	3. $3 + \frac{x}{5} = 1$ x = ______
4. $2t = \frac{t}{3} + 5$ t = ______	5. $d + \frac{3d}{2} = 2$ d = ______	6. $5 + \frac{7}{2h} = 3$ h = ______

7. The root of the equation at the right is reported to be $-\frac{3}{4}$. For $r = -\frac{3}{4}$, determine:

$$3r + 2 = \frac{r}{3}$$

(a) The value of the left side of the equation. ______

(b) The value of the right side of the equation. ______

(c) Is $-\frac{3}{4}$ the root? ______

ANSWERS:

1. $R = 27$
2. $w = \frac{3}{10}$
3. $x = -10$
4. $t = 3$
5. $d = \frac{4}{5}$
6. $h = -\frac{7}{4}$
7. (a) $-\frac{1}{4}$ (b) $-\frac{1}{4}$ (c) Yes

6-6 THE DISTRIBUTIVE PRINCIPLE OVER SUBTRACTION AND RELATED EQUATIONS

Here are some instances of the distributive principle which contain a subtraction in the grouping:

$$3(2x - 5) \qquad\qquad 4(7 - m)$$

Up to this point in the course, we have converted the subtractions to additions before multiplying by the distributive principle over addition, as shown here:

$$3(2x - 5) = 3[2x + (-5)] = 3(2x) + 3(-5)$$

$$4(7 - m) = 4[7 + (-m)] = 4(7) + 4(-m)$$

In this section, we will introduce the distributive principle over subtraction. The advantage of this principle is that we can multiply before converting the subtraction to addition. Using the distributive principle over subtraction makes the solution of equations more efficient.

56. Here is the pattern for the <u>distributive principle over subtraction</u>:

$$\triangle(\bigcirc - \square) = \triangle(\bigcirc) - \triangle(\square)$$

For example: $7(5 - 9) = 7(5) - 7(9) = 35 - 63$

$3(x - 2) = 3(x) - 3(2) = 3x - 6$

Using this principle, complete these:

(a) $5(2y - 3) = 5(2y) - 5(3) =$ _____ - _____

(b) $10(4 - m) = 10(4) - 10(m) =$ _____ - _____

---

a) $10y - 15$

b) $40 - 10m$

57. We can justify this principle with numerical examples.

Example:

$$3(2 - 7) = 3(2) - 3(7)$$
$$3\ (-5) = 6 - 21$$
$$-15 = -15$$

Show that both sides of the equation at the right are equal:

$$10(5 - 2) = 10(5) - 10(2)$$

$$10(5 - 2) = 10(5) - 10(2)$$
$$10\ (3) = 50 - 20$$
$$30 = 30$$

58. When multiplying by the distributive principle over subtraction, you should be able to write the simplified product in one step. <u>Be sure to multiply BOTH TERMS in the grouping by the factor.</u>

Example: $5(m - 7) = 5m - 35$

(a) $3(2d - 4) =$ ____________ (b) $8(7 - 3t) =$ ____________

---

a) $6d - 12$ b) $56 - 24t$

59. The left side of this equation contains an instance of the distributive principle over subtraction:

$$7(2x - 7) = 6$$

Do the multiplication on the left side. What new equation do you get? ____________

$14x - 49 = 6$

60. What new equation do you get when you multiply by the distributive principle over subtraction in these?

(a) $5 + 4(2t - 1) = 7$ (b) $3x + 2(x - 5) = 10(3 - x)$

____________ ____________

---

a) $5 + 8t - 4 = 7$

b) $3x + 2x - 10 = 30 - 10x$

61. $$10 - 2(y - 4) = 7$$

In the above equation, an instance of the distributive principle <u>over subtraction</u> follows a subtraction sign. To simplify this equation, first put brackets around the instance of the distributive principle, as follows:

$$10 - [2(y - 4)] = 7$$

Multiplying by the distributive principle <u>over subtraction</u>, we get:

$$10 - [2y - 8] = 7$$

Converting the subtraction within the grouping to addition, we get:

$$10 - [2y + (-8)] = 7$$

Now, converting the subtraction of a grouping to addition, we get:

$$10 + (-2y) + 8 = 7$$

Here's another equation of the same type: $3x - 5(7 - x) = 1$

(a) Put a bracket around the instance of the distributive principle over subtraction. Then multiply by the distributive principle over subtraction. The resulting equation is: ______________

(b) Now, show the two steps needed to convert the subtractions to additions: ______________ ______________

a) $3x - [35 - 5x] = 1$

b) $3x - [35 + (-5x)] = 1$
$3x + (-35) + 5x = 1$

---

62. When clearing the fraction in a fractional equation, we can encounter an instance of the distributive principle over subtraction. Here is an example:

$$\frac{d}{3} - 2 = 7$$

To clear the fraction, we multiply both sides by "3":

$$3\left(\frac{d}{3} - 2\right) = 3(7)$$

Now to simplify the left side, we multiply by the distributive principle over subtraction.

$$3\left(\frac{d}{3}\right) - 3(2) = 3(7)$$

$$d - 6 = 21$$

The root of the last equation is "27". Show that this root satisfies the original equation: $\frac{d}{3} - 2 = 7$

$$\frac{d}{3} - 2 = 7$$
$$\frac{27}{3} - 2 = 7$$
$$9 - 2 = 7$$
$$7 = 7$$

63. For each of the following, multiply by the distributive principle over subtraction and write the product in its simplest form:

(a) $10\left(\frac{2t}{10} - 3\right) = (\quad)(\quad) - (\quad)(\quad) =$ ____ - ____

(b) $7\left(4 - \frac{m}{7}\right) = (\quad)(\quad) - (\quad)(\quad) =$ ____ - ____

(c) $R\left(\frac{8}{R} - 1\right) = (\quad)(\quad) - (\quad)(\quad) =$ ____ - ____

(d) $2q\left(1 - \frac{3}{2q}\right) = (\quad)(\quad) - (\quad)(\quad) =$ ____ - ____

---

64. Show the steps for clearing the fraction in this equation. Use the distributive principle over subtraction: $7 = \frac{9}{x} - 5$

____

a) $10\left(\frac{2t}{10}\right) - 10(3) = 2t - 30$

b) $7(4) - 7\left(\frac{m}{7}\right) = 28 - m$

c) $R\left(\frac{8}{R}\right) - R(1) = 8 - R$

d) $2q(1) - 2q\left(\frac{3}{2q}\right) = 2q - 3$

---

65. When clearing the fraction in a fractional equation, you will frequently encounter an instance of the distributive principle. Whenever the grouping contains a subtraction, it is easier to multiply by the distributive principle over subtraction.

For each of these, write the non-fractional equation you obtain when the fraction is cleared:

(a) $\frac{x}{5} - 7 = 4x$

____

(b) $6 = \frac{11}{2R} - 3$

____

$x(7) = x\left(\frac{9}{x} - 5\right)$

$7x = x\left(\frac{9}{x}\right) - x(5)$

$7x = 9 - 5x$

---

66. What non-fractional equation do you obtain when you clear the fraction in each of these?

(a) $4 = 3x - \frac{4x}{5}$

____

(b) $10 - \frac{5}{2t} = 7$

____

a) $x - 35 = 20x$

b) $12R = 11 - 6R$

---

a) $20 = 15x - 4x$

b) $20t - 5 = 14t$

67. (a) What non-fractional equation do you obtain when you clear the fraction in the equation at the right? $3 = \frac{5}{4y} - 5$ (b) The root of this non-fractional equation is ______. (c) The root of the fractional equation must be ______.	
68. Find the root of each equation: (a) $5 - \frac{m}{3} = 2m$ (b) $2 = 5 - \frac{7}{4q}$ m = ______ q = ______	a) $12y = 5 - 20y$ b) $\frac{5}{32}$ c) $\frac{5}{32}$
	a) $m = \frac{15}{7}$ b) $q = \frac{7}{12}$

SELF-TEST 4 (Frames 56-68)

Multiply, using the distributive principle over subtraction:

1. $3(5 - 2a) =$ ______
2. $8\left(\frac{3h}{8} - 5\right) =$ ______
3. $4t\left(3 - \frac{1}{4t}\right) =$ ______

4. Solve: $3 - \frac{4h}{5} = 2h$

h = ______

5. Solve: $11 = 9 - \frac{1}{3r}$

r = ______

ANSWERS: 1. $15 - 6a$ 2. $3h - 40$ 3. $12t - 1$ 4. $h = \frac{15}{14}$ 5. $r = -\frac{1}{6}$

6-7 EQUATIONS CONTAINING ONE FRACTION WHOSE NUMERATOR OR DENOMINATOR IS AN ADDITION OR SUBTRACTION

In each of the following fractional equations, there is one fraction. In each case, either the numerator or denominator of the fraction is an addition or subtraction.

$$\frac{3y + 7}{5} = 9 - y \qquad 6 = \frac{9x}{x - 3}$$

In this section, we will show the method for solving equations of this type.

---

69. In this equation, the numerator of the fraction is an addition:

$$\frac{x + 3}{2} = 5$$

We can clear the fraction by multiplying both sides by "2", as follows:

$$2\left(\frac{x + 3}{2}\right) = 2(5)$$

$$\frac{2(x + 3)}{2} = 10$$

$$\left(\frac{2}{2}\right)(x + 3) = 10$$

$$(1)(x + 3) = 10$$

$$x + 3 = 10$$

The root of $x + 3 = 10$ is "7". Show that this root satisfies the original equation:

$$\frac{x + 3}{2} = 5$$

Answer to 69:

$$\frac{x + 3}{2} = 5$$

$$\frac{7 + 3}{2} = 5$$

$$\frac{10}{2} = 5$$

$$5 = 5$$

70. Here is another of the same type:

$$\frac{y + 6}{y} = 7$$

To clear the fraction, we multiply both sides by "y":

$$y\left(\frac{y + 6}{y}\right) = y(7)$$

$$\frac{y(y + 6)}{y} = 7y$$

$$\left(\frac{y}{y}\right)(y + 6) = 7y$$

$$(1)(y + 6) = 7y$$

$$y + 6 = 7y$$

(a) Solve the last equation. Its root is _______.

(b) The root of the original equation must be _______.

Answer to 70:

a) 1

b) 1

71. When clearing the fraction in equations of this type, you will encounter multiplications like the following:

$$5\left(\frac{x+3}{5}\right) \qquad t\left(\frac{t-7}{t}\right)$$

You must be able to perform these multiplications in order to clear the fraction. Here is the procedure:

Step 1: $5\left(\frac{x+3}{5}\right) = \frac{5(x+3)}{5}$

Step 2: $= \left(\frac{5}{5}\right)(x+3)$

Step 3: $= (1)(x+3)$

Step 4: $= x+3$

Notice how we made use of the two principles:

$$\frac{\square}{\square} = 1 \quad \text{and} \quad (1)(\square) = \square$$

Do these multiplications:

(a) $7\left(\frac{R+6}{7}\right) =$ ____ (b) $4\left(\frac{V+7}{4}\right) =$ ____ (c) $9\left(\frac{5d-4}{9}\right) =$ ____

a) R + 6
b) V + 7
c) 5d - 4

---

72. Watch the procedure in this multiplication:

$$t\left(\frac{t-7}{t}\right) = \frac{t(t-7)}{t}$$
$$= \left(\frac{t}{t}\right)(t-7)$$
$$= (1)(t-7)$$
$$= t-7$$

Multiply: (a) $x\left(\frac{x+5}{x}\right) =$ ____ (c) $Q\left(\frac{2Q-7}{Q}\right) =$ ____

(b) $3s\left(\frac{s+7}{3s}\right) =$ ____ (d) $4w\left(\frac{3w-5}{4w}\right) =$ ____

a) x + 5 c) 2Q - 7
b) s + 7 d) 3w - 5

---

73. Write the non-fractional equation which you obtain when the fraction is cleared in each of these equations. We have given the first step in each case:

(a) $t = \frac{t-5}{6}$ (b) $\frac{m+7}{3} = 2m - 6$ (c) $\frac{2V-5}{V} = 4$

$6(t) = 6\left(\frac{t-5}{6}\right)$ $3\left(\frac{m+7}{3}\right) = 3(2m-6)$ $V\left(\frac{2V-5}{V}\right) = V(4)$

____ ____ ____

a) 6t = t - 5
b) m + 7 = 6m - 18
c) 2V - 5 = 4V

74. Clear the fraction and write the resulting non-fractional equation for each of these equations:

(a) $\frac{3b - 5}{4} = 2b$ (b) $x + 5 = \frac{x - 5}{7}$ (c) $10 = \frac{y - 1}{2y}$

________ ________ ________

---

75. Solve each of these equations:

(a) $4 = \frac{S + 6}{5}$ (b) $\frac{4V - 15}{3V} = 2$

S = ______ V = ______

Answers to 74:

a) $3b - 5 = 8b$

b) $7x + 35 = x - 5$

c) $20y = y - 1$

---

76. (a) Solve the equation below:

$\frac{p - 5}{7} = 2p + 3$

The root is ______.

(b) Show that the root you obtained satisfies the original equation:

Answers to 75:

a) $S = 14$

b) $V = -\frac{15}{2}$

---

77. In the following equation, the denominator of the fraction is an addition:

$$\frac{12}{x + 2} = 2$$

To clear the fraction, we multiply both sides by "x + 2". Here are the steps:

Step 1: $(x + 2)\left(\frac{12}{x + 2}\right) = 2(x + 2)$

Step 2: $\frac{(x + 2)(12)}{x + 2} = 2x + 4$

Step 3: $\left(\frac{x + 2}{x + 2}\right)(12) = 2x + 4$

Step 4: $(1)(12) = 2x + 4$

Step 5: $12 = 2x + 4$

(Continued on following page.)

Answers to 76:

a) -2

b) Check:

$$\frac{p - 5}{7} = 2p + 3$$

$$\frac{(-2) - 5}{7} = 2(-2) + 3$$

$$\frac{(-2) + (-5)}{7} = (-4) + 3$$

$$\frac{-7}{7} = -1$$

$$-1 = -1$$

77. (Continued)

In Steps 3 and 4, notice that $\frac{x+2}{x+2} = 1$. Let's justify this:

If you plug in "3" for "x", you get: $\frac{3+2}{3+2} = \frac{5}{5} = 1$

If you plug in "10" for "x", you get: $\frac{10+2}{10+2} = \frac{12}{12} = 1$

If you plug in "-5" for "x", you get: $\frac{(-5)+2}{(-5)+2} = \frac{-3}{-3} = 1$

---

78. The point we are making is this:

Expressions like $\frac{x+2}{x+2} = 1$ and $\frac{2t-7}{2t-7} = 1$

are merely more complicated instances of the principle $\frac{\square}{\square} = 1$.

Complete these: (a) $\frac{3m+1}{3m+1} =$ ______ (b) $\frac{y-8}{y-8} =$ ______

Answer: Go to next frame.

---

79. To clear the fraction in this equation, you should multiply both sides by ______. $\frac{15}{y+3} = 3$

Answer: a) 1 b) 1

---

80. Let's solve the equation in the last frame:

$$\frac{15}{y+3} = 3$$

Step 1: $(y+3)\left(\frac{15}{y+3}\right) = 3(y+3)$

Step 2: $\frac{(y+3)(15)}{y+3} = 3y + 9$

Step 3: $\left(\frac{y+3}{y+3}\right)(15) = 3y + 9$

Step 4: $(1)(15) = 3y + 9$

Step 5: $15 = 3y + 9$

The root of $15 = 3y + 9$ is "2". Show that this root satisfies the original equation: $\frac{15}{y+3} = 3$

Answer: $y + 3$

Check:

$\frac{15}{y+3} = 3$

$\frac{15}{2+3} = 3$

$\frac{15}{5} = 3$

$3 = 3$

81. To clear the fraction in equations in which the denominator of the fraction is an addition or subtraction, you must be able to perform multiplications like the following. Watch the steps in each example:

Example: $(t - 3)\left(\frac{10}{t - 3}\right) = \frac{(t - 3)(10)}{t - 3} = \left(\frac{t - 3}{t - 3}\right)(10)$

$= (1)(10)$

$= 10$

Example: $(p + 7)\left(\frac{2p + 5}{p + 7}\right) = \frac{(p + 7)(2p + 5)}{p + 7} = \left(\frac{p + 7}{p + 7}\right)(2p + 5)$

$= (1)(2p + 5)$

$= 2p + 5$

Do these multiplications:

(a) $(5b + 6)\left(\frac{9}{5b + 6}\right) =$ ______

(c) $(D - 3)\left(\frac{7}{D - 3}\right) =$ ______

(b) $(N + 5)\left(\frac{N - 5}{N + 5}\right) =$ ______

(d) $(7s - 4)\left(\frac{3s + 1}{7s - 4}\right) =$ ______

Answers: a) 9 c) 7 b) $N - 5$ d) $3s + 1$

82. Clear the fraction in each equation and write the resulting non-fractional equation:

(a) $5 = \frac{7}{x + 1}$

(b) $\frac{V}{V - 3} = 9$

______ ______

Answers: a) $5x + 5 = 7$ b) $V = 9V - 27$

83. Clear the fraction in each equation and write the resulting non-fractional equation:

(a) $\frac{2x - 1}{2x + 1} = 3$

(b) $7 = \frac{y - 5}{y - 3}$

______ ______

Answers: a) $2x - 1 = 6x + 3$ b) $7y - 21 = y - 5$

84. Solve: (a) $\frac{2x}{x+1} = 4$ (b) $4 = \frac{w+3}{2w-3}$

x = ______ w = ______

a) $x = -2$ b) $w = \frac{15}{7}$

SELF-TEST 5 (Frames 69-84)

Do the following multiplications:

1. $6w\left(\frac{w-4}{6w}\right) =$ ______
2. $(h+1)\left(\frac{h-1}{h+1}\right) =$ ______
3. $(5x-2)\left(\frac{3x}{5x-2}\right) =$ ______

Solve the following equations:

4. $\frac{P-7}{4} = 2P$ P = ______
5. $V + 2 = \frac{3V-4}{7}$ V = ______
6. $\frac{4a-1}{3a} = 2$ a = ______
7. $\frac{3x}{4x-5} = 2$ x = ______
8. $\frac{2d+5}{d+4} = 1$ d = ______
9. $\frac{6-t}{t-2} = 5$ t = ______

ANSWERS:

1. $w - 4$
2. $h - 1$
3. $3x$
4. $P = -1$
5. $V = -\frac{9}{2}$
6. $a = -\frac{1}{2}$
7. $x = 2$
8. $d = -1$
9. $t = \frac{8}{3}$

6-8 EQUATIONS IN WHICH ONE SIDE IS "0" OR "1"

Students frequently have difficulty with equations like the following:

$$\frac{3x}{x+5} = 0 \qquad\qquad \frac{3x}{x+5} = 1$$

The difficulty is due to the fact that one side of each equation is either "0" or "1". Since the same principles apply to equations of this type, the difficulty can easily be overcome. We will show the method in this section.

---

85. When any quantity is multiplied by "0", the product is always "0". Therefore:

(a) $0(7) =$ ____ (c) $0(2p + 7) = 0(2p) + 0(7) =$ ____

(b) $0(2M) =$ ____ (d) $(3t - 5)(0) = 3t(0) - 5(0) =$ ____

---

86. In this equation, one side is "0": $\frac{x}{4} = 0$

To solve, we clear the fraction, multiplying both sides by "4":

Step 1: $4\left(\frac{x}{4}\right) = 4(0)$

Step 2: $\left(\frac{4}{4}\right)x = 0$

Step 3: $(1)x = 0$

Therefore $x =$ ____.

Answers: a) 0 b) 0 c) 0 d) 0

---

87. In this equation, one side is "1": $\frac{x}{4} = 1$

The steps in the solution are:

Step 1: $4\left(\frac{x}{4}\right) = 4(1)$

Step 2: $\left(\frac{4}{4}\right)(x) = 4$

Step 3: $(1)(x) = 4$

Therefore $x =$ ____.

Answer: $x = 0$

---

88. Watch this solution when one side is "0":

$$\frac{t}{t+2} = 0$$

Step 1: $(t+2)\left(\frac{t}{t+2}\right) = 0(t+2)$

Step 2: $\left(\frac{t+2}{t+2}\right)(t) = 0$

Step 3: $t = 0$

Show that "0" satisfies the original equation: $\frac{t}{t+2} = 0$

Answer: $x = 4$

89. Watch this solution when one side is "1": $$\frac{2y}{y+2} = 1$$ Step 1: $(y+2)\left(\frac{2y}{y+2}\right) = 1(y+2)$ Step 2: $\left(\frac{y+2}{y+2}\right)(2y) = 1(y+2)$ Step 3: $2y = y + 2$ The root of $2y = y + 2$ is "2". Show that this root satisfies the original equation: $\frac{2y}{y+2} = 1$	Check: $\frac{t}{t+2} = 0$ $\frac{0}{0+2} = 0$ $\frac{0}{2} = 0$ $0 = 0$
90. Solve: (a) $0 = \frac{N}{5}$ (b) $1 = \frac{N}{5}$ The root is _____. The root is _____.	Check: $\frac{2y}{y+2} = 1$ $\frac{2(2)}{2+2} = 1$ $\frac{4}{4} = 1$ $1 = 1$
91. In all of the equations below, the root is _____. $3Q = 0$ $-5t = 0$ $10m = 0$	a) 0 b) 5
92. Solve: (a) $\frac{3Q}{7} = 0$ (b) $\frac{3Q}{7} = 1$ The root is _____. The root is _____.	0
93. Solve: (a) $0 = \frac{3b}{b+5}$ (b) $1 = \frac{3b}{b+5}$ The root is _____. The root is _____.	a) 0 b) $\frac{7}{3}$
94. Solve: (a) $\frac{3d}{d-5} = 0$ (b) $\frac{3d}{d-5} = 1$ d = _____ d = _____	a) 0 b) $\frac{5}{2}$
	a) $d = 0$ b) $d = -\frac{5}{2}$

95. Watch this solution: $\frac{x+3}{x+5} = 0$ $(x+5)\left(\frac{x+3}{x+5}\right) = 0(x+5)$ $\left(\frac{x+5}{x+5}\right)(x+3) = 0$ $1(x+3) = 0$ $x + 3 = 0$ The root is _____.	
96. In the last frame, we decided that the root of the following equation is "-3": $\frac{x+3}{x+5} = 0$ Show that "-3" satisfies the equation:	-3
97. Solve: (a) $\frac{3d+10}{d-3} = 0$ (b) $\frac{3d+10}{d-3} = 1$ The root is _____. The root is _____.	Check: $\frac{x+3}{x+5} = 0$ $\frac{(-3)+3}{(-3)+5} = 0$ $\frac{0}{2} = 0$ $0 = 0$
	a) $-\frac{10}{3}$ b) $-\frac{13}{2}$

SELF-TEST 6 (Frames 85-97)

Complete: 1. $(1)(\square) =$ _____ 2. $(0)(\square) =$ _____

3. Solve: $\frac{4h}{5} = 0$ h = _____	4. Solve: $\frac{4h}{5} = 1$ h = _____	5. Solve: $\frac{2p}{p-5} = 0$ p = _____
6. Solve: $\frac{2p}{p-5} = 1$ p = _____	7. Solve: $\frac{2y+1}{6y-2} = 0$ y = _____	8. Solve: $\frac{2y+1}{6y-2} = 1$ y = _____

ANSWERS:

1. $\square$
2. 0
3. $h = 0$
4. $h = \frac{5}{4}$
5. $p = 0$
6. $p = -5$
7. $y = -\frac{1}{2}$
8. $y = \frac{3}{4}$

6-9 EQUATIONS WHICH CONTAIN TWO OR MORE FRACTIONS WHOSE DENOMINATORS ARE IDENTICAL

98. Here is an equation which contains two fractions:

$$\frac{x}{2} + 5 = \frac{7}{2}$$

The denominators are identical. Both are "2". Let's multiply both sides by "2":

$$2\left(\frac{x}{2} + 5\right) = 2\left(\frac{7}{2}\right)$$

$$2\left(\frac{x}{2}\right) + 2(5) = 2\left(\frac{7}{2}\right)$$

$$x + 10 = 7$$

When we multiplied both sides by "2", did we clear both fractions in one step? _______

Yes

99. Here is an equation containing three fractions whose denominators are identical:

$$\frac{R}{6} + 1 = \frac{2R}{6} - \frac{5}{6}$$

(a) Multiply both sides by "6".
What new equation do you get?

____

(b) Were all three fractions eliminated by this multiplication? _____

---

a) $R + 6 = 2R - 5$

b) Yes

100. (a) Multiply both sides of this equation by "y". Write the new equation: $\frac{3}{y} + 5 = \frac{2}{y}$

____

(b) Were all fractions eliminated by this multiplication? _______

---

a) $3 + 5y = 2$

b) Yes

101. (a) Multiply both sides of this equation by "3t". Write the new equation: $\frac{2}{3t} + 7 = \frac{5}{3t} - \frac{8}{3t}$

____

(b) Were all the fractions eliminated by this multiplication? _______

---

a) $2 + 21t = 5 - 8$

b) Yes

102. Here are two equations:

#1 $\frac{3}{7} + r = \frac{5r}{7} + \frac{5}{7}$

#2 $\frac{5}{V} - 1 = \frac{3}{V}$

You can eliminate all fractions in one step:

(a) if you multiply both sides of equation #1 by _____.

(b) if you multiply both sides of equation #2 by _____.

---

103. Solve: (a) $m - \frac{m}{10} = \frac{4m}{10} + \frac{9}{10}$ (b) $\frac{17}{5y} = 2 - \frac{11}{5y}$

m = _____ y = _____

Answers to 102: a) 7 b) V

Answers to 103: a) $m = \frac{9}{5}$ b) $y = \frac{14}{5}$

6-10 EQUATIONS WITH TWO OR MORE DIFFERENT NUMERICAL DENOMINATORS

When a fractional equation contains two or more different numerical denominators, we can "clear the fractions" in either of two ways.

(1) We can clear them in as many steps as there are different denominators.

(2) We can clear them in one step.

Though the first method is longer, it is a perfectly legitimate procedure. We will show both methods in this section.

---

104. There are two different numerical denominators in the following fractional equation:

$$\frac{x}{3} + 1 = \frac{x}{2}$$

We can clear the fractions in two steps:

Step 1: Multiplying both sides by "3", we can clear the fraction on the left side:

$$3\left(\frac{x}{3} + 1\right) = 3\left(\frac{x}{2}\right)$$

$$3\left(\frac{x}{3}\right) + 3(1) = \frac{3x}{2}$$

$$x + 3 = \frac{3x}{2}$$

Step 2: Now, multiplying both sides by "2", we can clear the remaining fraction on the right side:

$$2(x + 3) = 2\left(\frac{3x}{2}\right)$$

$$2x + 6 = 3x$$

The root of the last equation is "6". Show that this root satisfies the original equation:

$$\frac{x}{3} + 1 = \frac{x}{2}$$

105. In the equation at the right, we can "clear the fractions" in two steps: $\frac{2y}{5} = 1 + \frac{3y}{7}$ (a) Multiply both sides by "7" and write the new equation: __________ (b) Now multiply both sides of this new equation by "5" and write the resulting equation: __________	$\frac{x}{3} + 1 = \frac{x}{2}$ $\frac{6}{3} + 1 = \frac{6}{2}$ $2 + 1 = 3$ $3 = 3$
106. $\frac{R}{9} + 7 = \frac{3R}{4}$ If we multiply both sides of this equation by "4", which fraction would be cleared? _____	a) $\frac{14y}{5} = 7 + 3y$ b) $14y = 35 + 15y$
107. $\frac{t}{11} = 7 - \frac{t}{7}$ If we multiply both sides of this equation by "11", which fraction <u>would not be cleared</u>? _____	$\frac{3R}{4}$
108. Here is an equation with three fractions. Two of the denominators are identical. $\frac{5x}{7} + 1 = \frac{x}{2} + \frac{1}{2}$ (a) Multiply both sides by "2" and write the new equation: __________ (b) How many fractions were cleared by this multiplication? _____	The one whose denominator is "7". That is, $\frac{t}{7}$.
109. $\frac{x}{4} + 3 = \frac{x}{3} + \frac{1}{4}$ (a) If we multiply both sides of the equation above by "4", which fraction(s) will be cleared? _______ (b) If we multiply both sides of the equation above by "3", which fraction(s) will be eliminated? _______	a) $\frac{10x}{7} + 2 = x + 1$ b) Two. Both fractions with "2" as a denominator, namely $\frac{x}{2}$ and $\frac{1}{2}$.
	a) $\frac{x}{4}$ and $\frac{1}{4}$ b) $\frac{x}{3}$

110. We can clear all the fractions in the equation at the right in two steps:

$$\frac{m}{2} + \frac{3m}{5} = \frac{2m}{5} + 12$$

(a) Multiply both sides by "5" and write the resulting equation: ______

(b) Now multiply both sides of this new equation by "2" and write the resulting equation: ______

---

111. Here is a fractional equation with three different numerical denominators:

$$2 + \frac{y}{7} = \frac{2y}{5} + \frac{4y}{3}$$

If we multiply both sides by "7" and then by "3", would we clear all three fractions? ______

Answer to 110: a) $\frac{5m}{2} + 3m = 2m + 60$ b) $5m + 6m = 4m + 120$

112.

$$\frac{x}{2} = \frac{4}{3}$$

In the above equation, we can clear the fractions in two steps, multiplying by "2" and then by "3".

However, we can clear the fractions in one step by multiplying by both "2" and "3" at the same time. Here is the procedure:

$$(2)(3)\left(\frac{x}{2}\right) = (2)(3)\left(\frac{4}{3}\right)$$

$$\frac{(2)(3)(x)}{2} = \frac{(2)(3)(4)}{3}$$

$$\left(\frac{2}{2}\right)(3)(x) = \left(\frac{3}{3}\right)(2)(4)$$

$$(1)(3x) = (1)(8)$$

$$3x = 8$$

$$x = \frac{8}{3}$$

Show that $\frac{8}{3}$ is the root of the original equation: $\frac{x}{2} = \frac{4}{3}$ ______

Answer to 111: No. The fraction with denominator "5" would still remain.

113.

$$\frac{7}{8} = \frac{2R}{5}$$

To clear the fractions in one step, you must multiply by both ______ and ______ at the same time.

Answer to 112: Check:

$$\frac{x}{2} = \frac{4}{3}$$

$$\frac{\frac{8}{3}}{2} = \frac{4}{3}$$

$$\left(\frac{8}{3}\right)\left(\frac{1}{2}\right) = \frac{4}{3}$$

$$\frac{8}{6} = \frac{4}{3}$$

$$\frac{4}{3} = \frac{4}{3}$$

114. To clear fractions in one step, you must be able to perform multiplications like the following:

Step 1: $(7)(8)\left(\frac{2y}{7}\right) = \frac{(7)(8)(2y)}{7}$

Step 2: $= \left(\frac{7}{7}\right)(8)(2y)$

Step 3: $= (1)(8)(2y)$

Step 4: $= 16y$

Multiply: (a) $(10)(7)\left(\frac{V}{10}\right) =$ ____ (c) $(9)(4)\left(\frac{5s}{9}\right) =$ ____

(b) $(5)(6)\left(\frac{T}{6}\right) =$ ____ (d) $(3)(7)\left(\frac{12w}{7}\right) =$ ____

Answer: 8 and 5

---

115. What non-fractional equation do you get when you clear both fractions in this equation? $\frac{M}{4} = \frac{5}{7}$

____

Answers: a) 7V c) 20s b) 5T d) 36w

---

116. What non-fractional equation do you get when you clear both fractions in this equation? $\frac{3}{5} = \frac{5x}{6}$

____

Answer: 7M = 20

---

117. To clear the fractions in one step, we multiply by "3" and "2" at the same time. $\frac{x}{3} + 7 = \frac{5x}{2}$

To clear the fractions in one step, we multiply by "3" and "2" at the same time.

$$(3)(2)\left(\frac{x}{3} + 7\right) = (3)(2)\left(\frac{5x}{2}\right)$$

Notice that the left side is an instance of the distributive principle. Watch the multiplication of the left side in Step 1:

Step 1: $(3)(2)\left(\frac{x}{3}\right) + (3)(2)(7) = (3)(2)\left(\frac{5x}{2}\right)$

Step 2: $2x + 42 = 15x$

(a) Solve this last equation. Its root is ____.

(b) The root of the original equation must be ____.

Answer: 18 = 25x

Answers: a) $\frac{42}{13}$ b) $\frac{42}{13}$

118. To be able to clear the fraction in equations like the one in the last frame, you must be able to perform multiplications like the following:

$$(7)(5)\left(\frac{N}{7}+2\right)$$

This multiplication is an instance of the distributive principle. After multiplying by the distributive principle, we simplify each term as follows:

$$(7)(5)\left(\frac{N}{7}+2\right) = (7)(5)\left(\frac{N}{7}\right) + (7)(5)(2)$$
$$= \quad 5N \quad + \quad 70$$

Complete: (a) $(3)(4)\left(\frac{2t}{3}+5\right) = (3)(4)\left(\frac{2t}{3}\right) + (3)(4)(5)$
$= \_\_\_\_\_\_ + \_\_\_\_\_\_$

(b) $(2)(5)\left(\frac{V}{5}-\frac{3}{2}\right) = (2)(5)\left(\frac{V}{5}\right) - (2)(5)\left(\frac{3}{2}\right)$
$= \_\_\_\_\_\_ - \_\_\_\_\_\_$

---

119. Multiply: (a) $(7)(5)\left(\frac{3p}{7}+\frac{4}{5}\right) = \_\_\_\_\_\_$ (b) $(5)(3)\left(4-\frac{2y}{5}\right) = \_\_\_\_\_\_$

a) $8t + 60$

b) $2V - 15$

---

120. Write the non-fractional equation you obtain when you clear the fractions in each of these:

(a) $\frac{y}{6} - 1 = \frac{2y}{5}$ (b) $\frac{q}{4} + 3 = \frac{q}{3} - q$

__________ __________

a) $15p + 28$ b) $60 - 6y$

---

121. Clear the fractions and write the resulting non-fractional equation:

(a) $\frac{3m}{5} + \frac{6}{7} = 0$ (b) $1 = \frac{4d}{3} - \frac{5}{2}$

__________ __________

a) $5y - 30 = 12y$

b) $3q + 36 = 4q - 12q$

---

a) $21m + 30 = 0$

b) $6 = 8d - 15$

122. (a) Clear the fractions and write the resulting equation: $\frac{2x}{5} + 1 = \frac{x}{7}$

______

(b) The root of this new equation is ______.

(c) The root of the original equation must be ______.

---

123. Solve: (a) $1 = \frac{3t}{4} - \frac{2t}{3}$ (b) $\frac{R}{6} - \frac{1}{5} = 0$

t = ______ R = ______

Answers to 122: a) $14x + 35 = 5x$ b) $-\frac{35}{9}$ c) $-\frac{35}{9}$

---

124. $$\frac{5x}{7} + 1 = \frac{x}{2} + \frac{1}{2}$$

If we multiply both sides by "7", we eliminate the fraction on the left.

If we multiply both sides by "2", we eliminate both fractions on the right.

Therefore, we can eliminate all the fractions if we multiply by both ______ and ______ at the same time.

Answers to 123: a) $t = 12$ b) $R = \frac{6}{5}$

---

125. Here is the same equation: $\frac{5x}{7} + 1 = \frac{x}{2} + \frac{1}{2}$

(a) Clear all fractions in one step. What new equation do you get?

______

(b) The root of this new equation is ______.

(c) The root of the original equation must be ______.

Answer to 124: 7 and 2

Answers to 125: a) $10x + 14 = 7x + 7$ b) $-\frac{7}{3}$ c) $-\frac{7}{3}$

126. (a) To clear all fractions in one step, we multiply by both _____ and _____ at the same time.

$$\frac{b}{2} + \frac{3b}{5} = \frac{2b}{5} - 12$$

(b) Do so and write the new equation: ______________

(c) The root of the original equation must be _______.

a) 2 and 5

b) $5b + 6b = 4b - 120$

c) $-\frac{120}{7}$

---

127. Solve this one: $\frac{t}{3} - \frac{4t}{5} = 1 + \frac{t}{3}$

t = _______

$t = -\frac{5}{4}$

(Did you reduce to lowest terms?)

SELF-TEST 7 (Frames 98-127)

Solve the following equations:

1. $\frac{3}{4} + r = \frac{5r}{4}$ r = _______	2. $3 - \frac{7}{2x} = \frac{1}{2x} - 5$ x = _______	3. $\frac{5h}{3} + \frac{1}{2} = 0$ h = _______
4. $\frac{4i}{5} - 3 = \frac{i}{2}$ i = _______	5. $\frac{R}{5} - 1 = \frac{R}{3} + 1$ R = _______	6. $\frac{w}{4} + \frac{3}{5} = 2 - \frac{3w}{5}$ w = _______

ANSWERS: 1. $r = 3$ 2. $x = \frac{1}{2}$ 3. $h = -\frac{3}{10}$ 4. $i = 10$ 5. $R = -15$ 6. $w = \frac{28}{17}$

6-11 EQUATIONS WHICH CONTAIN TWO OR MORE DIFFERENT DENOMINATORS, ONE OF WHICH IS A LETTER

When one of the different denominators is a letter, we can still clear the fractions in one step. We will show in this section that the procedure is the same.

128.

$$\frac{7}{8} = \frac{3}{x}$$

(a) We can eliminate the fraction on the left if we multiply both sides by _____.

(b) We can eliminate the fraction on the right if we multiply both sides by _____.

(c) We can clear both fractions in one step if we multiply by both _____ and _____ at the same time.

---

129. Here is the same equation. Watch how we clear both fractions in one step:

$$\frac{7}{8} = \frac{3}{x}$$

$$(8)(x)\left(\frac{7}{8}\right) = (8)(x)\left(\frac{3}{x}\right)$$

$$7x = 24$$

(a) The root of this last equation is _____.

(b) The root of the original equation must be _____.

Answers to 128: a) 8 b) x c) 8 and x

---

130. (a) To clear the fractions in one step, we multiply by both _____ and _____ at the same time.

$$\frac{3}{y} + \frac{7}{5} = 1$$

(b) Do so. Write the new equation:

_____

(c) Solve the new equation. The root of the original equation must be _____.

_____

Answers to 129: a) $\frac{24}{7}$ b) $\frac{24}{7}$

---

131. Clear the fractions in each of the following in one step:

(a) $\frac{1}{3} = \frac{7}{x} - 4$ (b) $\frac{9}{m} - \frac{1}{2} = 0$

_____ _____

Answers to 130: a) y and 5 b) $15 + 7y = 5y$ c) $-\frac{15}{2}$

132. Solve: (a) $5 + \frac{2}{7} = \frac{3}{P}$ (b) $0 = \frac{7}{y} - \frac{2}{3}$

P = ______ y = ______

a) $x = 21 - 12x$

b) $18 - m = 0$

---

133. (a) We can clear all fractions in one step if we multiply both sides by _____ and _____ at the same time.

$\frac{2}{5} + \frac{3}{x} = \frac{6}{5} - 1$

(b) Do so and write the resulting equation: ______________

(c) Solve the resulting equation. The root of the original equation is _____. ______________

a) $P = \frac{21}{37}$

(From: $35P + 2P = 21$)

b) $y = \frac{21}{2}$

(From: $0 = 21 - 2y$)

---

134. (a) To clear the fractions in one step, we multiply by both _____ and _____.

$\frac{1}{m} - \frac{3}{4} = \frac{1}{4} + \frac{5}{m}$

(b) Do so and write the resulting equation: ______________

(c) The root of the original equation is _____. ______________

a) 5 and x

b) $2x + 15 = 6x - 5x$

c) -15

---

135. This equation contains three different denominators:

$\frac{x}{2} = \frac{2x}{3} + \frac{4}{5}$

(a) To clear all fractions in one step, you must multiply both sides by what? ___________

(b) Do so. Write the resulting equation: ______________

a) m and 4

b) $4 - 3m = m + 20$

c) -4

---

a) 2 and 3 and 5

b) $15x = 20x + 24$

136. (a) To clear all fractions in one step, you must multiply both sides by what? ____________

$$\frac{1}{m} - \frac{3}{2} = \frac{4}{5}$$

(b) Do so. Write the resulting equation: ____________________

137.
$$\frac{x+2}{3} = \frac{x}{4}$$

To clear all fractions in one step, we multiply both sides by 3 and 4. Here is the procedure:

$$(3)(4)\left(\frac{x+2}{3}\right) = (3)(4)\left(\frac{x}{4}\right)$$

$$\frac{(3)(4)(x+2)}{3} = \frac{(3)(4)(x)}{4}$$

$$\left(\frac{3}{3}\right)(4)(x+2) = \left(\frac{4}{4}\right)(3)(x)$$

$$1(4)(x+2) = 1(3)(x)$$

$$4x + 8 = 3x$$

(a) The root of $4x + 8 = 3x$ is _____.

(b) The root of the original equation must be _____.

Answers to 136: a) m and 2 and 5; b) $10 - 15m = 8m$

---

138. To solve equations like the one in the last frame, you must be able to perform multiplications like the following. Notice that we obtain an instance of the distributive principle in Step 4.

Step 1: $(5)(2)\left(\frac{x+3}{2}\right) = \frac{(5)(2)(x+3)}{2}$

Step 2: $= \left(\frac{2}{2}\right)(5)(x+3)$

Step 3: $= (1)(5)(x+3)$

Step 4: $= 5(x+3)$

Step 5: $= 5x + 15$

Multiply:

(a) $(5)(2)\left(\frac{t+7}{5}\right) =$ ____________

(b) $(3)(7)\left(\frac{2y+9}{7}\right) =$ ____________

(c) $(3)(4)\left(\frac{6d-5}{3}\right) =$ ____________

Answers to 137: a) -8; b) -8

Answers to 138: a) $2t + 14$; b) $6y + 27$; c) $24d - 20$

139. (a) Clear the fractions and write the resulting equation:	$\frac{x+9}{5} = \frac{x}{2}$ ________________	
(b) The root of the original equation must be _____.		
140. (a) Clear the fractions and write the resulting equation:	$\frac{2}{3} = \frac{V-2}{4}$ ________________	a) $2x + 18 = 5x$ b) 6
(b) The root of the original equation must be _____.		
141. (a) Clear the fractions in one step, and write the resulting equation:	$\frac{t}{5} - \frac{1}{2} = \frac{3t-1}{5}$ ________________	a) $8 = 3V - 6$ b) $\frac{14}{3}$
(b) The root of the original equation is _____.		
		a) $2t - 5 = 6t - 2$ b) $-\frac{3}{4}$

6-12 EQUATIONS CONTAINING TWO DIFFERENT DENOMINATORS, ONE OF WHICH IS AN ADDITION OR SUBTRACTION

When one of the denominators in a fractional equation is an addition or subtraction, we use the same procedure to clear the fractions. We will show the method in this section.

142. In the following equation, one denominator is an addition:

$$\frac{8}{x+3} = \frac{2}{5}$$

To clear the fraction on the left, we must multiply both sides by "x + 3".

To clear the fraction on the right, we must multiply both sides by "5".

To clear both fractions in one step, we must multiply both sides at the same time by _______ and _______.

---

143. Here is the same equation: $\frac{8}{x+3} = \frac{2}{5}$

To clear the fractions in one step, we multiply by both "x + 3" and "5" at the same time. The procedure is:

Step 1: $(5)(x+3)\left(\frac{8}{x+3}\right) = \left(\frac{2}{5}\right)(5)(x+3)$

Step 2: $\frac{(5)(x+3)(8)}{x+3} = \frac{(2)(5)(x+3)}{5}$

Step 3: $\left(\frac{x+3}{x+3}\right)(5)(8) = \left(\frac{5}{5}\right)(2)(x+3)$

Step 4: $1(5)(8) = (1)(2)(x+3)$

Step 5: $40 = 2(x+3)$

(a) The root of the last equation is _____.

(b) Show that this root satisfies the original equation: $\frac{8}{x+3} = \frac{2}{5}$

Answer: x + 3 and 5

---

144. To solve equations like the one in the last frame, you must be able to perform multiplications like the following:

$$7(x+5)\left(\frac{3}{x+5}\right) = \frac{(7)(x+5)(3)}{x+5}$$

$$= \left(\frac{x+5}{x+5}\right)(7)(3)$$

$$= (1)(7)(3)$$

$$= 21$$

Do these multiplications:

(a) $8(y+2)\left(\frac{7}{y+2}\right) =$ _______

(b) $5(R+6)\left(\frac{3R}{R+6}\right) =$ _______

(c) $7(2b-3)\left(\frac{6}{2b-3}\right) =$ _______

(d) $11(5N-7)\left(\frac{2N}{5N-7}\right) =$ _______

Answers:

a) 17

b) $\frac{8}{x+3} = \frac{2}{5}$

$\frac{8}{17+3} = \frac{2}{5}$

$\frac{8}{20} = \frac{2}{5}$

$\frac{2}{5} = \frac{2}{5}$

145. Here is another type of multiplication which occurs when clearing fractions in one step:

$$\left(\frac{5}{6}\right)(6)(t + 4) = \frac{(5)(6)(t + 4)}{6}$$

$$= \left(\frac{6}{6}\right)(5)(t + 4)$$

$$= (1)(5)(t + 4)$$

$$= 5(t + 4)$$

$$= 5t + 20$$

Notice that we obtained an instance of the distributive principle in the second-last step.

Do these multiplications:

(a) $\left(\frac{3}{7}\right)(7)(x + 6) =$ ______

(b) $\left(\frac{4}{5}\right)(5)(V + 2) =$ ______

(c) $\left(\frac{2}{9}\right)(9)(5t - 3) =$ ______

(d) $\left(\frac{3}{8}\right)(8)(7s - 6) =$ ______

a) 56 b) 15R c) 42 d) 22N

---

146. In each of the following, clear the fractions and write the resulting equation:

(a) $\frac{16}{x + 9} = \frac{4}{5}$

(b) $\frac{5}{2} = \frac{10}{P - 3}$

______ ______

a) $3x + 18$ b) $4V + 8$ c) $10t - 6$ d) $21s - 18$

---

147. Solve each of the following equations:

(a) $\frac{3w}{w + 5} = \frac{4}{3}$ $w =$ ______

(b) $\frac{3}{2} = \frac{3D}{D - 5}$ $D =$ ______

a) $80 = 4x + 36$ b) $5P - 15 = 20$

---

a) $w = 4$ b) $D = -5$

148.
$$\frac{7}{y+2} = \frac{5}{y}$$

To clear the fractions in one step, we multiply by both "y + 2" and "y". Here is the procedure:

Step 1: $(y)(y+2)\left(\frac{7}{y+2}\right) = \left(\frac{5}{y}\right)(y)(y+2)$

Step 2: $\frac{(y)(y+2)(7)}{y+2} = \frac{(5)(y)(y+2)}{y}$

Step 3: $\left(\frac{y+2}{y+2}\right)(y)(7) = \left(\frac{y}{y}\right)(5)(y+2)$

Step 4: $(1)(y)(7) = (1)(5)(y+2)$

Step 5: $7y = 5y + 10$

(a) The root of the last equation is _____.

(b) Show that this root satisfies the original equation: $\frac{7}{y+2} = \frac{5}{y}$

---

149. (a) Clear the fractions and write the resulting equation: $\frac{9}{2x+3} = \frac{3}{x}$

____

(b) The root of the original equation is _____.

Answers to 148:

a) 5

b) $\frac{7}{y+2} = \frac{5}{y}$

$\frac{7}{5+2} = \frac{5}{5}$

$\frac{7}{7} = 1$

$1 = 1$

---

150. (a) Clear the fractions and write the resulting equation: $\frac{7}{q} = \frac{10}{q-2}$

____

(b) The root of the original equation is _____.

Answers to 149:

a) $9x = 6x + 9$

b) 3

Answers to 150:

a) $7q - 14 = 10q$

b) $-\frac{14}{3}$

SELF-TEST 8 (Frames 128-150)

Solve the following equations:

1. $\frac{3}{7} = \frac{5}{V}$ V = ______	2. $\frac{7}{2} = \frac{3}{m} - 4$ m = ______	3. $\frac{1}{P} - 2 = \frac{5}{3}$ P = ______
4. $\frac{t}{2} - \frac{1}{3} = \frac{5t - 3}{2}$ t = ______	5. $\frac{5}{4} = \frac{1}{3a - 4}$ a = ______	6. $\frac{4}{d - 3} = \frac{2}{d}$ d = ______

ANSWERS: 1. $V = \frac{35}{3}$ 2. $m = \frac{2}{5}$ 3. $P = \frac{3}{11}$ 4. $t = \frac{7}{12}$ 5. $a = \frac{8}{5}$ 6. $d = -3$

6-13 A SHORTCUT WHEN ONE NUMERICAL DENOMINATOR IS A MULTIPLE OF ANOTHER

In the following equation, one denominator "10" is a multiple of the other denominator "5".

$$\frac{3x}{10} = \frac{7}{5}$$

When this is the case, there is a shortcut we can use when clearing the fractions. We will explain the shortcut in this section.

151. When a first number is divided by a second and the quotient is a whole number, we say that the first number is a "multiple" of the second.

Example: 10 is a multiple of 5, since:

$$\frac{10}{5} = 2, \text{ and 2 is a whole number.}$$

Example: 11 is not a multiple of 6, since:

$$\frac{11}{6} \text{ does not reduce to a whole number.}$$

(a) Is 16 a multiple of 8? _____ (b) Is 16 a multiple of 6? _____

---

152. In the following equation, 10 is a multiple of 5.

$$\frac{3x}{10} = \frac{7}{5}$$

We can clear the fractions in one step in the usual way. That is, we can multiply both sides by 10 and 5 at the same time. We get:

$$(10)(5)\left(\frac{3x}{10}\right) = (10)(5)\left(\frac{7}{5}\right)$$

$$\left(\frac{10}{10}\right)(5)(3x) = \left(\frac{5}{5}\right)(10)(7)$$

$$(1)(5)(3x) = (1)(10)(7)$$

$$15x = 70$$

However, we can also clear the fractions if we multiply both sides by 10. We get:

$$(10)\left(\frac{3x}{10}\right) = 10\left(\frac{7}{5}\right)$$

$$\left(\frac{10}{10}\right)(3x) = \left(\frac{10}{5}\right)(7)$$

$$(1)(3x) = (2)(7)$$

$$3x = 14$$

Let's examine the two non-fractional equations which we obtained:

(a) The root of $15x = 70$ is _____.

(b) The root of $3x = 14$ is _____.

(c) Are $15x = 70$ and $3x = 14$ equivalent equations? _____

Answers to 151:

a) Yes, since: $\frac{16}{8} = 2$

b) No, since: $\frac{16}{6} = \frac{8}{3}$

Answers to 152:

a) $\frac{14}{3}$

b) $\frac{14}{3}$

c) Yes, since they have the same root.

153. In this equation, 12 is a multiple of 4: $\frac{y}{12} = y + \frac{1}{4}$

(a) What non-fractional equation do you get if you multiply both sides by 12 and 4 at the same time? ________

(b) What non-fractional equation do you get if you multiply both sides by 12 alone? ________

(c) The two non-fractional equations are equivalent because the root of each is ____.

---

a) $4y = 48y + 12$

b) $y = 12y + 3$

c) $-\frac{3}{11}$

---

154. Here is another case in which the larger denominator is a multiple of the smaller:

$$\frac{t}{20} + 1 = \frac{t}{4}$$

We can clear the fractions in either of two ways:

(1) By multiplying both sides by 20 and 4 at the same time, we get:

$$4t + 80 = 20t$$

(2) By multiplying both sides by 20 alone, we get:

$$t + 20 = 5t$$

Both methods are legitimate since they give the same root. <u>But method 2 is a shortcut for this reason</u>: The non-fractional equation which it leads to is easier to solve because the numbers in it are smaller.

(a) The root of $t + 20 = 5t$ is ____.

(b) Show that this root satisfies the original equation: $\frac{t}{20} + 1 = \frac{t}{4}$

155. Equations in which one numerical denominator is a multiple of another are not very common. However, when they do occur, clearing the fractions by multiplying by the larger denominator is a shortcut. Use this shortcut whenever possible.

Can the shortcut be used for:

(a) $\frac{R}{18} - 1 = \frac{R}{3} + 1$? ______ (b) $\frac{2m}{7} - 1 = \frac{m}{3}$? ______

a) $t = 5$

b) Check: $\frac{t}{20} + 1 = \frac{t}{4}$

$\frac{5}{20} + 1 = \frac{5}{4}$

$\frac{1}{4} + \frac{4}{4} = \frac{5}{4}$

$\frac{5}{4} = \frac{5}{4}$

---

156. The shortcut can also be used when there are three different numerical denominators, provided that the larger is a multiple of the two smaller ones.

For example, we can clear all of the fractions in the following equation by multiplying both sides by 10, since 10 is a multiple of both 5 and 2:

What equation do you get when you use the shortcut to clear the fractions: $\frac{q}{2} + \frac{q}{5} = \frac{3q}{10} + 4$

______

a) Yes, since 18 is a multiple of 3.

b) No, since 7 is not a multiple of 3.

---

157. Use the shortcut to clear the fractions in each of these. Write the resulting non-fractional equation:

(a) $1 - \frac{t}{4} = \frac{3t}{8}$ (b) $\frac{2p - 1}{5} = \frac{11}{15}$ (c) $\frac{x}{2} - 1 = \frac{x}{3} - \frac{5x}{6}$

______ ______ ______

$5q + 2q = 3q + 40$

---

158. Solve: (a) $\frac{x+2}{5} = \frac{7}{10}$ (b) $\frac{R}{2} - \frac{R}{4} = \frac{R-1}{8}$

$x =$ ______ $R =$ ______

a) $8 - 2t = 3t$

b) $6p - 3 = 11$

c) $3x - 6 = 2x - 5x$

---

a) $x = \frac{3}{2}$ b) $R = -1$

6-14 EQUATIONS CONTAINING ONE FRACTION WHOSE DENOMINATOR IS AN INSTANCE OF THE DISTRIBUTIVE PRINCIPLE

In each equation below, there is one fraction whose denominator is an instance of the distributive principle:

$$\frac{7}{3(x+2)} = 5 \qquad 0 = \frac{3t}{2(t-9)}$$

In this section, we will solve equations of this type.

---

159. In the following equations, there are two factors in the denominator since 5x means: 5 times x or (5)(x).

$$\frac{7}{5x} = 3$$

If we multiply both sides by "5" alone, we do not clear the fraction since we get:

$$\frac{7}{x} = 15$$

If we multiply both sides by "x" alone, we do not clear the fraction since we get:

$$\frac{7}{5} = 3x$$

To clear the fraction <u>in</u> <u>one</u> <u>step</u>, we must multiply both sides of the original equation by _____ and _____ at the same time.

Answer to 159: 5 and x

160. In the following equation, there is an instance of the distributive principle in the denominator. This instance of the distributive principle is also two factors since 5(x + 1) means: 5 times (x + 1).

$$\frac{7}{5(x+1)} = 3$$

If we multiply both sides by "5" alone, we do not clear the fraction since we get:

$$5\left[\frac{7}{5(x+1)}\right] = 5(3)$$

$$\left(\frac{5}{5}\right)\left(\frac{7}{x+1}\right) = 15$$

$$\frac{7}{x+1} = 15$$

If we multiply both sides by (x + 1) alone, we do not clear the fraction since we get:

$$(x+1)\left[\frac{7}{5(x+1)}\right] = 3(x+1)$$

$$\left(\frac{x+1}{x+1}\right)\left(\frac{7}{5}\right) = 3x + 3$$

$$\frac{7}{5} = 3x + 3$$

To clear the fraction in one step, we must multiply both sides by _______ and _______ at the same time.

161. In this equation, the denominator is an instance of two factors in the distributive principle:

$$\frac{7}{3(t+2)} = 5$$

To clear the fraction, you should multiply both sides by ____________.

Answer: 5 and (x + 1)

---

162. Let's solve the equation in the last frame:

$$\frac{7}{3(t+2)} = 5$$

Step 1: $(3)(t+2)\left[\frac{7}{3(t+2)}\right] = (5)(3)(t+2)$

Step 2: $\frac{(3)(t+2)(7)}{(3)(t+2)} = 15(t+2)$

Step 3: $\left(\frac{3}{3}\right)\left(\frac{t+2}{t+2}\right)(7) = 15(t+2)$

Step 4: $(1)(1)(7) = 15(t+2)$

Step 5: $7 = 15(t+2)$

(a) The root of the last equation (Step 5) is ______.

(b) The root of the original equation must be ______.

Answer: (3)(t + 2)

---

163. To solve equations like the one in the last frame, you must be able to perform multiplications like the following:

$$5(x+7)\left[\frac{9}{5(x+7)}\right] = \frac{(5)(x+7)(9)}{(5)(x+7)} = \left(\frac{5}{5}\right)\left(\frac{x+7}{x+7}\right)(9)$$

$$= (1)(1)(9)$$

$$= 9$$

Do these multiplications:

(a) $9(b+2)\left[\frac{5}{9(b+2)}\right] =$ ______

(b) $6(M+3)\left[\frac{8}{6(M+3)}\right] =$ ______

(c) $11(V-5)\left[\frac{14}{11(V-5)}\right] =$ ______

Answers:

a) $-\frac{23}{15}$

b) $-\frac{23}{15}$

---

164. You must also be able to perform multiplications like the following. Though there are three factors, we multiply two at a time:

$$(4)(2)(x+5)$$

$$(8)\ (x+5) = 8x + 40$$

Multiply:

(a) $(5)(4)(y+3) =$ ____________

(b) $(7)(2)(s+2) =$ ____________

(c) $(4)(3)(t-6) =$ ____________

Answers:

a) 5

b) 8

c) 14

165. What non-fractional equation do you obtain when you clear the fractions in each of the following?

(a) $\frac{4}{2(p+3)} = 5$ (b) $4 = \frac{9}{2(q-3)}$

______ ______

a) $20y + 60$

b) $14s + 28$

c) $12t - 72$

---

166. Solve: (a) $2 = \frac{23}{5(x+2)}$ (b) $\frac{7}{3(y-5)} = 5$

x = ______ y = ______

a) $4 = 10p + 30$

b) $8q - 24 = 9$

---

167. In this one, one side of the equation is "0". Watch the steps:

$$\frac{5x}{3(x+3)} = 0$$

Step 1: $(3)(x+3)\left[\frac{5x}{3(x+3)}\right] = (0)(3)(x+3)$

Step 2: $\left(\frac{3}{3}\right)\left(\frac{x+3}{x+3}\right)(5x) = (0)\ (x+3)$

Step 3: $(1)(1)(5x) = 0$

Therefore $5x = 0$ and $x =$ ______

a) $x = \frac{3}{10}$

b) $y = \frac{82}{15}$

---

168. To solve equations like the one in the last frame, you must recognize that the product of any multiplication is "0" as long as one of the factors is "0".

Complete: (a) $(0)(5)(x+4)$

$(0)\ (x+4) =$ ______

(b) $(0)(3)(7)(y-1)$

$(0)\ (7)(y-1)$

$(0)\ (y-1) =$ ______

0

---

169. (a) Clear the fraction in this equation, and write the resulting equation: $\frac{3t}{2(t-5)} = 0$

______

(b) Therefore, the root of the original equation is ______.

a) 0 b) 0

170. Here is the same equation with the "0" replaced by "1":

$$\frac{3t}{2(t - 5)} = 1$$

(a) Clear the fraction, and write the resulting fraction:

____

(b) The root of the original equation is _______.

a) $3t = 0$

b) 0

---

171. Solve: (a) $\frac{5t}{3(t + 1)} = 0$ (b) $\frac{5t}{3(t + 1)} = 1$

$t =$ _____ $t =$ _____

a) $3t = 2(t - 5)$

or

$3t = 2t - 10$

b) -10

---

a) $t = 0$

b) $t = \frac{3}{2}$

SELF-TEST 9 (Frames 151-171)

Solve the following equations:

1. $\frac{5a - 3}{12} = \frac{5a}{3}$ $a =$ _____	2. $\frac{3y}{20} - 7 = \frac{y}{4}$ $y =$ _____	3. $\frac{t}{12} - \frac{t}{2} = \frac{1}{4} - t$ $t =$ _____
4. $\frac{9}{5(2E - 3)} = 3$ $E =$ _____	5. $\frac{4h}{3(2h + 2)} = 1$ $h =$ _____	6. $\frac{24R}{5(R - 7)} = 0$ $R =$ _____

ANSWERS: 1. $a = -\frac{1}{5}$ 2. $y = -70$ 3. $t = \frac{3}{7}$ 4. $E = \frac{9}{5}$ 5. $h = -3$ 6. $R = 0$

6-15 DETERMINING THE MULTIPLIER FOR MORE-COMPLICATED EQUATIONS WITH TWO-FACTOR DENOMINATORS

Our general strategy for solving a fractional equation is:

(1) Clear the fractions.
(2) Then solve the resulting non-fractional equation.

To clear the fractions in an equation, we use the multiplication axiom. The "multiplier" is determined by examining each denominator. We can clear all fractions in one step by using the proper "multiplier."

When an equation contains one or more two-factor denominators, it is sometimes difficult to determine the proper "multiplier." The purpose of this section is to develop some skill in choosing the correct "multiplier."

---

172. What "multiplier" would we use to clear the fractions <u>in one step</u> in each of the following equations?

(a) $\frac{y}{5} + 1 = \frac{3y}{7}$ ______

(c) $\frac{1}{R} + \frac{1}{3} = \frac{1}{2}$ ______

(b) $\frac{t+1}{t-1} = \frac{7}{8}$ ______

(d) $\frac{p}{2} - \frac{1}{3} = \frac{4p}{12}$ ______

---

173. To clear the fractions <u>in one step</u>, the "multiplier" must eliminate each denominator. If a denominator contains more than one factor, the "multiplier" must eliminate each factor.

In this equation, the denominator on the left contains two factors: $\frac{4}{3x} = \frac{1}{2}$

To clear the fractions, the "multiplier" must clear the "3", the "x", and the "2". What multiplier will do so? ______

Answers to 172:
a) $(5)(7)$
b) $(8)(t - 1)$
c) $(2)(3)(R)$
d) 12, since 12 is a multiple of both 2 and 3.

---

174. Here is another example: $\frac{4}{3m} = \frac{1}{m} + 2$

We can clear the "m" in each denominator by multiplying both sides by "m".

What multiplier can we use to clear both fractions in one step? ______

Answer to 173: $(3)(x)(2)$ or $(2)(3)(x)$

---

175. $\frac{1}{t} + 2 = \frac{2}{3} - \frac{2}{3t}$

We can clear both "t's" by multiplying by "t". We can clear both "3's" by multiplying by "3".

Therefore, we can clear all fractions in one step by multiplying both sides by ______.

Answer to 174: $(3)(m)$

---

176. $\frac{1}{x} + \frac{7}{8} = \frac{3}{4}$

Since 8 is a multiple of 4, we can clear both the 8 and the 4 by multiplying by 8.

Therefore, we can clear all fractions in one step by multiplying both sides by ______.

Answer to 175: $(3)(t)$

177. $$\frac{q+3}{5q} = \frac{3}{10q} + 4$$ We can clear both the 10 and the 5 by multiplying by 10. Therefore, we can clear all the fractions in one step by multiplying both sides by _________.	(8)(x)
178. $$\frac{4}{3V} - \frac{7}{4V} = \frac{5}{12}$$ (a) We can clear the 3, the 4, and the 12 by multiplying both sides by _____. (b) Therefore, what multiplier can we use to clear the fractions in one step? _________	(10)(q)
179. What multiplier can we use to clear all fractions in one step in each of the following? (a) $\frac{7}{5x} = \frac{9}{2}$ ______ (b) $\frac{5}{R} = \frac{R-1}{4R} + 3$ ______ (c) $\frac{2}{t} - \frac{1}{7t} = \frac{3}{2}$ ______	a) 12, since 12 is a multiple of 3 and 4. b) (12)(V)
180. What multiplier should we use for each of these? (a) $\frac{x-1}{7x} = \frac{13}{14}$ _____ (b) $\frac{1}{3D} - \frac{5}{D} = \frac{7}{9}$ _____ (c) $\frac{9}{2y} - \frac{7}{5y} = \frac{1}{10y} - 2$ _____	a) (2)(5)(x) b) (4)(R) c) (2)(7)(t)
181. In this case, the denominator on the right is an addition: $\frac{7}{3t} = \frac{5}{t+4}$ To clear the fraction on the left, we must multiply by (3)(t). To clear the fraction on the right, we must multiply by (t + 4). Therefore, to clear all fractions in one step, we must multiply by __________.	a) (14)(x) b) (9)(D) c) (10)(y)
182. In this case, the denominator on the left is an instance of the distributive principle. It contains two factors: 2 and (x + 3). $$\frac{5}{2(x+3)} = \frac{3}{7}$$ To clear the fraction on the left, we must multiply by 2(x + 3). Therefore, to clear both fractions in one step, we must multiply both sides by __________.	(3)(t)(t + 4)
183. $$\frac{3}{5} = \frac{7t}{5(t-1)}$$ We can clear both 5's by multiplying by 5. Therefore, we can clear <u>all</u> fractions by multiplying by __________.	(7)(2)(x + 3)
	5(t - 1)

184. $\dfrac{5}{2(x+3)} = \dfrac{3}{4}$ We can clear both the 4 and the 2 by multiplying by 4. Therefore, we can clear all fractions by multiplying by ________.	
185. $\dfrac{7}{3(m-1)} = \dfrac{5}{m-1}$ We can clear the (m - 1) in each denominator by multiplying by (m - 1). To clear all fractions in one step, we must multiply both sides by ______.	4(x + 3)
186. To clear all fractions in one step, we must multiply each of the following equations by what? (a) $\dfrac{m}{6(m-5)} = \dfrac{1}{6}$ ________ (b) $\dfrac{5}{9} = \dfrac{R-1}{3(R+1)}$ ________ (c) $\dfrac{2p-7}{5(p-1)} = \dfrac{1}{3}$ ________	3(m - 1)
187. What multiplier would we use to clear all fractions in each of these? (a) $\dfrac{3}{5(V-2)} = \dfrac{9}{5V}$ ________ (b) $\dfrac{m+1}{2(m-1)} = \dfrac{4}{m-1}$ ________	a) (6)(m - 5) b) (9)(R + 1) c) (3)(5)(p - 1)
	a) (5)(V)(V - 2) b) (2)(m - 1)

SELF-TEST 10 (Frames 172-187)

Directions: For each equation list the multiplier that should be used to clear the fractions in a single step. Do not actually clear the fractions or solve the equation. Just list the multiplier.

1. $\dfrac{a}{3} + 7 = \dfrac{2a}{5}$ ________
2. $\dfrac{5}{2} - \dfrac{1}{2y} = \dfrac{3}{y}$ ________
3. $\dfrac{1}{8} + \dfrac{4}{w} = \dfrac{2}{3}$ ________
4. $\dfrac{11}{12} - \dfrac{7}{3r} = \dfrac{9}{2r}$ ________
5. $\dfrac{3}{2(x+1)} = \dfrac{1}{x}$ ________
6. $\dfrac{d-3}{5d} = \dfrac{9}{4d}$ ________
7. $\dfrac{2}{E} - \dfrac{1}{9E} = \dfrac{7}{2}$ ________
8. $\dfrac{P-2}{4(P+2)} = \dfrac{3P}{P+2}$ ________

ANSWERS:

1. (3)(5)
2. (2)(y)
3. (3)(8)(w)
4. (12)(r)
5. (2)(x)(x + 1)
6. (4)(5)(d)
7. (2)(9)(E)
8. (4)(P + 2)

6-16 SOLVING MORE-COMPLICATED EQUATIONS WITH ONE OR MORE TWO-FACTOR DENOMINATORS

Having decided on the multiplier to use in order to clear the fractions in a more-complicated equation with one or more two-factor denominators, we must clear the fractions and solve the resulting non-fractional equation. In this section, we will solve equations of that type. Our main emphasis will be on clearing the fractions.

188. To clear the fractions in this one, we must multiply both sides by "3x":

$$\frac{4}{3x} = \frac{1}{x} + 2$$

Do so and write the resulting equation: ______________

189. To clear the fractions in this equation, we multiply by "4V":

$$\frac{1}{4} = 1 - \frac{3}{2V}$$

Do so and write the resulting equation: ______________

Answer to 188: $4 = 3 + 6x$

190. Clear the fractions in each of these and write the resulting non-fractional equation:

(a) $\frac{1}{x} + 2 = \frac{2}{3} - \frac{2}{3x}$ (b) $\frac{1}{m} = \frac{5}{3m} - \frac{1}{6}$

______________ ______________

Answer to 189: $V = 4V - 6$

191. To clear all fractions in this equation, we must multiply by (2)(3)(y):

$$\frac{4}{3y} = \frac{1}{2}$$

Do so and write the resulting equation: ______________

Answer to 190:
a) $3 + 6x = 2x - 2$
b) $6 = 10 - m$

192. Clear the fractions in this one. Write the resulting equation:

$$\frac{1}{5} = 3 - \frac{1}{2d}$$

____

Answer to 191: $8 = 3y$

Answer to 192: $2d = 30d - 5$

193.
$$\frac{t+4}{2t} = \frac{1}{3}$$

To clear the fractions in one step, we must multiply by (3)(2)(t). Here is the procedure:

$$(3)(2)(t)\left[\frac{t+4}{2t}\right] = (3)(2)(t)\left(\frac{1}{3}\right)$$

$$\frac{(3)(2)(t)(t+4)}{2t} = \frac{(3)(2)(t)(1)}{3}$$

$$\left(\frac{2}{2}\right)\left(\frac{t}{t}\right)(3)(t+4) = \left(\frac{3}{3}\right)(2)(t)$$

$$3(t+4) = 2t$$

(a) The root of the last equation is ______.

(b) Show that this root satisfies the original equation: $\frac{t+4}{2t} = \frac{1}{3}$

---

194. To solve equations like the one in the last frame, you must be able to perform multiplications like the following:

$$(2)(4)(x)\left(\frac{x+3}{4x}\right) = \frac{(2)(4)(x)(x+3)}{4x} = \left(\frac{4}{4}\right)\left(\frac{x}{x}\right)(2)(x+3)$$

$$= (1)(1)(2)(x+3)$$

$$= 2(x+3)$$

$$= 2x + 6$$

Do these multiplications:

(a) $(5)(3)(y)\left(\frac{y+2}{5y}\right) =$ ______

(b) $(3)(4)(t)\left(\frac{t+7}{4t}\right) =$ ______

(c) $(2)(7)(m)\left(\frac{m-5}{7m}\right) =$ ______

(d) $(7)(4)(V)\left(\frac{V-4}{7V}\right) =$ ______

Answers to 193:

a) -12

b) Check:

$$\frac{t+4}{2t} = \frac{1}{3}$$

$$\frac{(-12)+4}{2(-12)} = \frac{1}{3}$$

$$\frac{-8}{-24} = \frac{1}{3}$$

$$\frac{1}{3} = \frac{1}{3}$$

---

195. Clear the fractions and write the resulting equation for each:

(a) $\frac{x+2}{5x} = \frac{1}{10}$ ______

(b) $\frac{5}{3p} = \frac{p-3}{4p}$ ______

Answers to 194:

a) 3y + 6

b) 3t + 21

c) 2m - 10

d) 4V - 16

Answers to 195:

a) 2x + 4 = x
Multiplier is (10)(x)

b) 20 = 3p - 9
Multiplier is (3)(4)(p)

196. Clear the fractions and write the resulting equation:

$$\frac{q+3}{5q} = 1 - \frac{3}{10q}$$

____

197.

$$\frac{7}{3t} = \frac{5}{t-4}$$

To clear all fractions in one step, we multiply both sides by "3", "t", and "t - 4". Here is the procedure:

Step 1: $(3)(t)(t-4)\left(\frac{7}{3t}\right) = (3)(t)(t-4)\left(\frac{5}{t-4}\right)$

Step 2: $\frac{(3)(t)(t-4)(7)}{3t} = \frac{(3)(t)(t-4)(5)}{t-4}$

Step 3: $\left(\frac{3}{3}\right)\left(\frac{t}{t}\right)(7)(t-4) = \left(\frac{t-4}{t-4}\right)(3)(t)(5)$

Step 4: $(1)(1)(7)(t-4) = (1)(3)(t)(5)$

Step 5: $7t - 28 = 15t$

The root of the original equation is: t = ______

Answer (196): $2q + 6 = 10q - 3$

198. Clear the fractions and write the resulting equation for each:

(a) $\frac{3}{4(y-7)} = \frac{5}{y}$ ____________

(b) $\frac{4}{3} = \frac{R}{5(R+2)}$ ____________

Answer (197): $t = -\frac{7}{2}$

199. Watch how we clear the fractions in the following equation:

$$\frac{3x+2}{5(x+2)} = \frac{4}{7}$$

Step 1: $(5)(7)(x+2)\left(\frac{3x+2}{5(x+2)}\right) = \left(\frac{4}{7}\right)(5)(7)(x+2)$

Step 2: $\left(\frac{5}{5}\right)\left(\frac{x+2}{x+2}\right)(7)(3x+2) = \left(\frac{7}{7}\right)(4)(5)(x+2)$

Step 3: $(1)(1)(7)(3x+2) = (1)(4)(5)(x+2)$

Step 4: $7(3x+2) = 20(x+2)$

Step 5: $21x + 14 = 20x + 40$

(Continued on following page.)

Answer (198):
a) $3y = 20y - 140$
b) $20R + 40 = 3R$

199. (Continued)

Clear the fractions in this equation: $\frac{2}{5} = \frac{t - 2}{7(t + 3)}$

____

---

200. (a) What multiplier should be used to clear the fractions? ________

(b) Clear the fractions and write the resulting equation:

$\frac{m - 1}{2(m + 1)} = \frac{3}{4}$

____

Answer (199): $14t + 42 = 5t - 10$

---

201. Solve: (a) $\frac{7}{10} - \frac{1}{2m} = \frac{3}{5m}$ (b) $\frac{S - 4}{2S} = \frac{3}{5S}$

m = ______ S = ______

Answers (200): a) $4(m + 1)$ b) $2m - 2 = 3m + 3$

---

202. Solve: (a) $\frac{7}{4} = \frac{5}{3(V - 2)}$ (b) $\frac{k + 4}{5(k - 2)} = \frac{2}{3}$

V = ______ k = ______

Answers (201): a) $m = \frac{11}{7}$ b) $S = \frac{26}{5}$

Answers (202): a) $V = \frac{62}{21}$ b) $k = \frac{32}{7}$

SELF-TEST 11 (Frames 188-202)

Solve the following equations:

1. $\frac{1}{2R} - 3 = \frac{2}{R}$ R = ______	2. $\frac{3}{4w} - \frac{1}{20} = \frac{2}{5w}$ w = ______	3. $\frac{h+2}{3h} = \frac{4}{9h} + 1$ h = ______
4. $\frac{2}{3(d+1)} = \frac{1}{d}$ d = ______	5. $\frac{t-2}{4t} = \frac{1}{3t}$ t = ______	6. $\frac{x-1}{3(x+1)} = \frac{2x}{x+1}$ x = ______

ANSWERS: 1. $R = -\frac{1}{2}$ 2. $w = 7$ 3. $h = \frac{1}{3}$ 4. $d = -3$ 5. $t = \frac{10}{3}$ 6. $x = -\frac{1}{5}$

6-17 EQUATIONS IN WHICH A FRACTION WITH A COMPLEX NUMERATOR APPEARS WITH ANOTHER TERM ON ONE SIDE

In each of the following equations, a fraction with a complex numerator appears with another term on one side:

$$x - \frac{x+2}{3} = 5 \qquad 3 - \frac{4(m-1)}{7} = 2m$$

The main problem in solving equations of this type is "clearing the fractions." We will show the method in this section.

203. In the following equation, there are two terms ("1" and "$\frac{t+5}{4}$") on the left side:

$$1 + \frac{t+5}{4} = t$$

To clear the fraction, we multiply both sides by "4". We get:

<u>Step 1</u>: $4\left(1 + \frac{t+5}{4}\right) = 4(t)$

<u>Step 2</u>: $4(1) + 4\left(\frac{t+5}{4}\right) = 4t$

<u>Step 3</u>: $4 + \left(\frac{4}{4}\right)(t+5) = 4t$

<u>Step 4</u>: $4 + (t+5) = 4t$

<u>Step 5</u>: $4 + t + 5 = 4t$

<u>Notice these two points</u>:

(1) In Step 1, there is an instance of the distributive principle on the left side. We show the multiplication by the distributive principle in Step 2.

(2) In Step 4, there is a grouping after an addition symbol. As we showed before, the grouping can be dropped in this case. We did so in Step 5.

---

204. To clear the fractions in equations like those in the last frame, you must be able to do multiplications by the distributive principle like the following:

$$5\left(p + \frac{2(p-1)}{5}\right) = 5(p) + 5\left(\frac{2(p-1)}{5}\right)$$

$$= 5p + \left(\frac{5}{5}\right)(2)(p-1)$$

$$= 5p + 2(p-1)$$

$$= 5p + 2p - 2$$

Do these multiplications:

(a) $7\left(2 + \frac{5x-1}{7}\right) =$ ____________________

(b) $7\left(3d + \frac{5(d-4)}{7}\right) =$ ____________________

Go to next frame.

---

205. For each of these, clear the fraction and write the resulting equation:

(a) $V + \frac{3V+5}{2} = 4$ (b) $5S = 1 + \frac{3(S-2)}{4}$

____________________ ____________________

a) $14 + 5x - 1$

b) $21d + 5d - 20$

206. Solve: (a) $2F + \frac{F - 1}{5} = 3$ (b) $2x = 1 + \frac{7(x - 2)}{5}$

F = ______ x = ______

a) $2V + 3V + 5 = 8$

b) $20S = 4 + 3S - 6$

Answer to Frame 206: a) $F = \frac{16}{11}$ b) $x = -3$

From: $10F + F - 1 = 15$ From: $10x = 5 + 7x - 14$

207. In this equation, the two terms on the left side are separated by a subtraction symbol. Students frequently make sign errors in solving this type of equation:

$$y - \frac{y + 3}{4} = 1$$

Here are the steps for clearing the fraction:

Step 1: $4\left(y - \frac{y + 3}{4}\right) = 4(1)$

Step 2: $4(y) - 4\left(\frac{y + 3}{4}\right) = 4$

Step 3: $4y - \left(\frac{4}{4}\right)(y + 3) = 4$

Step 4: $4y - (y + 3) = 4$

Step 5: $4y + (-y) + (-3) = 4$

Notice these two points:

(1) In Step 1, there is an instance of the distributive principle over subtraction on the left side. We show the multiplication in Step 2.

(2) In Step 2, the grouping appears after a subtraction symbol. IMPORTANT: We cannot drop the grouping until we convert this subtraction to addition as we did in Step 5.

208. When a fraction with a complex numerator follows a subtraction symbol, we always meet the subtraction of a grouping when clearing the fraction. Let's review the conversion of the subtraction of a grouping to addition:

(a) Since "the opposite of (y + 5)" is "(-y) + (-5)",

$3y - (y + 5) = 3y +$ ______ $+$ ______

(b) Since "the opposite of [2d + (-1)]" is "(-2d) + 1",

$10 - [2d + (-1)] = 10 +$ ______ $+$ ______

Go to next frame.

209. When a grouping contains a subtraction, we convert the subtraction within the grouping to addition before converting the subtraction of the grouping to addition. For example:

(a) $m - (2m - 3) = m - [2m + (-3)]$

$= m + \_\_\_\_ + \_\_\_\_$

(b) $1 - (7 - x) = 1 - [7 + (-x)]$

$= 1 + \_\_\_\_ + \_\_\_\_$

a) $+ (-y) + (-5)$

b) $+ (-2d) + 1$

---

210. When an instance of the distributive principle follows a subtraction symbol, we multiply by the distributive principle before converting the subtraction to addition. Note: Always "bracket" the instance of the distributive principle first.

Complete: (a) $3t - 2(t + 4) = 3t - [2(t + 4)] = 3t - [2t + 8]$

$= 3t + \_\_\_\_ + \_\_\_\_$

(b) $4 - 5(y + 2) = 4 - [5(y + 2)] = 4 - [5y + 10]$

$= 4 + \_\_\_\_ + \_\_\_\_$

a) $+ (-2m) + 3$

b) $+ (-7) + x$

---

211. In the examples below, an instance of the distributive principle over subtraction follows a subtraction symbol. Notice the steps below:

(1) We bracket the instance of the distributive principle.
(2) We multiply by the distributive principle over subtraction.
(3) We convert the subtraction within the grouping to addition.
(4) We convert the subtraction of the grouping to addition.

Complete: (a) $2y - 4(y - 7) = 2y - [4(y - 7)] = 2y - [4y - 28]$

$= 2y - [4y + (-28)]$

$= 2y + \_\_\_\_ + \_\_\_\_$

(b) $1 - 3(1 - 2R) = 1 - [3(1 - 2R)] = 1 - [3 - 6R]$

$= 1 - [3 + (-6R)]$

$= 1 + \_\_\_\_ + \_\_\_\_$

a) $+ (-2t) + (-8)$

b) $+ (-5y) + (-10)$

---

a) $+ (-4y) + 28$

b) $+ (-3) + 6R$

212. Convert each of the following subtractions of an instance of the distributive principle to addition. Note that the first step is to put brackets around the instance of the distributive principle.

Complete: (a) $5 - 4(2 + 3t) = 5 - [4(2 + 3t)]$

$= 5 - [(\quad) + (\quad)]$

$= 5 + \_\_\_\_\_ + \_\_\_\_\_$

(b) $3x - 2(x - 5) = 3x - [2(x - 5)]$

$= 3x - [(\quad) - (\quad)]$

$= 3x - [(\quad) + (\quad)]$

$= 3x + \_\_\_\_\_ + \_\_\_\_\_$

Answers to 212:

a) $5 - [8 + 12t]$
$5 + (-8) + (-12t)$

b) $3x - [2x - 10]$
$3x - [2x + (-10)]$
$3x + (-2x) + 10$

213. In the equations we are solving, you will encounter instances of the distributive principle like this one. Watch the steps we use to multiply by the distributive principle and to convert all subtractions to additions:

$$5\left(3m - \frac{4(m-3)}{5}\right) = 5(3m) - 5\left(\frac{4(m-3)}{5}\right)$$

$$= 15m - \left(\frac{5}{5}\right)(4)(m-3)$$

$$= 15m - (4m - 12)$$

$$= 15m - [4m + (-12)]$$

$$= 15m + (-4m) + 12$$

For each of these, multiply by the distributive principle and convert all subtractions to additions:

(a) $7\left(9 - \frac{2p+5}{7}\right) = \_\_\_\_\_\_\_\_\_\_$

(b) $9\left(5x - \frac{3(x+1)}{9}\right) = \_\_\_\_\_\_\_\_\_\_$

(c) $3\left(1 - \frac{V-6}{3}\right) = \_\_\_\_\_\_\_\_\_\_$

(d) $5\left(2R - \frac{6(R-1)}{5}\right) = \_\_\_\_\_\_\_\_\_\_$

Answers to 213:

a) $63 + (-2p) + (-5)$

b) $45x + (-3x) + (-3)$

c) $3 + (-V) + 6$

d) $10R + (-6R) + 6$

214. Clear the fraction in each equation, convert all subtractions to additions, and write the resulting equation:

(a) $y - \frac{y+7}{3} = 5$

(b) $10 - \frac{5(t+2)}{4} = 3t$

__________ __________

215. Clear the fraction in each equation, convert all subtractions to additions, and write the resulting equation:

(a) $3m - \frac{m - 9}{5} = 1$ ____________

(b) $1 - \frac{5(f - 7)}{6} = f$ ____________

a) $3y + (-y) + (-7) = 15$

b) $40 + (-5t) + (-10) = 12t$

---

216. Solve: (a) $2y - \frac{5y + 6}{3} = 4$ (b) $4 - \frac{2(m - 3)}{5} = 3m$

y = ______ m = ______

a) $15m + (-m) + 9 = 5$

b) $6 + (-5f) + 35 = 6f$

---

a) $y = 18$

From:

$6y + (-5y) + (-6) = 12$

b) $m = \frac{26}{17}$

From:

$20 + (-2m) + 6 = 15m$

SELF-TEST 12 (Frames 203-216)

Solve the following equations:

1. $2w - \frac{w+4}{5} = 10$

w = ______

2. $i - \frac{5i-2}{3} = 4$

i = ______

3. $10 - \frac{2(3p+5)}{4} = 3p$

p = ______

4. $6E - \frac{3(4E-6)}{4} = 2$

E = ______

ANSWERS: 1. $w = 6$ 2. $i = -5$ 3. $p = \frac{5}{3}$ 4. $E = -\frac{5}{6}$

Chapter 7 INTRODUCTION TO GRAPHING

Mathematicians, scientists and technicians frequently communicate with each other by means of graphs. If you want to learn what someone else is communicating in a graph, you must be able to read his graph correctly. If you want to communicate something to someone else by means of a graph, you must be able to construct a graph correctly.

The purpose of this chapter is to develop some basic skills in reading and constructing graphs. In order to do so, the basic principles of the coordinate system used in graphing will be discussed. This chapter is merely an introduction to graphing. However, a knowledge of the basic principles which it contains are essential for an understanding of more advanced types of graphing.

7-1 EQUATIONS CONTAINING TWO DIFFERENT LETTERS

Up to this point, we have not encountered equations which contain two different letters. Equations of this type occur frequently in mathematics. We will examine equations containing two different letters in this section.

1. Here is an equation which contains two different letters: $$y = 2x$$ In this equation: (a) If we plug in 3 for "x", $y = 2(3) =$ ____ (b) If we plug in 10 for "x", $y = 2(10) =$ ____	
2. Here is another equation which contains two different letters: $$p = 3q - 1$$ (a) If we plug in 2 for q, $p = 3(2) - 1 =$ ____ (b) If we plug in -1 for q, $p = 3(-1) - 1 =$ ____	a) 6 b) 20
3. $$y = \frac{10}{x}$$ (a) If we plug in 2 for x, $y = \frac{10}{2} =$ ____ (b) If we plug in -5 for x, $y = \frac{10}{-5} =$ ____ (c) If we plug in 30 for x, $y = \frac{10}{30} =$ ____	a) 5 b) -4
	a) 5 b) -2 c) $\frac{1}{3}$

4. $$m = \frac{t}{20}$$

(a) If we plug in 40 for t, m = _____.

(b) If we plug in -60 for t, m = _____.

(c) If we plug in 5 for t, m = _____.

---

5. $$xy = 20$$

(a) If we plug in "5" for x, we get: 5y = 20. Then y = _____.

(b) If we plug in "-2" for x, we get: -2y = 20. Then y = _____.

a) 2
b) -3
c) $\frac{1}{4}$

---

6. In the equation below, we meet an expression like "x^2" for the first time:

$$y = x^2$$

"x^2" is a shorthand way of writing "x times x" or (x)(x).

(a) $(-7)^2$ means ( )( ) (b) p^2 means ( )( )

a) 4
b) -10

---

7. Mathematicians use either of the following names for "x^2":

(1) "x squared" or (2) "x to the <u>second</u> power."

Similarly: (a) 5^2 is called "5 ____________."

(b) t^2 is called "t to the ____________ power."

a) (-7)(-7)
b) (p)(p)

---

8. $$y = x^2$$

(a) If we plug in 2 for x, $y = 2^2 = (2)(2) =$ _____

(b) If we plug in -5 for x, $y = (-5)^2 = (-5)(-5) =$ _____

a) 5 <u>squared</u>
b) <u>second</u>

---

9. When a number is squared, the answer is called "the square" of the original number. For example:

25 is "the square" of both +5 and -5

since (5)(5) = 25

and (-5)(-5) = 25.

(a) The square of -6 is _____.

(b) 81 is the square of both _____ and _____.

a) 4
b) +25

---

10. The square of any number is <u>always positive</u>. That is:

(a) $(-1)^2 =$ _____ (b) $(-8)^2 =$ _____ (c) $(-10)^2 =$ _____

a) 36
b) +9 and -9

---

11. $$m = q^2 - 5$$

(a) If we plug in 4 for q, $m = 4^2 - 5 =$ ____________

(b) If we plug in -10 for q, $m = (-10)^2 - 5 =$ ____________

a) +1 b) +64 c) +100

a) $16 - 5 = 11$

b) $100 - 5 = 95$

---

12. $$y = 4x^2$$

In the equation above, $4x^2$ means "4 times x^2" or $(4)(x^2)$.

When plugging in a number for x, we perform the squaring <u>before</u> multiplying by 4.

(a) If $x = 3$, $y = 4\underbrace{(3)^2}$
$= 4(9) =$ _____

(b) If $x = -2$, $y = 4\underbrace{(-2)^2}$
$= 4\ (4) =$ _____

a) 36 b) 16

---

13. $$y = 3x^2 + 7$$

(a) If $x = 2$, $y = 3(2)^2 + 7$
$= 3(4) + 7 =$ ______________

(b) If $x = -1$, $y = 3(-1)^2 + 7$
$= 3(+1) + 7 =$ ______________

a) $12 + 7 = 19$

b) $3 + 7 = 10$

---

14. $$d = 2F^2 - 5$$

(a) If $F = 4$, $d =$ _____ (b) If $F = -1$, $d =$ _____

a) 27, since:
$2(4)^2 - 5 = 2(16) - 5$

b) -3, since:
$2(-1)^2 - 5 = 2(1) - 5$

---

15. In the equation below, we meet an expression like "x^3" for the first time:

$$y = x^3$$

"x^3" is a shorthand way of writing either: (1) x times x times x
or (2) (x)(x)(x).

Similarly:

(a) $(-2)^3 = (\ \ )(\ \ )(\ \ )$ (b) $F^3 = (\ \ )(\ \ )(\ \ )$

a) $(-2)(-2)(-2)$

b) $(F)(F)(F)$

---

16. Mathematicians use either of the following names for "x^3":

(1) "x cubed" or (2) "x to the <u>third</u> power."

Similarly: (a) 4^3 is called "4 ______________."

(b) b^3 is called "b to the ______________ power."

a) 4 <u>cubed</u>

b) <u>third</u>

---

17. $$y = x^3$$

(a) If we plug in 3 for x, $y = (3)(3)(3) =$ _____

(b) If we plug in -2 for x, $y = (-2)(-2)(-2) =$ _____

a) 27

b) -8

---

18. When a number is cubed, the answer is called "the cube" of the original number. For example:

8 is "the cube" of +2.
-8 is "the cube" of -2.

(a) The cube of -3 is _____. (b) The cube of +4 is _____.

a) -27

b) 64

19. The cube of any positive number is positive.
The cube of any negative number is negative.

(a) 1^3 = ____ (b) $(-1)^3$ = ____ (c) $(-5)^3$ = ____

---

20. $t = d^3 - 3$

(a) If we plug in +2 for d, $t = 2^3 - 3$ = ____________

(b) If we plug in -1 for d, $t = (-1)^3 - 3$ = ____________

a) +1 b) -1 c) -125

---

21. $y = 2x^3$

In this equation, the expression $2x^3$ means: "2 times x^3" or $(2)(x^3)$

When plugging in a number for x, we cube it before multiplying by 2.

(a) If $x = 3$, $y = 2(3)^3$
$= 2(27) =$ ____

(b) If $x = -2$, $y = 2(-2)^3$
$= 2(-8) =$ ____

a) $8 - 3 = 5$

b) $(-1) - 3 = -4$

---

22. $y = 3x^3 + 5$

(a) If $x = 2$, $y = 3(2)^3 + 5$
$= 3(8) + 5 =$ ____

(b) If $x = -1$, $y = 3(-1)^3 + 5$
$= 3(-1) + 5 =$ ____

a) 54 b) -16

---

23. $y = 4x^3 - 100$

(a) If $x = 5$, $y =$ ________ (b) If $x = -4$, $y =$ ________

a) 29

b) 2

---

24. $y = 10x$

Is the equation above true:

(a) If we plug in 2 for x and 20 for y? ______

(b) If we plug in -3 for x and -30 for y? ______

(c) If we plug in 1 for x and 12 for y? ______

a) $4(125) - 100 = 400$

b) $4(-64) - 100 = -356$

---

25. $t = 5F - 7$

Is the equation above true if we plug in:

(a) 6 for F and 23 for t? ______

(b) -3 for F and -8 for t? ______

(c) -1 for F and -12 for t? ______

a) Yes, since:
$20 = 10(2)$

b) Yes, since:
$-30 = 10(-3)$

c) No, since:
$12 \neq 10(1)$

---

a) Yes, since:
$23 = 30 - 7$

b) No, since:
$-8 \neq (-15) - 7$

c) Yes, since:
$-12 = (-5) - 7$

26. $xy = 50$

Is the equation above true: (a) If $x = 5$ and $y = 10$? ______

(b) If $x = -5$ and $y = 10$? ______

(c) If $x = -5$ and $y = -10$? ______

---

a) Yes, since: $(5)(10) = 50$

b) No, since: $(-5)(10) \neq 50$

c) Yes, since: $(-5)(-10) = 50$

27. $d = \frac{P}{4}$

Is the equation above true: (a) If $P = 24$ and $d = 8$? ______

(b) If $P = -16$ and $d = 4$? ______

(c) If $P = -20$ and $d = -5$? ______

---

a) No, since: $8 \neq \frac{24}{4}$

b) No, since: $4 \neq \frac{-16}{4}$

c) Yes, since: $-5 = \frac{-20}{4}$

28. When an equation contains two different letters, it is "true" when we plug in some pairs of values for the two letters. It is "false" when we plug in other pairs of values for them.

If a two-letter equation is true when we plug in a pair of values for the letters, we say that this pair of values "satisfies" the equation.

$$y = 3x^2$$

(a) Does this pair of values $(x = 2,\ y = 12)$ satisfy the equation above? ______

(b) Does this pair of values $(x = -1,\ y = -3)$ satisfy the equation above? ______

---

a) Yes, since: $12 = 3(4)$

b) No, since: $-3 = 3(1)$

29. $d = F^2 - 10$

Which of the following pairs of values satisfy the equation above? ______

(a) $F = -2,\ d = -14$ (b) $F = 5,\ d = 15$ (c) $F = -4,\ d = 6$

Both (b) and (c)

30. All of the following pairs of values satisfy the equation below:

$$\boxed{y = x^3}$$

$x = 1,\ y = 1$
$x = -1,\ y = -1$
$x = 2,\ y = 8$
$x = -2,\ y = -8$

How many more pairs of values would satisfy the same equation? ______

An infinite number

31. When a pair of values satisfies an equation, it is called a "solution" of the equation.

$$t = 2d^3$$

Which of the following pairs are "solutions" of the equation above? ______

(a) $d = 4,\ t = 128$ (b) $d = -1,\ t = -8$ (c) $d = -3,\ t = -54$

---

32. Each of the following pairs of values is a solution of the equation below:

Both (a) and (c)

| $x + y = 10$ |

$x = 0,\ y = 10$
$x = 1,\ y = 9$
$x = -1,\ y = 11$

How many solutions are there for the equation above? ______

---

33. Any solution for a two-letter equation contains a pair of values. One or both of these values can be a fraction.

An infinite number

$$y = 3x$$

Complete these solutions for the equation above:

(a) $x = \frac{1}{2}$, $y =$ ____ (b) $x = \frac{2}{3}$, $y =$ ____ (c) $x = -\frac{5}{4}$, $y =$ ____

---

34. $$P = \frac{10}{V}$$

a) $y = \frac{3}{2}$
b) $y = 2$
c) $y = -\frac{15}{4}$

If $V = \frac{1}{2}$, $P = \dfrac{10}{\frac{1}{2}} = 10(2) = 20$

(a) If $V = \frac{7}{4}$, $P =$ ______

(b) If $V = -\frac{9}{5}$, $P =$ ______

---

35. Just as $7^2 = (7)(7) = 49$, (a) $\left(\frac{1}{2}\right)^2 = \left(\frac{1}{2}\right)\left(\frac{1}{2}\right) =$ ____

(b) $\left(-\frac{3}{5}\right)^2 = \left(-\frac{3}{5}\right)\left(-\frac{3}{5}\right) =$ ____

a) $P = \frac{40}{7}$
b) $P = -\frac{50}{9}$

---

36. $$y = x^2$$

a) $\frac{1}{4}$ b) $\frac{9}{25}$

(a) If $x = \frac{1}{4}$, $y =$ ____ (b) If $x = -\frac{10}{3}$, $y =$ ____

---

37. Just as $4^3 = (4)(4)(4) = 64$, (a) $\left(\frac{1}{3}\right)^3 = \left(\frac{1}{3}\right)\left(\frac{1}{3}\right)\left(\frac{1}{3}\right) =$ ____

(b) $\left(-\frac{2}{5}\right)^3 = \left(-\frac{2}{5}\right)\left(-\frac{2}{5}\right)\left(-\frac{2}{5}\right) =$ ____

a) $\frac{1}{16}$ b) $\frac{100}{9}$

38. $$m = q^3$$

Complete these solutions: (a) $q = \frac{1}{2}$, $m =$ ______

(b) $q = -\frac{4}{3}$, $m =$ ______

a) $\frac{1}{27}$ b) $-\frac{8}{125}$

---

39. $$t = 4b^2$$

If $b = \frac{1}{2}$, $t = 4\left(\frac{1}{2}\right)^2 = 4\left(\frac{1}{4}\right) = 1$

If $b = \frac{3}{5}$, $t = 4\left(\frac{3}{5}\right)^2 = 4\left(\frac{\quad}{\quad}\right) =$ ______

a) $m = \frac{1}{8}$

b) $m = -\frac{64}{27}$

---

40. $$y = 3x^3$$

If $x = \frac{1}{3}$, $y = 3\left(\frac{1}{3}\right)^3 = 3\left(\frac{1}{27}\right) = \frac{3}{27} = \frac{1}{9}$

If $x = \frac{1}{2}$, $y = 3\left(\frac{1}{2}\right)^3 = 3\left(\frac{\quad}{\quad}\right) =$ ______

$4\left(\frac{9}{25}\right) = \frac{36}{25}$

---

41. Just as one or both values in a solution can be a fraction, one or both can be a decimal.

$$y = 2x$$

(a) If $x = 1.1$, $y =$ ______ (b) If $x = -10.5$, $y =$ ______

$3\left(\frac{1}{8}\right) = \frac{3}{8}$

---

42. $$y = x^2$$

If $x = 2.5$, $y = (2.5)^2 = (2.5)(2.5) = 6.25$

If $x = 0.5$, $y =$ ______

a) $y = 2.2$

b) $y = -21$

---

43. $$t = \frac{d}{10}$$

Complete these solutions. Write each answer as a decimal.

(a) $d = 7.5$, $t =$ ______ (b) $d = 15.5$, $t =$ ______

$(0.5)^2 = 0.25$

---

a) 0.75 b) 1.55

SELF-TEST 1 (Frames 1-43)

1. Plug in each of the following values for x and find the corresponding value of y:

$y = 2x - 5$

(a) If $x = 1$, $y =$ _____
(b) If $x = -1$, $y =$ _____
(c) If $x = \frac{1}{2}$, $y =$ _____
(d) If $x = 0.3$, $y =$ _____

2. $V = \frac{40}{P}$

(a) If $P = 20$, $V =$ _____
(b) If $P = 80$, $V =$ _____
(c) If $P = \frac{1}{2}$, $V =$ _____
(d) If $P = 0.4$, $V =$ _____

3. $y = 3x^2 - 2$

(a) If $x = 1$, $y =$ _____
(b) If $x = -2$, $y =$ _____
(c) If $x = \frac{2}{3}$, $y =$ _____
(d) If $x = -0.2$, $y =$ _____

4. $y = 2x - x^2$

(a) Is the equation true if we plug in 1 for x and -1 for y? _______
(b) Does $x = -1$, $y = -3$ satisfy the equation? _______
(c) If $x = -2$, $y =$ _____.

5. Given: $s = 2t^3$ Which of these pairs are solutions of the equation? _____________

(a) $t = 1$, $s = 2$
(b) $t = -1$, $s = 2$
(c) $t = -2$, $s = -16$
(d) $t = \frac{1}{2}$, $s = \frac{1}{4}$
(e) $t = -\frac{1}{2}$, $s = -\frac{1}{4}$
(f) $t = 0.1$, $s = 0.02$

ANSWERS:

1. (a) -3 (b) -7 (c) -4 (d) -4.4
2. (a) 2 (b) $\frac{1}{2}$ (c) 80 (d) 100
3. (a) 1 (b) 10 (c) $-\frac{2}{3}$ (d) -1.88
4. (a) No (b) Yes (c) -8
5. The solutions are: (a), (c), (d), (e)

7-2 TABLES AND SOLUTIONS OF TWO-LETTER EQUATIONS

In the last section, we saw that a two-letter equation is satisfied by various pairs of values called solutions. In fact, we saw that any two-letter equation has an infinite number of solutions, some of which include fractions and decimals. Mathematicians frequently list solutions for these equations in tables. We will show the method in this section.

44. In the last section, we used various letters in two-letter equations. However, mathematicians ordinarily use the letters "x" and "y". We will follow their convention in this section.

Here is a simple two-letter equation: $\boxed{y = 5x}$

At the right, we have given an example of a table of solutions for this equation. Notice that the x and y values are separated by a vertical line:

x	y
2	10
1	5
0	0
-1	-5
-2	-10

The first solution shown in the table is $x = 2$, $y = 10$. How many different solutions of the equation are listed in the table? _______

---

45. We can make up a similar table of solutions for any two-letter equation. To do so, mathematicians always plug in values for x and then find the corresponding values for y. We will also follow this convention.

Five

Here is another example of an equation with a table of solutions: $\boxed{y = x^2}$

x	y
5	25
3	9
1	1
0	0
-1	1
-3	9
-5	25

Notice these two points:

(1) The columns are labeled "x" and "y" at the top.
(2) The x-values are listed in some order. They are not listed haphazardly.

How many solutions for the equation are listed in the table? _______

---

46. $\boxed{y = 3x + 1}$

Seven

In the table on the left below, we have listed some solutions for the equation above. However, the table is clumsy because we have listed the x-values haphazardly. Using the table on the right, rewrite the solutions so that there is some order in the x-values.

x	y
2	7
-4	-11
4	13
0	1
-2	-5

x	y

47. When plugging in values of x to make a table, we do not necessarily have to plug in negative values for x corresponding to the positive values for x. That is, if we use $x = 5$, we do not necessarily have to use $x = -5$.

Complete the table at the right for the equation shown:

$y = x^3$

x	y
5	
3	
0	
-2	
-4	

Either of the following is acceptable:

x	y
4	13
2	7
0	1
-2	-5
-4	-11

x	y
-4	-11
-2	-5
0	1
2	7
4	13

---

48. When making up a table of solutions, we can plug in any values we want for x. Below, we have given three tables based on the following equation. Notice the different choices of values for x:

$y = 2x$

Table 1

x	y
1,000	2,000
700	1,400
500	1,000
350	700
180	360

Table 2

x	y
-800	-1,600
-600	-1,200
-450	-900
-250	-500
-100	-200

Table 3

x	y
750	1,500
375	750
0	0
-375	-750
-750	-1,500

(a) In which table did we ignore positive values of x? ________

(b) In which table did we use only small positive and negative values of x which are close to 0? ________

x	y
5	125
3	27
0	0
-2	-8
-4	-64

---

49. Here is another equation with a table of solutions:

$xy = 50$

x	y
50	1
25	3
10	5
-5	10
-25	-2

It is easy to make errors when constructing a table of solutions. In the table above, which pairs of values for x and y do not satisfy the equation? ________________

a) Table 2

b) None of them.

The following pairs do not satisfy $xy = 50$:

$x = 25, \ y = 3$

$x = -5, \ y = 10$

50. Below we have constructed three tables of solutions for the following equation:

$$y = x^2 - 5$$

Table 1

x	y
10	95
8	59
7	44
5	20
3	4

Table 2

x	y
5	20
2	-1
0	-5
-1	-4
-4	11

Table 3

x	y
0	-5
-1	-4
-3	4
-5	20
-7	44

When choosing values of x for Table 1, the values ranged from 10 to 3. Therefore, we say that "the range of values of x is from 3 to 10".

(a) The range of values of x in Table 2 is from _____ to _____.

(b) The range of values of x in Table 3 is from _____ to _____.

---

51. (a) How many solutions are there for any two-letter equations? _______

(b) Could we list all of the possible solutions for a two-letter equation in one table? __________

Answers: a) -4 to 5 b) -7 to 0

---

52. Here is a table of solutions for the equation shown. The range of x is from 0 to 10:

$$x + y = 40$$

x	y
10	30
8	32
6	34
4	36
2	38
0	40

There are all sorts of values between 0 and 10 which we could plug in for x. We could plug in:

(1) Fractions, like $\frac{1}{2}$, $\frac{11}{8}$, $\frac{139}{141}$, etc.

or (2) Decimals, like 0.7, 1.23, 9.667, etc.

(a) Including fractions and decimals, how many values between 0 and 10 could we plug in for x? ___________

(b) If we limited ourselves to the range of x from 0 to 10, could we list all of the solutions within that range in one table? _______

Answers: a) An infinite number. b) Obviously not.

Answers: a) An infinite number. b) No, obviously not.

Here are two conclusions we can make about tables and two-letter equations:

(1) Since the number of possible solutions is infinite, no table can contain all of them.

(2) Since the number of possible solutions within a given range of x is also infinite, no table can contain all of them.

SELF-TEST 2 (Frames 44-52)

In Problems 1 to 4, complete the table of solutions for each equation:

1. $y = 10 - 2x$

x	y
-4	
-2	
0	
$1\frac{1}{2}$	
3	
6	

2. $y = x^2 - 3x$

x	y
4	
3	
1	
0	
-1	
-2	

3. $y = x^3 - 1$

x	y
5	
2	
0	
-1	
-3	
-5	

4. $xy = 200$

x	y
-100	
-50	
-10	
20	
80	
200	

5. State the "range of values of x" for:

(a) Problem 1: _____ to _____

(b) Problem 2: _____ to _____

(c) Problem 3: _____ to _____

(d) Problem 4: _____ to _____

ANSWERS:

1.

x	y
-4	18
-2	14
0	10
$1\frac{1}{2}$	7
3	4
6	-2

2.

x	y
4	4
3	0
1	-2
0	0
-1	4
-2	10

3.

x	y
5	124
2	7
0	-1
-1	-2
-3	-28
-5	-126

4.

x	y
-100	-2
-50	-4
-10	-20
20	10
80	$2\frac{1}{2}$
200	1

5. (a) -4 to 6

(b) -2 to 4

(c) -5 to 5

(d) -100 to 200

7-3 THE COORDINATE SYSTEM - READING AND PLOTTING POINTS AT INTERSECTIONS

Mathematicians are not satisfied with the fact that they cannot list all of the solutions for a two-letter equation in a table. Therefore, they have developed another way of representing these solutions. This second way is by means of a graph. In this section and the next one, we will introduce the coordinate system which is used for graphing. Then, in a later section, we will show how two-letter equations are graphed.

53. The coordinate system is constructed by means of two number lines. We will briefly reexamine the number line and then show how the coordinate system is constructed.

Here is a diagram of the number line:

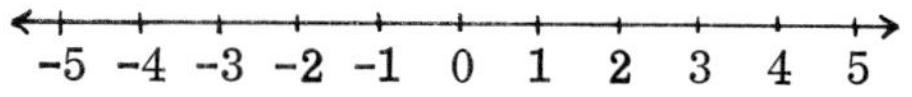

By convention:

(1) Positive numbers go to the right of "0"; negative numbers go to the left of "0".

(2) The absolute values of positive numbers increase as you move to the right.

(3) The absolute values of negative numbers increase as you move to the left.

What is the special name given to "point 0"? ____________

The origin.

---

54. Drawing the number line horizontally (sideways) is merely a convention. We can also draw it vertically (up and down). Here is the way it looks when drawn vertically:

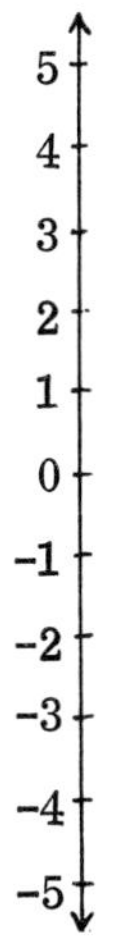

By convention:

(1) Positive numbers are above "0"; negative numbers are below "0".

(2) The absolute values of positive numbers increase as you move "up."

(3) The absolute values of negative numbers increase as you move "down."

When the number line is drawn vertically, "point 0" has the same special name. It is called the ____________.

origin

55. The "coordinate system" is constructed by putting together a horizontal and vertical number line as we have done below:

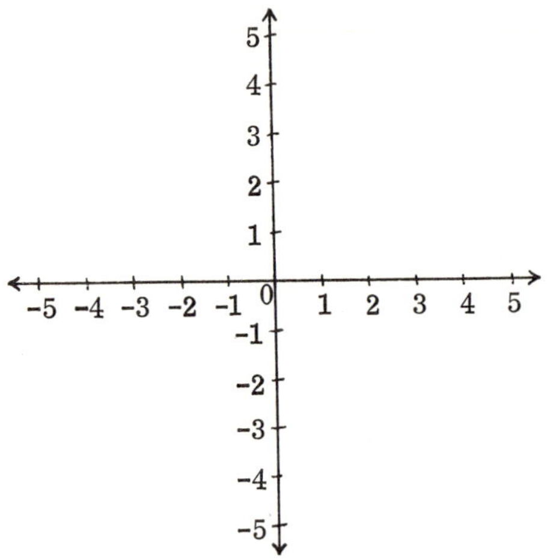

The vertical number line is perpendicular (at a 90° angle) to the horizontal number line.

The horizontal number line is called the horizontal axis.

The vertical number line is called the vertical axis.

The two number lines together are called coordinate axes.

The two coordinate axes (number lines) intersect at "point 0" on each. This special point in the coordinate system is called the ___________.

origin

56. A diagram of the coordinate system is shown at the right. Note that additional horizontal and vertical lines have been included, forming a "gridwork." Such lines make it easier to use the coordinate system.

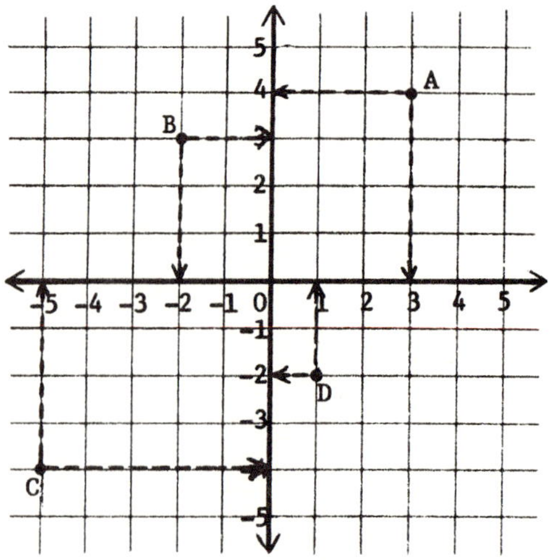

Each point in the coordinate system represents a pair of values, one related to the horizontal axis and one related to the vertical axis.

(Continued on following page.)

56. (Continued)

To find the pair of values for point A:

(1) We drew an arrow down to the horizontal axis. The tip of the arrow points to "3".
(2) We drew an arrow over to the vertical axis. The tip of the arrow points to "4".
(3) Therefore, point A represents 3 on the horizontal axis and 4 on the vertical axis.

Similarly: (a) Point B represents: _____ on the horizontal; _____ on the vertical

(b) Point C represents: _____ on the horizontal; _____ on the vertical

(c) Point D represents: _____ on the horizontal; _____ on the vertical

---

Answer to Frame 56: a) -2 (horizontal) 3 (vertical) b) -5 (horizontal) -4 (vertical) c) 1 (horizontal) -2 (vertical)

---

57. Mathematicians use the two axes to represent the values of "x" and "y", the letters they use in two-letter equations. They put:

x-values on the horizontal axis
y-values on the vertical axis

They label the two axes "x" and "y", as we have done on the coordinate system at the right:

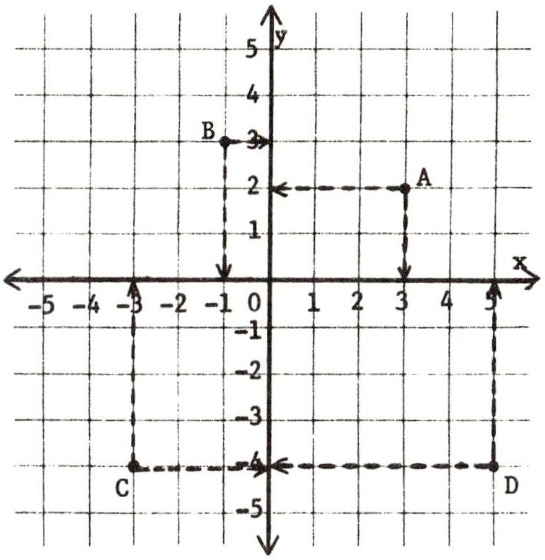

On the coordinate system shown: Point A represents: x = 3, y = 2

(a) Point B represents: x = _____, y = _____

(b) Point C represents: x = _____, y = _____

(c) Point D represents: x = _____, y = _____

Answer to Frame 57: a) B (x = -1, y = 3) b) C (x = -3, y = -4) c) D (x = 5, y = -4)

58. Each axis is a scale. When placing numbers at the calibration marks on the scale, we do not have to use consecutive whole numbers. On each axis at the right, we have counted "by two's."

Point A represents: x = 8, y = 4

(a) Point B represents: x = _____, y = _____

(b) Point C represents: x = _____, y = _____

(c) Point D represents: x = _____, y = _____

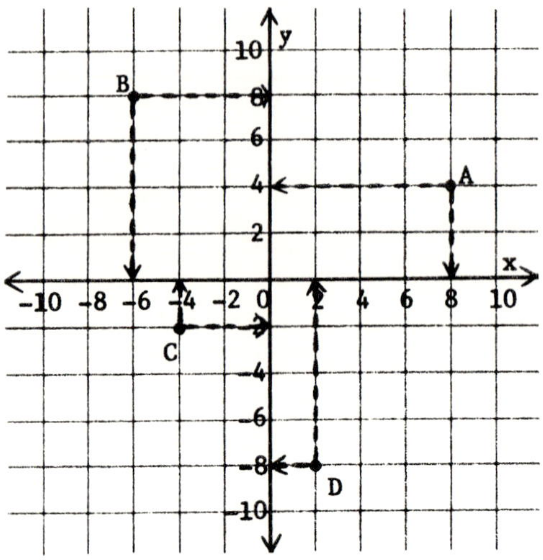

Answer to Frame 58:	a) B (x = -6, y = 8)	b) C (x = -4, y = -2)	c) D (x = 2, y = -8)

59. We also do not have to use the same scale for each axis. At the right, we have counted "by fives" on the horizontal axis and "by tens" on the vertical axis.

Point A represents: x = 20, y = 30

(a) Point B represents: x = _____, y = _____

(b) Point C represents: x = _____, y = _____

(c) Point D represents: x = _____, y = _____

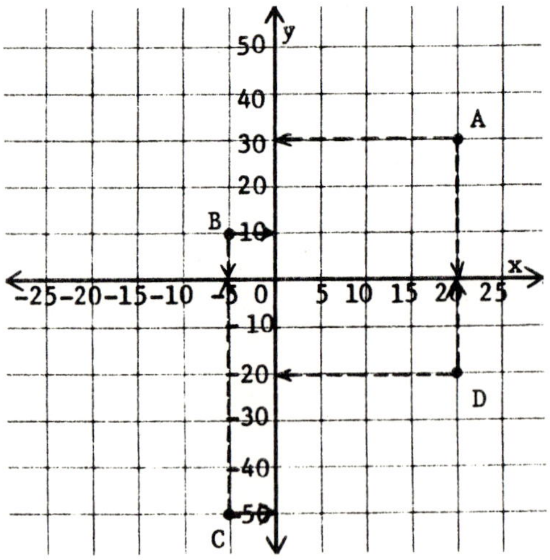

60. Since mathematicians use the scales on the axes to represent x-values and y-values, they call them the "x-axis" and the "y-axis."

(a) The "x-axis" is another name for the ______________ (horizontal/vertical) axis.

(b) The "y-axis" is another name for the ______________ (horizontal/vertical) axis.

Answer to Frame 59:

a) B (x = -5, y = 10)

b) C (x = -5, y = -50)

c) D (x = 20, y = -20)

a) horizontal

b) vertical

61. The coordinate axes divide the coordinate system into four parts. These four parts are called quadrants. We have labeled the four quadrants in Figure 1 below with Roman numerals. Notice that they are numbered in a counter-clockwise direction, beginning with the upper right quadrant.

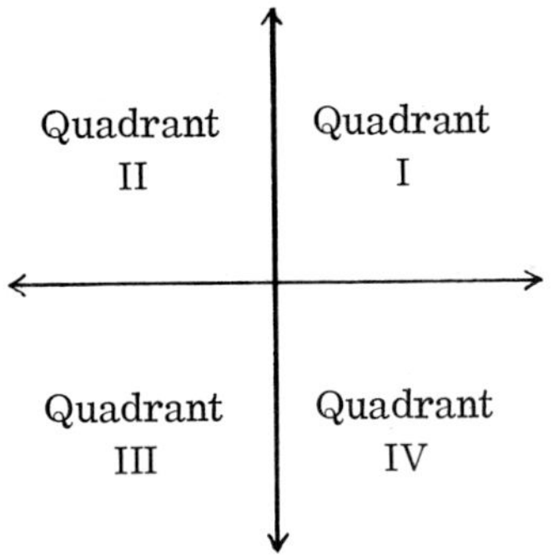

Figure 1

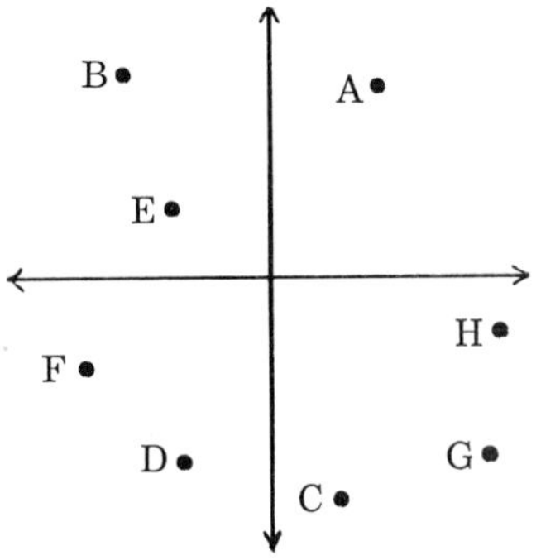

Figure 2

On the coordinate system in Figure 2 above, there is at least one point labeled in each quadrant.

(a) Points B and E lie in Quadrant _____. (b) Points C, G, and H lie in Quadrant _____.

Answer to Frame 61: a) Quadrant II b) Quadrant IV

62. Any single solution of an equation involving the letters x and y contains a pair of values, one for x and one for y. This pair of values can be represented by one point in the coordinate system. Locating this point is called "plotting the point." To do so, we simply reverse the procedure for reading the pair of values of a point.

At the right, for example, we have plotted two points:

A (x = 2, y = 5)

B (x = -4, y = -7)

We do so by drawing arrows from the axes.

Plot and label the following points on the coordinate system at the right:

C (x = 6, y = 7)

D (x = -2, y = 6)

E (x = -7, y = -2)

F (x = 8, y = -3)

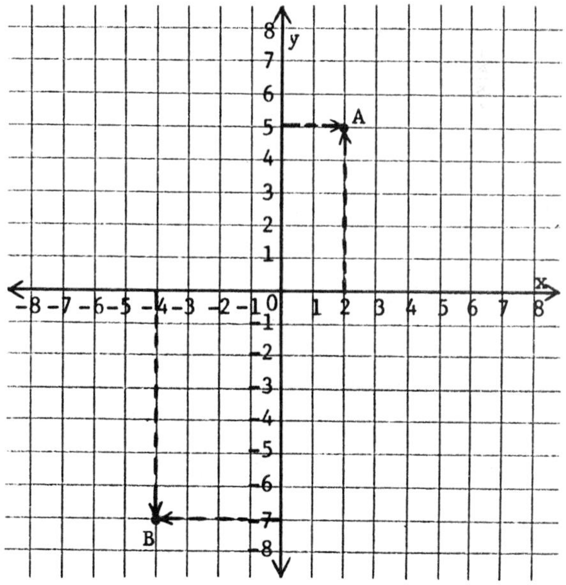

Answer to Frame 62: See next frame.

63. The correct plotting of the points from the last frame is shown at the right:

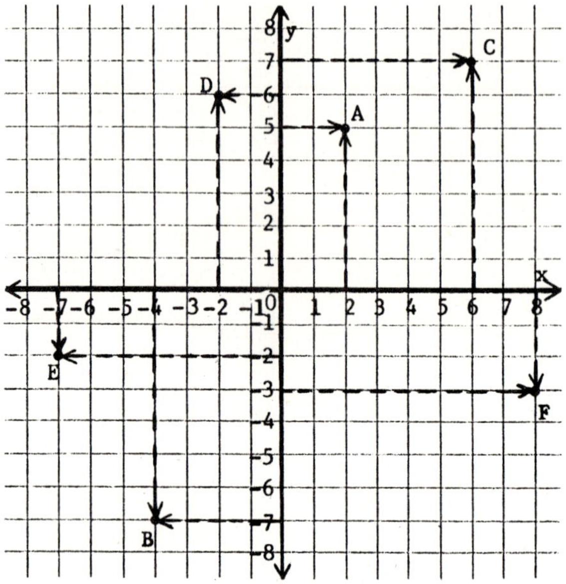

Of the four points (C, D, E, and F), which point lies in:

(a) Quadrant III? _____

(b) Quadrant IV? _____

Answer to Frame 63: a) E b) F

64. When the scales on the axes are different, as at the right, be careful when plotting points.

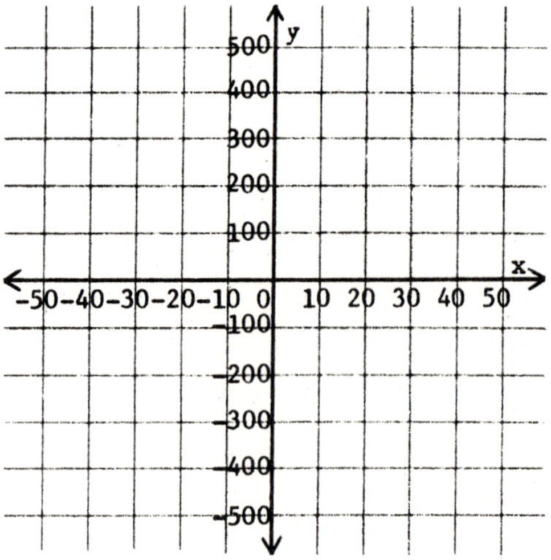

Plot and label the following points:

A (x = -50, y = -400)

B (x = -40, y = 200)

C (x = 30, y = -300)

D (x = -20, y = 500)

Answer to Frame 64: See next frame.

65. The points plotted in the last frame are shown at the left.

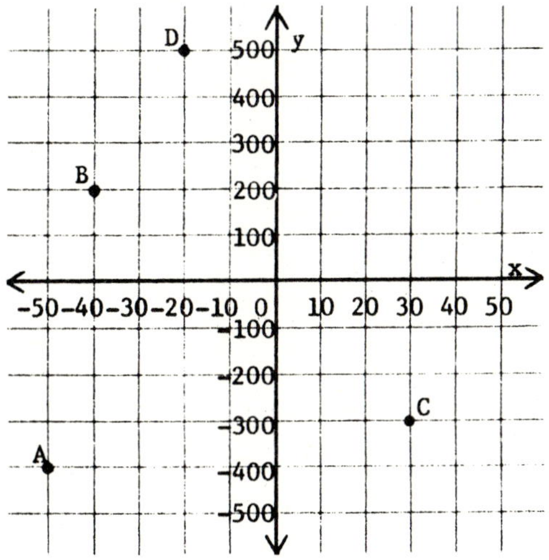

(a) Which point or points lie in Quadrant II? __________

(b) Which point or points lie in Quadrant IV? __________

Answer to Frame 65: a) B and D b) C

66. Point A represents these values: x = 6, y = 8.

"6" and "8" are called the coordinates of point A.

"6" is called the x-coordinate.

"8" is called the y-coordinate.

Next to point A, we have written its coordinate in parentheses like this: (6,8). The x-coordinate is always written first.

Write the coordinates for points B, C, and D in the parentheses next to them. (Write the x-coordinates first.)

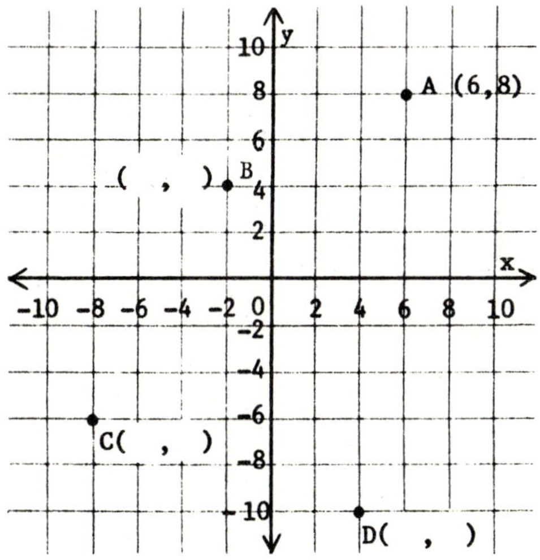

67. IMPORTANT: The coordinates of a point are always enclosed in parentheses.

When writing the coordinates of a point in parentheses, which coordinate is always written first? __________

B (-2,4)

C (-8,-6)

D (4,-10)

The x-coordinate.

68. For the point (1,3): (a) The x-coordinate is _____.

(b) The y-coordinate is _____.

For the point (3,1): (c) The x-coordinate is _____.

(d) The y-coordinate is _____.

Answer to Frame 68: a) 1 c) 3
b) 3 d) 1

---

69. Plot the following pairs of coordinates on the graph at the right. Write the coordinates in parentheses next to the plotted points.

(2,4) (-3,-5)

(4,2) (-5,-3)

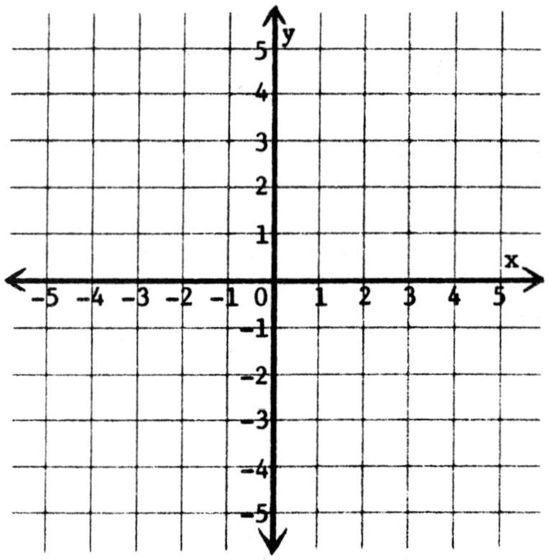

Answer to Frame 69: See next frame.

---

70. We have plotted the points given in the last frame on the coordinate system at the right:

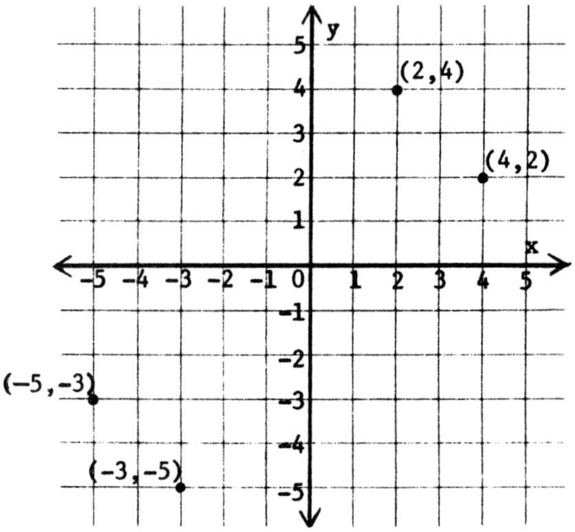

(a) Are points (2,4) and (4,2) identical? _______

(b) Are points (-5,-3) and (-3,-5) identical? _______

71. Write the following pairs of coordinates in parentheses:

(a) When x is 7, y is 10. ________

(b) When x is 10, y is 7. ________

(c) Does it make a difference which coordinate is written first? _____

a) No

b) No

---

72. When writing a pair of coordinates in parentheses, the order in which they are written makes a difference. Therefore, we call a pair of coordinates an ordered pair.

The first coordinate of an ordered pair is the x-coordinate. It is also called the abscissa.

The second coordinate of an ordered pair is the y-coordinate. It is also called the ordinate.

In (-4,8): (a) The abscissa is _____. (b) The ordinate is _____.

Note: Coordinates of points must be enclosed in parentheses, as shown in (a) and (b).

a) (7,10)

b) (10,7)

c) Yes, the x-coordinate is always written first.

---

73. (a) If the abscissa is 40 and the ordinate is 63, the ordered pair is _______.

(b) If the ordinate is 112 and the abscissa is 97, the ordered pair is _______.

a) -4 b) 8

---

74. Is the ordinate another name for the x-coordinate or the y-coordinate?

____

a) (40,63)

b) (97,112)

---

75. In an ordered pair, the __________ (abscissa/ordinate) is always written first.

The y-coordinate.

---

abscissa

---

76. Plot the following points on the graph at the right:

Point	Abscissa	Ordinate
A	6	20
B	-6	20
C	-4	-15
D	4	-15

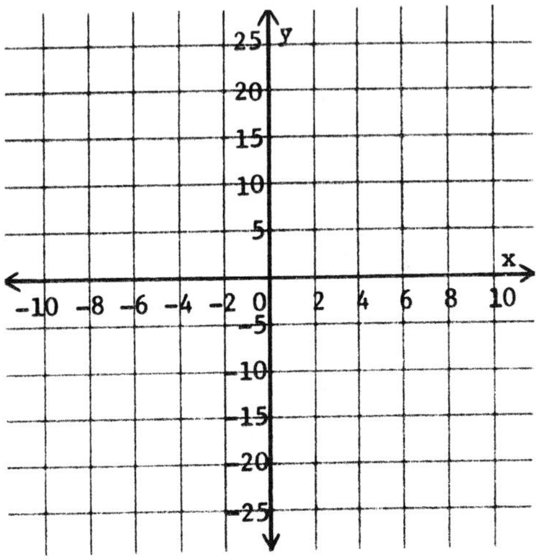

Answer to Frame 76: See next frame.

77.

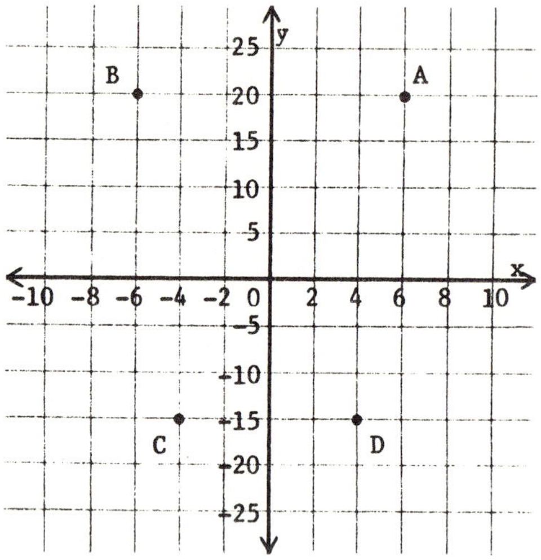

The points listed in the last frame are plotted on the coordinate axes at the left.

(a) Point B is in what quadrant?
Quadrant _____

(b) Point C is in what quadrant?
Quadrant _____

---

78. If you plotted the following points, in what quadrant would they be? You should be able to answer without actually plotting the points.

(a) (-3,6) _______ (c) (9,-10) _______

(b) (-5,-7) _______ (d) (4,12) _______

a) Quadrant II
b) Quadrant III

---

79. Answer "positive" or "negative" for each of the following:

(a) The abscissa of any point in Quadrant I is _____________.

(b) The ordinate of any point in Quadrant II is _____________.

(c) The x-coordinate of any point in Quadrant III is _____________.

(d) The y-coordinate of any point in Quadrant IV is _____________.

a) Quadrant II
b) Quadrant III
c) Quadrant IV
d) Quadrant I

---

80. Answer "positive" or "negative" for each of the following:

(a) The ordinate of any point in Quadrant III is _____________.

(b) The abscissa of any point in Quadrant IV is _____________.

(c) The x-coordinate of any point in Quadrant II is _____________.

(d) The y-coordinate of any point in Quadrant IV is _____________.

a) positive
b) positive
c) negative
d) negative

Note: The labels "horizontal," "x", and "abscissa" go together.

The labels "vertical," "y", and "ordinate" go together.

To help you remember what goes together, notice that alphabetically:

horizontal comes before vertical

x comes before y

abscissa comes before ordinate

a) negative
b) positive
c) negative
d) negative

SELF-TEST 3 (Frames 53-80)

1. Refer to the coordinate system at the right. Write the coordinates of the following points. Be sure to include parentheses in each answer.

A: _______ D: _______

B: _______ E: _______

C: _______ F: _______

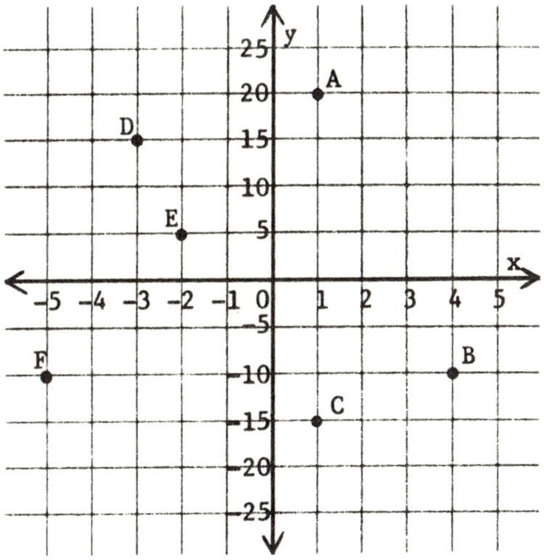

2. List the quadrant (I, II, III, or IV) in which each point lies:

(a) (-3,7) _____ (b) (8,-6) _____ (c) (25,82) _____ (d) (-5,-5) _____ (e) (-1,3) _____

3. Write the coordinates of the point whose ordinate is 2 and whose abscissa is -8: _______

4. List the quadrant (I, II, III, or IV) in which each point lies:

(a) Abscissa negative and ordinate positive. _______

(b) Ordinate negative and abscissa positive. _______

(c) Abscissa negative and ordinate negative. _______

ANSWERS:
1. A: (1,20) D: (-3,15)
B: (4,-10) E: (-2,5)
C: (1,-15) F: (-5,-10)
2. (a) II (b) IV (c) I (d) III (e) II
3. (-8,2)
4. (a) II (b) IV (c) III

7-4 THE COORDINATE SYSTEM - READING AND PLOTTING POINTS WHICH ARE NOT AT INTERSECTIONS

Many points in the coordinate system do not lie at an intersection of a horizontal and vertical line. Even though they do not lie at such an intersection, they do represent a pair of values. That is, these points also have coordinates. Reading and plotting points of this type is more difficult because we have to estimate the value of places on the two axes. We will show the method in this section.

81. On the coordinate system at the right, point A does not lie at an intersection. When we draw arrows to the two axes to determine its coordinates, as shown, the tips of the arrows do not fall at calibration marks. Therefore, we must estimate these two values. Its x-coordinate is approximately 25; its y-coordinate is approximately 32. The coordinates of point A, therefore, are approximately (25,32).

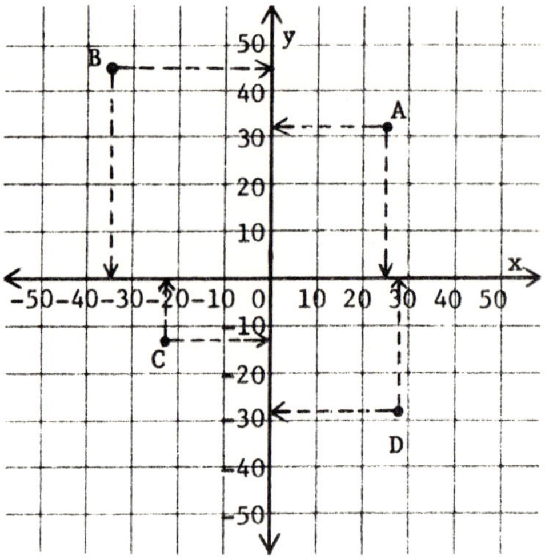

By the same type of estimation, the coordinates:

Of point B are approximately (,).

Of point C are approximately (,).

Of point D are approximately (,).

Answer to Frame 81: B (-35,45) C (-23,-13) D (28,-28)

---

82. In order to estimate the coordinates of points of this type, you must pay close attention to the scales on the axes. On the enlarged x-axis below, "hundreds" appear at the calibration marks. When estimating the value of points on this axis, you cannot read closer than the nearest "ten." For example, point A is approximately 170. To report 172 or 171.3 or 172.57 is ridiculous because the scale cannot be read that closely.

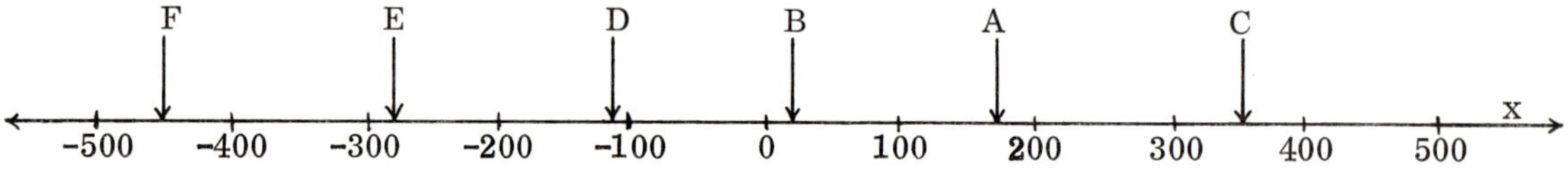

Estimate the numerical value of the following points:

B is approximately _______.

C is approximately _______.

D is approximately _______.

E is approximately _______.

F is approximately _______.

(B) 20 or 30

(C) 350

(D) -110 or -120

(E) -270 or -280

(F) -450

83.

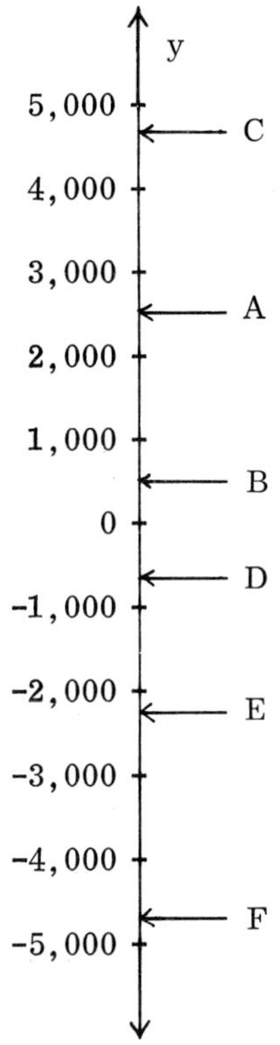

On the y-axis to the left, "thousands" appear at the calibration marks. When estimating values of points on this axis, we cannot read more closely than "hundreds." Therefore, A is approximately 2,500.

Estimate the value of the following points:

B is approximately _______.

C is approximately _______.

D is approximately _______.

E is approximately _______.

F is approximately _______.

<u>Answer to Frame 83:</u>

(B) 500

(C) 4,600 or 4,700

(D) -600 or -700

(E) -2,200 or -2,300

(F) -4,700 or -4,800

84. On the x-axis below, "tens" appear at the calibration marks. In this case, we can estimate to the "nearest unit." For example, point A is approximately 25.

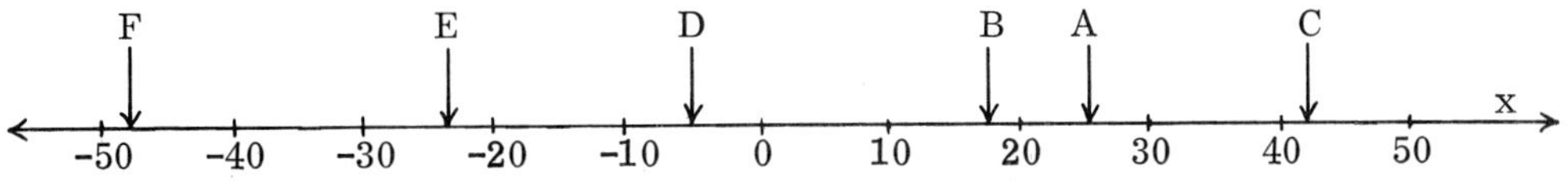

Estimate the values of the other points:

B is approximately _______.

C is approximately _______.

D is approximately _______.

E is approximately _______.

F is approximately _______.

(B) 17 or 18

(C) 42 or 43

(D) -5

(E) -23 or -24

(F) -47 or -48

85.

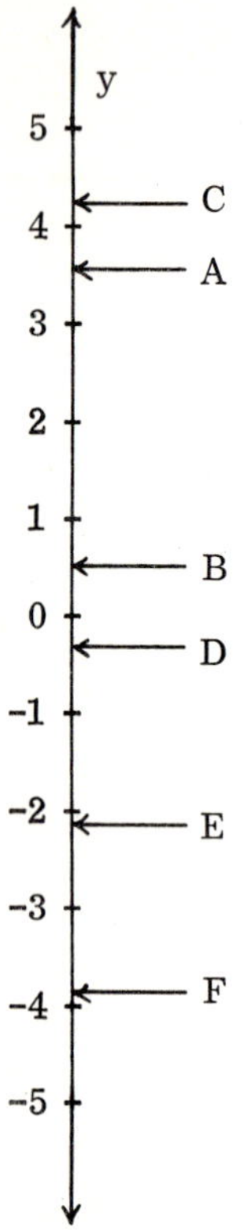

On the y-axis to the left, the scale is "units" or "whole numbers." When reading points on it, we can estimate to the "nearest tenth." For example, point A is approximately 3.6 or 3.7.

Estimate the value of the other points:

B is approximately __________.

C is approximately __________.

D is approximately __________.

E is approximately __________.

F is approximately __________.

---

86. Several points are shown on the coordinate system below. Write the coordinates for each point in the space provided. You will have to estimate the coordinates. <u>Notice that the scales on the two axes are different.</u>

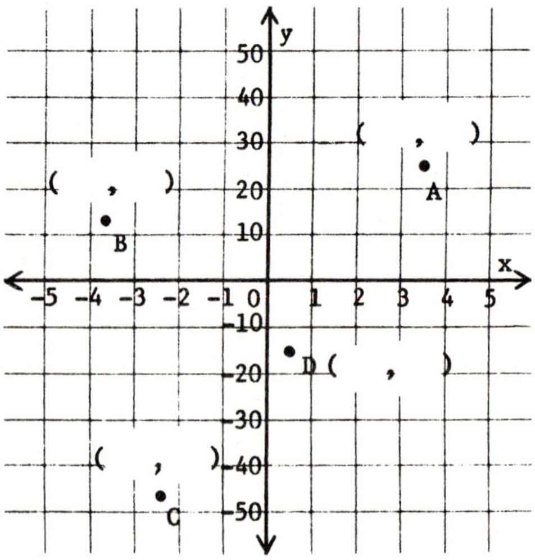

(B) 0.5

(C) 4.2 or 4.3

(D) -0.3 or -0.4

(E) -2.1 or -2.2

(F) -3.8 or -3.9

87. On the coordinate system below, write the coordinates of each point in the space provided. Notice that the scales on the axes are different.

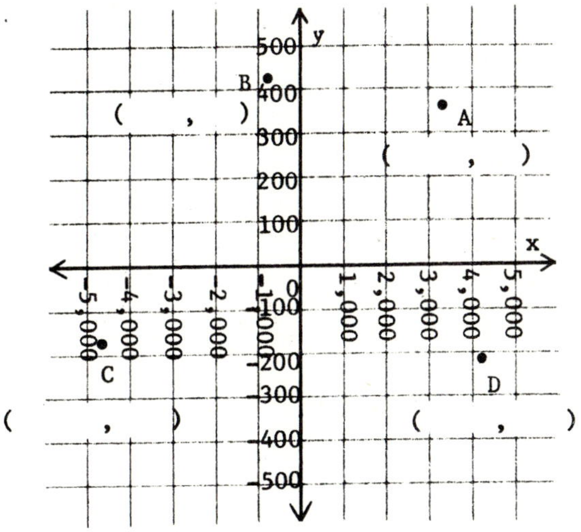

Your answers can be slightly different than ours. Check your signs!

A (3.5,25)

B (-3.7,12)

C (-2.4,-47)

D (0.5,-15)

A (3,300, 360) B (-800,420) C (-4,700,-170) D (4,200,-210) (Yours can differ slightly)

---

88. When plotting points on the coordinate system, we have to estimate the positions of the coordinates on the axes if the coordinates do not lie at calibration marks. On the coordinate system at the right, for example, we have plotted point A whose coordinates are (35,30). To do so, we had to estimate the position of 35 on the x-axis, since 35 does not lie at a calibration mark.

Plot the following points on the coordinate system at the right:

B (25,-40)

C (-15,25)

D (-45,-15)

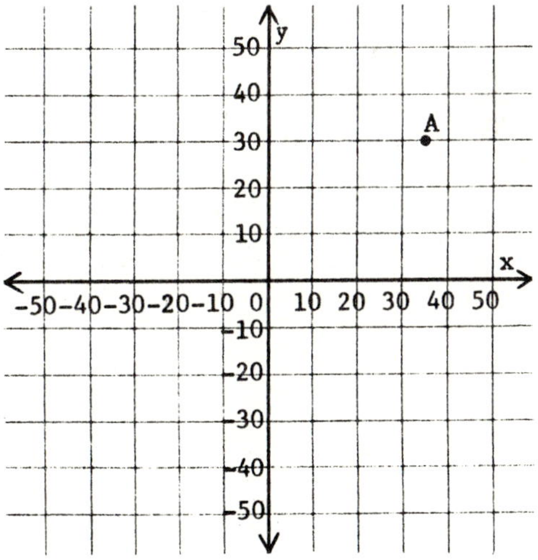

Answer to Frame 88: See next frame.

89.

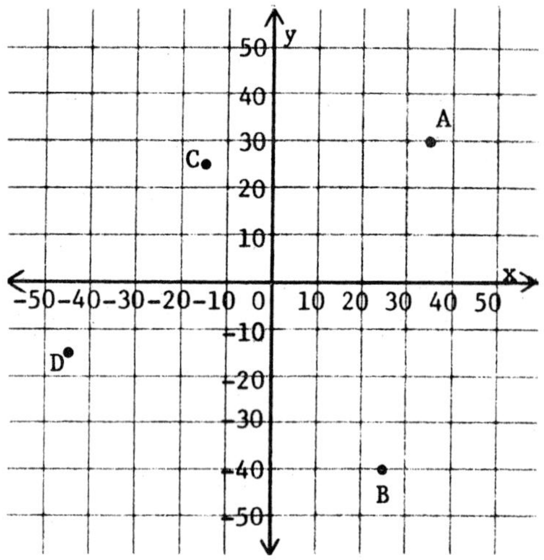

Your plotted points should be in the locations shown at the left.

Which point (A, B, C, or D) lies:

(a) In quadrant II? ______

(b) In quadrant IV? ______

Answer to Frame 89: a) C b) B

90. We have located 16 on each x-axis below by estimating its position. Use arrows to show the position of each of the following numbers on both axes. Label each arrow:

8, 22, -5, -17, -24

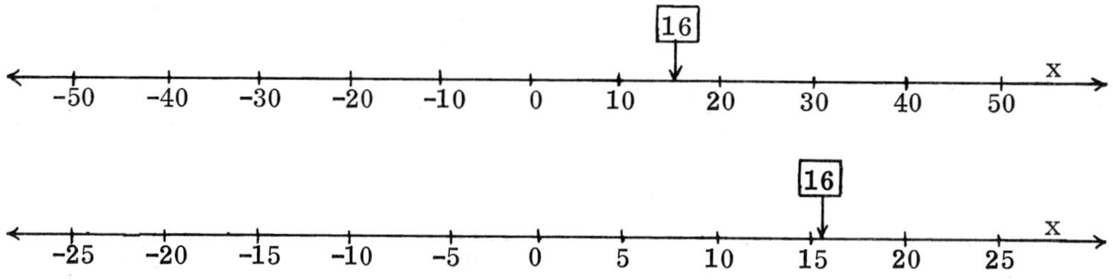

Answer to Frame 90: Your axes should look approximately like this:

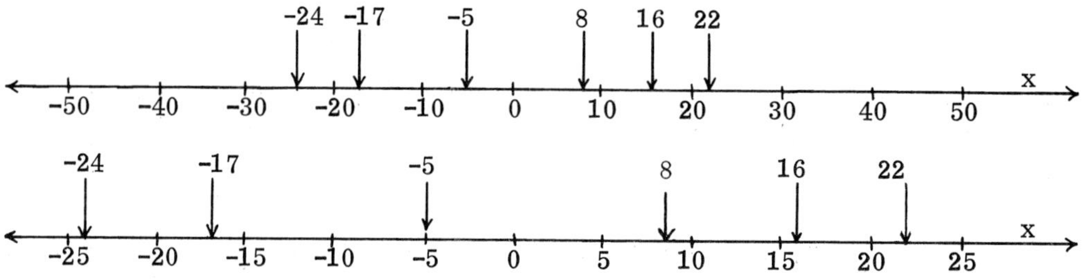

91.

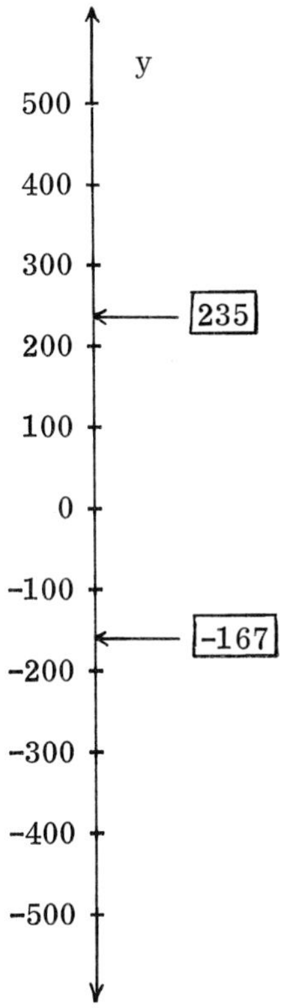

On the y-axis to the left, we have located 235 and -167 by estimating their positions. To do so, we rounded each to the nearest "ten" (240 and -170) since we can locate points only to the nearest "ten."

Use arrows to locate the following numbers on the y-axis at the left. Mentally round to the nearest "ten" before doing so:

16, 447, -32.6, -381.77

Note: Answers are shown on the vertical axis at the right.

Answer to Frame 91:

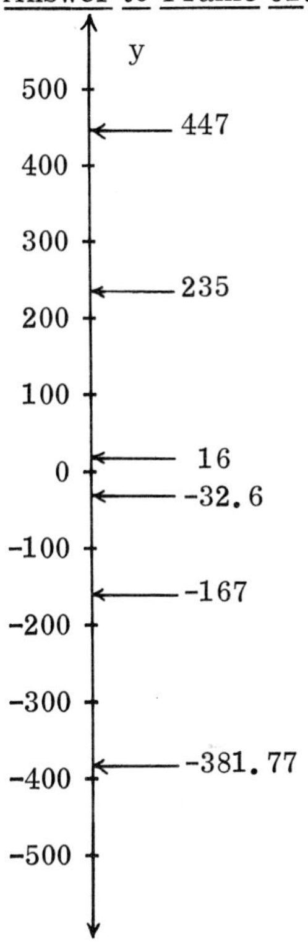

---

92. On the x-axis below, we have placed "thousands" at the calibration marks. When estimating the position of a number on this scale, we cannot locate more closely than the nearest "hundred." Therefore, when putting numbers on this scale, we round to the nearest "hundred" first.

Examples: 2,169 rounds to 2,200 -3,512.99 rounds to -3,500

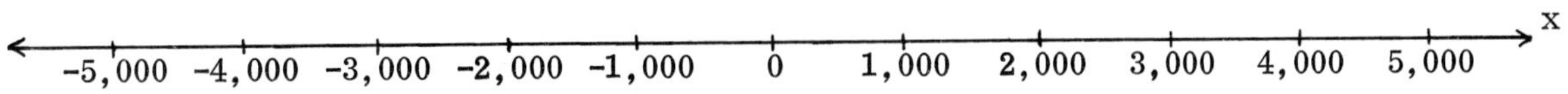

Locate the following numbers on the x-axis above. Mentally round to the nearest "hundred." Label the points with arrows:

787, 2,112, 3,642.9, -299, -4,621.08

Answer to Frame 92: Your x-axis should look approximately like this:

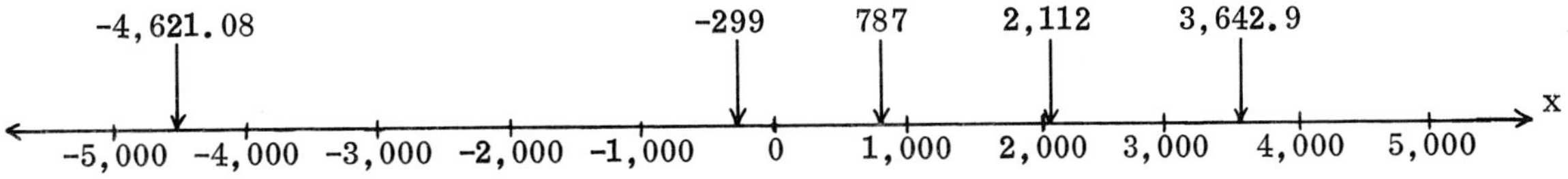

93. On the x-axis below, we have placed consecutive units or whole numbers on the calibration marks. When placing numbers on this scale, we cannot locate numbers closer than the nearest "tenth."

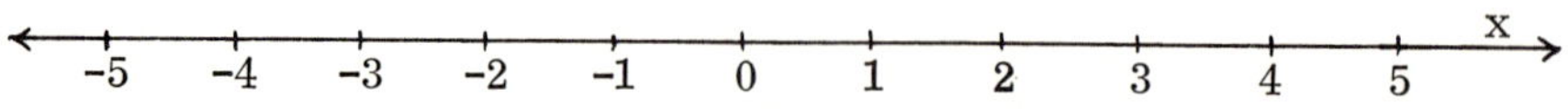

Locate the following numbers on the x-axis above. Mentally round to the nearest "tenth" if necessary. Label each point with an arrow.

$0.5,\ 1.9,\ 3.34,\ -\frac{1}{2},\ -2\frac{1}{2},\ -4.67$

Answer to Frame 93: Your x-axis should look approximately like this:

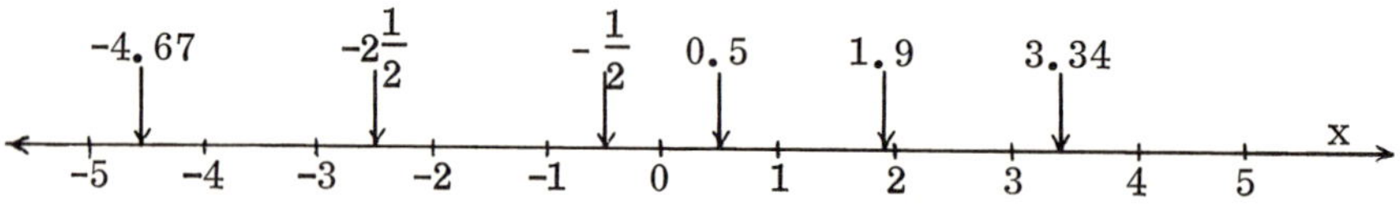

---

94. Plot and label the following points on the coordinate system at the right. Note: The scales on the axes are different.

A (2.5,45)

B (4.8,-37)

C (-1.1,-12)

D (-2.8,25)

E $(\frac{1}{2}, 18)$

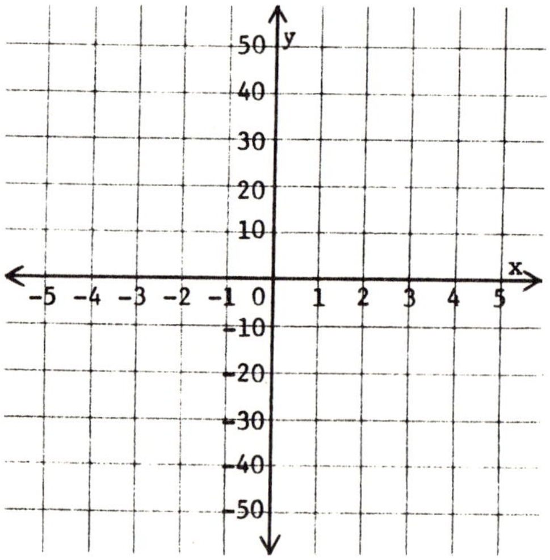

Answer to Frame 94: See next frame.

95.

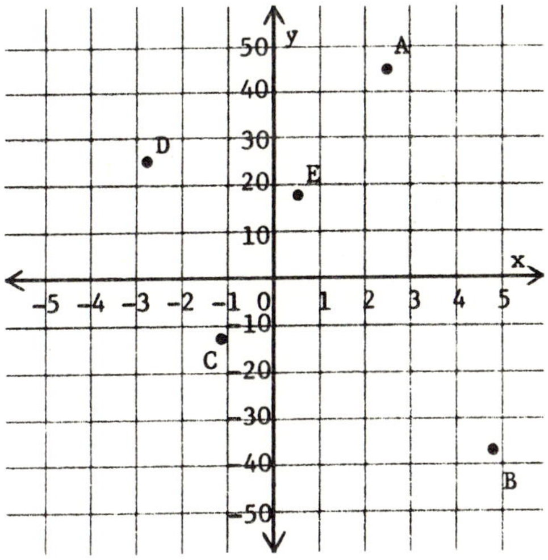

Your plotted points should be approximately in the locations shown at the left.

Complete: (a) The x-coordinate of A is _____.

(b) The ordinate of C is _____.

(c) The y-coordinate of B is _____.

(d) The abscissa of E is _____.

Answer to Frame 95: a) 2.5 b) -12 c) -37 d) $\frac{1}{2}$

96. Plot and label the following points on the coordinate system at the right. Note: The scales on the axes are different.

A (-150, 2,800)

B (312, -4,199)

C (259, 687)

D (-434, -1,912)

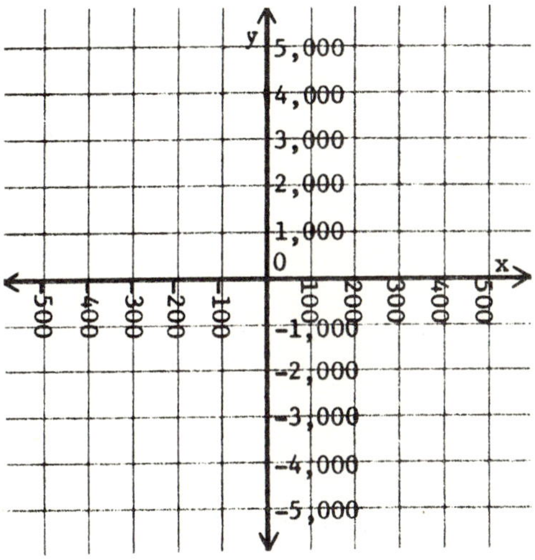

Answer to Frame 96: See next frame.

97.

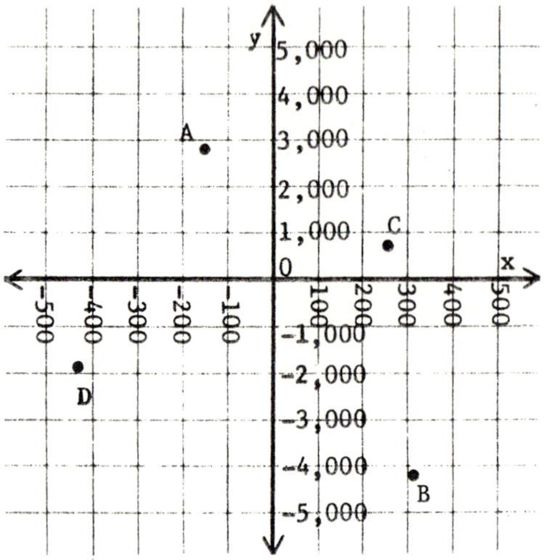

Your plotted points should be approximately in the locations shown at the left.

Complete: (a) The abscissa of D is _____.

(b) The y-coordinate of A is _____.

(c) The x-coordinate of C is _____.

(d) The ordinate of B is _____.

Answer to Frame 97: a) -434 b) 2,800 c) 259 d) -4,199

SELF-TEST 4 (Frames 81-97)

Examine the coordinate system at the right. Note that the two axes are calibrated differently.

Using care, write the coordinates of the following points:

1. A: _______ 5. E: _______

2. B: _______ 6. F: _______

3. C: _______ 7. G: _______

4. D: _______ 8. H: _______

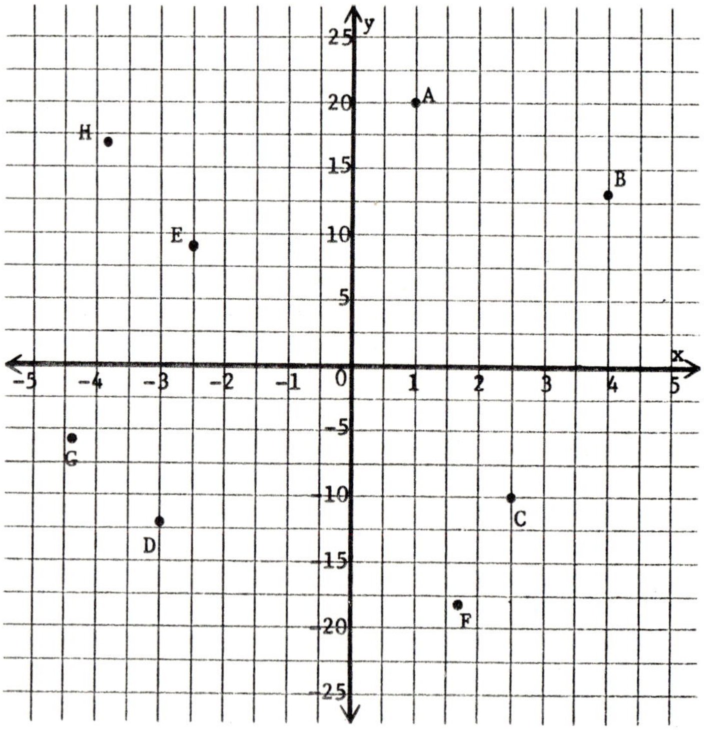

ANSWERS: Because estimation is used, your answers may deviate slightly from the answers shown. By using care, however, you should obtain the answers shown.

1. A: (1,20) 3. C: (2.5,-10) 5. E: (-2.5, 9) 7. G: (-4.4,-6)
2. B: (4,13) 4. D: (-3,-12) 6. F: (1.7,-18) 8. H: (-3.8, 17)

7-5 GRAPHING TWO-LETTER EQUATIONS

As we said before, mathematicians are not satisfied with a table of solutions for two-letter equations. They are uncomfortable with the fact that only a small part of the infinite number of solutions can be listed in a table. Therefore, they use graphs on the coordinate system to represent the solutions for these equations.

There are three steps in graphing any equation:

(1) Making up a table of solutions.

(2) Plotting the points which represent the solutions in the table.

(3) Drawing a straight or curved line through the plotted points.

The graph of an equation is either a straight line or some type of curved line. In this section, we will show the method for graphing equations whose graphs are straight lines. We will also consider the meaning of these graphs.

---

98. In this frame, we will show the method for graphing this equation:

$$\boxed{y = x + 1}$$

Step 1: We made up a table of solutions:

	x	y
A	4	5
B	2	3
C	-2	-1
D	-4	-3
E	-5	-4

Step 2: On the coordinate system at the right, we plotted and labeled each point.

Step 3: We drew a straight line through the plotted points.

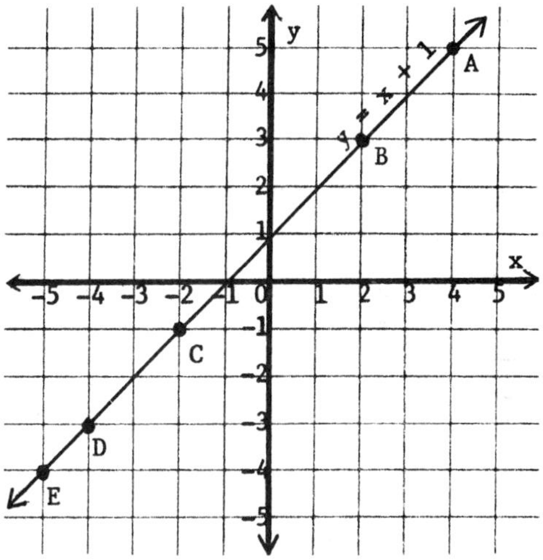

The straight line shown is the graph of the equation: $y = x + 1$

(a) Does the straight line (or graph) pass through the origin? _______

(b) The straight line (or graph) passes through what three quadrants? _______________

Answer to Frame 98: a) No b) I, II, and III

99. Below we have given a table of solutions for the following equation:

$y = 2x$

	x	y
A	4	8
B	2	4
C	-1	-2
D	-2	-4
E	-4	-8

Plot and label each point on the coordinate system at the right. Then draw the graph (a straight line) through the plotted points. Note: The scales on the axes are different.

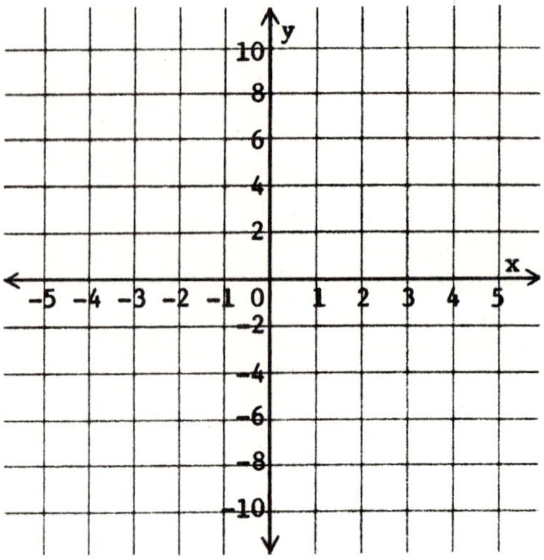

Answer to Frame 99: See next frame.

100.

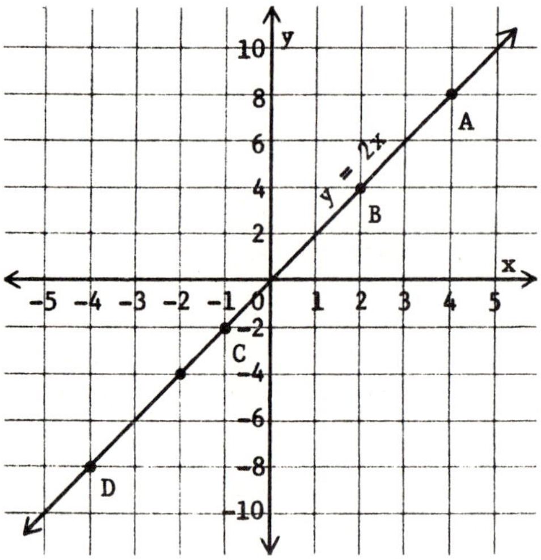

Your graphed line should look like the one shown at the left.

From the graph:

(a) Does the graphed line pass through the origin? ______

(b) The graphed line passes through what two quadrants? __________

Answer to Frame 100: a) Yes b) I and III

101. At the right we have graphed the equation:

$x + y = 3$

Any point on the graphed line represents one solution of the equation. Therefore, its coordinates satisfy the equation.

Any point off the graphed line does not represent a solution of the equation. Therefore, its coordinates do not satisfy the equation.

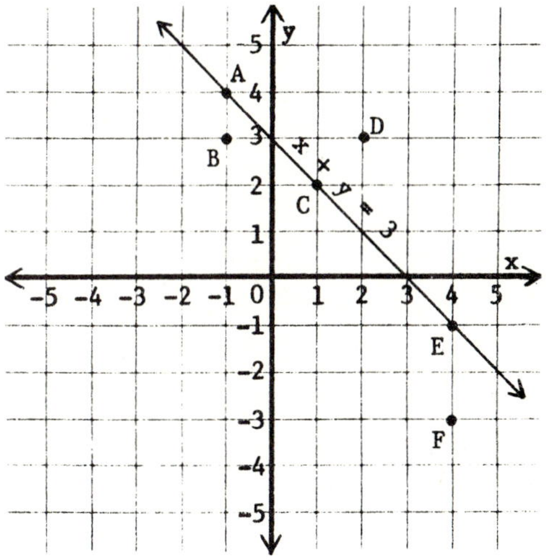

Let's examine the points on the line first.

(a) Write the coordinates of the three points on the graphed line:

A (,) C (,) E (,)

(b) Do the coordinates of these points satisfy the equation $x + y = 3$? _______

Now let's examine the points off the line.

(c) Write the coordinates of the three points off the graphed line:

B (,) D (,) F (,)

(d) Do the coordinates of these points satisfy the equation $x + y = 3$? _______

Answer to Frame 101: a) A (-1,4) C (1,2) E (4,-1) b) Yes, since in each case the sum of the coordinates is "3". c) B (-1,3) D (2,3) F (4,-3) d) No, since in each case the sum of the coordinates is not "3".

---

102. In principle, the coordinates of any point on a graphed line satisfy the equation. This is true even for points which do not lie at intersections on the coordinate system. However, since there might be some error in estimating the coordinates of points of this type, in practice our estimated coordinates might not satisfy the equation exactly.

At the right we have graphed the equation:

$y = 5x$

Points A and B on the graphed line each represent one solution of the equation: $y = 5x$. However, if we estimate the coordinates of these points, we get:

A (3.5, 18) B (-1.3, -7)

(a) Do these estimated coordinates for each point satisfy the equation exactly? _______

(b) Why don't they satisfy the equation exactly? ____________________

____

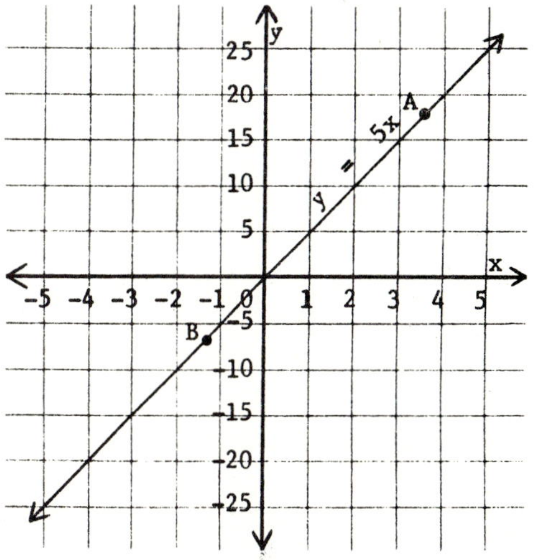

Answer to Frame 102:

a) No.

18 is only approximately equal to 5(3.5) or 17.5.

7 is only approximately equal to 5(-1.3) or -6.5.

b) Because our reported coordinates are only estimates.

103. When making up a table of solutions in order to graph an equation, we plug in some numbers for x and find the corresponding values of y. We can choose any values to plug in for x. The choice depends on the range of x which we want represented on the graph.

If we want a graph on which x ranges from -5 to 5, we plug in values of x between -5 and 5.

If we want a graph on which x ranges from -100 to 100, we plug in values of x between -100 to 100.

In both figures below, we have graphed the equation: $\boxed{y = x - 2}$

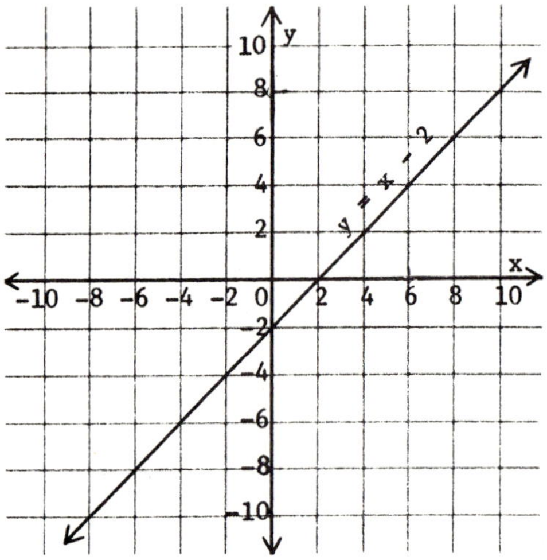

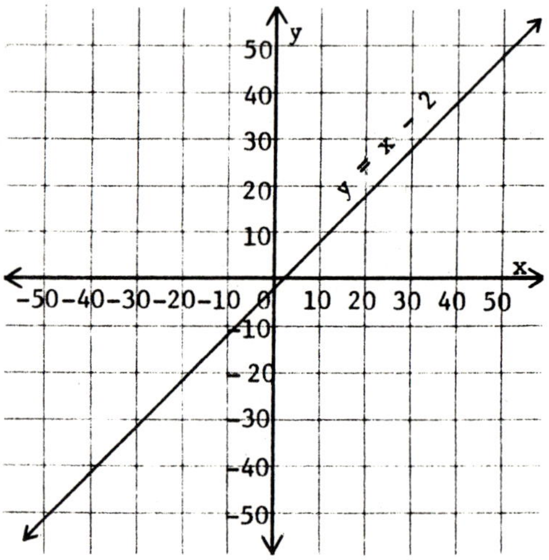

(a) On the coordinate system at the left, we have graphed the equation within what range for x?
From _____ to _____

(b) On the coordinate system at the right, we have graphed the equation within what range for x?
From _____ to _____

Answer to Frame 103: a) From -10 to 10 b) From -50 to 50

104. It should be obvious that all solutions for a given equation are not represented on one graph. When graphing $y = 5x$, for example:

If we use a range of x from -10 to 10, only solutions whose x-coordinates range from -10 to 10 will be represented.

If we use a range of x from -1,000 to 1,000, only solutions whose x-coordinates range from -1,000 to 1,000 will be represented.

One solution which satisfies the equation $y = 5x$ is (2,000, 10,000). Would this solution appear on the graph which uses either of the ranges above? _______

Answer to Frame 104: No, because 2,000 does not lie within either range.

105. On the two coordinate systems below, we have graphed the equation $x + y = 1$ from the table of solutions shown:

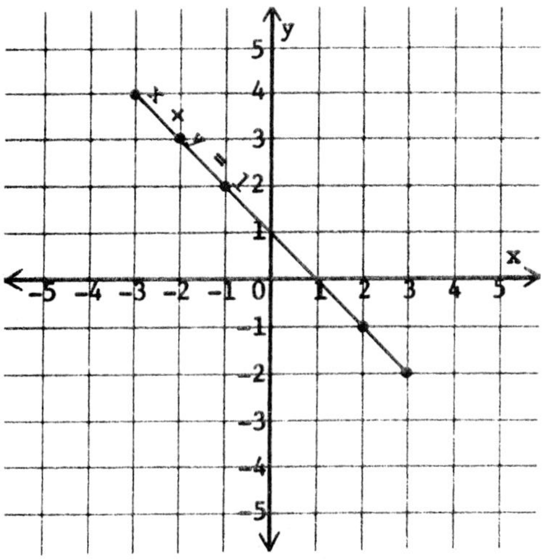

x + y = 1	
x	y
3	-2
2	-1
-1	2
-2	3
-3	4

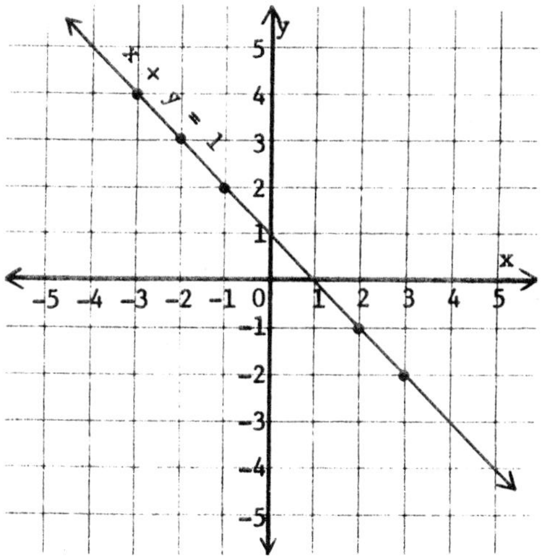

In the figure at the left, we stopped the graphed line at the last plotted point on each end.

In the figure at the right, we extended the graphed line to the edge of the coordinate system and put arrowheads at each end of it.

Stopping the graphed line at the extreme plotted points (as in the figure on the left) is wrong for this reason:

It implies that there are no solutions for the equation outside of the range -3 to 3 for x. This implication is false, since there are other solutions outside of this range like (4, -3), (-5, 6), (100, -99), (-78, 79).

Always extend your graphed line to the edge of the coordinate system and put arrowheads on the ends (as in the figure on the right). Doing so reflects these two facts:

(1) There are solutions within the full range -5 to 5 for x.

(2) There are solutions outside of that range which are not represented by this graph. The arrowheads indicate this fact.

---

106. The range of x in which you are interested determines how you should scale the x-axis.

If you are interested in values of x from -25 to 25, you would scale the x-axis this way:

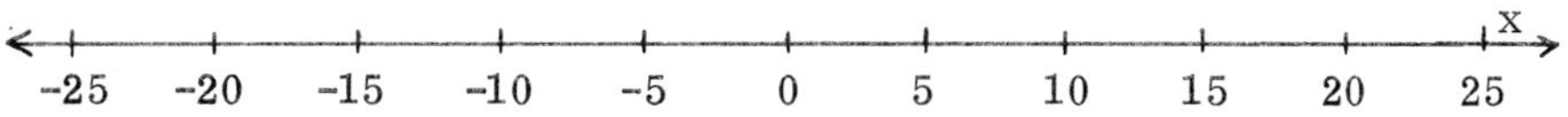

How would you scale the x-axis below if you were interested in values of x from -50 to 50?

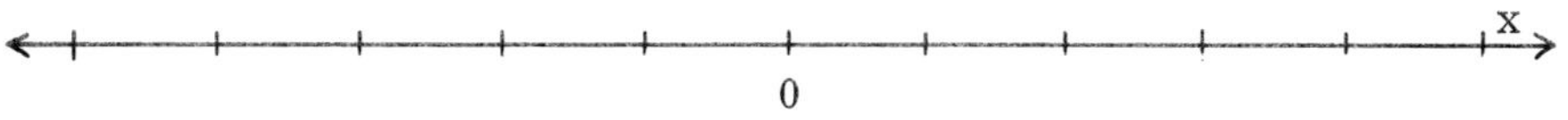

Answer to Frame 106:

107. How would you scale the x-axis below if you were interested in values of x from -100 to 100?

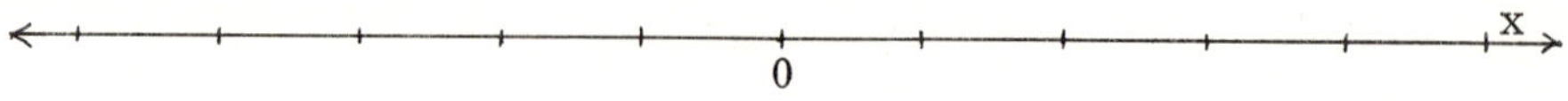

Answer to Frame 107:

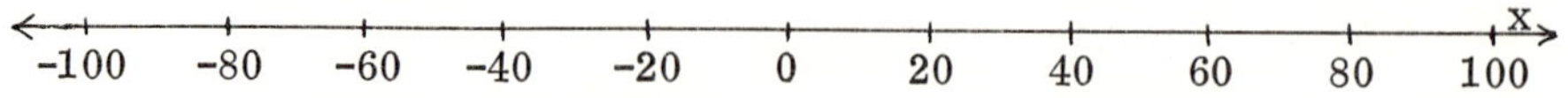

---

108. When graphing a linear equation, your scale on the x-axis depends on the range of values of x in which you are interested. If you want to graph $y = 5x$ within the range -5 to 5 for x, you plug in values of x from -5 to 5 when making up your table. A table of solutions for this range of x is shown below.

On the two coordinate systems below, we have graphed the equation $y = 5x$ using two different scales on the y-axis. Notice how different the positions of the graphed lines are.

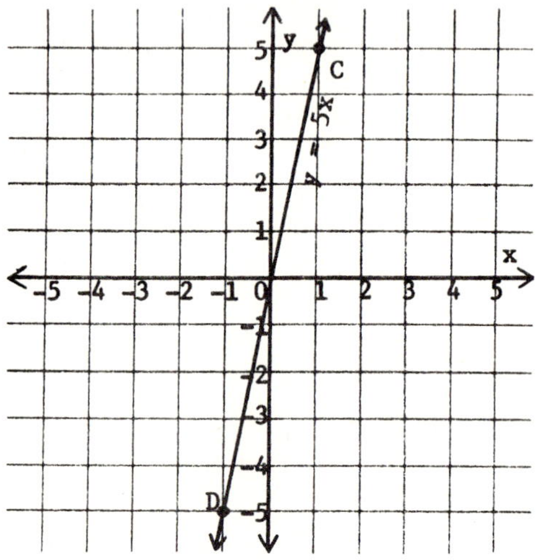

$y = 5x$

	x	y
A	5	25
B	3	15
C	1	5
D	-1	-5
E	-3	-15
F	-5	-25

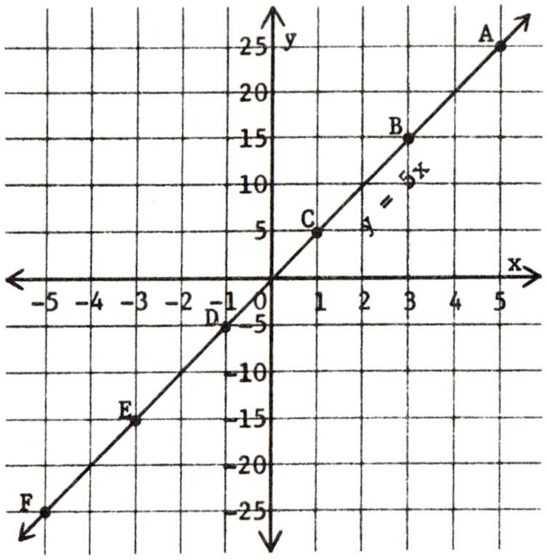

In the graph on the left, we used the same scale on each axis. Notice this point:

> Only two of the points (C and D) in the table could be plotted since the y-axis does not contain values like 25, 15, -15, -25. Therefore, the graphed line represents solutions only within the range of -1 to 1 for x.

In the graph on the right, the scale on the y-axis is different than that on the x-axis. All of the points in the table are plotted on the graph. Therefore, we have actually represented all of the solutions within the range -5 to 5 for x. For this reason, the graph on the right is preferable.

Go to next frame.

109. The table below contains solutions for $y = 10x$ with the range -10 to 10 for x. On the two coordinate systems below, we have graphed the equation above using two different scales on the y-axis. Notice again how different the positions of the graphed lines are.

y = 10x

	x	y
A	10	100
B	5	50
C	-5	-50
D	-10	-100

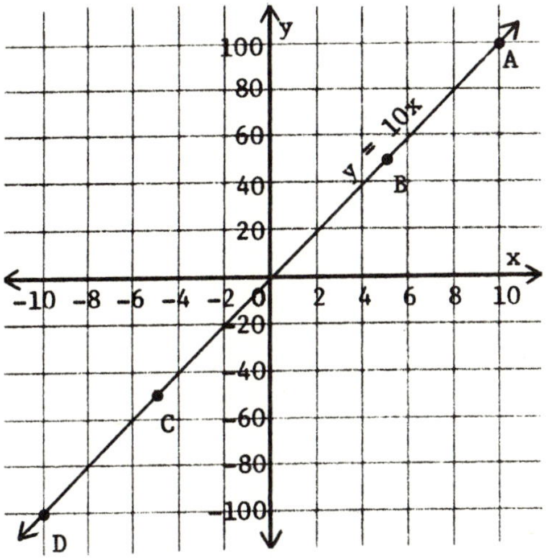

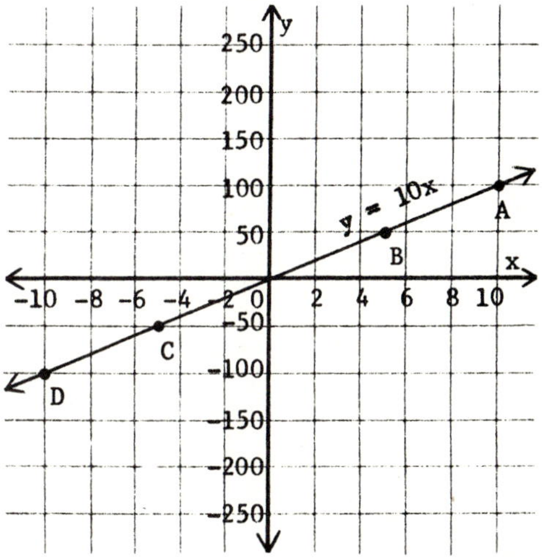

All of the points (A, B, C, D) in the table are plotted on both graphs. Therefore, all of the solutions within the range -10 to 10 for x are represented on both graphs.

However, notice these facts about the graph on the right.

(1) Though there are values on the y-axis from 100 to 250 and from -100 to -250, these values are not needed.

(2) Therefore, we are not using the space at the top and bottom of the coordinate system.

Therefore, the graph on the left is preferable.

Go to next frame.

110. Follow this procedure for determining the scales on the axes:

> The <u>scale on the x-axis</u> is determined <u>by the range of x we wish to represent.</u>
>
> The <u>scale on the y-axis</u> is determined by:
>
> (1) Making up a table of solutions. <u>Include the two extreme values of x from the range.</u>
>
> (2) Then examine the range of y-values in the table and scale the axes so that:
>
> (a) All of the y-values are represented.
> (b) <u>Unnecessarily large or small values of y are not included.</u>

Below we have constructed a table of solutions for: y = 4x. We want to represent the solutions within the range -25 to 25 for x. Notice that we have included the two extreme values (-25 and 25) for x:

y = 4x

x	y
25	100
10	40
-10	-40
-25	-100

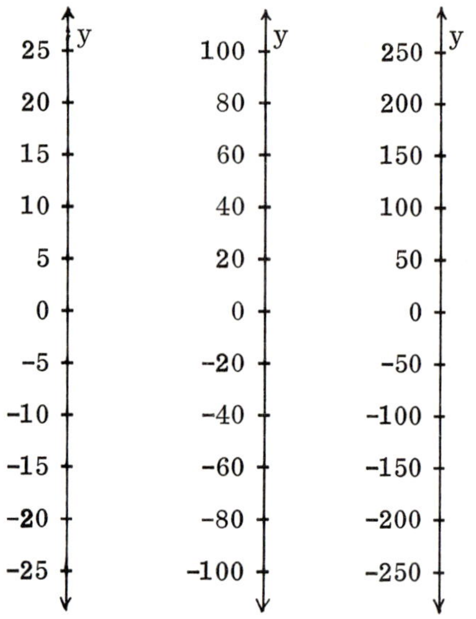

Now examine the range of y-values in the table. They range from -100 to 100. Which of the y-axes at the right would we use? ______________

The one in the middle.

The one <u>on the left</u> does not contain large enough values.

The one <u>on the right</u> contains unnecessary large values.

---

111. If you do not include the two extreme values of x, you cannot determine the range of y. For example, if you want to graph the equation y = 10x - 5 within the range -10 to 10 for x:

(a) Two solutions which should appear in your table are:

x = _____, y = _____

x = _____, y = _____

(b) Therefore, the values on your y-axis must range at least from _______ to _______.

---

a) x = 10, y = 95
x = -10, y = -105

b) -105 to 95

Probably you would use:

-110 to 100

or

-120 to 120

112. What is wrong with the graph for $y = x - 2$ shown at the right? ____________

____

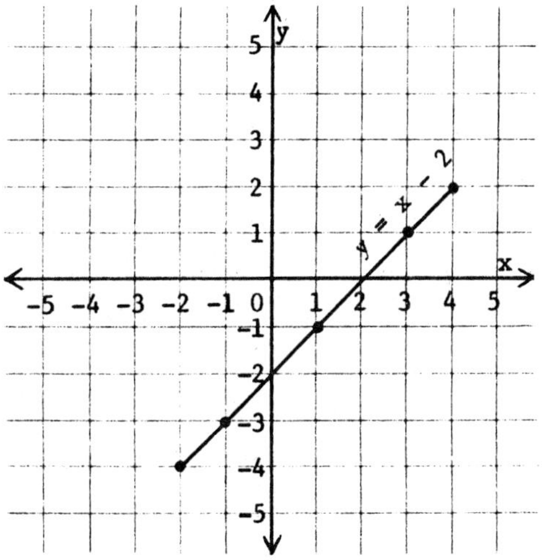

	Answer to Frame 112:	The graphed line should be extended in both directions to the edges of the coordinate system.

SELF-TEST 5 (Frames 98-112)

Using the table below and the coordinate system at the right, graph the equation: $y = 50x$. Graph it within the range -10 to 10 for x. Put your own scales on the axes.

y = 50x

x	y

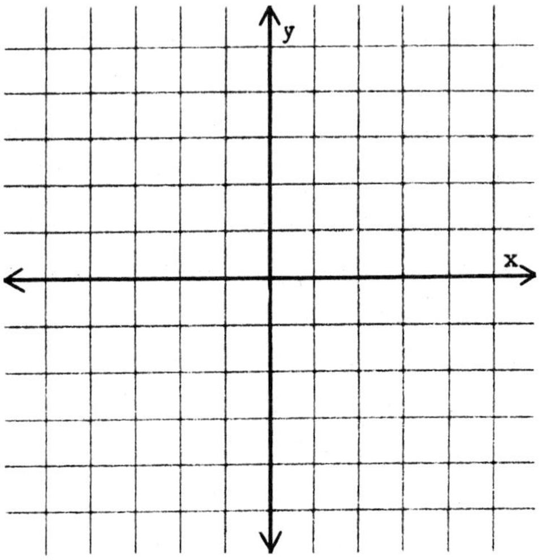

(Answers to Self-Test are on following page.)

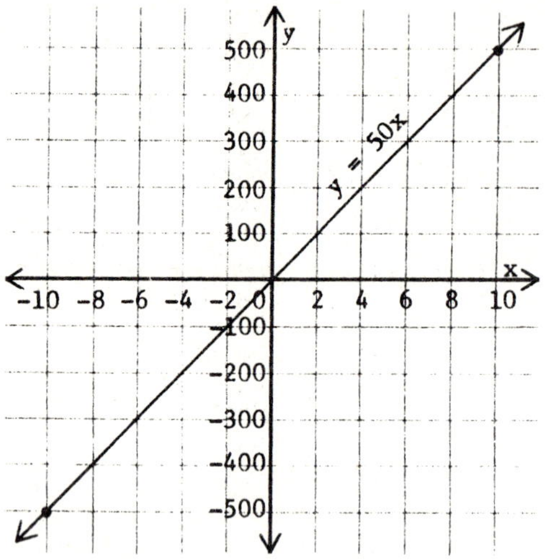

ANSWER:

In addition to other solutions, you should have included the following solutions in your table:

x = 10, y = 500

x = -10, y = -500

Your graph should look like the one at the left.

7-6 POINTS ON THE AXES AND INTERCEPTS

Up to this time, we have avoided reading and plotting points on the coordinate axes. In this section, we will examine the coordinates of points of this type and show the meaning of intercepts for graphed lines.

---

113. We have plotted various points on the coordinate system at the right. Points I, J, K, L lie on the horizontal axis.

(a) Points A, B, C, and D have the same y-coordinate (or ordinate). It is ______.

(b) Points E, F, G, and H have the same y-coordinate (or ordinate). It is ______.

(c) Points I, J, K, and L lie on the horizontal axis. They have the same y-coordinate (or ordinate). It is ______.

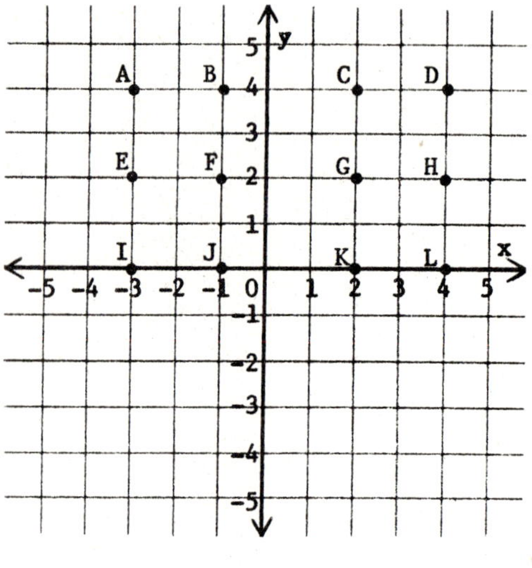

---

114. The y-coordinate of any point on the horizontal axis is ______.

a) 4 b) 2 c) 0

0

115. Eight points have been plotted on the coordinate system at the right.

Write the coordinates of each point:

A ________ E ________

B ________ F ________

C ________ G ________

D ________ H ________

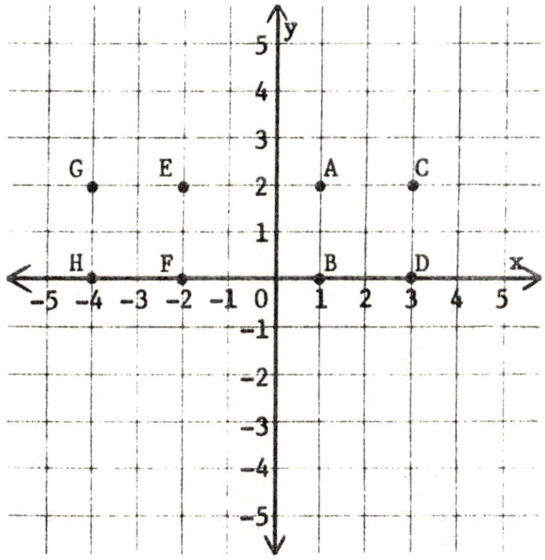

	Answer to Frame 115:	A (1,2)	C (3,2)	E (-2,2)	G (-4,2)
		B (1,0)	D (3,0)	F (-2,0)	H (-4,0)

116. We have plotted various points on the coordinate system at the right. Some of the points lie on the vertical axis.

(a) Points A, B, C, and D have the same x-coordinate (or abscissa). It is ____.

(b) Points E, F, G, and H have the same x-coordinate (or abscissa). It is ____.

(c) Points I, J, K, and L lie on the vertical axis. Each point has the same x-coordinate (or abscissa). It is ____.

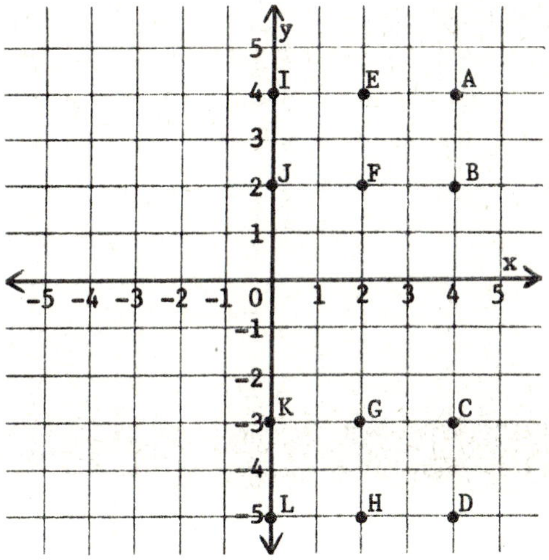

117. The x-coordinate of any point on the vertical axis is ____.

a) 4 b) 2 c) 0

0

118. Various points have been plotted on the coordinate system at the right.

Write the coordinates of each point:

A ________ E ________

B ________ F ________

C ________ G ________

D ________ H ________

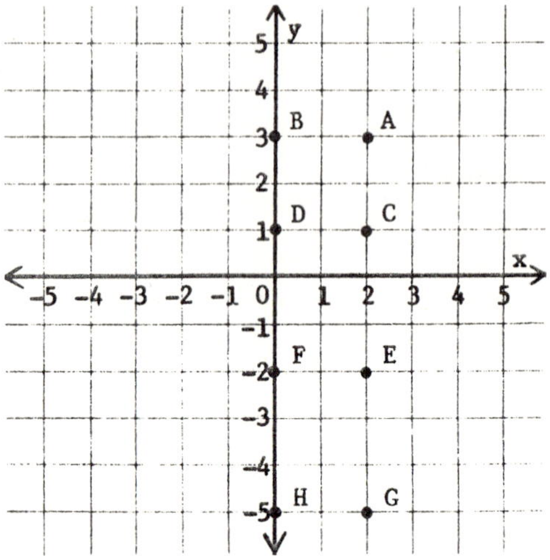

A (2,3)	E (2,-2)
B (0,3)	F (0,-2)
C (2,1)	G (2,-5)
D (0,1)	H (0,-5)

119. (a) If a point lies on the horizontal axis, is its x-coordinate or y-coordinate 0? ________

(b) If a point lies on the vertical axis, is its x-coordinate or y-coordinate 0? ________

a) y-coordinate

b) x-coordinate

120. Four points have been plotted on the coordinate system at the right.

Write the coordinates of each point:

A ________

B ________

C ________

D ________

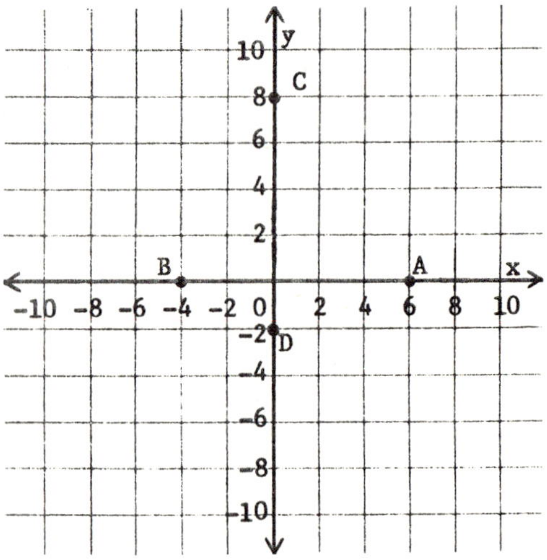

A (6,0)	C (0,8)
B (-4,0)	D (0,-2)

121. (a) Another name for the <u>horizontal</u> axis is the ____-axis.

(b) Another name for the <u>vertical</u> axis is the ____-axis.

122. State whether each of the following points lies on the horizontal or vertical axis:

(a) (0,-15) ____________ (c) (-15,0) ____________

(b) (7,0) ____________ (d) (0,23) ____________

a) x-axis

b) y-axis

123. State whether each of the following points lies on the x-axis or y-axis:

(a) (77,0) ____________ (c) (-100,0) ____________

(b) (0,53) ____________ (d) (0,-99) ____________

a) Vertical
b) Horizontal
c) Horizontal
d) Vertical

Answer to Frame 123: a) x-axis b) y-axis c) x-axis d) y-axis

124. Plot the following points on the coordinate system at the right:

A (4,0)

B (0,4)

C (-3,0)

D (0,-3)

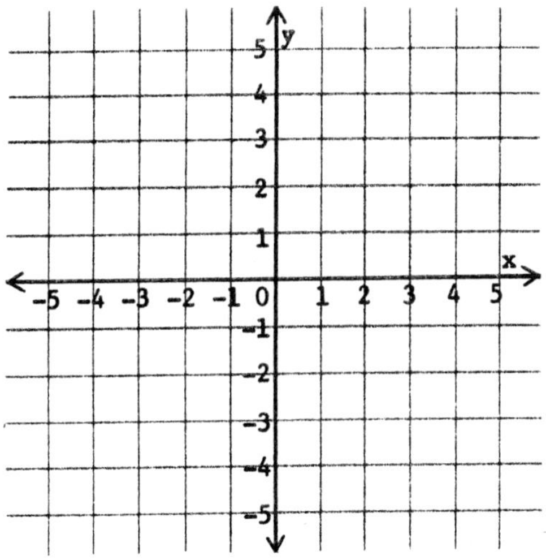

Answer to Frame 124:

The plotted points are shown at the right.

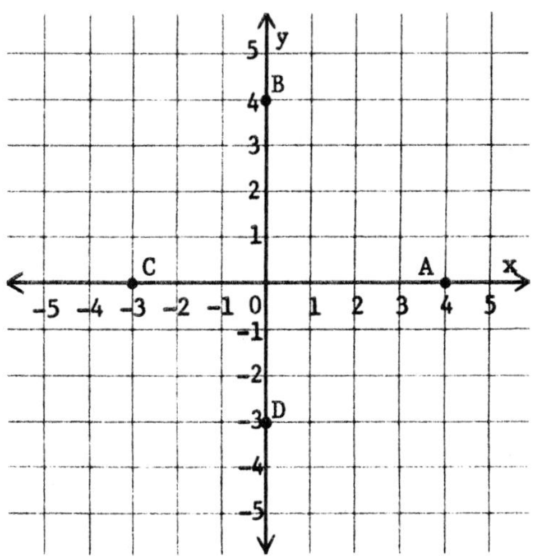

125. The origin lies on both axes.

(a) Its x-coordinate is _____.

(b) Its y-coordinate is _____.

(c) Its coordinates are __________.

---

126. (a) The "abscissa" is another name for which coordinate? ____________

(b) The "ordinate" is another name for which coordinate? ____________

a) 0

b) 0

c) (0,0)

---

127. (a) If the abscissa of a point is "0", the point lies on which axis? ____________

(b) If the ordinate of a point is "0", the point lies on which axis? ____________

a) x-coordinate

b) y-coordinate

---

128. If both the abscissa and ordinate of a point are "0", what is the name of the point? ____________

a) Vertical or y-axis

b) Horizontal or x-axis

The origin

---

129. Two lines have been graphed at the right.

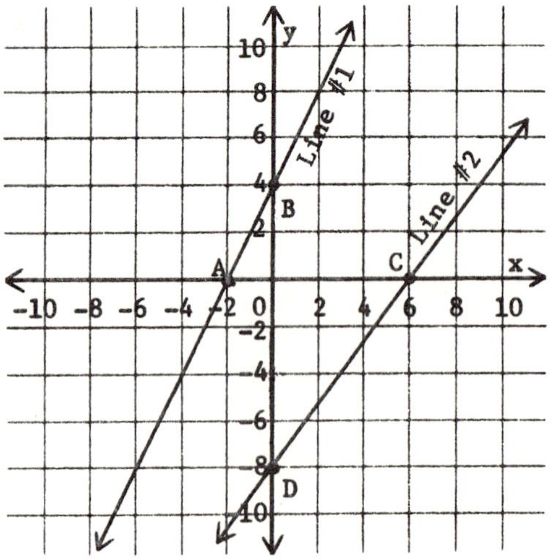

Look at Line #1:

Line #1 cuts the horizontal axis (x-axis) at point A. Point A is called the x-intercept of Line #1. Its coordinates are (-2,0).

Line #1 cuts the vertical axis (y-axis) at point B. Point B is called the y-intercept of Line #1. Its coordinates are (0,4).

Look at Line #2:

(a) Point C is the x-intercept of Line #2. Its coordinates are __________.

(b) Point D is the y-intercept of Line #2. Its coordinates are __________.

---

a) (6,0)

b) (0,-8)

130. Two more lines have been graphed at the right. The intercepts are labelled A, B, C, and D.

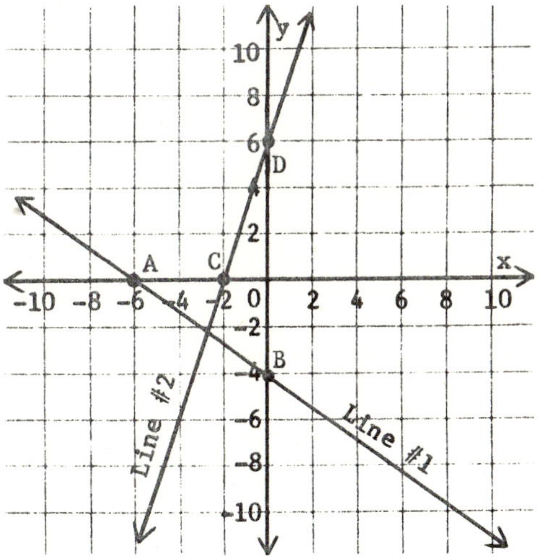

What point is the y-intercept for:

(a) Line #1 ? _____ (b) Line #2 ? _____

What point is the x-intercept for:

(c) Line #1 ? _____ (d) Line #2 ? _____

For Line #1, write the coordinates of:

(e) The x-intercept _________

(f) The y-intercept _________

For Line #2, write the coordinates of:

(g) The x-intercept _________

(h) The y-intercept _________

Answer to Frame 130: a) Point B b) Point D c) Point A d) Point C e) (-6,0) f) (0,-4) g) (-2,0) h) (0,6)

---

131. Here is another line graphed on the "xy" coordinate system:

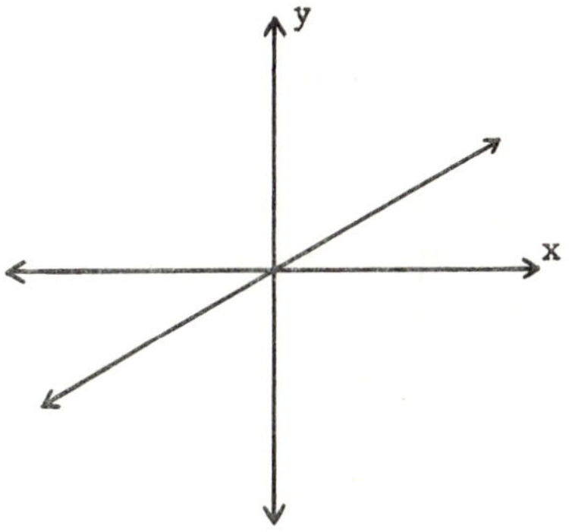

The graphed line passes through the origin.

The x-intercept is at the origin.
The y-intercept is also at the origin.

(a) What are the coordinates of the x-intercept? _____

(b) What are the coordinates of the y-intercept? _____

Answers: a) (0,0) b) (0,0)

132. When a graphed line passes through the __________, the x-intercept and y-intercept are identical.

Answer: origin

133. We have graphed the following equation at the right.

$$\boxed{y = x + 3}$$

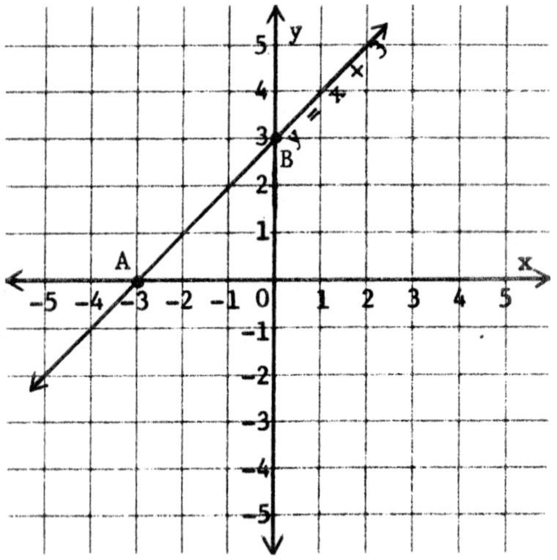

(a) Point A is the x-intercept. Its coordinates are ________.

(b) Point B is the y-intercept. Its coordinates are ________.

(c) Show that the coordinates of point A satisfy the equation.

(d) Show that the coordinates of point B satisfy the equation.

<u>Answer to Frame 133</u>: a) (-3, 0) b) (0, 3) c) For A: (-3, 0) d) For B: (0, 3)

c)
$$y = x + 3$$
$$0 = (-3) + 3$$
$$0 = 0$$

d)
$$y = x + 3$$
$$3 = 0 + 3$$
$$3 = 3$$

---

134. Since the coordinates of any point on a graphed line satisfy the equation of the line, <u>the coordinates of its intercepts satisfy the equation also.</u>

At the right we have graphed this equation:

$$\boxed{y = 2x}$$

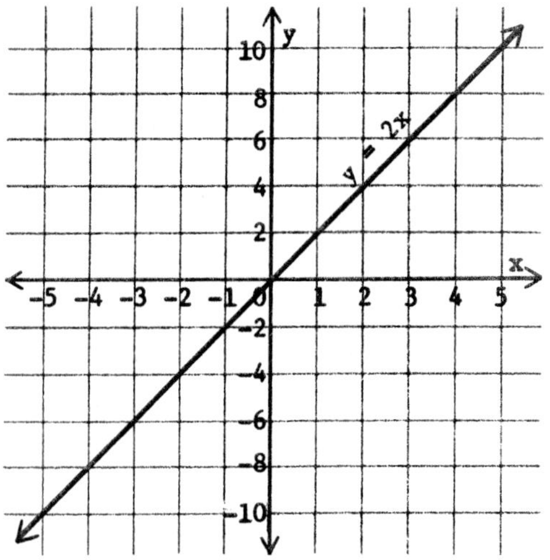

Since the graphed line passes through the origin, the coordinates of both its intercepts are (0, 0).

Do the coordinates (0, 0) satisfy the equation of the line? ______

<u>Answer to Frame 134</u>: Yes, since:
$$y = 2x$$
$$0 = 2(0)$$
$$0 = 0$$

SELF-TEST 6 (Frames 113-134)

Write the coordinates of:

1. A point on the x-axis three units to the right of the origin. __________
2. A point on the y-axis five units below the origin. __________

Complete: 3. If a point lies on the x-axis, its ______________ (abscissa/ordinate) is 0.

4. If a point lies on the y-axis, its ______________ (abscissa/ordinate) is 0.

5. Which of the following points lie on the x-axis? __________

(a) (0, -5) (b) (-2, 0) (c) (0, 0) (d) (9, 1) (e) (0, 7)

Refer to the two lines graphed at the right:

6. For Line #1, write the coordinates of:
   (a) The x-intercept: __________
   (b) The y-intercept: __________
7. For Line #2, write the coordinates of:
   (a) The x-intercept: __________
   (b) The y-intercept: __________

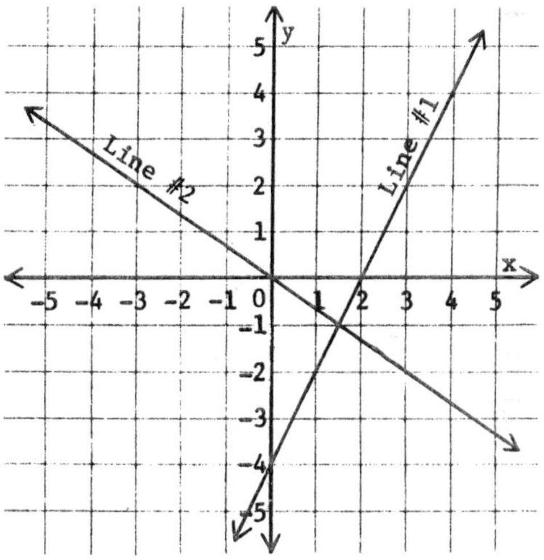

ANSWERS: 1. (3, 0) 2. (0, -5) 3. ordinate 4. abscissa 5. (b), (c) 6. (a) (2, 0) (b) (0, -4) 7. (a) (0, 0) (b) (0, 0)

7-7 CURVILINEAR GRAPHS AND EQUATIONS

When graphing an equation, the graph does not have to be a straight line. Frequently, the graph is a curved line. Such graphs are called "curvilinear" graphs. In this section, we will construct curvilinear graphs and examine them.

135. We have given some examples of equations with rough sketches of their graphs immediately below them. In each case, the graph is curvilinear. Notice the different shapes that curved graphs can have.

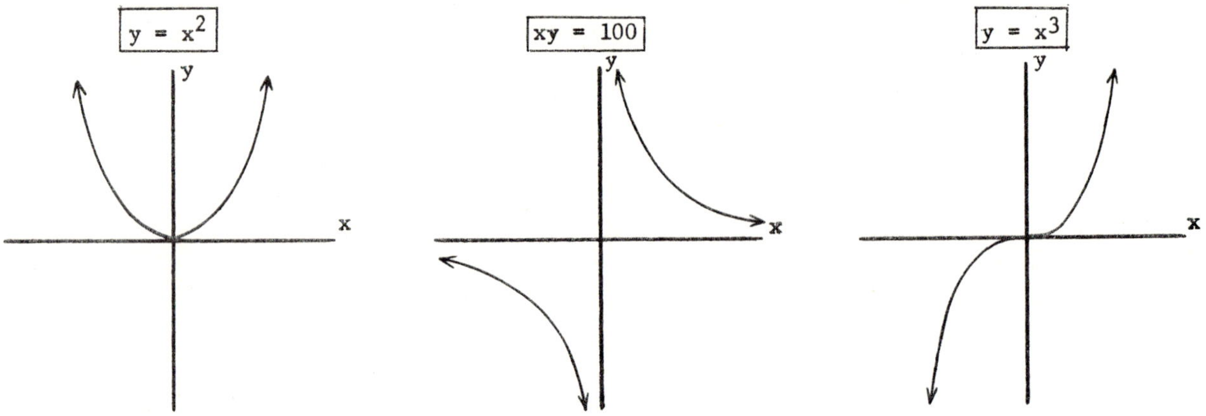

Note also that no arrows are shown at each end of the x-axis and the y-axis, as was done in earlier sections. Such arrows are usually omitted, and will be omitted in our further work in graphing.

136. When the graph of an equation is a curve, it still represents the solutions for the equation. Any point on the curved line stands for one solution. Therefore, the coordinates of each point on the line satisfy the equation. The coordinates of points not on the line do not satisfy the equation.

At the right is the graph of this equation:

$y = 2x^2$

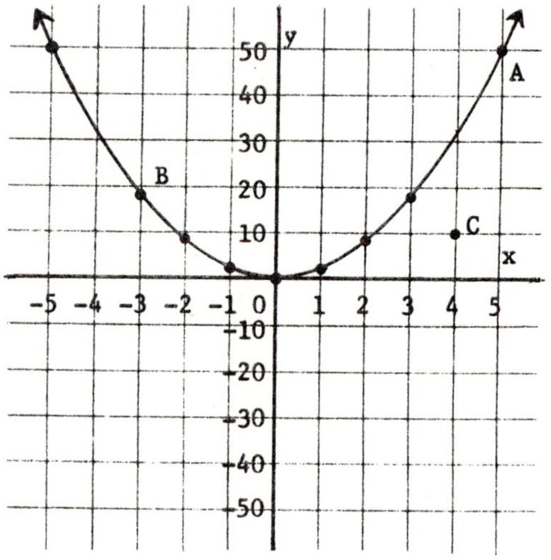

(a) Point A (5,50) is a point on the curve. Show that its coordinates satisfy the equation:

(b) Point B (-3,18) is a point on the curve. Show that its coordinates satisfy the equation:

(c) Point C (4,10) is a point which is not on the curve. Show that its coordinates do not satisfy the equation:

(d) The origin (0,0) is a point on the curve. Show that its coordinates satisfy the equation:

Answer to Frame 136:	a)	b)	c)	d)
	$y = 2x^2$	$y = 2x^2$	$y = 2x^2$	$y = 2x^2$
	$50 = 2(5)^2$	$18 = 2(-3)^2$	$10 \neq 2(4)^2$	$0 = 2(0)^2$
	$50 = 2(25)$	$18 = 2(9)$	$10 \neq 2(16)$	$0 = 2(0)$
	$50 = 50$	$18 = 18$	$10 \neq 32$	$0 = 0$

137. A graph is supposed to represent the solutions of an equation within a given range for x. The purpose of this frame is to show why we draw a smooth curve through the plotted points when the graph is not a straight line. We draw a curve because otherwise the points for some solutions will not be on the graph.

We will graph the following equation as an example: $y = 4x^2$

Here is a table of solutions:

	x	y
(A)	-5	100
(B)	-3	36
(C)	0	0
(D)	3	36
(E)	5	100

We have plotted the five points above on each coordinate system below. On the left, we drew the graph by connecting the points with a series of straight lines. On the right, we drew the graph by drawing a curved line through the points.

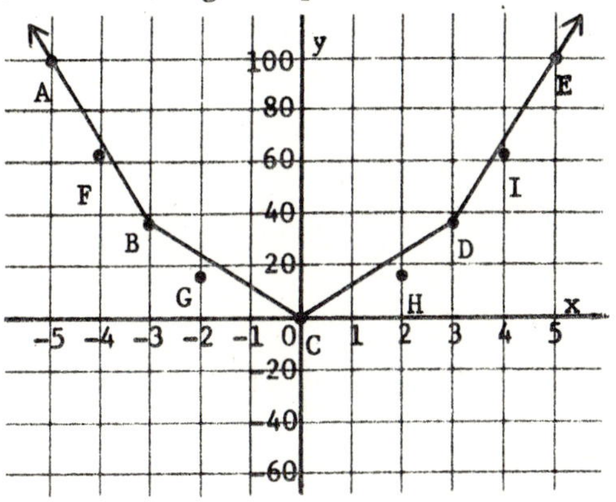

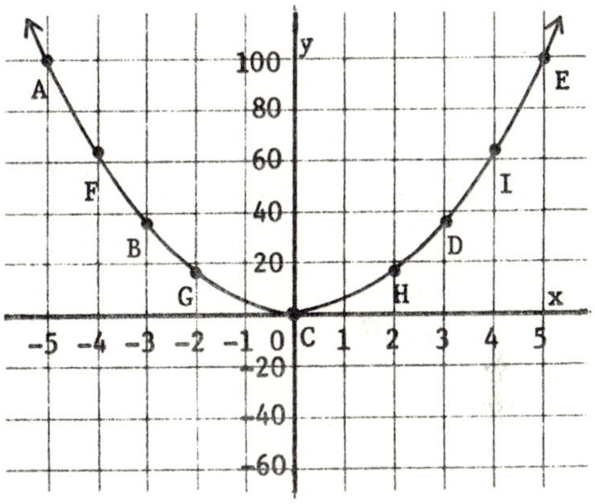

On each coordinate system, we have also plotted the following four solutions which were not in the original table:

$y = 4x^2$

	x	y
(F)	-4	64
(G)	-2	16
(H)	2	16
(I)	4	64

On the left diagram, these four points do not lie on the graph, even though they are supposed to.

On the right diagram, these four points do lie on the graph.

WE DRAW A SMOOTH CURVE THROUGH THE PLOTTED POINTS BECAUSE THE SOLUTIONS LIE IN A SMOOTH CURVE. THE MORE POINTS WE PLOT, THE CLEARER THIS FACT IS.

Go to next frame.

138. To graph the equation xy = 24 we have made up a table of solutions, shown below.

Plot the points on the coordinate system at the right. The graph is curvilinear, and it has two parts - one part in Quadrant I and one part in Quadrant III. The two parts are not connected. Draw the two curved lines.

xy = 24

	x	y
(A)	-6	-4
(B)	-4	-6
(C)	-3	-8
(D)	-2	-12
(E)	-1	-24
(F)	1	24
(G)	2	12
(H)	3	8
(I)	4	6
(J)	6	4

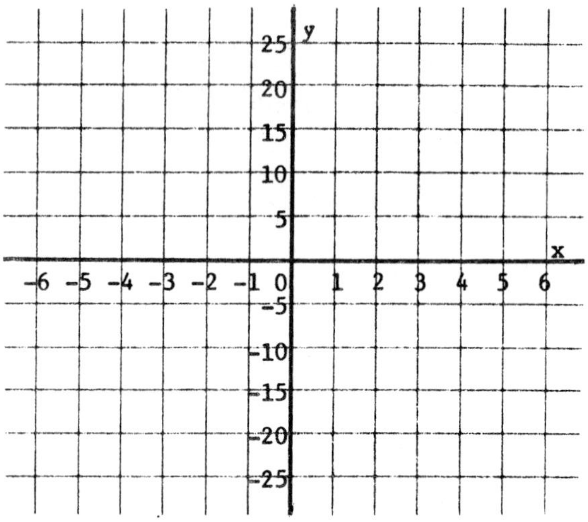

Answer to Frame 138: See next frame.

139. Here is the graph of xy = 24. Compare this graph with yours. Be sure you drew smooth curves. Notice that the curves do not touch the axes.

It should be obvious to you that you can get a better picture of the outline of the graph by plotting more points. Plotting more points is especially useful when two consecutive points are fairly far apart. For example, you could have drawn the graph better by:

(1) Plotting a point ($1\frac{1}{2}$ or $\frac{3}{2}$, 16) between points F and G.

(2) Plotting a point (5, $\frac{24}{5}$ or $4\frac{4}{5}$) between points I and J.

(3) Plotting a point ($-1\frac{1}{2}$ or $-\frac{3}{2}$, -16) between points D and E.

(4) Plotting a point (-5, $-\frac{24}{5}$ or $-4\frac{4}{5}$) between points A and B.

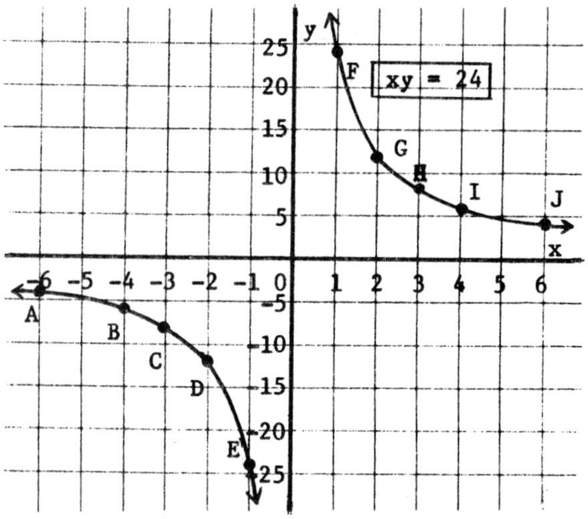

Go to next frame.

140. When an equation is graphed and the graphed relationship is a straight line, there is usually one x-intercept and one y-intercept. (If the line passes through the origin, the x-intercept and y-intercept are identical.)

But when the graphed relationship is curvilinear:

(1) There may be more than one x-intercept or y-intercept, or
(2) there may be no x-intercept or y-intercept.

Two curvilinear graphs are shown in Figures 1 and 2 at the right:

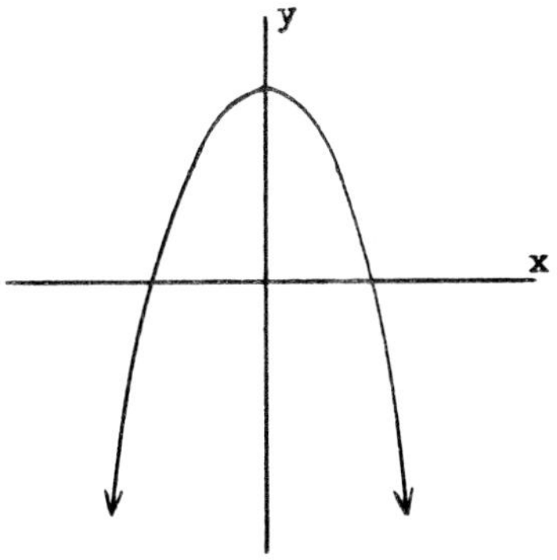

Figure 1

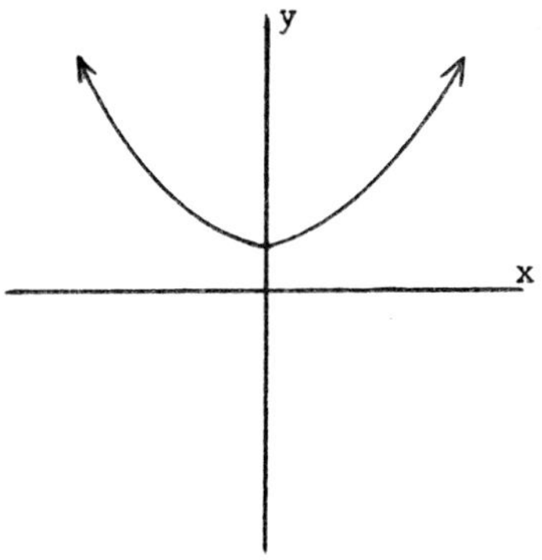

Figure 2

(a) In Figure 1: How many x-intercepts are there? _____

How many y-intercepts are there? _____

(b) In Figure 2: How many x-intercepts are there? _____

How many y-intercepts are there? _____

Answer to Frame 140: a) 2 x-intercepts, 1 y-intercept b) No x-intercepts, 1 y-intercept

---

141. Two more curvilinear graphs are shown in Figures 1 and 2 at the right:

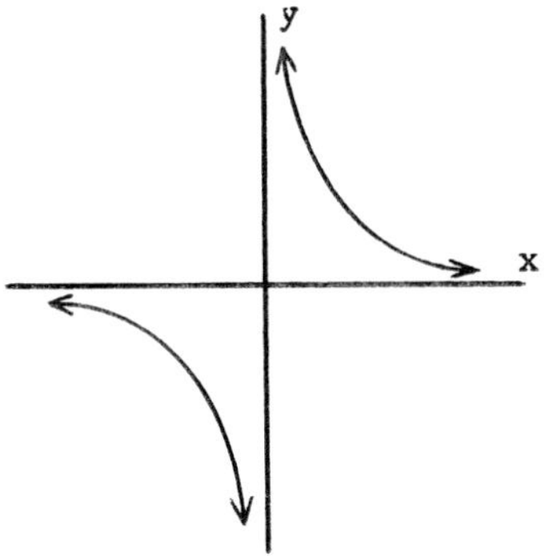

Figure 1

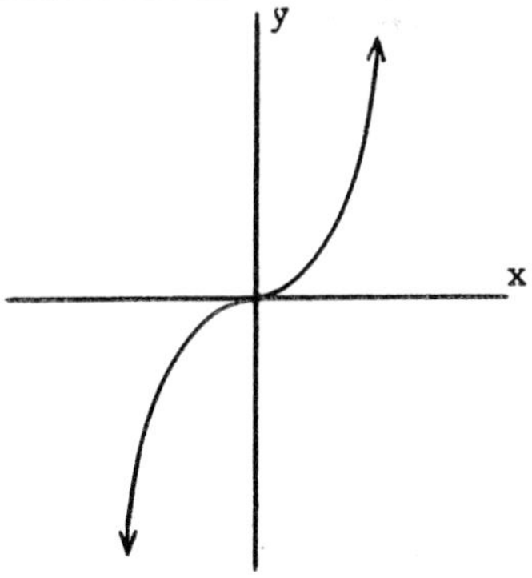

Figure 2

(a) In Figure 1: How many x-intercepts are there? _____ How many y-intercepts are there? _____

(b) In Figure 2: How many x-intercepts are there? _____ How many y-intercepts are there? _____

Answer to Frame 141: a) No x-intercepts, No y-intercepts (No matter how far the graphed curve is extended, it never touches the axes.) b) 1 x-intercept, 1 y-intercept (Identical intercepts, at the origin)

142. We want to graph the equation $y = 4x - x^2$. To do so, proceed as follows:

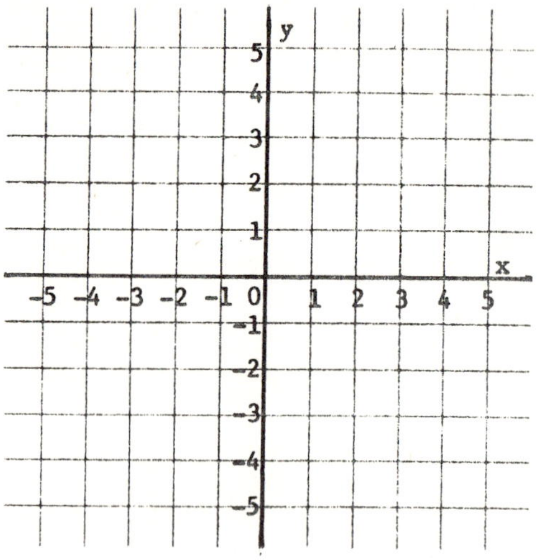

(1) Complete this table of solutions:

$y = 4x - x^2$

	x	y
(A)	-1	
(B)	0	
(C)	1	
(D)	2	
(E)	3	
(F)	4	
(G)	5	

(2) Plot the points on the coordinate system at the left, and then carefully draw the curve

Answer to Frame 142: See next frame.

143. The table of solutions and the graph for $y = 4x - x^2$ are shown below:

$y = 4x - x^2$

	x	y
(A)	-1	-5
(B)	0	0
(C)	1	3
(D)	2	4
(E)	3	3
(F)	4	0
(G)	5	-5

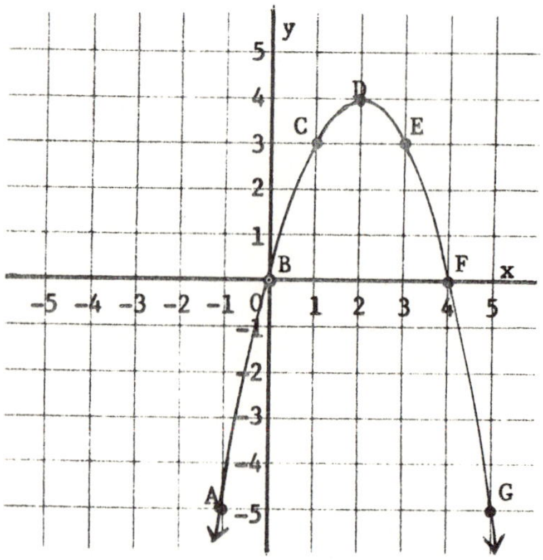

Referring to the graph, answer the following:

(a) There are two x-intercepts. Their coordinates are _______ and _______.

(b) There is one y-intercept. Its coordinates are _______.

Answer to Frame 143: a) B (0,0) F (4,0) b) B (0,0)

Note: When graphing a curvilinear equation, be sure that you plot enough points so that the outline of the curve becomes clear.

7-8 GRAPHING TWO-VARIABLE FORMULAS

Up to this point, we have graphed only pure mathematical equations containing x and y. In this section, we will show that the same procedure is used to graph formulas containing two letters or variables. Before doing so, we will discuss the meaning of variables and formulas.

144. Here is a typical mathematical equation: $y = 2x + 1$

In this equation, the two letters "x" and "y" are called variables. They are called variables because we can satisfy the equation by plugging in different or various pairs of values for them.

In the equation $xy = 100$, "x" and "y" are called ______________.

---

145. In the equation: $y = 2x + 1$

"y" and "x" are called variables.

"2" and "1" are called constants.

(They are called constants because they do not change no matter what numbers are plugged in for "y" and "x".)

Name the constants in each of the following equations:

(a) $y = 3x^2$ _____ (b) $xy = 50$ _____ (c) $y = 2x^2 - 7$ _____

Answer: variables

---

146. Here is a formula by which we can convert temperature from degrees-Centigrade (C) to degrees-Fahrenheit (F).

$$F = \frac{9}{5}C + 32°$$

This formula is an equation. There are pairs of values of "F" and "C" which satisfy it. For example:

$$\text{If } C = 10°,\ F = \frac{9}{5}(10°) + 32°$$
$$= 18° + 32° \quad = 50°$$

(a) If $C = 50°$, $F =$ _______ (b) If $C = -30°$, $F =$ _______

(c) How many pairs of values for "C" and "F" satisfy this formula? ____________________

Answers:
a) 3
b) 50
c) 2 and 7

Answers:
a) 122°
b) -22°
c) An infinite number

147. Here is a formula showing the relationship between pressure (measured in pounds per square inch) and volume (measured in cubic inches) of a gas under certain conditions. This formula is like any pure mathematical equation of the same type (for example, $xy = 100$).

$\boxed{PV = 100}$

(a) If $P = 5$, $V =$ ______

(b) If $P = 10$, $V =$ ______

(c) If $P = 25$, $V =$ ______

(d) How many pairs of values for "P" and "V" satisfy this equation? ______

---

148. Though the letters in any formula stand for a measurement of some physical quantity (like temperature or pressure or volume), mathematically we treat them just like any other letters.

Therefore, in the formula: $F = \frac{9}{5}C + 32°$

F and C are called variables because there are various pairs of values for them which satisfy the formula.

$\frac{9}{5}$ and 32° are called constants because they do not change no matter what values are plugged in for F and C.

In the formula $PV = 100$: (a) Name the variables ______

(b) Name the constants ______

Answers to 147: a) 20 b) 10 c) 4 d) An infinite number

---

149. In general: (a) The letters in either equations or formulas are called ______.

(b) The numbers in either equations or formulas are called ______.

Answers to 148: a) P and V b) 100

---

150. When graphing a pure mathematical equation which contains the two letters "x" and "y":

(1) We always scale "x" on the horizontal axis.
(2) We always scale "y" on the vertical axis.

The "x-axis" is another name for the "horizontal axis."
The "y-axis" is another name for the "vertical axis."

(a) Another name for the horizontal axis is the ______.

(b) When graphing an equation which contains the letters "x" and "y", we always put the y-scale on the ______ (horizontal/vertical) axis.

Answers to 149: a) variables b) constants

Answers to 150: a) x-axis b) vertical

151. Here is the formula showing the relationship between degrees-Fahrenheit and degrees-Centigrade:

$$F = \frac{9}{5}C + 32°$$

Where F stands for degrees-Fahrenheit, and C stands for degrees-Centigrade.

When graphing a formula which contains two variables (two letters), there is no rigid rule about "which variable goes on which axis." This decision is somewhat arbitrary. However, there are accepted conventions with certain formulas. These conventions will be learned in your science and technology courses.

In the formula above:

(a) If you are told to "graph C as x," you should put the C-scale on the ______________ (horizontal/vertical) axis.

(b) If you are told to "graph C as y," you should put the C-scale on the ______________ (horizontal/vertical) axis.

Answer to Frame 151: a) horizontal b) vertical

152. Let's graph the formula: $F = \frac{9}{5}C + 32°$

(We will graph C as x. That is, we will put the C-scale on the horizontal axis.)

Here is a table of pairs of values of C and F which satisfy the formula. Notice that we plugged in multiples of 5 for C so that "$\frac{9}{5}C$" will be a whole number. When graphing a formula, always plug in numbers which will make your calculations easy.

Each pair of values is labeled with a letter. Plot each point on the graph at the right and draw the straight line through them.

	C	F
(A)	30°	86°
(B)	20°	68°
(C)	10°	50°
(D)	-10°	14°
(E)	-40°	-40°

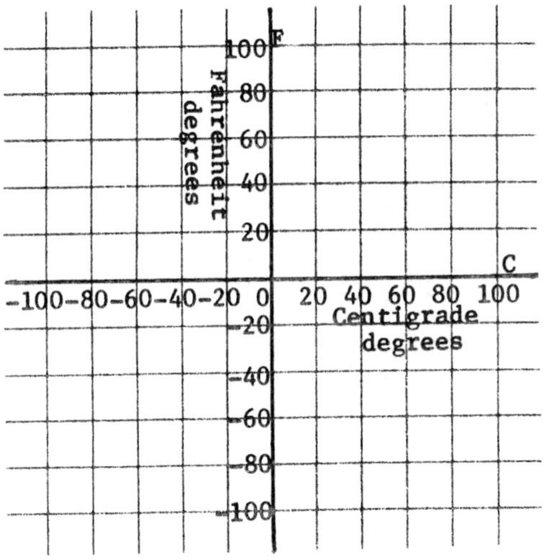

Answer to Frame 152: See next frame.

153. At the right is the graph of:

$$F = \frac{9}{5}C + 32°$$

(a) There is a point M in the second quadrant. It is on the graphed line. It represents this pair of values: C = -10°, F = 14°. Does this pair of values satisfy the formula? _______

(b) There is a point N in the second quadrant. It represents the pair of values: C = -60°, F = 20°. Does this pair of values satisfy the formula? _______

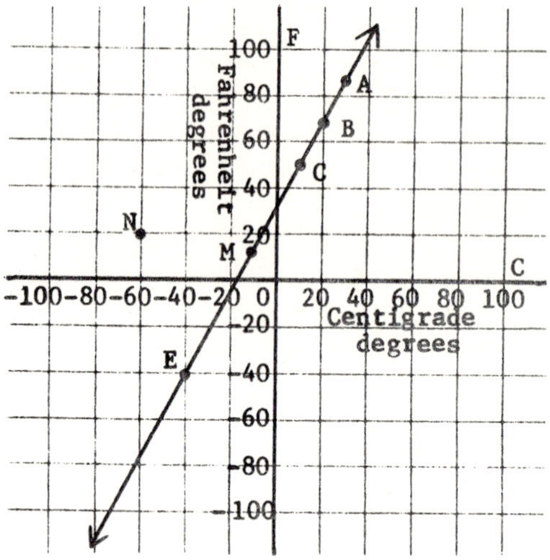

---

154. In a pair of coordinates:

(a) Another name for the x-coordinate is the ____________.

(b) Another name for the y-coordinate is the ____________.

a) Yes, since:

$$F = \frac{9}{5}C + 32°$$
$$14° = \frac{9}{5}(-10°) + 32°$$
$$14° = -18° + 32°$$
$$14° = 14°$$

b) No, since:

$$F = \frac{9}{5}C + 32°$$
$$20° \neq \frac{9}{5}(-60°) + 32°$$
$$20° \neq -108° + 32°$$
$$20° \neq -76°$$

---

155. $d = 35t$

The formula above states the relationship between "distance traveled" and "time" at a constant speed or velocity of 35 miles per hour.

The phrase "plot t as the abscissa" means "plot t as the x-coordinate."

(a) If you are told to "plot t as the x-coordinate," you should put the t-scale on the ______________ (horizontal/vertical) axis.

(b) If you are told to "plot t as the abscissa," you should put the t-scale on the ______________ (horizontal/vertical) axis.

a) abscissa

b) ordinate

---

156. $E = 80I$

This formula shows the relationship between "E" (voltage) and "I" (current) with a constant "R" (resistance) of 80 ohms.

(a) If we tell you to "plot I as the abscissa," you should put the I-scale on the ______________ (horizontal/vertical) axis.

(b) If we tell you to "plot I as the ordinate," you should put the I-scale on the ______________ (horizontal/vertical) axis.

a) horizontal

b) horizontal

---

a) horizontal

b) vertical

157. When measuring temperatures, we use negative values for both F (degrees-Fahrenheit) and C (degrees-Centigrade). However, negative values are usually not used when measuring most physical quantities. For example, we usually do not talk about negative distance or pressure or weight, and so on.

The following formula represents the distance a dropped object will fall in a given period of time:

$$\boxed{s = 16t^2}$$ where s is distance (in feet) and t is time (in seconds)

We do not use negative values when measuring either distance (s) or time (t). Since only positive coordinates will appear on the graph, the graph of this formula will appear only in what quadrant?

____

Answer to Frame 157: Quadrant I

158. Let's graph the formula: $\boxed{s = 16t^2}$ The graph will be curvilinear.

Plot the points represented in the following table on the graph at the right. Then draw a smooth curved line through them.

$$\boxed{s = 16t^2}$$

	t (in seconds)	s (in feet)
(A)	0	0
(B)	1	16
(C)	2	64
(D)	3	144
(E)	4	256
(F)	5	400

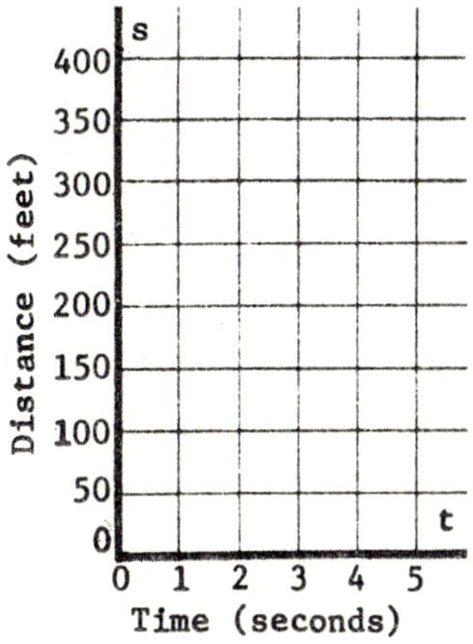

(Since we use only positive values for t and s, we only need Quadrant I for the graph.)

Answer to Frame 158: See next frame.

159. Your graph of $s = 16t^2$ should look like the one below, at the left.

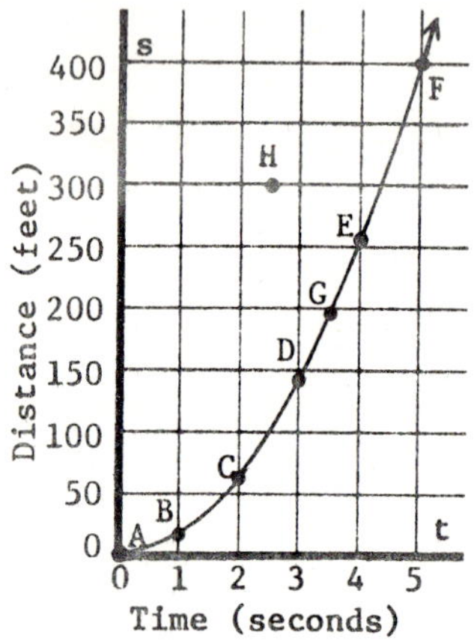

(a) There is a point G on the curved line. It represents these values: $3\frac{1}{2}$ seconds and 196 feet. Do these values satisfy the formula? ________

(b) There is a point H which is not on the curved line. It represents these values: $2\frac{1}{2}$ seconds and 300 feet. Do these values satisfy the formula? ________

(c) Since we put the t-scale on the horizontal axis, we plotted t as the ______________ (abscissa/ordinate).

---

a) Yes, since:

$s = 16t^2$

$196 = 16(3\frac{1}{2})^2$

$196 = 16(\frac{49}{4})$

$196 = 196$

b) No, since:

$s = 16t^2$

$300 \neq 16(2\frac{1}{2})^2$

$300 \neq 16(\frac{25}{4})$

$300 \neq 100$

c) abscissa

---

160. Let's graph the following specific instance of the principle known as Boyle's Law, showing the relationship between the pressure and volume of a gas.

$PV = 100$ where P is pressure (pounds per square inch) and V is volume (cubic inches).

This graph is also curvilinear. A table of solutions, sometimes called a "table of values," is shown below at the right.

$PV = 100$

	P	V
(A)	1	100
(B)	2	50
(C)	4	25
(D)	5	20
(E)	10	10
(F)	-1	-100
(G)	-2	-50
(H)	-4	-25
(I)	-5	-20
(J)	-10	-10

(a) Since negative values for pressure and volume do not usually make sense, which points in the table should not be plotted?

____

(b) Since only positive values of P and V are plotted, the curved line will appear in only one quadrant. Which quadrant is needed? ______________

---

a) F, G, H, I, J

b) Quadrant I

---

161. Plot the appropriate points (from the last frame) on the graph at the right, and draw the curved line through them.

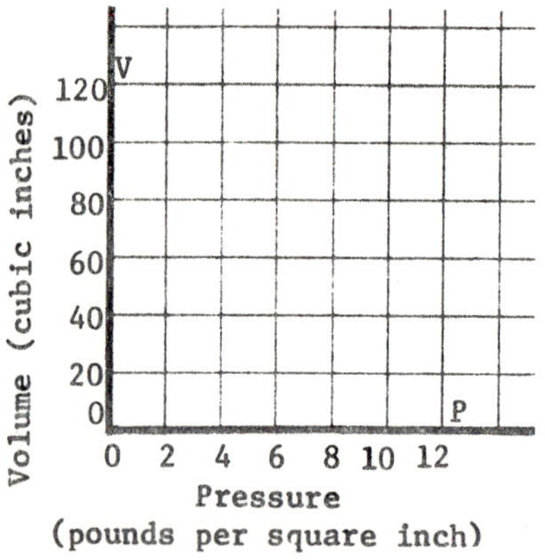

Answer to Frame 161:

See next frame.

162. The graph of PV = 100 is shown below, at the right.

Carefully examine the graph:

(a) As the pressure increases, does the volume increase or decrease? ________

(b) As the volume increases, does the pressure increase or decrease? ________

(c) Which variable was plotted as x? ____

(d) Which variable was plotted as the ordinate? ____

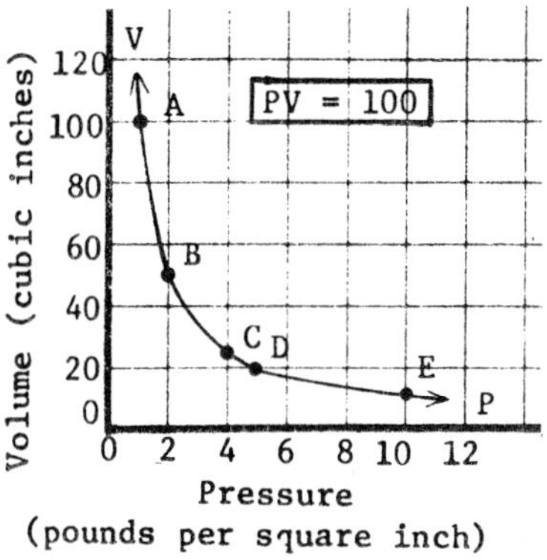

Answer to Frame 162: a) Decrease b) Decrease c) P d) V

163. Graphs of formulas can have intercepts on the coordinate axes. For example, in the graph at the right, the graphed line intercepts both axes. The graph shows the relationship between degrees-Centigrade and degrees-Fahrenheit.

(a) Which variable is plotted as the abscissa? ____

(b) Which variable is plotted as the ordinate? ____

(c) The approximate coordinates of the horizontal or F-intercept are (,).

(d) The approximate coordinates of the vertical or C-intercept are (,).

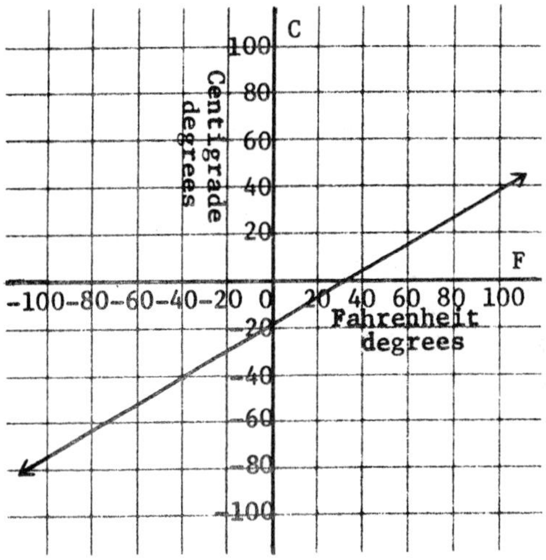

a) F (degrees-Fahrenheit)

b) C (degrees-Centigrade)

c) (30°, 0°)

d) (0°, -18°)

164. Here is a formula related to the concept of "load line" in transistor electronics:

$\boxed{E + 8I = 12}$ where E is voltage (in volts) and I is current (in milliamperes).

The graph of this formula is shown at the right. Notice these points:

(1) Since negative values for E and I are not needed in this situation, points are plotted only in the first quadrant.

(2) There is an E-axis and an I-axis.

(3) Even though the graphed line only touches the two axes, we still talk about an E-intercept and an I-intercept.

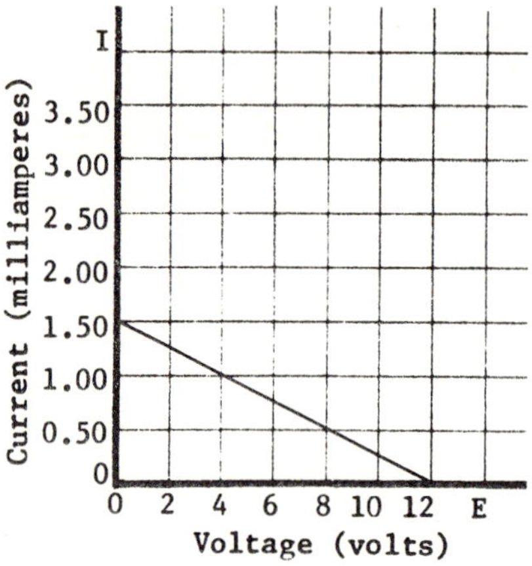

Referring to the graph, answer the following:

(a) Which variable is plotted as the ordinate? _______

(b) The coordinates of the E-intercept are (,).

(c) The coordinates of the I-intercept are (,).

---

165. The graphed line below shows the relationship between distance and time for an object whose velocity is 75 miles per hour.

Even though the graphed line only touches the origin, we still speak of intercepts.

(a) The coordinates of both the t-intercept and the d-intercept are identical. They are (,).

(b) Why don't we graph the other half of the line in the third quadrant? ________________

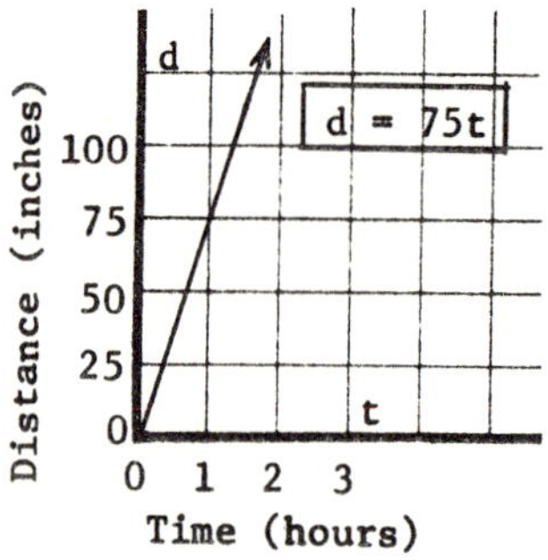

Answers to 164:

a) I (current)

b) (12,0)

c) (0, 1.50)

Answers to 165:

a) (0,0)

b) Because negative values for "d" and "t" are usually not used.

166. Shown below is the graph of the formula: $S = 0.5F$ This formula is an instance of the scientific principle known as Hooke's Law. This principle describes the amount of stretch S (in inches) that will occur in a spring when a force F (in pounds) is applied to the spring.

Each point on a graphed line represents a fact. Point A represents this fact: It takes 2 lbs. of force to stretch the spring 1 in.

(a) Point B represents this fact: It takes 4 lbs. of force to stretch the spring _______ in.

(b) Point C represents this fact: It takes _______ lbs. of force to stretch the spring $2\frac{1}{2}$ in.

(c) Point D represents this fact: It takes _______ lbs. of force to stretch the spring _______ in.

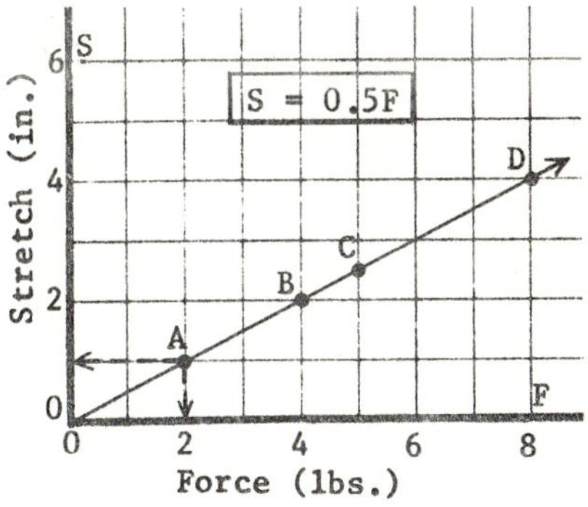

Answer to Frame 166: a) 2 in. b) 5 lbs. c) 8 lbs., 4 in.

7-9 READING TECHNICAL AND SCIENTIFIC GRAPHS

Any technical or scientific graph contains a wealth of information. Technicians and scientists have to be skilled in reading graphs in order to obtain this information. The purpose of this section is to introduce graph-reading. We will discuss the following:

(1) Any point on a graph represents some scientific or technical fact.
(2) Graphs can be used to solve problems.
(3) Graphs show the basic relationship between two variables.

167. Below we have graphed the distance a dropped object will fall in a given amount of time. The graphed relationship is curvilinear.

When reading a graph, you frequently have to estimate the values of your "readings" on the axes.

Point A represents this fact: An object falls approximately 80 ft. in approximately 2.3 sec.

(a) Point B represents this fact: An object falls approximately ____ ft. in approximately ____ sec.

(b) Point C represents this fact: An object falls approximately ____ ft. in approximately ____ sec.

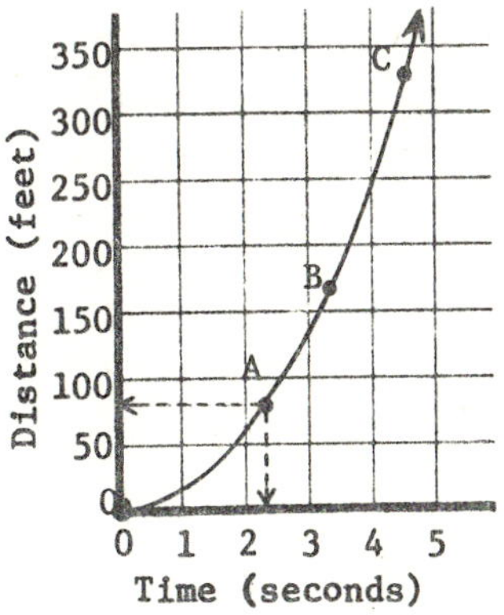

a) 170 ft. 3.3 sec.
b) 330 ft. 4.5 sec.

168. The graph below shows the relationship between temperature measured on a Fahrenheit scale and temperature measured on a Centigrade scale. Since there are negative values of temperature, the graph appears in other quadrants besides Quadrant I.

Point A represents this fact: A Fahrenheit temperature of -80° is approximately equivalent to a Centigrade temperature of -61° or -62°.

(a) Point B represents this fact: A Fahrenheit temperature of -40° is approximately equivalent to a Centigrade temperature of _______.

(b) Point C represents this fact: A Fahrenheit temperature of 10° is approximately equivalent to a Centigrade temperature of _______.

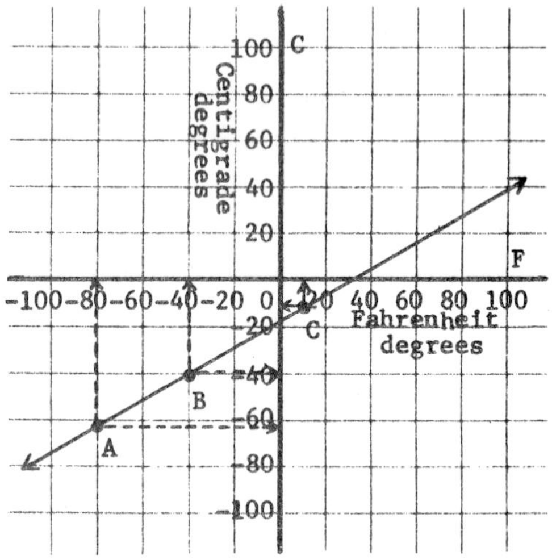

Answer to Frame 168: a) about -40° b) about -10°

169. Here is a graph relating force and stretch of a spring (Hooke's Law):

We can use the graph to solve the following problem: If you apply 3 lbs. of force, how much stretch do you get in this spring? To solve it, we do the following:

(1) Draw a vertical arrow up from "3" on the "force" axis until it intersects the graphed line at point A.

(2) Now draw a horizontal arrow from point A to the "stretch" axis.

(3) Since this horizontal arrow points to $1\frac{1}{2}$ on the "stretch" axis, we know that we will get $1\frac{1}{2}$ in. of stretch if we apply a 3 lb. force.

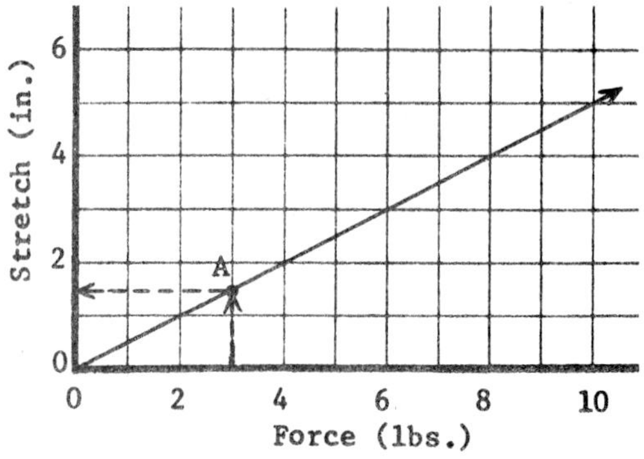

How many inches of stretch, approximately, would we get in this spring if the following forces were applied?

	Force	Stretch		Force	Stretch		Force	Stretch		Force	Stretch
(a)	4 lbs.	_____ in.	(b)	7 lbs.	_____ in.	(c)	2 lbs.	_____ in.	(d)	9 lbs.	_____ in.

Answer to Frame 169: a) 2 in. b) $3\frac{1}{2}$ in. c) 1 in. d) $4\frac{1}{2}$ in.

170. Some springs are stretched more easily than others. In graphing Hooke's Law, the graphed line is different for different springs. Below we show the graphs for two different springs.

We can use the graph to decide which of the two springs stretches more easily.

(a) When Spring #1 has a force of 3 lbs. applied to it, the spring is stretched _____ in.

(b) When Spring #2 has a force of 3 lbs. applied to it, the spring is stretched _____ in.

(c) When Spring #1 has a force of 6 lbs. applied to it, the spring is stretched _____ in.

(d) When Spring #2 has a force of 6 lbs. applied to it, the spring is stretched _____ in.

(e) Which of the two springs stretches more easily? ____________

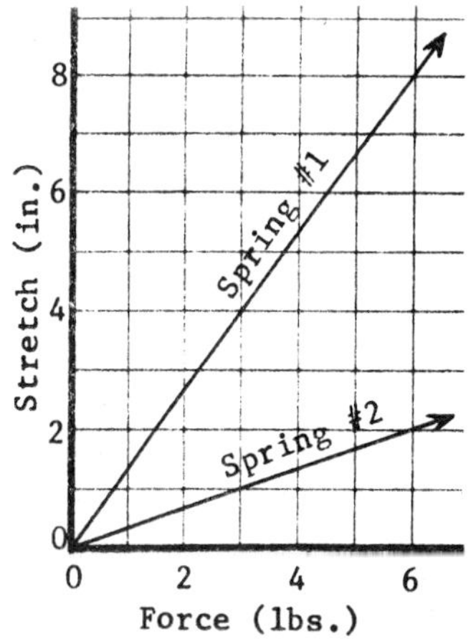

Answer to Frame 170: a) 4 in. b) 1 in. c) 8 in. d) 2 in. e) Spring #1

171. Another graph of Hooke's Law for a particular spring is given below on the right.

We can use the graph to solve this problem: "How much force do you need to get a 10 in. stretch?" To answer this question, do the following:

(1) Draw a horizontal arrow from 10 in. on the "stretch" axis to the graphed line.

(2) Draw a vertical arrow from this point of intersection down to the "force" scale.

(3) Read this point on the force scale. The answer is 400 lbs.

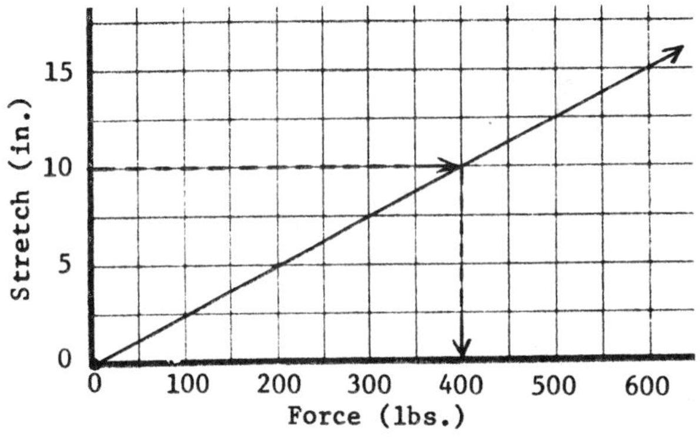

Using the same procedure: (a) How much force do you need to get a stretch of 15 in. ? ________

(b) How much force do you need to get a stretch of 5 in. ? ________

Answer to Frame 171: a) 600 lbs. b) 200 lbs.

172. The graph at the right relates Fahrenheit and Centigrade temperatures. If a specific Fahrenheit temperature is known, the corresponding Centigrade temperature can be determined from the graph, as follows:

(1) Draw a vertical arrow from the Fahrenheit axis to the graphed line.

(2) Draw a horizontal arrow from this point of intersection to the Centigrade axis.

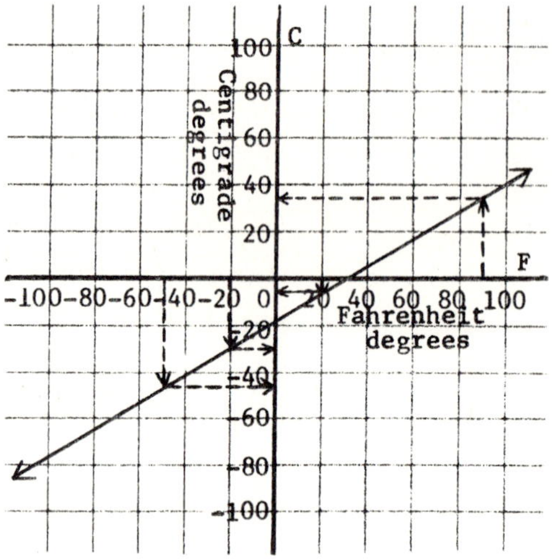

What Centigrade temperature, approximately, corresponds to each of the following Fahrenheit temperatures?

(a) -50° Fahrenheit
____ Centigrade

(b) -20° Fahrenheit
____ Centigrade

(c) 20° Fahrenheit
____ Centigrade

(d) 90° Fahrenheit
____ Centigrade

Answer to Frame 172: (approximately) a) -45° b) -30° c) -5° d) 35°

173. The Fahrenheit-Centigrade graph below at the right is the same as in the previous frame.

For the questions below, we have drawn the appropriate arrows on the graph. This time, the arrows go from the Centigrade axis to the graphed line and from there to the Fahrenheit axis, since we are converting Centigrade to Fahrenheit.

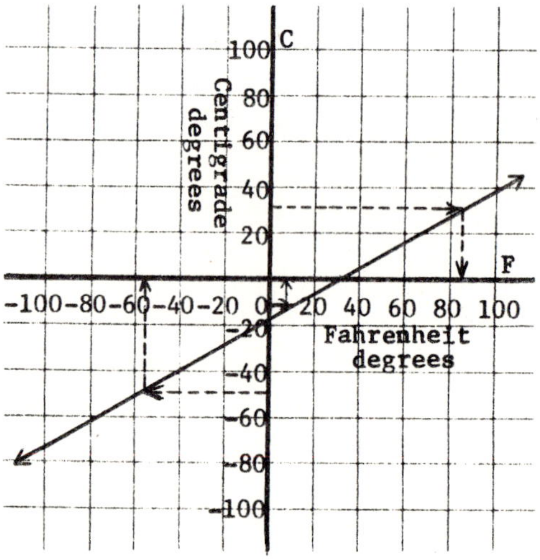

What Fahrenheit temperature, approximately, corresponds to each of the following Centigrade temperatures?

(a) 30° Centigrade
____ Fahrenheit

(b) -15° Centigrade
____ Fahrenheit

(c) -50° Centigrade
____ Fahrenheit

Answer to Frame 173: (approximately) a) 85° b) 5° c) -55°

174. At the right is a graph of the relationship between voltage and current in a circuit when the resistance of the circuit is constant. This relationship is called Ohm's Law. Voltage is measured in "volts," and current is measured in "amperes."

(a) If a voltage of 30 volts is applied to this circuit, how much current do you get? ________

(b) If you want to get 7 amperes of current in the circuit, how much voltage must you apply? ________

(c) If you apply a voltage of 55 volts to the circuit, how much current do you get? ________

(d) If you want to get 2.5 amperes of current in the circuit, how much voltage must you apply? ________

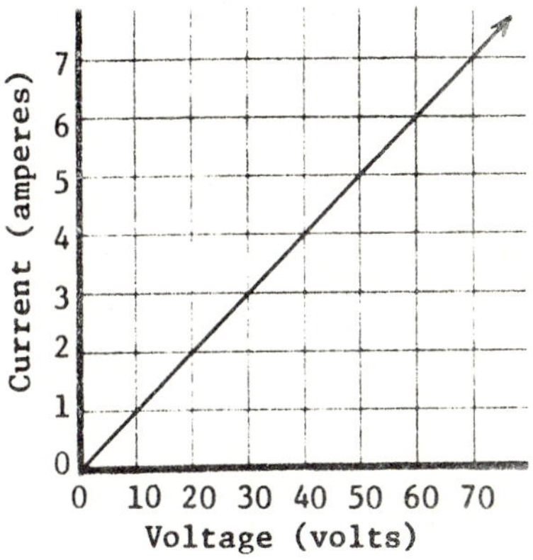

Answer to Frame 174: a) 3 amperes b) 70 volts c) 5.5 amperes d) 25 volts

---

175. At the right below is a graph relating voltage and current for two circuits with different resistances. We can use the graph to decide which circuit has the most resistance.

To get 4 amperes of current, how much voltage must you apply:

(a) In circuit #1 ? ______ (b) In circuit #2 ? ______

If you apply a voltage of 50 volts, how much current will you get:

(c) In circuit #1 ? ______ (d) In circuit #2 ? ______

(e) The more resistance there is in a circuit, the less current you get for the amount of voltage applied. Which circuit has more resistance, #1 or #2 ? ______

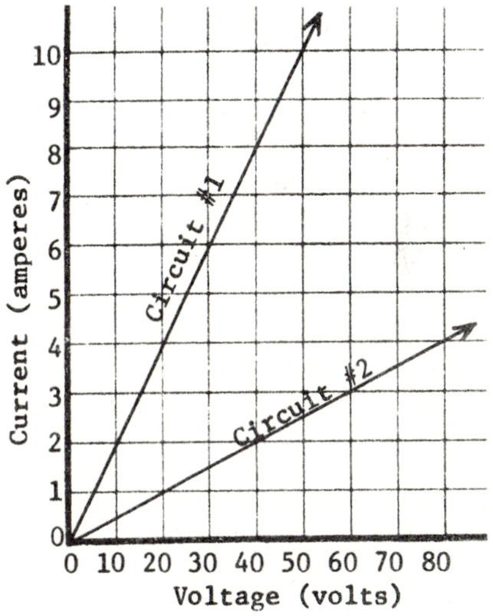

---

a) 20 volts

b) 80 volts

c) 10 amperes

d) 2.5 amperes

e) #2

176. When graphing the relationship between the Fahrenheit and Centigrade temperature scales in earlier frames, we put the Fahrenheit scale on the horizontal axis and the Centigrade scale on the vertical axis. This choice was arbitrary. On the graph below, we have reversed the scales and put Centigrade on the horizontal axis. This reversal does not change the relationship.

Use the graph to answer the following:

(a) What Fahrenheit temperature is equivalent to a Centigrade temperature of 20°? _______

(b) What Fahrenheit temperature is equivalent to a Centigrade temperature of -10°? _______

(c) What Centigrade temperature is equivalent to a Fahrenheit temperature of -60°? _______

(d) What Centigrade temperature is equivalent to a Fahrenheit temperature of 55°? _______

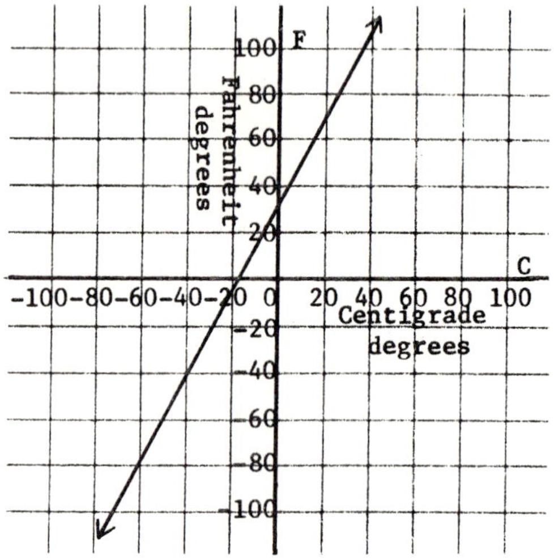

Answer to Frame 176: (approximately)	a) 70° F	b) 15° F	c) -50° C	d) 15° C

177. Many technical and scientific graphs are based on a formula. When solving a problem by reading one of these graphs, we obtain only an approximate solution. If a formula is available, we can obtain an exact solution by plugging into the formula.

The formula for the distance a dropped object falls in a given amount of time is:

$$\boxed{s = 16t^2}$$

where s is the distance in feet and t is the time in seconds.

The graphed relationship is shown at the right.

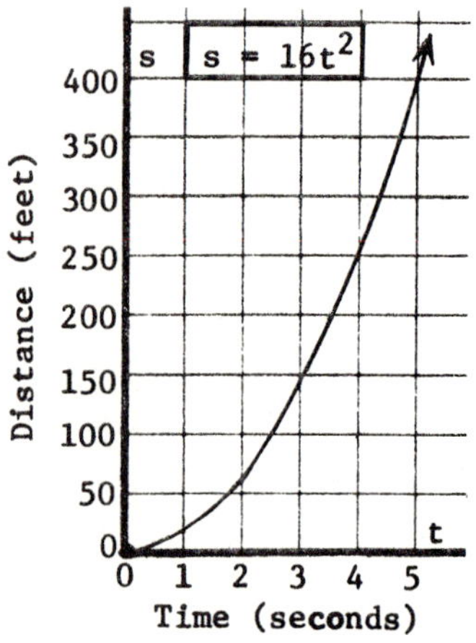

Read the graph to get approximate answers for these:

(a) In 2 seconds, an object falls approximately _______ feet.

(b) In 4 seconds, an object falls approximately _______ feet.

Now plug into the formula $s = 16t^2$ to get exact answers:

(c) In 2 seconds, an object falls exactly _______ feet.

(d) In 4 seconds, an object falls exactly _______ feet.

Answer to Frame 177:	a) Approximately 65 feet	b) Approximately 260 feet	c) Exactly 64 feet	d) Exactly 256 feet

178. The graph at the right below shows the relationship between pressure and volume of a gas at a constant temperature. It is based on the formula shown, which is an instance of Boyle's Law:

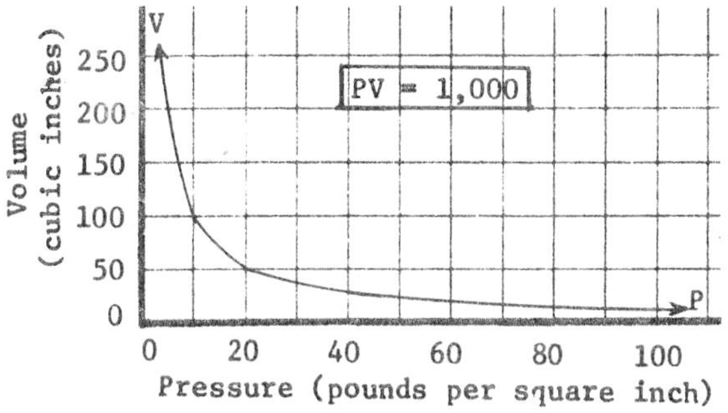

Use the graph to get approximate solutions:

(a) When P is 60 pounds per square inch, what is the approximate value of V?
_______ cubic inches

(b) When V is 150 cubic inches, what is the approximate value of P?
_______ pounds per square inch

Now plug values into the formula PV = 1,000 to get exact solutions:

(c) When P is 60 pounds per square inch, the exact value of V is _______ cubic inches.

(d) When V is 150 cubic inches, the exact value of P is _______ pounds per square inch.

Answer to Frame 178: a) 20 cubic inches b) 7 pounds per square inch c) $16\frac{2}{3}$ or 16.7 cubic inches d) $6\frac{2}{3}$ pounds per square inch

---

179. Below at the right we have graphed the relationship between distance and time when traveling at a constant velocity of 50 miles per hour. The graph is based on the formula shown.

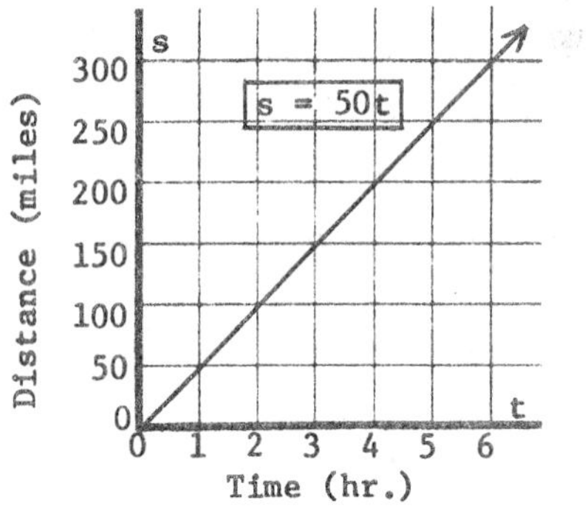

Use the graph to get approximate answers:

(a) In 3.3 hours, the distance traveled is approximately _______ miles.

(b) The object travels 264 miles in approximately _______ hours.

Now plug into the formula s = 50t to get exact answers:

(c) In 3.3 hours, the distance traveled is exactly _______ miles.

(d) The object travels 264 miles in exactly _______ hours.

Answer to Frame 179: a) 160 or 170 miles b) 5.2 or 5.3 hrs. c) 165 miles d) 5.28 hrs.

180. Sometimes the same value of a variable can be produced by either of two values of the other variable. We can use the graph below as an example. It shows the relationship between output power (in watts) and output resistance (in kilohms) in an electric circuit.

The two dashed lines on the graph show that we can get an output power of 3.5 watts with either of two output resistances: Approximately 2.7 kilohms or 7.4 kilohms.

(a) We can get 2.5 watts of output power with either of two output resistances, approximately _______ kilohms or _______ kilohms.

(b) The maximum output power we can get is approximately 3.8 or 3.9 watts. At what value of output resistance do we get this maximum value?

_______ kilohms

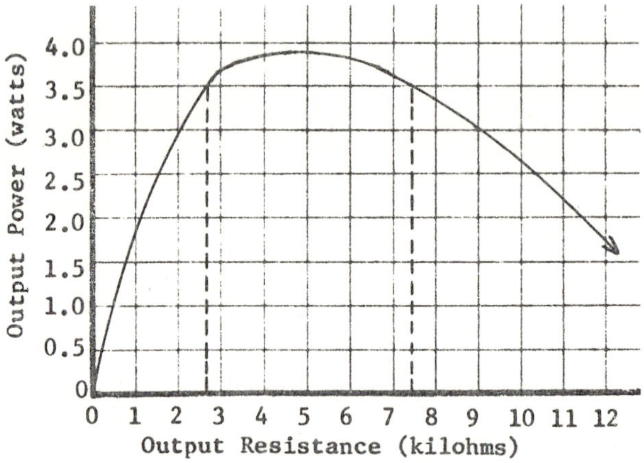

Answer to Frame 180: a) 1.5 or 10.3 kilohms b) 4.8 kilohms

181. Here is a graph showing the relationship between voltage and current in an electronic device called a "tunnel diode." Voltage is measured in volts; current is measured in milliamperes.

(a) Which variable, voltage or current, is plotted as the ordinate? _______

(b) We can obtain 1.0 milliampere of current by applying either of two voltages, approximately _______ volts or _______ volts.

(c) We can obtain 0.5 milliampere of current by applying any of three voltages, approximately _______ volts or _______ volts or _______ volts.

(d) What is the maximum amount of current produced? _______ milliamperes

(e) What voltage produces this maximum amount of current? _______ volts

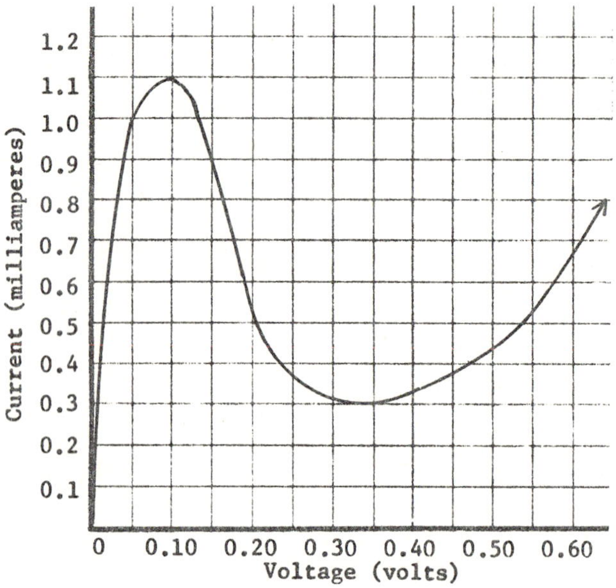

Answer to Frame 181:

a) Current b) 0.05 or 0.13 volts c) 0.02 or 0.20 or 0.54 volts d) 1.1 milliamperes e) 0.10 volts

182. Below is another graph of Boyle's Law, showing the relationship between the pressure and volume of a gas at a fixed temperature. By examining the graph, we can see how volume changes as pressure changes, or vice versa.

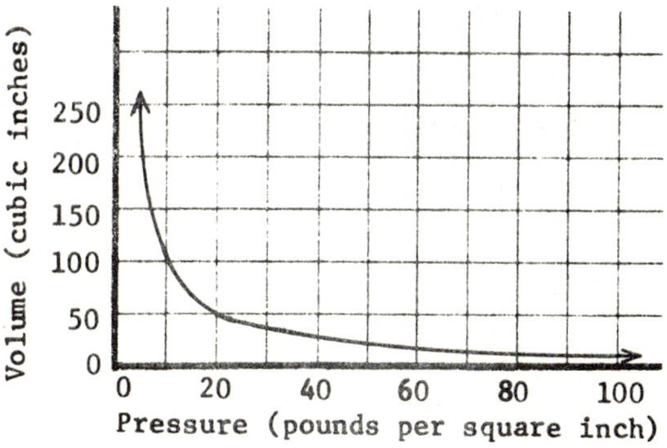

(a) If the pressure is 20 pounds per square inch, the volume is approximately _______ cubic inches.

(b) If the pressure is 40 pounds per square inch, the volume is approximately _______ cubic inches.

(c) On the basis of (a) and (b), as the pressure increases, does the volume increase or decrease? _______________

(d) As you increase the volume from 50 cubic inches to 100 cubic inches, the pressure changes from 20 pounds per square inch to _______ pounds per square inch.

(e) On the basic of (d), as the volume increases, what happens to the pressure? ________________

Answer to Frame 182: a) 50 b) 25 c) Decreases d) 10 e) Decreases

183. Here is the same graph we looked at in an earlier frame, showing the relationship between voltage and current in a tunnel diode.

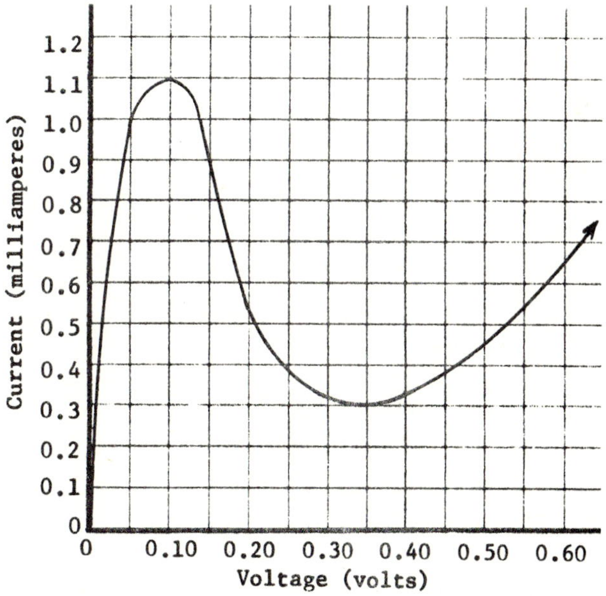

(a) As the voltage increases from 0 to 0.05 volts, does the current increase or decrease?

____

(b) As the voltage increases from 0.10 to 0.30 volts, does the current increase or decrease?

____

(c) As the voltage increases from 0.40 to 0.60 volts, does the current increase or decrease?

____

Answer to Frame 183: a) Increases b) Decreases c) Increases

7-10 GRAPHING FORMULAS WHICH CONTAIN THREE VARIABLES

Most formulas and many equations contain more than two variables. In this section, we will show a method for graphing formulas which contain three variables.

---

184. In general: (1) Any letter in an equation or formula is a variable. (There are some exceptions.)

(2) Any number in an equation or formula is a constant.

Here is an electronics formula called "Ohm's Law":

$$\boxed{E = IR}$$ where "E" is voltage
"I" is current
"R" is resistance

(a) Name the variables: ________ (b) Name the constants: ________

---

185. Here is the formula for the area of a triangle:

$$\boxed{A = \frac{1}{2}bh}$$ where "A" is area
"b" is the length of the base
"h" is the length of the altitude

(a) Name the constants: ________ (b) Name the variables: ________

Answers to Frame 184: a) E, I, R b) There are none.

---

186. In ordinary mathematical equations, letters usually stand for variables.

In $\boxed{y = 7xz + 19}$ list the: (a) Variables ______ (b) Constants ______

Answers to Frame 185: a) $\frac{1}{2}$ b) A, b, h

Answer to Frame 186: a) y, x, z b) 7, 19

---

187. Ordinarily, letters stand for variables. However, some letters in formulas are constants.

Here is the formula for the circumference of a circle:

$$\boxed{C = \pi d}$$ where "C" is length of the circumference
and "d" is length of the diameter.

In this case, "π" is a constant. Its value is 3.1416...

The string of three periods means that we could carry the value out to more decimal places. Of course, we use the number of decimal places that makes sense from the standpoint of the accuracy needed in a specific situation.

Here is the formula for the distance a body falls from re.. in space:

$$\boxed{s = \frac{1}{2}gt^2}$$ where "s" stands for distance
and "t" stands for time.

"g" is the acceleration due to gravity. At sea level, this value is always 32.17... "g" is therefore a constant.

What do the three periods "..." mean after 32.17... ? ____________

That we can carry this value out to more decimal places.

188. In scientific formulas, most letters stand for variables, but an occasional letter is actually a constant. You will learn the specific letters that stand for constants in your science and technology courses.

---

189. Here is the formula showing the relationship between voltage (E), current (I), and resistance (R). The formula contains three variables.

$$E = IR$$

We cannot graph the formula as it stands because it contains three variables and there are only two coordinate axes. That is, if we put the E-scale on the horizontal axis and the I-scale on the vertical axis, there is no axis left for the R-scale.

On the coordinate axes, we can only graph equations or formulas which contain how many variables? __________

Go to next frame.

---

190. On the coordinate system, we can only graph the relationship between two variables at one time.

Here is the same formula: $E = IR$

Even though this formula contains three variables, we can graph the relationship between any two of them on the coordinate system IF THE THIRD VARIABLE IS SET EQUAL TO A CONSTANT.

If we set "R" equal to 50 ohms in the formula $E = IR$,

we get $E = I(50)$ or $E = 50I$.

We can graph the new formula $E = 50I$ because it contains only _______ variables.

Two

---

191. Here is the same formula: $E = IR$

To graph the relationship between voltage (E) and current (I), we must set R equal to some constant. Since resistance (R) can be any of a wide range of values, we have a choice of values to use as the constant. Depending upon our choice of a constant, we get different equations.

If we set R equal to 20 ohms, we get: $E = 20I$

If we set R equal to 100 ohms, we get: $E = 100I$

How many different equations of this type are possible?

____

two

As many as there are different values of R (resistance). That is, an infinite number.

192. Here is the same equation: E = IR

By setting R equal to 20 ohms, 50 ohms, and 100 ohms respectively, we get a family of equations:

E = 20I E = 50I E = 100I

Each of these equations can be plotted on the same graph. To do so, we prepare a table of values for each equation, as follows:

E = 20I

E	I
20	1
40	2
60	3
80	4
100	5

E = 50I

E	I
50	1
100	2
150	3
200	4
250	5

E = 100I

E	I
100	1
200	2
300	3
400	4
500	5

Using these values, all three equations are plotted on the graph below. Notice that we have labeled each graphed relationship with a specific resistance-value.

E stands for "voltage."
I stands for "current."

(a) Which variable is plotted as the abscissa? _______

(b) Which variable is plotted as x? _______

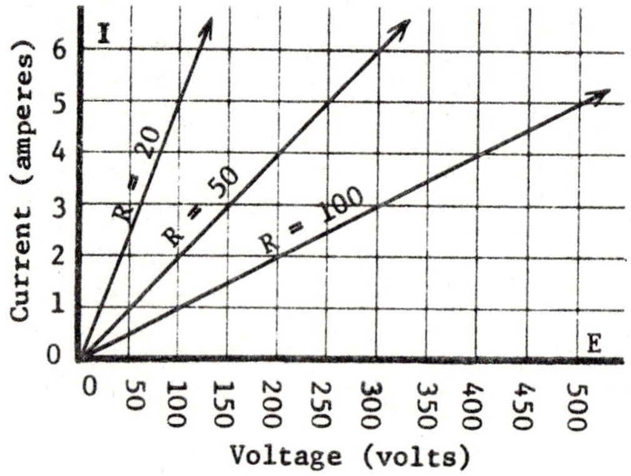

---

a) E (voltage)

b) E (voltage)

193. Here is another three-variable formula: $P = EI$ where "P" is power (watts)
"E" is voltage (volts)
"I" is current (amperes)

In this case, to graph the relationship between E and I, we must set P equal to a constant or constants. If we set P equal to 2 watts, 4 watts, and 8 watts respectively, we get the following family of equations:

$$EI = 2 \qquad EI = 4 \qquad EI = 8$$

When these equations are graphed with E (voltage) plotted as the abscissa, we get the following family of curves.

(a) When P is 2 watts, approximately how much current do we get with a voltage of 5 volts?
_______ amperes

(b) When P is 4 watts, approximately how much current do we get with a voltage of 3 volts?
_______ amperes

(c) When P is 8 watts, approximately how much voltage do we need to get a current of 3 amperes?
_______ volts

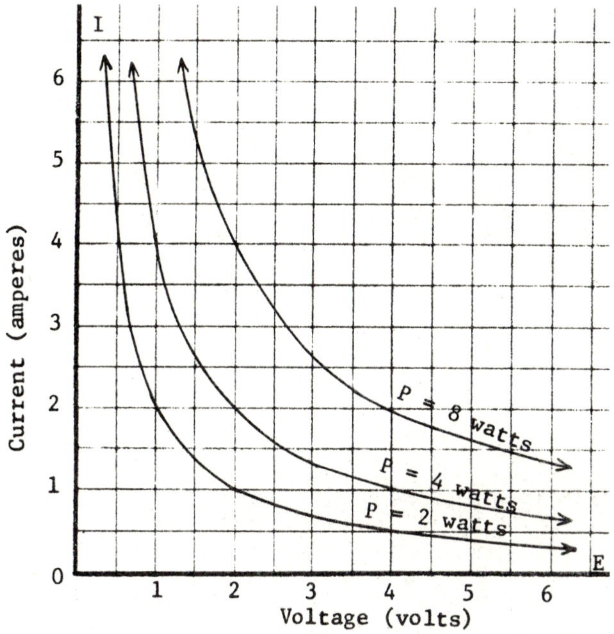

---

a) 0.4 amperes

b) 1.3 amperes

c) 2.7 volts

194. Here is another three-variable formula: $\boxed{E = \frac{1}{2}mv^2}$ where "E" is kinetic energy (in ergs)
"m" is mass (in grams)
"v" is velocity (in centimeters per second)

Let's graph the relationship between E (kinetic energy) and v (velocity). If we set m (mass) equal to 50 grams, 100 grams, 200 grams, and 500 grams respectively, we get the following family of equations.

$E = 25v^2$ $\quad E = 50v^2$ $\quad E = 100v^2$ $\quad E = 250v^2$

The family of curves is given on the graph at the right:

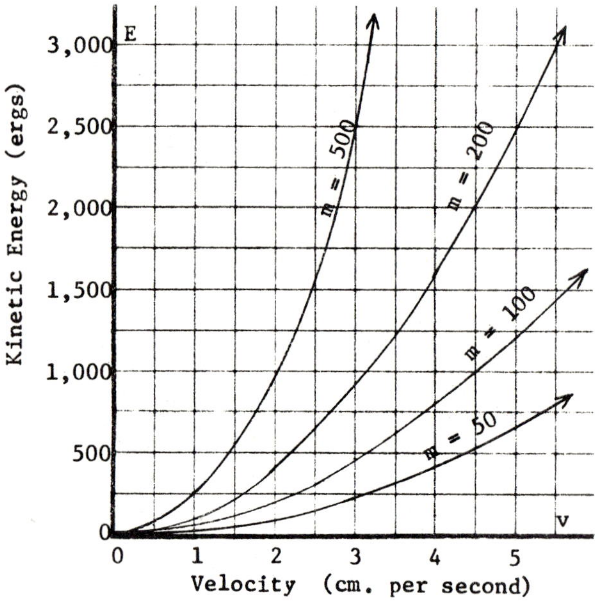

(a) With a mass of 50 grams, approximately how much kinetic energy is produced with a velocity of 4.5 centimeters per second? ________

(b) With a mass of 200 grams, approximately how much kinetic energy is produced with a velocity of 4.5 centimeters per second? ________

(c) With a mass of 500 grams, what velocity is required to produce a kinetic energy of 1,000 ergs? ________

(d) With a mass of 100 grams, what velocity is required to produce a kinetic energy of 1,000 ergs? ________

---

a) 510 ergs

b) 2,000 ergs

c) 2 centimeters per second

d) 4.5 centimeters per second

Chapter 8 LITERAL FRACTIONS

Any scientist or technician must be totally fluent in manipulations with literal fractions. This fluency with literal fractions is needed both for formula rearrangement and formula derivations. Formula rearrangement will be discussed in Chapter 9; some basic formula derivation will be discussed in Chapter 10.

The purpose of this chapter is to examine operations with literal fractions. The two major goals of the chapter are:

(1) to show that the principles used in operations with literal fractions are the same as the principles used with numerical fractions.

(2) to show when literal fractions can or cannot be written in equivalent forms.

8-1 MULTIPLICATIONS INVOLVING LITERAL FRACTIONS

In this section, we will show the method for multiplying two literal fractions or a literal fraction and a non-fraction. The procedures are the same as those used with numerical fractions.

---

1. Any fraction stands for a division in which the numerator is divided by the denominator. A fraction stands for a division even when it contains more than one letter.

Just as $\frac{6}{3}$ means $6 \div 3$ and $\frac{2x}{7}$ means $2x \div 7$

(a) $\frac{V}{P}$ means ____ $\div$ ____ (b) $\frac{3x}{2y}$ means ____ $\div$ ____

---

2. Write each of the following divisions as a fraction:

(a) $t \div m =$ ____ (b) $5z \div 4y =$ ____ (c) $3q \div r =$ ____

Answers to 1: a) $V \div P$ b) $3x \div 2y$

---

3. The expression "7x" means "7 times x" or (7)(x).

Note: We always write the numerical factor first.

The expression "mp" means "m times p" or (m)(p).

When each factor is a letter, we usually write the letters in alphabetical order.

Therefore, instead of "td", we would write ____.

Answers to 2: a) $\frac{t}{m}$ b) $\frac{5z}{4y}$ c) $\frac{3q}{r}$

Answer to 3: dt

4. "7xy" and "PRV" are three-factor multiplications.

"7xy" means "7 times x times y" or (7)(x)(y).

"PRV" means "P times R times V" or (P)(R)(V).

Note: We usually write the literal factors in alphabetical order.

(a) Instead of "3tm", we would write _______.

(b) Instead of "bda", we would write _______.

---

5. Here is the pattern for multiplying two fractions:

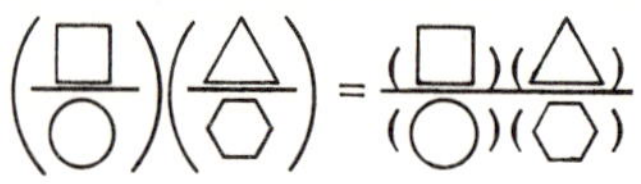

To get the product, we: (1) Multiply the two numerators.
(2) Multiply the two denominators.

This same pattern is used with literal fractions. For example:

$$\left(\frac{3}{4}\right)\left(\frac{a}{b}\right) = \frac{3a}{4b} \qquad \left(\frac{x}{d}\right)\left(\frac{y}{h}\right) = \frac{xy}{dh}$$

Complete these. Write the literal factors in alphabetical order.

(a) $\left(\frac{F}{7}\right)\left(\frac{g}{h}\right) =$ _______ (b) $\left(\frac{m}{t}\right)\left(\frac{p}{q}\right) =$ _______

Answers to 4: a) 3mt b) abd

---

6. Some fractions contain two or more literal factors in either the numerator or denominator, or both. The same process is used for multiplying fractions of this type. For example:

$$\left(\frac{7}{8}\right)\left(\frac{ab}{3c}\right) = \frac{(7)(ab)}{(8)(3c)} = \frac{7ab}{24c}$$

$$\left(\frac{cF}{n}\right)\left(\frac{R}{VT}\right) = \frac{(cF)(R)}{(n)(VT)} = \frac{cFR}{nVT}$$

Complete these. Write the literal factors in alphabetical order.

(a) $\left(\frac{ms}{r}\right)\left(\frac{xy}{ab}\right) =$ _______ (b) $\left(\frac{3D}{t}\right)\left(\frac{A}{2h}\right) =$ _______

Answers to 5: a) $\frac{Fg}{7h}$ b) $\frac{mp}{qt}$ (not $\frac{mp}{tq}$)

---

7. Any non-fractional literal expression can be converted to a fraction by the "principle of dividing a quantity by +1". That is:

Since $\square = \frac{\square}{1}$, $t = \frac{t}{1}$ and $ab = \frac{ab}{1}$

Because of this fact, any multiplication of a non-fraction and a fraction can be converted into a multiplication of two fractions.

Complete: (a) $b\left(\frac{V}{R}\right) = \left(\frac{b}{1}\right)\left(\frac{V}{R}\right) =$ _____ (b) $pq\left(\frac{1}{m}\right) = \left(\frac{pq}{1}\right)\left(\frac{1}{m}\right) =$ _____

Answers to 6: a) $\frac{msxy}{abr}$ b) $\frac{3AD}{2ht}$

Answers to 7: a) $\frac{bV}{R}$ b) $\frac{pq}{m}$

8. Do these multiplications:

(a) $m\left(\frac{7}{y}\right) =$ ______ (c) $AV\left(\frac{Z}{m}\right) =$ ______

(b) $c\left(\frac{d}{5}\right) =$ ______ (d) $Q\left(\frac{R}{ST}\right) =$ ______

---

a) $\frac{7m}{y}$ c) $\frac{AVZ}{m}$

b) $\frac{cd}{5}$ d) $\frac{QR}{ST}$

9. Since we can use the principle of dividing a number by +1 to convert the number "1" to a fraction $(1 = \frac{1}{1})$, we can use the same method to show that the identity principle of multiplication, $(1)(\square) = \square$, is true for literal fractions.

(a) $1\left(\frac{a}{b}\right) = \left(\frac{1}{1}\right)\left(\frac{a}{b}\right) =$ ______ (c) $1\left(\frac{M}{N}\right) =$ ______

(b) $1\left(\frac{PQ}{R}\right) = \left(\frac{1}{1}\right)\left(\frac{PQ}{R}\right) =$ ______ (d) $1\left(\frac{pt}{qr}\right) =$ ______

Answer to Frame 9: a) $\frac{a}{b}$ b) $\frac{PQ}{R}$ c) $\frac{M}{N}$ d) $\frac{pt}{qr}$

8-2 FACTORING LITERAL FRACTIONS

Just as numerical fractions can be factored, literal fractions can be factored in the same way. We will show the method in this section.

---

10. Factoring is done by reversing the procedure for multiplication. When factoring, we look upon the original fraction as if it were a product. Here is an example:

$$\frac{3t}{4V} = \frac{(3)(t)}{(4)(V)} = \left(\frac{3}{4}\right)\left(\frac{t}{V}\right) \text{ or } \left(\frac{t}{4}\right)\left(\frac{3}{V}\right)$$

Of course, we can commute the order of the factors and get:

$$\left(\frac{t}{V}\right)\left(\frac{3}{4}\right) \text{ or } \left(\frac{3}{V}\right)\left(\frac{t}{4}\right)$$

Note: <u>When writing factors, we always put parentheses around them.</u>

Factor the following fraction in four different ways:

$$\frac{aF}{bG} = \left(\frac{\quad}{\quad}\right)\left(\frac{\quad}{\quad}\right) \text{ or } \left(\frac{\quad}{\quad}\right)\left(\frac{\quad}{\quad}\right) \text{ or } \left(\frac{\quad}{\quad}\right)\left(\frac{\quad}{\quad}\right) \text{ or } \left(\frac{\quad}{\quad}\right)\left(\frac{\quad}{\quad}\right)$$

You can have the following factors in any order:

$\left(\frac{a}{b}\right)\left(\frac{F}{G}\right)$ or $\left(\frac{F}{G}\right)\left(\frac{a}{b}\right)$

$\left(\frac{a}{G}\right)\left(\frac{F}{b}\right)$ or $\left(\frac{F}{b}\right)\left(\frac{a}{G}\right)$

11. To factor a fraction, the numerator and denominator must contain two factors. If either does not, we can always factor it by $\square = (1)(\square)$ first.

$$\frac{p}{q} = \frac{(p)(1)}{(1)(q)} = \left(\frac{p}{1}\right)\left(\frac{1}{q}\right) = (p)\left(\frac{1}{q}\right)$$

Note: (1) Since $\frac{p}{1}$ is an instance of $\frac{\square}{1}$, we replaced it with "p", a non-fraction.

(2) We can again commute the factors and write $\left(\frac{1}{q}\right)(p)$.

Factor each into a fraction and a non-fraction:

(a) $\frac{R}{m} =$ ______ (b) $\frac{a}{b} =$ ______

---

12. If the numerator and denominator of a fraction already contain two factors, we can factor it in all sorts of ways. Look at the following examples:

$$\frac{rs}{tv} = \frac{(r)(s)}{(t)(v)} = \left(\frac{r}{t}\right)\left(\frac{s}{v}\right) \text{ or } \left(\frac{s}{t}\right)\left(\frac{r}{v}\right)$$

$$= \frac{(rs)(1)}{(t)(v)} = \left(\frac{rs}{t}\right)\left(\frac{1}{v}\right) \text{ or } \left(\frac{1}{t}\right)\left(\frac{rs}{v}\right)$$

$$= \frac{(rs)(1)}{(1)(tv)} = (rs)\left(\frac{1}{tv}\right)$$

$$= \frac{(r)(s)}{(1)(tv)} = r\left(\frac{s}{tv}\right) \text{ or } s\left(\frac{r}{tv}\right)$$

The point of this frame was to show the flexibility we have in factoring. We can factor into two fractions or a fraction and a non-fraction in various ways.

Answers to 11:

a) $(R)\left(\frac{1}{m}\right)$ or $\left(\frac{1}{m}\right)(R)$

b) $(a)\left(\frac{1}{b}\right)$ or $\left(\frac{1}{b}\right)(a)$

---

13. The test for the correctness of a factoring is: THE PRODUCT OF TWO FACTORS MUST EQUAL THE ORIGINAL FRACTION. Always use this test.

(a) Are $\left(\frac{m}{q}\right)$ and (P) a pair of factors for $\frac{mP}{q}$? ______

(b) Are (ab) and $\left(\frac{1}{c}\right)$ a pair of factors for $\frac{ac}{b}$? ______

Answer to 12: Go to next frame.

---

14. Which of the following are pairs of factors for $\frac{FG}{H}$? ______

(a) $\left(\frac{F}{H}\right)(G)$ (b) $(FG)\left(\frac{1}{H}\right)$ (c) $FH\left(\frac{1}{G}\right)$ (d) $F\left(\frac{G}{H}\right)$

Answers to 13:

a) Yes, since $\left(\frac{m}{q}\right)(P) = \frac{mP}{q}$

b) No, since $(ab)\left(\frac{1}{c}\right) \neq \frac{ac}{b}$

15. Which of the following are pairs of factors for $\frac{t}{ms}$? ________ (a) $t\left(\frac{1}{ms}\right)$ (b) $t\left(\frac{m}{s}\right)$ (c) $\left(\frac{t}{s}\right)\left(\frac{1}{m}\right)$ (d) $\left(\frac{1}{s}\right)\left(\frac{t}{m}\right)$	(a), (b), and (d) Not (c), since: $FH\left(\frac{1}{G}\right) = \frac{FH}{G}$
16. Which of the following are pairs of factors for $\frac{3x}{2y}$? ________ (a) $3x\left(\frac{1}{2y}\right)$ (b) $3\left(\frac{x}{2y}\right)$ (c) $\left(\frac{x}{2}\right)\left(\frac{3}{y}\right)$ (d) $3x\left(\frac{2}{y}\right)$	(a), (c), and (d) Not (b), since: $t\left(\frac{m}{s}\right) = \frac{tm}{s}$
17. In each case, insert the missing factor: (a) $\frac{3m}{t} = 3(\quad)$ (b) $\frac{d}{bR} = d(\quad)$ (c) $\frac{5a}{c} = 5a(\quad)$	(a), (b), and (c) Not (d), since: $3x\left(\frac{2}{y}\right) = \frac{6x}{y}$
18. Insert the missing factors: (a) $\frac{pq}{dv} = p(\quad)$ (b) $\frac{ab}{cd} = ab(\quad)$ (c) $\frac{5x}{3y} = \frac{5}{y}(\quad)$	a) $\left(\frac{m}{t}\right)$ b) $\left(\frac{1}{bR}\right)$ c) $\left(\frac{1}{c}\right)$
19. Insert the missing factors: (a) $\frac{R}{TV} = \left(\frac{R}{T}\right)(\quad)$ (b) $\frac{mq}{p} = \left(\frac{1}{p}\right)(\quad)$ (c) $\frac{pqr}{xy} = \left(\frac{q}{y}\right)\left(\frac{r}{x}\right)(\quad)$	a) $\left(\frac{q}{dv}\right)$ b) $\left(\frac{1}{cd}\right)$ c) $\left(\frac{x}{3}\right)$
	a) $\left(\frac{1}{V}\right)$ b) (mq) c) (p)

SELF-TEST 1 (Frames 1-19)

Do the following multiplications:

1. $\left(\frac{1}{2}\right)\left(\frac{3h}{r}\right) =$ ____ 2. $2P\left(\frac{1}{dw}\right) =$ ____ 3. $\left(\frac{2}{c}\right)3t =$ ____ 4. $\left(\frac{h}{2d}\right)\left(\frac{gr}{a}\right) =$ ____

Insert the missing factor in each problem:

5. $\frac{2k}{t} = \left(\frac{2}{t}\right)(\quad)$ 6. $\frac{E}{PR} = \frac{E}{R}(\quad)$ 7. $\frac{6w}{kx} = \left(\frac{2w}{k}\right)(\quad)$

8. Which of the following are pairs of factors for $\frac{4a}{rw}$? ________

(a) $\left(\frac{2a}{w}\right)\left(\frac{2}{r}\right)$ (b) $\left(\frac{4}{rw}\right)a$ (c) $\left(\frac{4}{w}\right)\left(\frac{r}{a}\right)$ (d) $\left(\frac{1}{w}\right)\left(\frac{4a}{r}\right)$

ANSWERS: 1. $\frac{3h}{2r}$ 2. $\frac{2P}{dw}$ 3. $\frac{6t}{c}$ 4. $\frac{ghr}{2ad}$ 5. k 6. $\frac{1}{P}$ 7. $\frac{3}{x}$ 8. (a), (b), (d)

8-3 GENERATING EQUIVALENT FRACTIONS

There is a family of fractions which is equivalent (or equal) to any fraction or non-fraction. This family of equivalent fractions can be generated by means of the two principles:

$$(\square)(1) = \square \quad \text{and} \quad \frac{\square}{\square} = 1$$

In this section, we will review the procedure for generating a family of equivalent fractions. We will also discuss the method for converting any literal fraction or non-fraction to an equivalent fraction.

---

20. In each fraction below, a quantity is divided by itself:

$$\frac{m}{m} \qquad \frac{tv}{tv} \qquad \frac{7xy}{7xy}$$

Whenever a quantity is divided by itself, the quotient is +1. Therefore, the value of each fraction above is _______.

Answer: +1

21. If any quantity is multiplied by +1, the product equals the original quantity. This principle is true even if the original quantity is a literal fraction or non-fraction. Therefore:

(a) $(ab)(1) =$ _____ (b) $\left(\frac{R}{S}\right)(1) =$ _____ (c) $\left(\frac{pq}{3d}\right)(1) =$ _____

Answers: a) ab b) $\frac{R}{S}$ c) $\frac{pq}{3d}$

22. Any instance of $\frac{n}{n}$ equals "1". When we multiply a fraction or non-fraction by an instance of $\frac{n}{n}$, we are really multiplying it by "1". Since we are multiplying by "1", the resulting fraction is equivalent to the original fraction or non-fraction. Here are two examples:

$$\left(\frac{a}{b}\right)(1) = \left(\frac{a}{b}\right)\left(\frac{c}{c}\right) = \frac{ac}{bc}$$

$\frac{ac}{bc}$ is equivalent to $\frac{a}{b}$ since we obtained

$\frac{ac}{bc}$ by multiplying $\frac{a}{b}$ by $\frac{c}{c}$ which equals 1.

$$(m)(1) = (m)\left(\frac{t}{t}\right) = \frac{mt}{t}$$

$\frac{mt}{t}$ is equivalent to "m" since we obtained

$\frac{mt}{t}$ by multiplying "m" by $\frac{t}{t}$ which equals _____.

Answer: 1

23. It should be obvious that we can generate a whole family of equivalent fractions by multiplying any literal fraction or non-fraction by various instances of $\frac{n}{n}$.

Here are various examples of generating fractions equivalent to the fraction $\frac{c}{d}$:

$$\left(\frac{c}{d}\right)\left(\frac{F}{F}\right) = \frac{cF}{dF} \qquad \left(\frac{c}{d}\right)\left(\frac{7}{7}\right) = \frac{7c}{7d} \qquad \left(\frac{c}{d}\right)\left(\frac{3y}{3y}\right) = \frac{3cy}{3dy} \qquad \left(\frac{c}{d}\right)\left(\frac{xy}{xy}\right) = \frac{cxy}{dxy}$$

Here are various examples of generating fractions equivalent to the non-fraction "a":

$$(a)\left(\frac{b}{b}\right) = \frac{ab}{b} \qquad (a)\left(\frac{4}{4}\right) = \frac{4a}{4} \qquad (a)\left(\frac{5m}{5m}\right) = \frac{5am}{5m} \qquad (a)\left(\frac{xy}{xy}\right) = \frac{axy}{xy}$$

---

24. Using the same procedure, we can convert any literal fraction to an equivalent fraction which has a specific denominator, provided that the specific denominator contains the original denominator as a factor or factors.

If we want to convert $\frac{m}{p}$ to an equivalent fraction whose denominator is 6p, we multiply $\frac{m}{p}$ by $\frac{6}{6}$ and get:

$$\frac{m}{p} = \left(\frac{m}{p}\right)\left(\frac{6}{6}\right) = \frac{6m}{6p}$$

If we want to convert $\frac{m}{p}$ to an equivalent fraction whose denominator is pv, we multiply $\frac{m}{p}$ by $\frac{v}{v}$ and get:

$$\frac{m}{p} = \left(\frac{m}{p}\right)\left(\frac{v}{v}\right) = \_\_\_\_\_\_$$

Answer to frame 23: Go to next frame.

---

25. When converting a literal fraction to an equivalent literal fraction, you can determine the instance of $\frac{n}{n}$ to use by examining the additional factor or factors in the new denominator.

(a) In the conversion below, we are converting the original fraction to an equivalent fraction whose denominator is "as". Since the only additional factor in the new denominator is "s", we multiply by $\frac{s}{s}$:

$$\frac{t}{a} = \left(\frac{t}{a}\right)\left(\frac{s}{s}\right) = \frac{(\ \ )}{as}$$

(b) In the conversion below, we are converting the original fraction to an equivalent fraction whose denominator is "2cf". Since the additional factors in the new denominator are "2" and "f", we multiply by $\frac{2f}{2f}$:

$$\frac{ab}{c} = \left(\frac{ab}{c}\right)\left(\frac{2f}{2f}\right) = \frac{(\ \ )}{2cf}$$

Answer to frame 24: $\frac{mv}{pv}$

Answer to frame 25: a) $\frac{st}{as}$ b) $\frac{2abf}{2cf}$

26. Complete each of these conversions:

(a) $\frac{b}{d} = \left(\frac{b}{d}\right)\left(\frac{\quad}{\quad}\right) = \frac{(\quad)}{dm}$

(b) $\frac{R}{F} = \left(\frac{R}{F}\right)\left(\frac{\quad}{\quad}\right) = \frac{(\quad)}{7F}$

(c) $\frac{ct}{m} = \left(\frac{ct}{m}\right)\left(\frac{\quad}{\quad}\right) = \frac{(\quad)}{amx}$

(d) $\frac{a}{bc} = \left(\frac{a}{bc}\right)\left(\frac{\quad}{\quad}\right) = \frac{(\quad)}{5bc}$

---

a) $\left(\frac{b}{d}\right)\left(\frac{m}{m}\right) = \frac{bm}{dm}$

b) $\left(\frac{R}{F}\right)\left(\frac{7}{7}\right) = \frac{7R}{7F}$

c) $\left(\frac{ct}{m}\right)\left(\frac{ax}{ax}\right) = \frac{actx}{amx}$

d) $\left(\frac{a}{bc}\right)\left(\frac{5}{5}\right) = \frac{5a}{5bc}$

27. Similarly, we can convert any non-fraction to an equivalent literal fraction which has a specific denominator.

To do so, the instance of $\frac{n}{n}$ is determined by the denominator.

To convert "5" to an equivalent fraction whose denominator is "p", we must multiply 5 by $\frac{p}{p}$ and get:

$$5 = 5\left(\frac{p}{p}\right) = \frac{5p}{p}$$

To convert "m" to an equivalent fraction whose denominator is "xy", we must multiply "m" by $\frac{xy}{xy}$ and get:

$$m = m\left(\frac{xy}{xy}\right) = \frac{mxy}{xy}$$

Complete:

(a) $7 = 7\left(\frac{\quad}{\quad}\right) = \frac{(\quad)}{d}$

(b) $5 = 5\left(\frac{\quad}{\quad}\right) = \frac{(\quad)}{3x}$

(c) $t = t\left(\frac{\quad}{\quad}\right) = \frac{(\quad)}{vz}$

(d) $ab = ab\left(\frac{\quad}{\quad}\right) = \frac{(\quad)}{5c}$

---

a) $7\left(\frac{d}{d}\right) = \frac{7d}{d}$

b) $5\left(\frac{3x}{3x}\right) = \frac{15x}{3x}$

c) $t\left(\frac{vz}{vz}\right) = \frac{tvz}{vz}$

d) $ab\left(\frac{5c}{5c}\right) = \frac{5abc}{5c}$

28. Complete:

(a) $\frac{t}{bs} = \frac{(\quad)}{abs}$

(b) $\frac{qr}{p} = \frac{(\quad)}{abp}$

(c) $3 = \frac{(\quad)}{7R}$

(d) $mt = \frac{(\quad)}{2a}$

---

a) $\frac{at}{abs}$

b) $\frac{abqr}{abp}$

c) $\frac{21R}{7R}$

d) $\frac{2amt}{2a}$

8-4 REDUCING LITERAL FRACTIONS TO LOWEST TERMS

The procedure for reducing fractions to lower or lowest terms is merely the reverse of the procedure for converting a fraction to an equivalent fraction. We will review the method in this section.

29. A fraction can be reduced to lower terms only if we can factor in such a way that one of the factors is an instance of $\frac{n}{n}$.

$\frac{as}{bs}$ can be reduced to lower terms since we can factor it and obtain $\left(\frac{s}{s}\right)$ as one of the factors. We get:

$$\frac{as}{bs} = \left(\frac{a}{b}\right)\left(\frac{s}{s}\right) = \left(\frac{a}{b}\right)(1) = \frac{a}{b}$$

$\frac{ab}{cd}$ cannot be reduced to lower terms since we cannot obtain a factor which is an instance of $\frac{n}{n}$.

Complete this reduction: $\frac{ab}{bc} = \left(\frac{\quad}{\quad}\right)\left(\frac{a}{c}\right) = (\quad)\left(\frac{a}{c}\right) = $ ______

Answer: $\left(\frac{b}{b}\right)\left(\frac{a}{c}\right) = (1)\left(\frac{a}{c}\right) = \frac{a}{c}$

30. When reducing a fraction to lowest terms, the result may be a non-fraction. Here is an example:

$$\frac{4d}{d} = 4\left(\frac{d}{d}\right) = 4(1) = 4$$

Complete this one: $\frac{as}{a} = s\left(\frac{\quad}{\quad}\right) = s(\quad) = $ ______

Answer: $s\left(\frac{a}{a}\right) = s(1) = s$

31. Complete this one: $\frac{3F}{10F} = \left(\frac{3}{10}\right)\left(\frac{F}{F}\right) = \left(\frac{\quad}{\quad}\right)(\quad) = $ ______

Answer: $\left(\frac{3}{10}\right)(1) = \frac{3}{10}$

32. If one of the factors is a numerical fraction, be sure that it is also reduced to lowest terms.

Complete this one: $\frac{15ab}{10ab} = \left(\frac{15}{10}\right)\left(\frac{ab}{ab}\right) = \left(\frac{\quad}{\quad}\right)(\quad) = $ ______

Answer: $\left(\frac{3}{2}\right)(1) = \frac{3}{2}$

33. Reduce the following to lowest terms:

(a) $\frac{mn}{pm} = $ ______

(b) $\frac{qr}{vt} = $ ______

(c) $\frac{7ab}{b} = $ ______

(d) $\frac{14VT}{7VT} = $ ______

Answers: a) $\frac{n}{p}$ b) It cannot be reduced. c) 7a d) 2

34. In the following cases, we must factor the numerators first since they contain only one factor. We factor them by means of the principle: $\square = (\square)(1)$

Example: $\frac{b}{bd} = \frac{(b)(1)}{bd} = \left(\frac{b}{b}\right)\left(\frac{1}{d}\right) = (1)\left(\frac{1}{d}\right) = \frac{1}{d}$

Complete: $\frac{m}{mpq} = \frac{(m)(1)}{mpq} = \left(\frac{m}{m}\right)\left(\frac{}{\quad}\right) = (\quad)(\quad) = \underline{\qquad}$

---

35. Complete: $\frac{7}{21ac} = \frac{(7)(1)}{21ac} = \left(\frac{7}{21}\right)\left(\frac{1}{ac}\right) = (\quad)(\quad) = \underline{\qquad}$

Answer (34): $\left(\frac{m}{m}\right)\left(\frac{1}{pq}\right) = (1)\left(\frac{1}{pq}\right) = \frac{1}{pq}$

---

36. Reduce the following to lowest terms:

(a) $\frac{T}{VT} = \underline{\qquad}$ (b) $\frac{R}{3RD} = \underline{\qquad}$ (c) $\frac{b}{abc} = \underline{\qquad}$

Answer (35): $= \left(\frac{1}{3}\right)\left(\frac{1}{ac}\right) = \frac{1}{3ac}$

---

37. In this one, we must also add a "1" as a factor in the numerator so that there are as many factors in the numerator as there are in the denominator.

$$\frac{5c}{10bc} = \frac{(5c)(1)}{10bc} = \left(\frac{5}{10}\right)\left(\frac{c}{c}\right)\left(\frac{1}{b}\right)$$

$$= \left(\frac{1}{2}\right)(1)\left(\frac{1}{b}\right) = \left(\frac{1}{2}\right)\left(\frac{1}{b}\right) = \underline{\qquad}$$

Answer (36): a) $\frac{1}{V}$ b) $\frac{1}{3D}$ c) $\frac{1}{ac}$

---

38. Reduce each of these to lowest terms:

(a) $\frac{7}{7PQ} = \underline{\qquad}$ (b) $\frac{4}{8ab} = \underline{\qquad}$ (c) $\frac{10df}{20cdfK} = \underline{\qquad}$

Answer (37): $\frac{1}{2b}$

---

39. When multiplying literal fractions, the product should always be written in lowest terms.

Example: $\left(\frac{ab}{c}\right)\left(\frac{c}{bd}\right) = \frac{abc}{bcd} = \left(\frac{bc}{bc}\right)\left(\frac{a}{d}\right) = \frac{a}{d}$

Complete: $\left(\frac{3x}{2}\right)\left(\frac{4}{y}\right) = \frac{12x}{2y} = \underline{\qquad}$

Answer (38): a) $\frac{1}{PQ}$ b) $\frac{1}{2ab}$ c) $\frac{1}{2cK}$

---

40. Do the following multiplications. Reduce each product to lowest terms.

(a) $\left(\frac{M}{T}\right)\left(\frac{T}{R}\right) = \underline{\qquad}$ (b) $\left(\frac{Pq}{r}\right)\left(\frac{r}{q}\right) = \underline{\qquad}$ (c) $\left(\frac{10m}{s}\right)\left(\frac{st}{5m}\right) = \underline{\qquad}$

Answer (39): $\frac{6x}{y}$

---

41. Multiply and reduce each product to lowest terms:

(a) $\left(\frac{y}{x}\right)\left(\frac{x}{y}\right) = \underline{\qquad}$ (b) $\left(\frac{a}{bc}\right)\left(\frac{c}{a}\right) = \underline{\qquad}$ (c) $\left(\frac{13t}{21z}\right)\left(\frac{7r}{26rt}\right) = \underline{\qquad}$

Answer (40): a) $\frac{M}{R}$ b) P c) 2t

Answer (41): a) 1 b) $\frac{1}{b}$ c) $\frac{1}{6z}$

SELF-TEST 2 (Frames 20-41)

Complete: 1. $\frac{t}{a} = \frac{(\ )}{ar}$ 2. $\frac{2h}{c} = \frac{(\ )}{2ct}$ 3. $w = \frac{(\ )}{p}$ 4. $3d = \frac{(\ )}{4V}$

Reduce each fraction to lowest terms:

5. $\frac{hk}{kw} =$ ______ 6. $\frac{8at}{6a} =$ ______ 7. $\frac{4r}{12rs} =$ ______ 8. $\frac{2ht}{3bd} =$ ______

Multiply. Write each product in lowest terms:

9. $\left(\frac{3}{br}\right)\left(\frac{b}{9a}\right) =$ ______ 10. $\left(\frac{4w}{d}\right)\left(\frac{d}{2w}\right) =$ ______ 11. $\left(\frac{3P}{H}\right)\left(\frac{2H}{9R}\right) =$ ______

ANSWERS:

1. rt
2. 4ht
3. pw
4. 12dV
5. $\frac{h}{w}$
6. $\frac{4t}{3}$
7. $\frac{1}{3s}$
8. Cannot be reduced.
9. $\frac{1}{3ar}$
10. 2
11. $\frac{2P}{3R}$

8-5 DIVISIONS INVOLVING LITERAL FRACTIONS

To handle divisions which contain fractions, we always convert the division to a multiplication first. To do so, we multiply by the reciprocal of the denominator. In this section, we will briefly examine the meaning of reciprocals for literal fractions and non-fractions. Then we will perform some divisions involving literal fractions.

42. Two quantities are a pair of reciprocals if their product is +1.

Since $(b)\left(\frac{1}{b}\right) = \frac{b}{b} = 1$:

(a) The reciprocal of b is ______.

(b) The reciprocal of $\frac{1}{b}$ is ______.

43. Write the reciprocals of the following:

(a) m ______ (b) $\frac{1}{t}$ ______ (c) $\frac{1}{ab}$ ______ (d) Pq ______

Answers to Frame 42: a) $\frac{1}{b}$ b) b

44. Since $\left(\frac{m}{q}\right)\left(\frac{q}{m}\right) = \frac{mq}{mq} = 1$:

(a) The reciprocal of $\frac{m}{q}$ is ______.

(b) The reciprocal of $\frac{q}{m}$ is ______.

Answers to Frame 43: a) $\frac{1}{m}$ b) t c) ab d) $\frac{1}{Pq}$

45. Write the reciprocals of each of the following:

(a) $\frac{t}{s}$ ______ (b) $\frac{m}{Pq}$ ______ (c) $\frac{VT}{R}$ ______ (d) $\frac{abc}{df}$ ______

Answers to Frame 44: a) $\frac{q}{m}$ b) $\frac{m}{q}$

46. Any fraction stands for a division. This is true even when the numerator or denominator or both are literal fractions. The main fraction line divides the numerator and denominator. For example:

$\dfrac{\frac{t}{s}}{\frac{w}{2}}$ means: divide $\frac{t}{s}$ by $\frac{w}{2}$ $\dfrac{d}{\frac{b}{c}}$ means: divide d by $\frac{b}{c}$

Complete: $\dfrac{\frac{m}{p}}{q}$ means: divide ____ by ____

Answers: a) $\frac{s}{t}$ b) $\frac{Pq}{m}$ c) $\frac{R}{VT}$ d) $\frac{df}{abc}$

---

47. To perform a division of this type, we convert it to a multiplication. That is, we multiply the numerator by the reciprocal of the denominator.

Complete the following divisions:

(a) $\dfrac{\frac{t}{s}}{\frac{w}{z}} = \left(\frac{t}{s}\right)\left(\text{the reciprocal of } \frac{w}{z}\right) = \left(\frac{t}{s}\right)(\quad) =$ ____

(b) $\dfrac{d}{\frac{b}{c}} = (d)\left(\text{the reciprocal of } \frac{b}{c}\right) = (d)(\quad) =$ ____

(c) $\dfrac{\frac{m}{p}}{q} = \left(\frac{m}{p}\right)(\text{the reciprocal of } q) = \left(\frac{m}{p}\right)(\quad) =$ ____

Answer: $\frac{m}{p}$ by q

---

48. Complete the following divisions:

(a) $\dfrac{\frac{p}{q}}{\frac{r}{s}} = (\quad)(\quad) =$ ____ (b) $\dfrac{\frac{w}{tv}}{\frac{mp}{q}} = (\quad)(\quad) =$ ____

Answers: a) $\left(\frac{t}{s}\right)\left(\frac{z}{w}\right) = \frac{tz}{sw}$ b) $d\left(\frac{c}{b}\right) = \frac{cd}{b}$ c) $\left(\frac{m}{p}\right)\left(\frac{1}{q}\right) = \frac{m}{pq}$

---

49. Complete the following. (Be sure to reduce to lowest terms.)

(a) $\dfrac{\frac{3mp}{q}}{\frac{pr}{t}} =$ ____ (b) $\dfrac{\frac{T}{VZ}}{\frac{4T}{RZ}} =$ ____

Answers: a) $= \left(\frac{p}{q}\right)\left(\frac{s}{r}\right) = \frac{ps}{qr}$ b) $= \left(\frac{w}{tv}\right)\left(\frac{q}{mp}\right) = \frac{qw}{mptv}$

---

50. (a) $\dfrac{m}{\frac{s}{t}} = (\quad)(\quad) =$ ____ (b) $\dfrac{\frac{2a}{b}}{4c} = (\quad)(\quad) =$ ____

Answers: a) $\frac{3mt}{qr}$ b) $\frac{R}{4V}$

Answers: a) $m\left(\frac{t}{s}\right) = \frac{mt}{s}$ b) $\left(\frac{2a}{b}\right)\left(\frac{1}{4c}\right) = \frac{a}{2bc}$

51. Divide, and reduce each answer to lowest terms:

(a) $\dfrac{mt}{\dfrac{d}{mn}} =$ ____________ (b) $\dfrac{R}{\dfrac{R}{T}} =$ ____________

Answers: a) $\dfrac{t}{dn}$ b) T

---

52. Reduce each answer to lowest terms:

(a) $\dfrac{VT}{\dfrac{V}{S}} =$ ____________ (b) $\dfrac{\dfrac{5a}{b}}{5b} =$ ____________

Answers: a) ST b) $\dfrac{a}{b^2}$

---

53. Watch this one in which the numerator is "1":

$$\frac{1}{\dfrac{F}{G}} = 1\,(\text{the reciprocal of } \tfrac{F}{G}) = 1\left(\frac{G}{F}\right) = \frac{G}{F}$$

Complete: (a) $\dfrac{1}{\dfrac{p}{q}} =$ ____________ (b) $\dfrac{1}{\dfrac{a}{bc}} =$ ____________

Answers: a) $\dfrac{q}{p}$ b) $\dfrac{bc}{a}$

SELF-TEST 3 (Frames 42-53)

Divide. Report answers in lowest terms.

1. $\dfrac{\dfrac{k}{a}}{\dfrac{b}{2s}}$	2. $\dfrac{\dfrac{d}{2}}{\dfrac{h}{4}}$	3. $\dfrac{\dfrac{1}{at}}{\dfrac{1}{a}}$
4. $\dfrac{r}{\dfrac{2r}{w}}$	5. $\dfrac{\dfrac{bc}{w}}{3b}$	6. $\dfrac{1}{\dfrac{v}{2t}}$

ANSWERS: 1. $\dfrac{2ks}{ab}$ 2. $\dfrac{2d}{h}$ 3. $\dfrac{1}{t}$ 4. $\dfrac{w}{2}$ 5. $\dfrac{c}{3w}$ 6. $\dfrac{2t}{v}$

8-6 ADDITION OF LITERAL FRACTIONS

Fractions can be added only if they have identical or "like" denominators. When they do have identical denominators, they are added by means of the following pattern:

$$\frac{\square}{\bigcirc} + \frac{\triangle}{\bigcirc} = \frac{\square + \triangle}{\bigcirc}$$

In this section, we will show that this pattern for addition also holds for literal fractions. Here are two examples:

(1) $\frac{2x}{3} + \frac{y}{3} = \frac{2x + y}{3}$ (2) $\frac{m}{pq} + \frac{n}{pq} = \frac{m + n}{pq}$

Note: In these two examples, the numerators of the sums cannot be simplified further.

---

54. Do these additions: (a) $\frac{3t}{7} + \frac{5v}{7} =$ ______ (b) $\frac{R}{VT} + \frac{BP}{VT} =$ ______

---

55. In the addition below, the denominators are not identical. However, one denominator "6" is a multiple of the other denominator "3". Before the addition is possible, we must convert $\frac{4y}{3}$ into an equivalent fraction whose denominator is "6". Notice the steps:

$$\frac{5x}{6} + \frac{4y}{3} = \frac{5x}{6} + \left(\frac{4y}{3}\right)\left(\frac{2}{2}\right)$$

$$= \frac{5x}{6} + \frac{8y}{6} = \text{______}$$

Answers: a) $\frac{3t + 5v}{7}$ b) $\frac{R + BP}{VT}$

---

56. In each addition below, one denominator is a multiple of the other.

(a) $\frac{7x}{12} + \frac{y}{3} =$ ______

(b) $\frac{5t}{8} + \frac{3m}{4} =$ ______

Answer: $\frac{5x + 8y}{6}$

---

57. As we saw earlier, we can generate a family of multiples for any number by multiplying the numbers by various whole numbers. For example, we have generated a series of "multiples of 4" below:

2(4) = 8	4(4) = 16	10(4) = 40
3(4) = 12	5(4) = 20	20(4) = 80

A multiple of a number is always larger than that number. To test whether a larger number is a multiple of a smaller number, we divide the larger by the smaller. The larger number is a multiple of the smaller only if the quotient is a whole number. For example:

24 is a multiple of 8, since $\frac{24}{8} = 3$.

15 is not a multiple of 7, since $\frac{15}{7}$ does not equal a whole number.

Which of these numbers (10, 17, 30, 45, 53) are multiples of 5? ______

Answers: a) $\frac{7x + 4y}{12}$ b) $\frac{5t + 6m}{8}$

58. Letters also have multiples. We can generate a family of multiples for any letter by multiplying it by either (1) various <u>whole numbers</u>, or (2) various <u>letters or letter-terms.</u> We have generated various multiples of the letter "x" below:

$3(x) = 3x$	$a(x) = ax$	$3b(x) = 3bx$
$7(x) = 7x$	$d(x) = dx$	$ab(x) = abx$

To test whether an expression is a multiple of a particular letter, we divide the expression by the letter. The expression is a multiple of the letter <u>only if the quotient is a non-fraction.</u> For example:

3ab is a multiple of "a" since $\frac{3ab}{a}$ can be reduced to the non-fraction "3b".

4xy is not a multiple of "d" since $\frac{4xy}{d}$ cannot be reduced to a non-fraction.

Which of these expressions (ax, cy, 3mx, pqx, 4R) are multiples of the letter "x"? ____________________

Answer: 10, 30, and 45

---

59. We can also generate multiples of any letter-term by multiplying it by either (1) various whole numbers, or (2) various letters or letter-terms.

We have generated various multiples of "3y" below:

$5(3y) = 15y$	$d(3y) = 3dy$	$7x(3y) = 21xy$	$ab(3y) = 3aby$

We have generated various multiples of "mx" below:

$7(mx) = 7mx$	$a(mx) = amx$	$4b(mx) = 4bmx$	$cd(mx) = cdmx$

To test whether an expression is a multiple of a letter-term, we again divide the expression by the letter-term. The expression is a multiple of the letter-term <u>only if the quotient is a non-fraction.</u> For example:

"4xyz" is a multiple of "xy", since $\frac{4xyz}{xy}$ reduces to the non-fraction "4z".

"10abx" is a multiple of "2b", since $\frac{10abx}{2b}$ reduces to the non-fraction "5ax".

(a) Which of these expressions (12dt, 6ax, 9adt) are multiples of "3d"? ____________________

(b) Which of these expressions (pdm, pqm, 5pqR) are multiples of "pq"? ____________________

Answer: ax, 3mx, and pqx

---

60. "3abc" is a multiple of which of the expressions below? ______________

ab, 3ac, b, 3c, ac

Answer:
a) 12dt and 9adt
b) pqm and 5pqR

Answer: All of them.

61. In the addition below, one denominator "pq" is a multiple of the other denominator "p". Before the addition is possible, we must convert $\frac{t}{p}$ to an equivalent fraction whose denominator is "pq". Notice the steps:

Example:
$$\frac{m}{pq} + \frac{t}{p} = \frac{m}{pq} + \left(\frac{t}{p}\right)\left(\frac{q}{q}\right)$$
$$= \frac{m}{pq} + \frac{tq}{pq} = \frac{m + tq}{pq}$$

Complete this one:
$$\frac{R}{VT} + \frac{Q}{T} = \frac{R}{VT} + \left(\frac{Q}{T}\right)\left(\frac{V}{V}\right)$$
$$= \frac{R}{VT} + \frac{QV}{VT} = \underline{\qquad\qquad}$$

Answer: $\frac{R + QV}{VT}$

62. Complete:

(a) $\frac{s}{p} + \frac{q}{pt} = \left(\frac{\quad}{\quad}\right) + \frac{q}{pt} = \underline{\qquad\qquad}$

(b) $\frac{n}{pqr} + \frac{m}{pq} = \frac{n}{pqr} + \left(\frac{\quad}{\quad}\right) = \underline{\qquad\qquad}$

(c) $\frac{x}{2a} + \frac{y}{a} = \frac{x}{2a} + \left(\frac{\quad}{\quad}\right) = \underline{\qquad\qquad}$

(d) $\frac{1}{a} + \frac{1}{ab} = \left(\frac{\quad}{\quad}\right) + \frac{1}{ab} = \underline{\qquad\qquad}$

Answers:

a) $\left(\frac{st}{pt}\right) + \frac{q}{pt} = \frac{st + q}{pt}$

b) $\frac{n}{pqr} + \left(\frac{mr}{pqr}\right) = \frac{n + mr}{pqr}$

c) $\frac{x}{2a} + \left(\frac{2y}{2a}\right) = \frac{x + 2y}{2a}$

d) $\left(\frac{b}{ab}\right) + \frac{1}{ab} = \frac{b + 1}{ab}$

63. A fraction can only be added to another fraction with an identical denominator. To add the whole number and the fraction below, we must convert "2" to an equivalent fraction whose denominator is "b". Notice the steps:

$$2 + \frac{a}{b} = 2\left(\frac{b}{b}\right) + \frac{a}{b} = \frac{2b}{b} + \frac{a}{b} = \underline{\qquad\qquad}$$

Answer: $\frac{2b + a}{b}$ or $\frac{a + 2b}{b}$

64. Complete: $1 + \frac{m}{pq} = 1\left(\frac{pq}{pq}\right) + \frac{m}{pq} = \frac{pq}{pq} + \frac{m}{pq} = \underline{\qquad\qquad}$

Answer: $\frac{pq + m}{pq}$ or $\frac{m + pq}{pq}$

65. Complete:

(a) $1 + \frac{V}{T} = \left(\frac{\quad}{\quad}\right) + \frac{V}{T} = \underline{\qquad\qquad}$

(b) $\frac{c}{d} + 4 = \frac{c}{d} + \left(\frac{\quad}{\quad}\right) = \underline{\qquad\qquad}$

Answers:

a) $\left(\frac{T}{T}\right) + \frac{V}{T} = \frac{T + V}{T}$

b) $\frac{c}{d} + \left(\frac{4d}{d}\right) = \frac{c + 4d}{d}$

66. To add "h" and the fraction $\frac{"p"}{t}$ below, we must convert "h" to an equivalent fraction whose denominator is "t". Notice the steps:

$$h + \frac{p}{t} = h\left(\frac{t}{t}\right) + \frac{p}{t} = \frac{ht}{t} + \frac{p}{t} = \underline{\hspace{3cm}}$$

Answer: $\frac{ht + p}{t}$

---

67. Complete: $\frac{T}{VR} + M = \frac{T}{VR} + M\left(\frac{VR}{VR}\right) = \frac{T}{VR} + \frac{MVR}{VR} = \underline{\hspace{3cm}}$

Answer: $\frac{T + MVR}{VR}$

---

68. Do these: (a) $p + \frac{q}{r} = \underline{\hspace{4cm}}$

(b) $\frac{f}{gh} + 2d = \underline{\hspace{4cm}}$

Answer: a) $\frac{pr + q}{r}$ b) $\frac{f + 2dgh}{gh}$

---

69. In many additions of literal fractions, neither denominator is a multiple of the other. When this is the case, we can obtain identical denominators in two steps:

(1) <u>We convert one fraction into an equivalent fraction whose denominator is a multiple of the other denominator.</u> Below, for example, we converted $\frac{"r"}{p}$ into an equivalent fraction whose denominator is "pt". "pt" is a multiple of "t":

$$\frac{m}{t} + \frac{r}{p} = \frac{m}{t} + \frac{r}{p}\left(\frac{t}{t}\right) = \frac{m}{t} + \frac{rt}{pt}$$

(2) <u>Then we convert the other original fraction to an equivalent fraction whose denominator is that multiple.</u> Below, we converted $\frac{"m"}{t}$ to an equivalent fraction whose denominator is "pt":

$$\frac{m}{t} + \frac{rt}{pt} = \frac{m}{t}\left(\frac{p}{p}\right) + \frac{rt}{pt} = \frac{mp}{pt} + \frac{rt}{pt}$$

Having obtained identical denominators, we can now add. The sum is

____

Answer: $\frac{mp + rt}{pt}$

---

70. When neither denominator is a multiple of the other, we can obtain identical denominators <u>in one step</u>. We have done so for the addition below:

$$\frac{1}{R} + \frac{1}{T} = \frac{1}{R}\left(\frac{T}{T}\right) + \frac{1}{T}\left(\frac{R}{R}\right) = \frac{T}{RT} + \frac{R}{RT}$$

<u>Note</u>: Each fraction was multiplied by an instance of $\frac{n}{n}$ which was determined by the denominator of the other fractions.

Insert the instances of $\frac{n}{n}$ which you would need to get identical denominators in these additions:

(a) $\frac{1}{c} + \frac{1}{d} = \frac{1}{c}\left(\frac{\quad}{\quad}\right) + \frac{1}{d}\left(\frac{\quad}{\quad}\right)$

(b) $\frac{V}{ab} + \frac{t}{cd} = \frac{V}{ab}\left(\frac{\quad}{\quad}\right) + \frac{t}{cd}\left(\frac{\quad}{\quad}\right)$

71. Complete: (a) $\frac{c}{d}+\frac{f}{h}=\frac{c}{d}\left(\frac{h}{h}\right)+\frac{f}{h}\left(\frac{d}{d}\right)=$ _____ + _____ = __________

(b) $\frac{x}{2}+\frac{3}{ms}=\frac{x}{2}\left(\frac{ms}{ms}\right)+\frac{3}{ms}\left(\frac{2}{2}\right)=$ _____ + _____ = __________

Answers: a) $\frac{d}{d}$ and $\frac{c}{c}$ b) $\frac{cd}{cd}$ and $\frac{ab}{ab}$

---

72. Add: (a) $\frac{V}{P}+\frac{T}{S}=$ __________

(b) $\frac{1}{m}+\frac{1}{q}=$ __________

Answers: a) $\frac{ch}{dh}+\frac{df}{dh}=\frac{ch+df}{dh}$ b) $\frac{msx}{2ms}+\frac{6}{2ms}=\frac{msx+6}{2ms}$

---

73. Add: (a) $\frac{1}{2p}+\frac{2}{q}=$ __________

(b) $\frac{x}{my}+\frac{1}{p}=$ __________

Answers: a) $\frac{SV+PT}{PS}$ b) $\frac{q+m}{mq}$

---

74. Before adding fractions with non-identical denominators, always check first to see whether one denominator is a multiple of the other.

In which addition below is one denominator a multiple of the other? __________

(a) $\frac{m}{ab}+\frac{t}{abc}$ (b) $\frac{c}{d}+\frac{f}{h}$ (c) $\frac{1}{pq}+\frac{1}{t}$

Answers: a) $\frac{q+4p}{2pq}$ b) $\frac{px+my}{mpy}$

---

75. Before adding, check first to see whether one denominator is a multiple of the other:

(a) $\frac{5x}{12}+\frac{y}{4}=$ __________

(b) $\frac{a}{5}+\frac{b}{4}=$ __________

(c) $\frac{t}{sy}+\frac{x}{t}=$ __________

Answer: Only in (a), where "abc" is a multiple of "ab".

---

76. Add: (a) $1+\frac{m}{t}=$ __________

(b) $\frac{v}{r}+2s=$ __________

(c) $\frac{5a}{b}+\frac{4c}{d}=$ __________

Answers: a) $\frac{5x+3y}{12}$ b) $\frac{4a+5b}{20}$ c) $\frac{t^2+sxy}{sty}$

Answers: a) $\frac{t+m}{t}$ or $\frac{m+t}{t}$ b) $\frac{v+2sr}{r}$ c) $\frac{5ad+4bc}{bd}$

77. We can add fractions with identical denominators even if the denominators are more complicated.

Examples: $\frac{m}{a+b} + \frac{t}{a+b} = \frac{m+t}{a+b}$

$\frac{1}{2(m+q)} + \frac{R}{2(m+q)} = \frac{1+R}{2(m+q)}$

Note: In each case, we simply added the numerators.

Add: (a) $\frac{1}{p-q} + \frac{t}{p-q} =$ ______ (b) $\frac{V}{3(a-b)} + \frac{T}{3(a-b)} =$ ______

Answer to Frame 77: a) $\frac{1+t}{p-q}$ b) $\frac{V+T}{3(a-b)}$

78. We can also add fractions when one or both numerators are more complicated as long as they have identical denominators. Here is an example:

$$\frac{a+b}{R} + \frac{a}{R} = \frac{(a+b)+a}{R} = \frac{2a+b}{R}$$

Notice that we merely added the two numerators.

Add: (a) $\frac{3x+y}{5} + \frac{y}{5} =$ ______

(b) $\frac{m+t}{V} + \frac{m-t}{V} =$ ______

(c) $\frac{a}{b+c} + \frac{2a+b}{b+c} =$ ______

(d) $\frac{3a+2b}{p-q} + \frac{3a-2b}{p-q} =$ ______

Answer to Frame 78: a) $\frac{3x+2y}{5}$ b) $\frac{2m}{V}$ c) $\frac{3a+b}{b+c}$ d) $\frac{6a}{p-q}$

8-7 SUBTRACTION OF LITERAL FRACTIONS

When fractions have identical denominators, we use the following pattern for subtractions:

$$\frac{\square}{\bigcirc} - \frac{\triangle}{\bigcirc} = \frac{\square - \triangle}{\bigcirc}$$

This pattern also holds for literal fractions. Here are two examples:

(1) $\frac{3y}{5} - \frac{2x}{5} = \frac{3y-2x}{5}$ (2) $\frac{V}{R} - \frac{T}{R} = \frac{V-T}{R}$

---

79. Complete: (a) $\frac{s}{7} - \frac{2t}{7} =$ ______ (b) $\frac{a}{x} - \frac{1}{x} =$ ______ (c) $\frac{d}{pq} - \frac{3f}{pq} =$ ______

Answer to Frame 79: a) $\frac{s-2t}{7}$ b) $\frac{a-1}{x}$ c) $\frac{d-3f}{pq}$

80. In these subtractions, you will have to obtain identical denominators first. In each case, one denominator is a multiple of the other.

(a) $\frac{t}{ab} - \frac{v}{b} =$ ______

(b) $\frac{1}{c} - \frac{t}{cd} =$ ______

(c) $\frac{4m}{9} - \frac{2p}{3} =$ ______

Answers to 80: a) $\frac{t - av}{ab}$ b) $\frac{d - t}{cd}$ c) $\frac{4m - 6p}{9}$

81. When getting identical denominators for subtractions, always check to see whether one denominator is a multiple of the other.

In which subtractions below is one denominator a multiple of the other? ______

(a) $\frac{V}{7} - \frac{2}{3}$ (b) $\frac{1}{k} - \frac{1}{w}$ (c) $\frac{a}{RS} - \frac{b}{S}$

Answer to 81: Only in (c), where "RS" is a multiple of "S".

82. Complete: (a) $\frac{V}{7} - \frac{h}{3} =$ ______

(b) $\frac{2x}{y} - \frac{a}{bc} =$ ______

(c) $\frac{1}{m} - \frac{1}{mp} =$ ______

Answers to 82: a) $\frac{3V - 7h}{21}$ b) $\frac{2bcx - ay}{bcy}$ c) $\frac{p - 1}{mp}$

83. Of course, if one of the terms is a non-fraction, we must convert it to a fraction before a subtraction can be performed. Here is an example:

$$1 - \frac{p}{q} = 1\left(\frac{q}{q}\right) - \frac{p}{q} = \frac{q}{q} - \frac{p}{q} = \frac{q - p}{q}$$

Complete: (a) $\frac{a}{b} - 3 =$ ______

(b) $m - \frac{V}{T} =$ ______

Answers to 83: a) $\frac{a - 3b}{b}$ b) $\frac{mT - V}{T}$

84. Complete: (a) $\frac{c}{d} - 2a =$ ______

(b) $2m - \frac{RS}{TV} =$ ______

Answers to 84: a) $\frac{c - 2ad}{d}$ b) $\frac{2mTV - RS}{TV}$

85. Some of the following are additions:

(a) $\frac{1}{k} + \frac{1}{w} =$ ______

(b) $\frac{3p}{t} - \frac{q}{ty} =$ ______

(c) $\frac{R}{MQ} + \frac{2T}{M} =$ ______

(d) $\frac{1}{R} - \frac{1}{S} =$ ______

86. We can perform subtractions by means of the same pattern even if the denominators are more complicated. Do these:

(a) $\frac{2R}{a+b} - \frac{T}{a+b} =$ ________ (b) $\frac{m}{3(p-q)} - \frac{0.8d}{3(p-q)} =$ ________

a) $\frac{w+k}{kw}$ b) $\frac{3py-q}{ty}$ c) $\frac{R+2QT}{MQ}$ d) $\frac{S-R}{RS}$

---

87. The same pattern is used for subtraction even when one or both numerators are more complex. Here is an example:

$$\frac{x+7}{y} - \frac{3}{y} = \frac{(x+7)-3}{y} = \frac{x+4}{y}$$

Complete: (a) $\frac{y+5x}{t} - \frac{3x}{t} =$ ________

(b) $\frac{R-V}{R+V} - \frac{2T}{R+V} =$ ________

a) $\frac{2R-T}{a+b}$ b) $\frac{m-0.8d}{3(p-q)}$

Answer to Frame 87: a) $\frac{y+2x}{t}$ b) $\frac{R-V-2T}{R+V}$

SELF-TEST 4 (Frames 54-87)

Perform the following additions and subtractions:

1. $\frac{t}{12r} + \frac{2t}{3r} =$ ________	2. $\frac{1}{h} + \frac{1}{w} =$ ________
3. $1 + \frac{d}{s} =$ ________	4. $\frac{2a}{pr} - \frac{c}{p} =$ ________
5. $1 - \frac{d}{kt} =$ ________	6. $\frac{h-1}{R-r} + \frac{h+1}{R-r} =$ ________

ANSWERS: 1. $\frac{3t}{4r}$ (from $\frac{9t}{12r}$) 2. $\frac{w+h}{hw}$ 3. $\frac{s+d}{s}$ 4. $\frac{2a-cr}{pr}$ 5. $\frac{kt-d}{kt}$ 6. $\frac{2h}{R-r}$

8-8 THE PATTERN FOR THE SUM OF FRACTIONS

In many equations and formulas, you will meet fractions like those below. In each one, the numerator of the fraction is an addition.

$$\frac{x+3}{4} \qquad \frac{5x+3y}{7} \qquad \frac{dm+t}{p} \qquad \frac{a+b}{c+d}$$

Any fraction whose numerator in an addition is the sum of two simpler fractions. In this section, we will show how fractions like those above can be "broken up" into two simpler fractions and written in equivalent forms.

---

88. Let's examine some fractions whose numerators are additions. <u>All fractions of this type are sums of simpler fractions.</u>

$\frac{x+2}{3}$ is the sum of $\frac{x}{3}$ and $\frac{2}{3}$.

(a) $\frac{7x+8}{y}$ is the sum of ____ and ____.

(b) $\frac{m+p}{q}$ is the sum of ____ and ____.

Answers: a) $\frac{7x}{y}$ and $\frac{8}{y}$ b) $\frac{m}{q}$ and $\frac{p}{q}$

89. Break up each sum into two simpler fractions:

(a) $\frac{2x+5y}{9} = \_\_\_\_ + \_\_\_\_$ (b) $\frac{mt+3R}{5q} = \_\_\_\_ + \_\_\_\_$

Answers: a) $\frac{2x}{9}+\frac{5y}{9}$ b) $\frac{mt}{5q}+\frac{3R}{5q}$

90. When we break up a sum into two fractions, sometimes one or both of the fractions can be reduced to lowest terms.

Here is a case where neither fraction can be reduced:

$$\frac{5x+6}{7} = \frac{5x}{7}+\frac{6}{7}$$

Here is a case where one of the fractions can be reduced:

$$\frac{3y+7}{14} = \frac{3y}{14}+\frac{7}{14} = \frac{3y}{14}+\frac{1}{2}$$

Here is a case where both fractions can be reduced:

$$\frac{4x+5y}{20} = \frac{4x}{20}+\frac{5y}{20} = \frac{x}{5}+\frac{y}{4}$$

In the following: (1) break up each sum into two simpler fractions
(2) then reduce each simpler fraction to lowest terms, if possible:

(a) $\frac{x+6}{4} = \_\_\_\_ + \_\_\_\_ = \_\_\_\_ + \_\_\_\_$

(b) $\frac{5d+15p}{10} = \_\_\_\_ + \_\_\_\_ = \_\_\_\_ + \_\_\_\_$

(c) $\frac{3x+7}{11} = \_\_\_\_ + \_\_\_\_ = \_\_\_\_ + \_\_\_\_$

91. When a sum of two fractions is broken up into two simpler fractions, sometimes one or both simpler fractions reduces to a non-fraction. For example:

$$\frac{x+4}{4} = \frac{x}{4} + \frac{4}{4} = \frac{x}{4} + 1$$

(a) $\frac{t+10}{5} = \frac{t}{5} + \frac{10}{5} =$ ______ + ______

(b) $\frac{4p+8q}{2} = \frac{4p}{2} + \frac{8q}{2} =$ ______ + ______

a) $\frac{x}{4} + \frac{6}{4} = \frac{x}{4} + \frac{3}{2}$

b) $\frac{5d}{10} + \frac{15p}{10} = \frac{d}{2} + \frac{3p}{2}$

c) $\frac{3x}{11} + \frac{7}{11}$ (Neither fraction can be reduced.)

Answer to Frame 91: a) $\frac{t}{5} + 2$ b) $2p + 4q$

92. Let's examine more closely the case where a sum is broken up into two fractions and one of them can be reduced to lower terms.

$$\underbrace{\frac{x+2}{2}}_{(A)} = \underbrace{\frac{x}{2} + \frac{2}{2}}_{(B)} = \underbrace{\frac{x}{2} + 1}_{(C)}$$

A, B, and C are called equivalent expressions. That is, no matter what number you plug in for "x" in each, the three are still equal.

Let's plug in a 4 for "x" in each:

(A) $\frac{x+2}{2} = \frac{4+2}{2} = \frac{6}{2} = 3$

(B) $\frac{x}{2} + \frac{2}{2} = \frac{4}{2} + \frac{2}{2} = 2 + 1 = 3$

(C) $\frac{x}{2} + 1 = \frac{4}{2} + 1 = 2 + 1 = 3$

Let's plug in a 7 for "x" in each:

(A) $\frac{x+2}{2} = \frac{7+2}{2} = \frac{9}{2}$

(B) $\frac{x}{2} + \frac{2}{2} = \frac{7}{2} + \frac{2}{2} = \frac{9}{2}$

(C) $\frac{x}{2} + 1 = \frac{7}{2} + 1 = \frac{7}{2} + \frac{2}{2} = \frac{9}{2}$

You can plug in any number you want and the above three expressions will be equal. Since this is the case, these three expressions are called ______________ expressions.

equivalent

93. Write each fraction below in two equivalent forms:

(a) $\frac{x+6}{6} =$ ____________ or ____________

(b) $\frac{2x+6}{3} =$ ____________ or ____________

a) $\frac{x}{6} + \frac{6}{6}$ or $\frac{x}{6} + 1$

b) $\frac{2x}{3} + \frac{6}{3}$ or $\frac{2x}{3} + 2$

94. Write each of these in an equivalent form with both parts reduced to lowest terms:

(a) $\frac{3y+5}{5} =$ ______________ (b) $\frac{6x+8y}{4} =$ ______________

a) $\frac{3y}{5} + 1$

b) $\frac{3x}{2} + 2y$

95. Write each of the following in two equivalent forms:

(a) $\frac{2x}{7} + \frac{7}{7} =$ __________ or __________

(b) $\frac{4x}{5} + \frac{10y}{5} =$ __________ or __________

Answer: a) $\frac{2x + 7}{7}$ or $\frac{2x}{7} + 1$

b) $\frac{4x + 10y}{5}$ or $\frac{4x}{5} + 2y$

96. Write each of these in two equivalent forms:

(a) $\frac{y}{4} + 3 =$ __________ or __________

(b) $\frac{5x}{4} + y =$ __________ or __________

Answer: a) $\frac{y}{4} + \frac{12}{4}$ or $\frac{y + 12}{4}$

b) $\frac{5x}{4} + \frac{4y}{4}$ or $\frac{5x + 4y}{4}$

97. Which of the following expressions are equivalent to $\frac{3x + 4}{4}$? __________

(a) $3x + 1$ (b) $\frac{3x}{4} + 1$ (c) $\frac{3x}{4} + \frac{4}{4}$ (d) $3x$

Answer: (b) and (c)

98. Which of the following expressions are equivalent to $\frac{5x + 6}{6}$? __________

(a) $\frac{11x}{6}$ (b) $\frac{5x}{6} + \frac{6}{6}$ (c) $5x + 1$ (d) $\frac{5x}{6} + 1$

Answer: (b) and (d)

99. Which of the following expressions are equivalent to $\frac{3x}{2} + 2$? __________

(a) $\frac{3x + 4}{2}$ (b) $3x$ (c) $\frac{3x}{2} + \frac{4}{2}$ (d) $\frac{5x}{2}$

Answer: (a) and (c)

100. Whenever a fraction contains an addition in its numerator, it is the sum of two simpler fractions. This is true even if the fraction contains only letters.

$\frac{a + b}{c}$ is the sum of $\frac{a}{c}$ and $\frac{b}{c}$.

$\frac{2V + T}{R}$ is the sum of ___ and ___.

Answer: $\frac{2V}{R}$ and $\frac{T}{R}$

101. If a complicated literal fraction is the sum of two simpler fractions, it can also be broken down into two parts. Sometimes the parts can be reduced to lower terms. For example:

$$\frac{a + bc}{b} = \frac{a}{b} + \frac{bc}{b} = \frac{a}{b} + c$$

Each of the two expressions on the right are equivalent to the complicated fraction on the left.

Write $\frac{Q + CT}{C}$ in two equivalent forms: __________ or __________

Answer: $\frac{Q}{C} + \frac{CT}{C}$ or $\frac{Q}{C} + T$

102. Write each of the following in two equivalent forms:

(a) $\frac{aV + T}{V} =$ ________ or ________

(b) $\frac{LH + AKTt_1}{AKT} =$ ________ or ________

---

103. Which of the following expressions are equivalent to $\frac{A + DF}{D}$? ________

(a) $A + F$ (b) $\frac{A}{D} + F$ (c) $A + \frac{F}{D}$

a) $\frac{aV}{V} + \frac{T}{V}$ or $a + \frac{T}{V}$

b) $\frac{LH}{AKT} + \frac{AKTt_1}{AKT}$ or $\frac{LH}{AKT} + t_1$

---

104. Which of the following expressions are equivalent to $\frac{ABC + X}{AB}$? ________

(a) $C + X$ (b) $\frac{C}{AB} + X$ (c) $C + \frac{X}{AB}$

Only (b)

---

105. Which of the following expressions are equivalent to $\frac{Q}{C} + T$? ________

(a) $\frac{Q + CT}{C}$ (b) $\frac{Q + T}{C}$ (c) $\frac{CQ + T}{C}$

Only (c)

---

106. Which of the following expressions are equivalent to $\frac{33,000H}{\pi dR} + F$? ________

(a) $\frac{33,000H + F}{\pi dR}$ (b) $\frac{33,000\pi dRH + F}{\pi dR}$ (c) $\frac{33,000H + \pi dRF}{\pi dR}$

Only (a)

---

107. As long as a fraction contains an addition in its numerator, it is the sum of two fractions even if its denominator is more complicated.

$\frac{a + b}{c + d}$ is the sum of $\frac{a}{c + d}$ and $\frac{b}{c + d}$.

(a) $\frac{V + t}{x - y}$ is the sum of ________ and ________.

(b) $\frac{2m + t}{c(p + q)}$ is the sum of ________ and ________.

Only (c)

---

108. If the addition in its numerator contains <u>three</u> terms, the complicated fraction is the sum of <u>three</u> simpler fractions.

$$\frac{x + y + 2}{7} = \frac{x}{7} + \frac{y}{7} + \frac{2}{7}$$

$\frac{a + b + c}{d} =$ ____ + ____ + ____

a) $\frac{V}{x - y}$ and $\frac{t}{x - y}$

b) $\frac{2m}{c(p + q)}$ and $\frac{t}{c(p + q)}$

$\frac{a}{d} + \frac{b}{d} + \frac{c}{d}$

109. When we break up the sum of three fractions into simpler fractions, sometimes one or more parts can be reduced to lowest terms. For example:

$$\frac{x+y+3}{3} = \frac{x}{3} + \frac{y}{3} + \frac{3}{3} = \frac{x}{3} + \frac{y}{3} + 1$$

Write these in equivalent forms with each simpler fraction reduced to lowest terms:

(a) $\frac{a+b+c}{b} =$ ______________________

(b) $\frac{mx+ky+m}{m} =$ ______________________

Answer to Frame 109: a) $\frac{a}{b} + 1 + \frac{c}{b}$ b) $x + \frac{ky}{m} + 1$

8-9 THE PATTERN FOR THE SUBTRACTION OF FRACTIONS

The numerator of each fraction below is a subtraction:

$$\frac{x-5}{7} \qquad \frac{2y-3t}{6} \qquad \frac{a-bc}{d} \qquad \frac{R-V}{Q+T}$$

Any fraction whose numerator is a subtraction can be looked upon as resulting from the subtraction of two simpler fractions. In this section, we will show how fractions of this type can be "broken up" into two simpler fractions and written in equivalent forms.

110. Whenever you see a fraction whose numerator is a "subtraction," you should immediately recognize it as a fraction obtained by subtracting two simpler fractions.

$\frac{3x-2y}{7}$ was obtained from this subtraction $\frac{3x}{7} - \frac{2y}{7}$.

$\frac{m-p}{d}$ was obtained from this subtraction ___ - ___.

111. When the numerator of a fraction is a subtraction, we can write the fraction in equivalent forms. For example:

$\frac{m}{d} - \frac{p}{d}$

$$\frac{x-2}{2} = \frac{x}{2} - \frac{2}{2} = \frac{x}{2} - 1$$

$\frac{a-b}{b} =$ ___ - ___ = ___ - ___

112. Write each of the following fractions in two equivalent forms:

$\frac{a}{b} - \frac{b}{b} = \frac{a}{b} - 1$

(a) $\frac{2-x}{2} =$ ______________ or ______________

(b) $\frac{a-x}{a} =$ ______________ or ______________

113. Write each of these in two equivalent forms:

(a) $\frac{bc - d}{b} =$ ______ or ______

(b) $\frac{LH - TCX}{CX} =$ ______ or ______

Answer: a) $\frac{2}{2} - \frac{x}{2}$ or $1 - \frac{x}{2}$ b) $\frac{a}{a} - \frac{x}{a}$ or $1 - \frac{x}{a}$

---

114. Which of the following expressions are equivalent to $\frac{tAKT - LH}{AKT}$? ______

(a) $tAKT - \frac{LH}{AKT}$ (b) $t - \frac{LH}{AKT}$ (c) $t - LH$

Answer: a) $\frac{bc}{b} - \frac{d}{b}$ or $c - \frac{d}{b}$ b) $\frac{LH}{CX} - \frac{TCX}{CX}$ or $\frac{LH}{CX} - T$

---

115. Which of the following expressions are equivalent to $\frac{abt - V}{ab}$? ______

(a) $t - V$ (b) $abt - \frac{V}{ab}$ (c) $\frac{t}{ab} - V$

Answer: Only (b)

---

116. Which of the following expressions are equivalent to $m - \frac{QZ}{PX}$? ______

(a) $\frac{mPX - QZ}{PX}$ (b) $\frac{m - QZ}{PX}$ (c) $\frac{mPX - QZPX}{PX}$

Answer: None of them are equivalent.

---

117. You should recognize the fact that fractions with more complicated denominators can also result from the subtraction of two fractions. <u>The clue is the fact that there is a subtraction in the numerator.</u> For example:

$\frac{a - b}{c + d}$ was obtained from this subtraction: $\frac{a}{c + d} - \frac{b}{c + d}$

(a) $\frac{x - 2}{y - 3}$ was obtained from this subtraction: ______ - ______

(b) $\frac{m - t}{a(b + c)}$ was obtained from this subtraction: ______ - ______

Answer: Only (a)

---

118. If a complicated fraction contains both an addition and a subtraction in its numerator, it can be broken down into an addition and subtraction. For example:

$$\frac{x + y - 5}{7} = \frac{x}{7} + \frac{y}{7} - \frac{5}{7}$$

(a) $\frac{m - p + q}{t} =$ ____ - ____ + ____ (b) $\frac{V - T - R}{m} =$ ____ - ____ - ____

Answer: a) $\frac{x}{y - 3} - \frac{2}{y - 3}$ b) $\frac{m}{a(b + c)} - \frac{t}{a(b + c)}$

Answer: a) $\frac{m}{t} - \frac{p}{t} + \frac{q}{t}$ b) $\frac{V}{m} - \frac{T}{m} - \frac{R}{m}$

SELF-TEST 5 (Frames 88-118)

Write each fraction in an equivalent form which involves the addition or subtraction of two simpler fractions. Reduce each part to lowest terms.

1. $\frac{3a + 4h}{12} =$ ________	2. $\frac{E + 2}{6} =$ ________	3. $\frac{t - kw}{kt} =$ ________
4. $\frac{2s - p}{2p} =$ ________	5. $\frac{b - 3h}{h} =$ ________	6. $\frac{R + 2r}{R - r} =$ ________

ANSWERS: 1. $\frac{a}{4} + \frac{h}{3}$ 2. $\frac{E}{6} + \frac{1}{3}$ 3. $\frac{1}{k} - \frac{w}{t}$ 4. $\frac{s}{p} - \frac{1}{2}$ 5. $\frac{b}{h} - 3$ 6. $\frac{R}{R - r} + \frac{2r}{R - r}$

8-10 CONTRASTING TWO PATTERNS OF COMPLICATED FRACTIONS

The two patterns of complicated fractions below are different. The pattern on the left stands for fractions with either an addition or subtraction in their numerators. The pattern on the right stands for fractions with either an addition or subtraction in their denominators.

$$\frac{\square \pm \triangle}{\bigcirc} \quad \text{and} \quad \frac{\square}{\triangle \pm \bigcirc}$$

Note: We use the symbol "±" to show that the operation can be either an addition or a subtraction.

In this section, we will discuss the difference between these two patterns. We will show that a fraction can be broken down into two simpler fractions only if its numerator is an addition or subtraction.

119. If a fraction contains an addition or subtraction in its numerator, it is the result of adding or subtracting two fractions. Consequently, it can be broken down into two simpler fractions. For example:

$$\frac{x + 4}{4} = \frac{x}{4} + \frac{4}{4} = \frac{x}{4} + 1$$

$$\frac{y - 5}{5} = \frac{y}{5} - \frac{5}{5} = \frac{y}{5} - 1$$

However, if a fraction contains an addition or subtraction only in its denominator, it is not the result of adding or subtracting two fractions. Therefore, it cannot be broken down into two simpler fractions. That is:

$$\frac{8}{x + 8} \neq \frac{8}{x} + \frac{8}{8}$$

We can show this fact numerically by plugging in a "4" for "x" and evaluating each expression above.

(a) $\frac{8}{x + 8} = \frac{8}{4 + 8} =$ ________ (b) $\frac{8}{x} + \frac{8}{8} = \frac{8}{4} + \frac{8}{8} =$ ________

a) $\frac{8}{12} = \frac{2}{3}$ b) $2 + 1 = 3$

120. Does $\frac{a}{a+1} = \frac{a}{a} + \frac{a}{1}$? ______ (Plug in "4" for "a" and check it.)	
121. Does $\frac{m}{m-1} = \frac{m}{m} - \frac{m}{1}$? ______ (Plug in "3" for "m" and check it.)	No. $\frac{a}{a+1} = \frac{4}{4+1} = \frac{4}{5}$ $\frac{a}{a} + \frac{a}{1} = \frac{4}{4} + \frac{4}{1} = 1 + 4 = 5$
122. A complicated fraction is an addition or subtraction of fractions only if it has an addition or subtraction <u>in its numerator.</u> Which of the following are additions or subtractions of two fractions? ______ (a) $\frac{C+1}{C}$ (b) $\frac{d}{d-1}$ (c) $\frac{mn}{m+n}$ (d) $\frac{p-q}{pq}$	No, since: $\frac{m}{m-1} = \frac{3}{3-1} = \frac{3}{2}$ $\frac{m}{m} - \frac{m}{1} = \frac{3}{3} - \frac{3}{1} = 1 - 3 = -2$
123. A complicated fraction can be written in an equivalent form <u>by breaking it down into two simpler fractions</u> only it it is the addition or subtraction of two fractions. Write each of the following in an equivalent form by breaking it into two simpler fractions, if this is possible: (a) $\frac{B}{B+1} =$ ______ (c) $\frac{T+1}{T} =$ ______ (b) $\frac{B-1}{B} =$ ______ (d) $\frac{V}{V-S} =$ ______	Only (a) and (d)
124. If a fraction contains an addition (or subtraction) <u>in both numerator and denominator, it is</u> the result of adding (or subtracting) two fractions. Consequently, it can be broken down into the original fractions. For example: $\frac{M+P}{T+V} = \frac{M}{T+V} + \frac{P}{T+V}$ $\frac{I_c - I_{co}}{I_b - I_{co}} = \frac{I_c}{I_b - I_{co}} - \frac{I_{co}}{I_b - I_{co}}$ Complete these breakdowns: (a) $\frac{CI+B}{M+LV} =$ ______ + ______ (b) $\frac{I-VW}{C-D} =$ ______ - ______	a) Impossible b) $\frac{B}{B} - \frac{1}{B}$ or $1 - \frac{1}{B}$ c) $\frac{T}{T} + \frac{1}{T}$ or $1 + \frac{1}{T}$ d) Impossible
125. A complicated fraction is an addition or subtraction of two fractions <u>only if it has an addition or subtraction in its</u> ______ (numerator/denominator).	a) $\frac{CI}{M+LV} + \frac{B}{M+LV}$ b) $\frac{I}{C-D} - \frac{VW}{C-D}$
	numerator

126. Which of the following are additions or subtractions of two fractions? ____________

(a) $\frac{m}{p+q}$ (b) $\frac{x-y}{7}$ (c) $\frac{2b+a}{3x+y}$ (d) $\frac{t}{2m-d}$

Answer to Frame 126: Only (b) and (c) (The others do not have an addition or subtraction in the numerator.)

8-11 REDUCING COMPLICATED FRACTIONS TO LOWEST TERMS

The patterns below represent fractions with an addition or subtraction in their numerators, denominators, or both.

$$\frac{\square \pm \triangle}{\bigcirc} \qquad \frac{\square}{\triangle \pm \bigcirc} \qquad \frac{\square \pm \triangle}{\bigcirc \pm \hexagon}$$

Note: We use the symbol "±" to show that the operation can be either an addition or a subtraction.

In this section, we will begin by showing how we can generate equivalent forms of these fractions. Then we will show how fractions of these types can be reduced to lowest terms.

127. Beginning with any fraction, we can generate an equivalent fraction by multiplying it by various instances of $\frac{n}{n}$. This is true even if the original fraction has an addition or subtraction in its numerator or denominator. An example is given below.

Since $\frac{p}{p} = 1$, if we multiply the fraction below by $\frac{p}{p}$, we are multiplying it by "1". Therefore, the new fraction is equivalent to the original one.

Step 1: $\frac{b+c}{d} = \left(\frac{p}{p}\right)\left(\frac{b+c}{d}\right)$

Step 2: $= \frac{p(b+c)}{pd}$

Step 3: $= \frac{pb+pc}{pd}$

Note: (1) In Step 2, there is an instance of the distributive principle in the numerator of the fraction.

(2) In going from Step 2 to Step 3, we multiplied by the distributive principle in the numerator.

128. Complete this one: $\frac{m}{p-q} = \left(\frac{a}{a}\right)\left(\frac{m}{p-q}\right) = \frac{am}{a(p-q)} =$ ____________

Go to next frame.

129. In this case, you must multiply by the distributive principle in both numerator and denominator:

$\frac{v-t}{x+y} = \left(\frac{f}{f}\right)\left(\frac{v-t}{x+y}\right) = \frac{f(v-t)}{f(x+y)} =$ ____________

$\frac{am}{ap-aq}$

130. In each case below, the fraction on the right is equivalent to the one on the left. What instance of $\frac{n}{n}$ did we use to generate it?

(a) $\frac{x - y}{z} = \frac{3x - 3y}{3z}$ ____ (b) $\frac{y}{a + b} = \frac{my}{ma + mb}$ ____ (c) $\frac{x + 5}{x - 3} = \frac{4x + 20}{4x - 12}$ ____

Answer: $\frac{fv - ft}{fx + fy}$

---

131. When we multiply a complicated fraction by an instance of $\frac{n}{n}$, the new equivalent fraction has "n" as a factor in each term of its numerator and denominator. Here is an example:

$$\frac{t}{m + q} = \left(\frac{c}{c}\right)\left(\frac{t}{m + q}\right) = \frac{ct}{cm + cq}$$

In the fraction at the right, each term in the numerator and denominator has ____ as a factor.

Answer: a) $\frac{3}{3}$ b) $\frac{m}{m}$ c) $\frac{4}{4}$

---

132. Identify the common factor in each term of the numerator and denominator of each fraction below:

(a) $\frac{sv - st}{sx}$ ____ (b) $\frac{dm - dt}{dp + dq}$ ____ (c) $\frac{2x}{2y - 2}$ ____ (d) $\frac{3x - 6}{3y + 9}$ ____

Answer: c

---

133. Identify the common factor in each term of the numerator and denominator of each fraction, if it has a common factor:

(a) $\frac{mt}{bq + br}$ ____ (b) $\frac{ax - ay}{am}$ ____ (c) $\frac{7x - 3}{7y}$ ____ (d) $\frac{4}{4x + 12}$ ____

Answer: a) s b) d c) 2 d) 3

---

134. If a complicated fraction has a common factor in each term of its numerator and denominator, we can reduce it to lower terms by factoring out an instance of $\frac{n}{n}$. Here is an example:

Step 1: $\frac{CT + CS}{CM} = \frac{C(T + S)}{CM}$

Step 2: $= \left(\frac{C}{C}\right)\left(\frac{T + S}{M}\right)$

Step 3: $= (1)\left(\frac{T + S}{M}\right)$

Step 4: $= \frac{T + S}{M}$

Note: (1) In Step 1, we factored the numerator by the distributive principle.

(2) Then we factored out $\left(\frac{C}{C}\right)$ which can be dropped since it equals "1".

Complete this one: $\frac{am}{bm + dm} = \frac{am}{m(b + d)} = \left(\frac{m}{m}\right)\left(\frac{a}{b + d}\right) =$ ________

Answer: a) None b) a c) None d) 4

Answer: $\frac{a}{b + d}$

135. To reduce fractions by this method, we must factor by the distributive principle in either the numerator or denominator or both. Let's review this type of factoring.

In order to factor a complex product, there must be a common factor in each term. For example:

DE + FE can be factored because E is a common factor in each term.

CD + EF cannot be factored because there is no common factor in each term.

Factor the following if they can be factored:

(a) aT - bT = ________ (c) $V_1P_1 - V_2P_2$ = ________

(b) VX + ST = ________ (d) $V_1T_1 + V_2T_1$ = ________

---

a) T(a - b)

b) Cannot be factored.

c) Cannot be factored.

d) $T_1(V_1 + V_2)$

136. To factor the expression below, it is helpful to substitute "1T" for "T" so that the coefficient "1" is not forgotten. We get:

$$ET + T = ET + 1T = T(E + 1)$$

Factor the following by the distributive principle:

(a) ab + b = ________ (c) ST - S = ________

(b) c + cd = ________ (d) V - XV = ________

---

a) b(a + 1) c) S(T - 1)

b) c(1 + d) d) V(1 - X)

137. Factor the following if they can be factored:

(a) MN - PN = ________ (d) $V_1 - V_1P_1$ = ________

(b) mM + m = ________ (e) $C_1C_2 - C_1$ = ________

(c) $V_1P_1 - V_2$ = ________

---

a) N(M - P)

b) m(M + 1)

c) Cannot be factored.

d) $V_1(1 - P_1)$

e) $C_1(C_2 - 1)$

138. We can factor both the numerator and denominator of the fraction below by the distributive principle. Do so, and then reduce the fraction to lowest terms.

$$\frac{ax - ay}{am + ap} = \_\_\_\_\_\_\_\_ = \_\_\_\_\_\_\_\_$$

$$\frac{a(x - y)}{a(m + p)} = \frac{x - y}{m + p}$$

139. We cannot always reduce a fraction to lower terms just because we can factor its numerator or denominator by the distributive principle. Here is an example:

$$\frac{md + pd}{tv} = \frac{d(m + p)}{tv}$$

The fraction cannot be reduced to lower terms since there is no common factor in both numerator and denominator. Therefore, we cannot get an instance of $\frac{n}{n}$.

In each case below, we have factored by the distributive principle. In which one can we reduce to lower terms? ________

(a) $\frac{4m}{8x - 4} = \frac{4m}{4(2x - 1)}$ (b) $\frac{7x - 7y}{3a - 6} = \frac{7(x - y)}{3(a - 2)}$

140. If a complicated fraction is a sum of two simple fractions, and is reducible, it can be reduced to lowest terms in either of two ways: (1) By breaking the fraction up into the two original fractions, reducing each of them to lowest terms, and recombining: Example: $\frac{pq + fq}{mq} = \frac{pq}{mq} + \frac{fq}{mq} = \frac{p}{m} + \frac{f}{m} = \frac{p + f}{m}$ (2) By factoring out an instance of $\frac{n}{n}$, and then reducing the fraction to lowest terms: Example: $\frac{pq + fq}{mq} = \frac{q(p + f)}{mq} = \left(\frac{q}{q}\right)\left(\frac{p + f}{m}\right) = \frac{p + f}{m}$	Only in (a).
141. If a complicated fraction is not a sum of two simpler fractions, it can be reduced to lower terms in only one way. It can be reduced only by eliminating a common factor in both numerator and denominator. For example: $\frac{CT}{CV + CS} = \frac{CT}{C(V + S)} = \left(\frac{C}{C}\right)\left(\frac{T}{V + S}\right) = \frac{T}{V + S}$ Complete: $\frac{m}{bm + cm} = \frac{(m)(1)}{m(b + c)} =$ __________	Go to next frame.
142. Though the following fraction cannot be reduced to lower terms, it can be written in an equivalent form since it is a sum of two fractions. We get: $\frac{A + BC}{A} = \frac{A}{A} + \frac{BC}{A} = 1 + \frac{BC}{A}$ Write the following in an equivalent or simpler form, if this is possible: (a) $\frac{wr}{w - 2fw} =$ __________ (c) $\frac{E - IR}{I} =$ __________ (b) $\frac{A}{1 - AB} =$ __________ (d) $\frac{Vv}{V - v} =$ __________	$\frac{1}{b + c}$
143. In the formula: $T = \frac{am}{a + m}$, can the right side be written in an equivalent or simpler form? ______	a) $\frac{r}{1 - 2f}$ b) Impossible c) $\frac{E}{I} - R$ d) Impossible
144. In the formula: $A = \frac{B}{B + 1}$, can the right side be written in an equivalent or simpler form? ______	No
145. Which of the following expressions are equivalent to $\frac{P + XH}{X}$? ______ (a) $P + H$ (b) $\frac{P + H}{X}$ (c) $\frac{P}{X} + \frac{H}{X}$ (d) $\frac{P}{X} + H$	No
	Only (d)

146. Which of the following are equivalent to $\dfrac{AKTv_1 - Rv_2}{AKT}$? ______ (a) $\dfrac{v_1 - Rv_2}{AKT}$ (b) $v_1 - \dfrac{Rv_2}{AKT}$ (c) $v_1 - Rv_2$ (d) $\dfrac{v_1}{AKT} - \dfrac{Rv_2}{AKT}$	
147. Which of the following are equivalent to $\dfrac{Q}{Z} - r_1$? ______ (a) $\dfrac{Q - r_1Z}{Z}$ (b) $\dfrac{Q - r_1}{Z}$ (c) $\dfrac{QZ - r_1Z}{Z}$ (d) $\dfrac{Q}{Z} - \dfrac{r_1}{Z}$	Only (b)
148. Which of the following are equivalent to $\dfrac{PQ}{PD - DQ}$? ______ (a) $\dfrac{Q}{D - DQ}$ (b) $\dfrac{1}{D - D}$ (c) $\dfrac{P}{PD - D}$ (d) $\dfrac{Q}{D} - \dfrac{P}{Q}$	Only (a)
149. Which of the following are equivalent to $\dfrac{0.24IR - R}{R}$? ______ (a) $0.24IR - 1$ (b) $\dfrac{0.24I - 1}{R}$ (c) $0.24I - 1$ (d) $0.24I - R$	None of them.
150. Which of the following are equivalent to $\dfrac{vs}{vt_1 - vt_2}$? ______ (a) $\dfrac{s}{t_1 - t_2}$ (b) $\dfrac{s}{t_1} - \dfrac{s}{t_2}$ (c) $\dfrac{s}{t_1 - vt_2}$ (d) $\dfrac{s}{t_1} - \dfrac{1}{t_2}$	Only (c)
151. Which of the following are equivalent to $\dfrac{R_1 - R_2}{R_1 + R_t}$? ______ (a) $-\dfrac{R_2}{R_t}$ (b) $\dfrac{1 - R_2}{1 + R_t}$ (c) $\dfrac{1}{R} - \dfrac{R_2}{R_1 + R_t}$ (d) $\dfrac{R_1}{R_1 + R_t} - \dfrac{R_2}{R_1 + R_t}$	Only (a)
	Only (d)

SELF-TEST 6 (Frames 119-151)

1. Which of the following are additions or subtractions of two fractions? ________

(a) $\frac{r + 2w}{3}$ (b) $\frac{h - 2t}{d + a}$ (c) $\frac{k}{E + e}$ (d) $\frac{4w}{3h - 2c}$ (e) $\frac{s + v}{p}$

In Problems 2 and 3 below, the fraction on the right is equivalent to the fraction on the left. What instance of $\frac{n}{n}$ was used to generate the fraction on the right?

2. $\frac{c}{b - at} = \frac{ch}{bh - aht}$ ________

3. $\frac{d + 2}{2d - 3} = \frac{4d + 8}{8d - 12}$ ________

Reduce each of the following to lowest terms. Write each result as a single fraction:

4. $\frac{2t}{tw - 3t} =$ ________

5. $\frac{6}{2h + 4} =$ ________

6. $\frac{AB}{B - AB} =$ ________

7. $\frac{4k - 2}{2} =$ ________

8. $\frac{2ar + arw}{2ar} =$ ________

9. $\frac{rR}{IR - Ir} =$ ________

ANSWERS:

1. (a), (b), (e)
2. $\frac{h}{h}$
3. $\frac{4}{4}$
4. $\frac{2}{w - 3}$
5. $\frac{3}{h + 2}$
6. $\frac{A}{1 - A}$
7. $\frac{2k - 1}{1}$ or $2k - 1$
8. $\frac{2 + w}{2}$
9. Cannot be reduced.

8-12 SIMPLER FORMS OF LITERAL FRACTIONS WHOSE NUMERATOR OR DENOMINATOR CONTAINS A FRACTION

When dealing with formulas, you will encounter complicated fractions which contain fractions in their numerators or denominators or both. Here are some examples:

$$\frac{1 + \frac{a}{b}}{2a} \qquad \frac{\frac{m}{q}}{1 - m} \qquad \frac{1 + \frac{1}{a}}{1 + \frac{1}{b}}$$

Any fraction of this type can be written as a single fraction which does not contain a fraction in its numerator or denominator. We will show the method in this section.

152. The complex fraction below contains a fraction in its numerator. To write it as a single fraction without a fraction in its numerator, we need two steps:

Step 1: Performing the addition in the numerator:

$$\frac{1+\frac{m}{q}}{t} = \frac{\frac{q+m}{q}}{t}$$

Step 2: Dividing the new numerator by the denominator: (Note: We multiply the new numerator by the reciprocal of the denominator.)

$$\frac{1+\frac{m}{q}}{t} = \frac{\frac{q+m}{q}}{t} = \left(\frac{q+m}{q}\right)\left(\frac{1}{t}\right) = \underline{\qquad\qquad}$$

Answer: $\frac{q+m}{qt}$

153. When simplifying to eliminate a fraction in a numerator or denominator, you must frequently perform divisions like the following:

$$\frac{\frac{m}{t}}{\frac{a+b}{c}} \qquad \frac{\frac{a}{m}}{1-q}$$

When performing these divisions, we convert to multiplication by multiplying the numerator by the reciprocal of the denominator. Let's examine the reciprocals of each denominator above.

The reciprocal of $\frac{a+b}{c}$ is $\frac{c}{a+b}$, since:

$$\left(\frac{a+b}{c}\right)\left(\frac{c}{a+b}\right) = \frac{c(a+b)}{c(a+b)} = 1$$

The reciprocal of $1-q$ is $\frac{1}{1-q}$, since:

$$(1-q)\left(\frac{1}{1-q}\right) = \frac{1-q}{1-q} = 1$$

Write the reciprocals of each of these expressions:

(a) $\frac{m-p}{R}$ ________ (c) $\frac{1}{V-2d}$ ________

(b) $a-c$ ________ (d) $\frac{F-S}{T+V}$ ________

Answers: a) $\frac{R}{m-p}$ b) $\frac{1}{a-c}$ c) $V-2d$ d) $\frac{T+V}{F-S}$

154. Convert each of the following divisions to a multiplication:

(a) $\frac{\frac{V}{d}}{\frac{p+q}{r}} = (\quad)(\quad)$ (c) $\frac{\frac{t}{m}}{\frac{1}{b-c}} = (\quad)(\quad)$

(b) $\frac{\frac{a-c}{b}}{x-y} = (\quad)(\quad)$

155. When performing multiplications like those in the last frame, we can write the product in either of two forms:

$$\left(\frac{V}{d}\right)\left(\frac{r}{p+q}\right) = \frac{Vr}{d(p+q)} \quad \text{or} \quad \frac{Vr}{dp+dq}$$

In the middle fraction above, the denominator is an instance of the distributive principle. We multiplied by the distributive principle to get the denominator on the extreme right.

Ordinarily we leave the denominator in factored form. That is, we do not multiply by the distributive principle. Therefore, when performing the multiplication above, which of the two forms of the product would we use? ________

Answers to frame 154:

a) $\left(\frac{V}{d}\right)\left(\frac{r}{p+q}\right)$

b) $\left(\frac{a-c}{b}\right)\left(\frac{1}{x-y}\right)$

c) $\left(\frac{t}{m}\right)(b-c)$

---

156. Perform these multiplications and write each product in the preferred form:

(a) $\left(\frac{a-c}{b}\right)\left(\frac{1}{x-y}\right) =$ ________

(b) $\left(\frac{t}{m}\right)(b-c) =$ ________

(c) $\left(\frac{R}{1-Q}\right)\left(\frac{1-T}{M}\right) =$ ________

Answer to frame 155: $\frac{Vr}{d(p+q)}$

---

157. Perform each of the following divisions. Write each answer in the preferred form:

(a) $\dfrac{m+q}{\frac{a}{b}} = (\quad)(\quad) =$ ________

(b) $\dfrac{\frac{c+d}{m}}{\frac{c-d}{t}} = (\quad)(\quad) =$ ________

(c) $\dfrac{\frac{c+d}{c}}{1+q} = (\quad)(\quad) =$ ________

Answers to frame 156:

a) $\frac{a-c}{b(x-y)}$

b) $\frac{t(b-c)}{m}$

c) $\frac{R(1-T)}{M(1-Q)}$

---

158. Here is the general strategy for simplifying complex fractions to obtain a single fraction which does not contain a fraction in either its numerator or denominator.

Step 1: If the complex fraction has a fractional addition or subtraction in either the numerator or denominator, perform this addition or subtraction first.

Step 2: Then perform the division.

Example: $\dfrac{a}{1+\frac{c}{b}} = \dfrac{a}{\frac{b+c}{b}} =$ ________

Answers to frame 157:

a) $(m+q)\left(\frac{b}{a}\right) = \frac{b(m+q)}{a}$

b) $\left(\frac{c+d}{m}\right)\left(\frac{t}{c-d}\right) = \frac{t(c+d)}{m(c-d)}$

c) $\left(\frac{c+d}{c}\right)\left(\frac{1}{1+q}\right) = \frac{c+d}{c(1+q)}$

Answer to frame 158: $\frac{ab}{b+c}$

159. To simplify the fraction below, you must perform the subtraction in the denominator first. Do so, and then perform the division:

$$\frac{g}{d - \frac{g}{h}} = \underline{\qquad\qquad} = \underline{\qquad\qquad\qquad\qquad}$$

$$\frac{g}{\frac{dh - g}{h}} = \frac{gh}{dh - g}$$

160. To simplify the fraction below, you must perform the addition in the numerator and denominator first. Do so, and then perform the division.

$$\frac{1 + \frac{1}{a}}{1 + \frac{1}{b}} = \underline{\qquad\qquad} = \underline{\qquad\qquad\qquad\qquad}$$

$$\frac{\frac{a + 1}{a}}{\frac{b + 1}{b}} = \frac{b(a + 1)}{a(b + 1)}$$

161. To simplify the fraction below, you do not have to perform any addition or subtraction. You can perform the division immediately. Do so:

$$\frac{\frac{a}{b}}{1 - a} = \underline{\qquad\qquad\qquad\qquad}$$

$$\frac{a}{b(1 - a)}$$

162.

Simplify: (a) $\frac{m - \frac{V}{T}}{R} = \underline{\qquad\qquad\qquad\qquad}$

(b) $\frac{p + \frac{x}{y}}{p - \frac{q}{r}} = \underline{\qquad\qquad\qquad\qquad}$

---

a) $\frac{mT - V}{RT}$

b) $\frac{r(py + x)}{y(pr - q)}$

SELF-TEST 7 (Frames 152-162)

Simplify each of the following fractions. Write each result as a single fraction which does not contain any fraction in its numerator or denominator:

1. $\dfrac{\frac{h}{2} + \frac{r}{d}}{t} =$ ______

2. $\dfrac{\frac{a}{2} - 1}{2a} =$ ______

3. $\dfrac{\frac{d}{h}}{1 - d} =$ ______

4. $\dfrac{r}{1 - \frac{b}{w}} =$ ______

5. $\dfrac{\frac{p-2}{s}}{\frac{1}{s}} =$ ______

6. $\dfrac{\frac{h}{k} + 1}{\frac{h}{k}} =$ ______

7. $\dfrac{b - \frac{b}{2}}{a - \frac{a}{2}} =$ ______

8. $\dfrac{1 + \frac{w}{r}}{1 - \frac{w}{r}} =$ ______

ANSWERS:

1. $\dfrac{dh + 2r}{2dt}$
2. $\dfrac{a - 2}{4a}$
3. $\dfrac{d}{h(1 - d)}$
4. $\dfrac{rw}{w - b}$
5. $\dfrac{p - 2}{1}$ or $p - 2$
6. $\dfrac{h + k}{h}$
7. $\dfrac{b}{a}$
8. $\dfrac{r + w}{r - w}$

8-13 "CANCELING" AND REDUCING TO LOWEST TERMS

When confronted with a complicated fraction, many students have an overwhelming temptation to "cancel." They do so in order to simplify, even when the "canceling" is not legitimate. In this section, we will briefly review the principles which must be used to simplify a complicated fraction or write it in an equivalent form.

163. Here is one of the most common errors with fractions. When dealing with a fraction whose numerator is a sum (like the one below), students frequently "cancel" the 2's and get:

$$\frac{x+2}{2} = \frac{x+\overset{1}{\cancel{2}}}{\cancel{2}} = x+1$$

Of course, x + 1 is simpler than $\frac{x+2}{2}$, and anyone likes to change something complicated to something simpler. However in this case, the following statement is clearly false:

$$\frac{x+2}{2} = x+1$$

We can easily show that it is false by plugging in the same number for "x" on each side. If we plug in "4" for "x" on each side:

The value of $\frac{x+2}{2}$ (the left side) is $\frac{\boxed{4}+2}{2}$, which is 3.

The value of x + 1 (the right side) is $\boxed{4}+1$, which is 5.

Since $3 \neq 5$, obviously $\frac{x+2}{2} \neq x+1$.

---

164. Here is another example of the same error. The student merely canceled the a's to get a simpler expression:

$$\frac{a+b}{a} = \frac{\overset{1}{\cancel{a}}+b}{\underset{1}{\cancel{a}}} = 1+b$$

Again we can easily show that this type of canceling is an error by plugging in some numbers for the letters. If we plug in 3 for "a" and 9 for "b":

(a) $\frac{a+b}{a} =$ ________ (b) $1+b =$ ________

Answers: a) $\frac{3+9}{3} = \frac{12}{3} = \underline{\underline{4}}$ b) $1+9 = \underline{\underline{10}}$

165. The point we are trying to make is this:

> When you see a complicated fraction, you cannot always "cancel" to make things simpler for yourself. Sometimes you just have to live with the complicated fraction.

Though we cannot "cancel" to simplify the fractions in the last two frames, we can break them apart and write them in equivalent forms.

$$\frac{x+2}{2} = \frac{x}{2} + \frac{2}{2} = \frac{x}{2} + 1$$

$$\frac{a+b}{a} = \_\_\_ + \_\_\_ = \_\_\_ + \_\_\_$$

Answer: $\frac{a}{a} + \frac{b}{a} = 1 + \frac{b}{a}$

166. Students are tempted to "cancel" the 2's in the fraction below. They get:

$$\frac{x-2}{2} = \frac{x - \overset{1}{\not{2}}}{\underset{1}{\not{2}}} = x - 1$$

By plugging in a 10 for "x" on both the left side and the right side, we can again see that this type of "canceling" is an error:

$$\frac{x-2}{2} = \frac{10-2}{2} = \frac{8}{2} = \underline{4} \qquad x - 1 = 10 - 1 = \underline{9}$$

Though we cannot cancel the 2's, we can write the original fraction in equivalent forms by breaking it into two fractions:

$$\frac{x-2}{2} = \frac{x}{2} - \frac{2}{2} = \frac{x}{2} - 1$$

Write this fraction in two equivalent forms:

$$\frac{a-b}{b} = \underline{\qquad\qquad} = \underline{\qquad\qquad}$$

$\frac{a}{b} - \frac{b}{b} = \frac{a}{b} - 1$

167. Students are also tempted to cancel the 3's in the fraction below.

$$\frac{x+y+3}{3} = \frac{x + y + \overset{1}{\not{3}}}{\underset{1}{\not{3}}} = x + y + 1$$

By plugging in some number for "x" and "y", however, we can see that this type of "canceling" <u>is an error.</u>

If we plug in 6 for "x" and 9 for "y":

(a) $\frac{x+y+3}{3} = \frac{6+9+3}{3} = \underline{\qquad}$ (b) $x + y + 1 = 6 + 9 + 1 = \underline{\qquad}$

---

a) 6 b) 16

168. Though we cannot "cancel" the 3's in the fraction below, we can write it in equivalent forms by breaking it into three fractions:

$$\frac{x+y+3}{3} = \frac{x}{3} + \frac{y}{3} + \frac{3}{3} = \frac{x}{3} + \frac{y}{3} + 1$$

Write these in equivalent forms with each simpler fraction reduced to lowest terms:

(a) $\frac{a+b+c}{b} = \underline{\qquad\qquad}$ (b) $\frac{mx+ky+m}{m} = \underline{\qquad\qquad}$

---

a) $\frac{a}{b} + 1 + \frac{c}{b}$

b) $x + \frac{ky}{m} + 1$

169. With some fractions which contain an addition or subtraction as their denominators, there is also a temptation to "cancel." We have done so below:

$$\frac{5}{x+5} = \frac{\overset{1}{\cancel{5}}}{x+\underset{1}{\cancel{5}}} = \frac{1}{x+1}$$

$$\frac{p}{p-q} = \frac{\overset{1}{\cancel{p}}}{\underset{1}{\cancel{p}}-q} = \frac{1}{1-q}$$

We can see that this type of "canceling" is an error if we plug in numbers for the letters.

If we plug in "10" for the "x" in the fractions above:

$$\frac{5}{x+5} = \frac{5}{10+5} = \frac{5}{15} = \frac{1}{3} \qquad \frac{1}{x+1} = \frac{1}{10+1} = \frac{1}{11}$$

If we plug in 10 for "p" and 5 for "q" in the fractions above:

$$\frac{p}{p-q} = \frac{10}{10-5} = \frac{10}{5} = 2 \qquad \frac{1}{1-q} = \frac{1}{1-5} = \frac{1}{-4} = -\frac{1}{4}$$

Each of the fractions above contains an addition or subtraction <u>only in its denominator.</u> Can we break fractions of this type apart and write them in equivalent forms? ______

Answer: No. Only fractions with an addition or subtraction in their numerators can be "broken apart."

170. The examples of "canceling" shown in the last few frames <u>were all incorrect.</u> There is, however, a correct method of "canceling." This correct method will now be discussed.

"Canceling" is merely a short way of reducing a fraction to lowest terms. Since reducing to lowest terms is only possible when we can factor out an instance of $\frac{n}{n}$, "canceling" is merely a shortcut for "factoring out an instance of $\frac{n}{n}$". This type of factoring is possible <u>only when each term in both the numerator and denominator contains a common factor.</u> For example:

$$\frac{a+ab}{a} = \frac{a(1+b)}{a} = \left(\frac{a}{a}\right)(1+b) = 1+b$$

$$\frac{m}{cm-m} = \frac{m(1)}{m(c-1)} = \left(\quad\right)\left(\quad\right) = \underline{\qquad\qquad}$$

Answer: $\left(\frac{m}{m}\right)\left(\frac{1}{c-1}\right) = \frac{1}{c-1}$

171. We can "cancel" only when we can factor out an instance of $\frac{n}{n}$. Here is a case in which "canceling" of this type is possible:

$$\frac{c(p+q)}{c(v-t)} = \left(\frac{c}{c}\right)\left(\frac{p+q}{v-t}\right) = \frac{p+q}{v-t}$$

Complete this one in which "canceling" is possible:

$$\frac{a(x-y)}{b(x-y)} = \left(\frac{a}{b}\right)\left(\frac{x-y}{x-y}\right) = \underline{\qquad}$$

Answer: $\frac{a}{b}$

172. Here is a case in which "canceling" is possible because we can factor the numerator and obtain an instance of $\frac{n}{n}$.

$$\frac{bt + bR}{b(m - p)} = \frac{b(t + R)}{b(m - p)} = \frac{t + R}{m - p}$$

Here is a case in which "canceling is not possible even though you might be tempted to cancel the "c's".

$$\frac{c + d}{c(1 + a)}$$

We cannot cancel since "c" is not a factor in the numerator, and therefore, we cannot obtain an instance of $\frac{n}{n}$.

Simplify each, if possible: (a) $\frac{m(p + q)}{m(p - q)} =$ ________________

(b) $\frac{tx + tz}{t(1 - z)} =$ ________________

(c) $\frac{V + R}{V(R - 1)} =$ ________________

---

173. Here is a case in which students are tempted to cancel the "h's":

$$\frac{h + \frac{r}{a}}{h}$$

Since "h" is not a factor in the numerator, this "canceling" is an error. However, we can simplify the fraction above and write it without a fraction in its numerator. Do so by performing the indicated division:

Answers to 172:

a) $\frac{p + q}{p - q}$

b) $\frac{x + z}{1 - z}$

c) Not possible. The fraction is in simplest form.

---

174. When simplifying a fraction, be alert for opportunities to reduce to lowest terms or "cancel." The fraction at the left below can be reduced to lowest terms. Do so:

$$\frac{m}{m - \frac{m}{p}} = \frac{m}{\frac{mp - m}{p}} = \frac{mp}{mp - m} = \_\_\_\_\_\_\_\_\_\_\_\_$$

Answer to 173: $1 + \frac{r}{ah}$ or $\frac{ah + r}{ah}$

Answer to 174: $\frac{p}{p - 1}$

175. Simplify each of these. Reduce to lowest terms whenever possible:

(a) $\dfrac{\dfrac{c+d}{a}}{\dfrac{c-d}{a}} =$ ____________

(b) $\dfrac{\dfrac{y-1}{x}}{\dfrac{y-1}{z}} =$ ____________

(c) $\dfrac{y - \dfrac{2y}{a}}{3y} =$ ____________

Answer to Frame 175: a) $\dfrac{c+d}{c-d}$ b) $\dfrac{z}{x}$ c) $\dfrac{a-2}{3a}$

SELF-TEST 8 (Frames 163-175)

Simplify each of the following, if possible:

1. $\dfrac{kv}{v(1-a)} =$ ____________	2. $\dfrac{p+2r}{p(r+1)} =$ ____________
3. $\dfrac{t - \dfrac{w}{p}}{t} =$ ____________	4. $\dfrac{w(f+2)}{w+2} =$ ____________
5. $\dfrac{2r}{r - \dfrac{1}{2}} =$ ____________	6. $\dfrac{\dfrac{h+1}{m}}{\dfrac{1}{h}+1} =$ ____________

ANSWERS: 1. $\dfrac{k}{1-a}$ 2. Cannot be simplified. 3. $\dfrac{pt-w}{pt}$ 4. Cannot be simplified. 5. $\dfrac{4r}{2r-1}$ 6. $\dfrac{h}{m}$

Chapter 9 FORMULA REARRANGEMENT

Formulas are a mathematical way of representing the laws of the physical world. Any scientist or technician must be skilled in formula rearrangement since formula rearrangement is frequently used in science and technology. Three of the common uses are:

(1) to make formula evaluations easier,
(2) to eliminate variables from systems of formulas and thereby derive new formulas,
(3) to write a formula in an alternate way so that a different meaning can be emphasized.

Formula rearrangement is an algebraic process by which we rearrange a formula so that a new variable is "solved for." For example:

We rearrange $I = \frac{E}{R}$ to get $E = IR$

We rearrange $P_1V_1 = P_2V_2$ to get $V_2 = \frac{P_1V_1}{P_2}$

We rearrange $F = \frac{mv^2}{r}$ to get $r = \frac{mv^2}{F}$

The same algebraic principles that were used in solving equations are used in formula rearrangement. In this chapter, we will review those principles, generalize them to literal expressions, and then use them to rearrange formulas.

9-1 THE MEANING OF LETTERS IN FORMULAS

In this section, we will review the fact that letters in formulas are usually variables representing physical quantities. We will discuss the use of capital and small letters in formulas and the use of subscripts on letters in formulas.

---

1. The laws of the physical world express relationships among physical quantities. Formulas are a mathematical way of stating these laws. Most letters which appear in formulas are abbreviations for physical quantities. Since these physical quantities can have various numerical values, the letters which represent them in formulas are <u>variables</u> mathematically.

   For example, the formula $\boxed{s = vt}$ is a concise way of stating the relationship between distance traveled (s), velocity (v), and time (t) for a moving object. Distance traveled, velocity, and time are physical quantities whose numerical values can vary. Since each of these three physical quantities is a variable quantity, the three letters which represent them in the formula are called ______________.

---

variables

2. When a formula is written, letters are used as abbreviations for physical quantities. Both capital and small letters are used. For example:

In $\boxed{P = \frac{Fs}{t}}$ "P" stands for "power"
"F" stands for "force"
"s" stands for "distance"
"t" stands for "time"

The formula above contains two capital letters and two small letters. The use of a capital or small letter to represent a physical quantity is purely arbitrary. However, once scientists or technicians decide to use a capital or small letter for a physical quantity, you must follow their convention. That is, capital letters in a formula should not be changed to small letters, and vice versa. Otherwise, what you write will be unclear and incorrect.

We have written the formula above in various ways below. Only one of them is correct. Which one is it? _______

(a) $p = \frac{Fs}{t}$ (b) $p = \frac{fs}{t}$ (c) $P = \frac{Fs}{t}$ (d) $P = \frac{Fs}{T}$

Answer to Frame 2: (c)

3. The formula below represents Newton's law of gravitational attraction between two bodies. Both the capital symbol and the small symbol for the same letter appear in the formula.

$\boxed{F = \frac{gMm}{r^2}}$ where "M" stands for the mass of one body
and "m" stands for the mass of the second body.

In this case, "M" and "m" stand for different physical quantities. In using this formula, you must be careful not to change the size of either letter or you will eliminate one of these quantities from the formula.

Go to next frame.

4. The formula below shows the load voltage of a battery or other power supply. The "e" and "E" stand for two different physical quantities.

$\boxed{e = E - Ir}$ where "e" stands for load voltage
and "E" stands for no-load voltage.

Could we rewrite the formula in either of these two ways? _______

$e = e - Ir$ or $E = E - Ir$

No. Otherwise one variable is eliminated.

5. Sometimes subscripts are used instead of using the same capital and small letters to distinguish two different variables. Here is an example:

$\boxed{R_t = R_1 + R_2}$

The formula above represents the total resistance of two resistors connected in series.

R_t stands for the total resistance.
R_1 stands for the resistance of one resistor.
R_2 stands for the resistance of the second resistor.

Note: The subscripts which distinguish the variables are written to the right and slightly below "R". Numbers or letters can be used as subscripts.

How many variables are there in the formula above? _________

6. This formula represents Boyle's law for gases: $\boxed{\dfrac{P_1}{P_2} = \dfrac{V_2}{V_1}}$

P_1 and V_1 stand for the initial pressure and volume.

P_2 and V_2 stand for the final pressure and volume.

The subscripts are used to distinguish the initial and final conditions. Therefore, how many variables are there in the formula? ______

Three

---

7. Formulas contain both subscripts and exponents. A subscript is written slightly below a letter. An exponent is written slightly above a letter.

In $\boxed{\dfrac{Kq_1q_2}{s^2}}$

(1) Both "q's" have subscripts. They are used to distinguish between two variables.

(2) "s" has an exponent. s^2 means: s times s or (s)(s).

Are there any exponents in $\dfrac{F_1}{F_2} = \dfrac{d_1}{d_2}$? ______

Four

---

8. In $\boxed{H = 0.24I^2Rt}$ is the little "2" in the formula a subscript or an exponent? ______

No. All of the numbers are subscripts.

---

9. In $\boxed{\dfrac{P_1V_1}{T_1} = \dfrac{P_2V_2}{T_2}}$ which variables are squared? ______

An exponent

---

10. How many variables are there in each formula below:

(a) $I_E = I_C + I_B$ ______

(b) $R_t = R_1 + R_2 + R_3$ ______

(c) $\dfrac{P_1V_1}{T_1} = \dfrac{P_2V_2}{T_2}$ ______

(d) $e = E - Ir$ ______

None of them. All of the numbers are subscripts.

Answer to Frame 10: a) 3 ("E", "C", and "B" are subscripts) b) 4 c) 6 d) 4 ("r" is not a subscript)

9-2 THE NEED FOR FORMULA REARRANGEMENT IN FORMULA EVALUATION

As mentioned in the last section, there are various needs for formula rearrangement in science and technology. We cannot talk about all of these needs at once. The need for formula rearrangement in derivations, for example, will not be discussed until the next chapter. In this section, we will discuss the need for formula rearrangement in formula evaluation.

11. If we know specific numerical values for all variables except one in a formula, we can find the corresponding numerical value of that one variable. This process of finding the corresponding value of that one variable is called FORMULA EVALUATION. For example:

The formula for the area of a rectangle is $\boxed{A = LW}$

If we know that the length (L) = 10 in. and the width (W) = 5 in., we can find the area (A) by plugging these numbers in the formula. We get:

$$A = (10 \text{ in.})(5 \text{ in.}) = 50 \text{ sq. in.}$$

What is the value of A if L = 20 in. and W = 10 in.? A = ________

Answer: A = 200 sq. in.

---

12. Let's do an evaluation with the formula for the amount of current in an electrical circuit: $\boxed{I = \frac{E}{R}}$

If the voltage (E) is 12 volts and the resistance (R) is 4 ohms, we can find the amount of current (I) by plugging in these values. We get:

$$I = \frac{12 \text{ volts}}{4 \text{ ohms}} = 3 \text{ amperes}$$

What is the value of I if E = 120 volts and R = 12 ohms?
I = ______ amperes

Answer: 10 amperes

---

13. Let's do one with the formula for centripetal force: $\boxed{F = \frac{mv^2}{r}}$
(From this point on, we will drop the "units" for each variable, since we are only interested in the calculation required.)

If m = 100, v = 5, and r = 20, we can find the value of F by plugging in these numbers. We get:

$$F = \frac{(100)(5^2)}{20} = \frac{(100)(25)}{20} = \frac{2,500}{20} = 125$$

What is the value of F if m = 200, v = 10, and r = 25? F = ________

Answer: F = 800

---

14. The evaluations done in the last three frames were easy since we were looking for the value of the "solved for" variable.

A variable is "solved for" in a formula if:

(1) It is alone on one side.
(2) All of the other variables and constants are on the other side.

In A = LW, for example, A is "solved for" since:

(1) "A" is alone on the left side.
(2) The other variables (L and W) are on the right side.

(a) Which variable is solved for in $H = 0.24I^2Rt$? ______

(b) Which variable is solved for in $P = \frac{Fs}{t}$? ______

Answer:
a) H
b) P

15. (a) In $H = ms(t_2 - t_1)$, is H "solved for?" ________

(b) In $P_1V_1 = P_2V_2$, is V_1 "solved for?" ________

(c) In $0.24R = \frac{H}{I^2t}$, is R "solved for?" ________

---

16. When doing a formula evaluation, we sometimes have to find the value of a variable which is not "solved for." Here is an example using the formula: $\boxed{A = LW}$

If we know that $A = 100$ and $L = 20$, we can find the value of W by plugging in these values. We get:

$$100 = 20W$$

To find the value of W, we have to solve the basic equation above. We get: W = ______

Answers to 15:

a) Yes

b) No, since P_1 also appears on the left side.

c) No, since 0.24 also appears on the left side.

---

17. Here is another case in which we must find the value of a variable which is not "solved for."

Using the formula $\boxed{I = \frac{E}{R}}$: If we know that $I = 12$ and $E = 120$, we can find the value of R by plugging in these values. We get:

$$12 = \frac{120}{R}$$

To find the value of R, we must solve the fractional equation above. We get: R = ______

Answer to 16: $W = 5$

---

18. Here is another example using the formula: $\boxed{F = \frac{mv^2}{r}}$

If we know that $F = 500$, $v = 10$, and $r = 5$, we can find the value of "m" by plugging in these values. We get:

$$500 = \frac{m(10^2)}{5} \quad \text{or} \quad 500 = \frac{100m}{5}$$

To find the value of "m", we must solve the fractional equation on the right. We get: m = ______

Answer to 17: $R = 10$

Answer to 18: $m = 25$

19. If we are looking for the value of a "solved for" variable, we do not have to solve an equation to find its value. We simply evaluate one side of the equation. For example:

$$A = (10)(5) \qquad \text{or} \qquad F = \frac{(100)(5^2)}{20}$$

If we are looking for the value of a variable which is not "solved for," we must solve an equation to find its value. For example:

$$100 = 20W \qquad \text{or} \qquad 500 = \frac{m(10^2)}{5}$$

Here is a case in which a rather difficult fractional equation must be solved in order to find the value of a variable which is not "solved for." The formula is from transistor electronics.

In $\boxed{A = \frac{B}{B+1}}$ find the value of B if $A = 0.98$.

B = ________

B = 49, since:

$$0.98 = \frac{B}{B+1}$$

$$0.98B + 0.98 = B$$

$$0.98 = 0.02B$$

$$B = \frac{0.98}{0.02} = 49$$

20. To avoid the necessity of solving equations when doing evaluations, scientists and technicians <u>always</u> rearrange formulas first. The purpose of the rearrangement is to solve for a variable which is not "solved for" in the original formula. For example:

Before finding the value of "W" in $A = LW$, they rearrange $A = LW$ to get $W = \frac{A}{L}$. (Note: W is "solved for" in the new formula.)

Before finding the value of "B" in $A = \frac{B}{B+1}$, they rearrange $A = \frac{B}{B+1}$ to $B = \frac{A}{1-A}$. (Note: B is "solved for" in the new formula.)

After a variable is "solved for," we can find its value without solving an equation. Let's solve the problem from the last frame with the new formula above:

In $\boxed{B = \frac{A}{1-A}}$, find the value of B if $A = 0.98$.

Answer to Frame 20: $B = \frac{0.98}{1 - 0.98} = \frac{0.98}{0.02} = 49$

9-3 REVIEW OF BASIC ALGEBRAIC PRINCIPLES

The purpose of this chapter is to teach formula rearrangement. The same principles and axioms which were used to solve one-letter equations are used to rearrange formulas. In this section, we will review some of these basic algebraic principles and begin to make the transition to using them with letters.

21. THE IDENTITY PRINCIPLE OF MULTIPLICATION: If any quantity is multiplied by +1, the product is the original quantity.

In symbols, we write the principle this way: $\square(1) = \square$

The $\square$ means that this principle is true for any expression.

For example: $(x + y)(1) = x + y$, and $(1)\left(\frac{p}{q}\right) = \frac{p}{q}$.

Complete the following:

(a) $(1)(x - 1) =$ _____ (b) $(1)(x + y + z) =$ _____ (c) $\frac{a}{m}(1) =$ _____

---

22. IF ANY QUANTITY IS DIVIDED BY ITSELF, THE QUOTIENT IS "+1".

Answers to 21: a) $x - 1$ b) $x + y + z$ c) $\frac{a}{m}$

In symbols, we write the principle this way: $\frac{\square}{\square} = 1$

The $\square$ again means that this principle is true for any expression or quantity.

Complete this one: $\frac{ab}{ab} = \left(\frac{a}{a}\right)\left(\frac{b}{b}\right) = (\quad)(\quad) =$ _____

---

23. Complete: (a) $\frac{V(P + Q)}{V(P + Q)} = \left(\frac{V}{V}\right)\left(\frac{P + Q}{P + Q}\right) = (\quad)(\quad) =$ _____

(b) $\left(\frac{cd}{cd}\right)\left(\frac{t + 1}{t + 1}\right) = (\quad)(\quad) =$ _____

Answer to 22: $(1)(1) = 1$

---

24. Complete: (a) $\frac{a + b + c}{a + b + c} =$ _____ (b) $\frac{m + n - 7}{m + n - 7} =$ _____

Answers to 23: a) $(1)(1) = 1$ b) $(1)(1) = 1$

---

25. Complete: (a) $\left(\frac{R}{R}\right)(S) = (\quad)(S) =$ _____

(b) $\left(\frac{x + 1}{x + 1}\right)(d) = (\quad)(d) =$ _____

Answers to 24: a) 1 b) 1

---

26. Complete:

(a) $\left(\frac{a + b}{a + b}\right)(xy) =$ _____ (b) $\left(\frac{mn}{mn}\right)\left(\frac{1}{c}\right) =$ _____ (c) $\left(\frac{FG}{D}\right)\left(\frac{m - p}{m - p}\right) =$ _____

Answers to 25: a) $(1)(S) = S$ b) $(1)(d) = d$

Answers to 26: a) xy b) $\frac{1}{c}$ c) $\frac{FG}{D}$

27. In formulas, you will encounter expressions like v^2 and I^2.

Just as 5^2 means (5)(5) and 5^3 means (5)(5)(5),
v^2 means (v)(v) and v^3 means (v)(v)(v).

Complete: (a) (m)(m) = ____ (b) (p)(p)(p) = ____ (c) (t)(t) = ____

Answers: a) m^2 b) p^3 c) t^2

28. (a) $1(x^2) =$ ____ (c) $(v^2 - 1)(1) =$ ____ (e) $\left(\frac{F^2}{F^2}\right)(k) =$ ____

(b) $\frac{R^2}{R^2} =$ ____ (d) $\frac{b^2 - 1}{b^2 - 1} =$ ____ (f) $\left(\frac{1}{t}\right)\left(\frac{a^2 + 4}{a^2 + 4}\right) =$ ____

Answers: a) x^2 b) 1 c) $v^2 - 1$ d) 1 e) k f) $\frac{1}{t}$

29. Here is the definition for the multiplication of two fractions:

$$\left(\frac{\square}{\bigcirc}\right)\left(\frac{\triangle}{\square}\right) = \frac{(\square)(\triangle)}{(\bigcirc)(\square)}$$

Example: $\left(\frac{3h}{2r}\right)\left(\frac{t + 2}{w}\right) = \frac{3h(t + 2)}{2rw}$

Perform the following multiplications of fractions:

(a) $\left(\frac{m}{n}\right)\left(\frac{y^2}{x}\right) =$ ____ (b) $\left(\frac{VT}{FG}\right)\left(\frac{R}{Q}\right) =$ ____ (c) $\left(\frac{D}{BE}\right)\left(\frac{T + R}{S^2}\right) =$ ____

Answers: a) $\frac{my^2}{nx}$ b) $\frac{VTR}{FGQ}$ or $\frac{RTV}{FGQ}$ c) $\frac{D(T + R)}{BES^2}$

30. Do the following. Be sure to reduce to lowest terms if possible.

(a) $\left(\frac{cd}{e}\right)\left(\frac{f}{ch}\right) =$ ____

(b) $\left(\frac{p(q + 1)}{r}\right)\left(\frac{r}{p}\right) = \frac{p(q + 1)(r)}{(r)(p)} =$ ____

(c) $\left(\frac{1}{s(t - 3)}\right)\left(\frac{t - 3}{m}\right) =$ ____

Answers: a) $\frac{df}{eh}$ b) $q + 1$ c) $\frac{1}{sm}$

31. <u>IF ANY EXPRESSION IS DIVIDED BY "+1", THE QUOTIENT IS THE ORIGINAL EXPRESSION.</u>

In symbols, we write the principle this way: $\frac{\square}{1} = \square$

The $\square$ means that this principle is true for <u>any</u> expression or quantity.

(a) $\frac{T + C}{1} =$ ____ (c) $\frac{a(b + c)}{1} =$ ____

(b) $\frac{x - y}{1} =$ ____ (d) $\frac{a(b + c)(d - e)}{1} =$ ____

Answers: a) $T + C$ b) $x - y$ c) $a(b + c)$ d) $a(b + c)(d - e)$

32. The principle $\frac{\boxed{}}{1} = \boxed{}$ is useful in multiplying a fractional expression by a non-fractional expression. We use it to convert such a multiplication into a multiplication of two fractions.

For example: $\frac{1}{ab}(cd) = \frac{1}{ab}\left(\frac{cd}{1}\right) = \frac{cd}{ab}$

Complete: $\frac{md^2}{p}(r) = \left(\frac{md^2}{p}\right)\left(\frac{r}{1}\right) =$ ______

Answer: $\frac{md^2r}{p}$

33. Multiply. Be sure all answers are reduced to lowest terms.

(a) $\frac{1}{VT}(ST) =$ ______ (b) $\frac{r}{pq}(qt) =$ ______ (c) $(xy)\left(\frac{av}{xw}\right) =$ ______

Answer: a) $\frac{S}{V}$ b) $\frac{rt}{p}$ c) $\frac{avy}{w}$

34. Complete: $\left(\frac{1}{a(b + c)}\right)(x + y) = \left(\frac{1}{a(b + c)}\right)\left(\frac{x + y}{1}\right) =$ ______

Answer: $\frac{x + y}{a(b + c)}$

35. Multiply: (a) $\frac{1}{pq}(c + d) =$ ______ (c) $\frac{1}{v(t + 1)}[v(s - 2)] =$ ______

(b) $\frac{1}{s + t}[m(s + t)] =$ ______

Answer: a) $\frac{c + d}{pq}$ c) $\frac{s - 2}{t + 1}$ b) m

36. Multiply: (a) $\frac{1}{2}(bh) =$ ______ (c) $\left(\frac{1}{\pi d(F_1 - F_2)}\right)(33{,}000H) =$ ______

(b) $\left(\frac{1}{R - r}\right)(2ab) =$ ______ (d) $\left(\frac{1}{AK(t_2 - t_1)}\right)(LH) =$ ______

Answer: a) $\frac{bh}{2}$ c) $\frac{33{,}000H}{\pi d(F_1 - F_2)}$ b) $\frac{2ab}{R - r}$ d) $\frac{LH}{AK(t_2 - t_1)}$

37. TWO EXPRESSIONS ARE A PAIR OF RECIPROCALS IF THEIR PRODUCT IS +1.

Since $\frac{1}{LW}(LW) = \frac{LW}{LW} = 1$: (a) The reciprocal of LW is ______.

(b) The reciprocal of $\frac{1}{LW}$ is ______.

Answer: a) $\frac{1}{LW}$ b) LW

38. Since $(c + d)\left(\frac{1}{c + d}\right) = \left(\frac{c + d}{1}\right)\left(\frac{1}{c + d}\right) = \frac{c + d}{c + d} = 1$

(a) The reciprocal of $(c + d)$ is ______.

(b) The reciprocal of $\frac{1}{c + d}$ is ______.

Answer: a) $\frac{1}{c + d}$ b) $c + d$

39. Since $s(m + n)\left(\frac{1}{s(m + n)}\right) = \left(\frac{s(m + n)}{1}\right)\left(\frac{1}{s(m + n)}\right) = \frac{s(m + n)}{s(m + n)} = 1$

(a) The reciprocal of $s(m + n)$ is ______.

(b) The reciprocal of $\frac{1}{s(m + n)}$ is ______.

Answer: a) $\frac{1}{s(m + n)}$ b) $s(m + n)$

40. In symbols: (1) The reciprocal of $\square$ is $\dfrac{1}{\square}$.

(2) The reciprocal of $\dfrac{1}{\square}$ is $\square$.

(Where $\square$ stands for an algebraic expression.)

Write the reciprocals of the following:

(a) LW _____ (b) abd _____ (c) $\dfrac{1}{ef}$ _____ (d) $\dfrac{1}{x^2}$ _____

---

41. Write the reciprocals of the following:

(a) $x - d$ _____ (c) $3a(s + t)$ _____

(b) $\dfrac{1}{x + y}$ _____ (d) $w(R - r)$ _____

Answers to 40: a) $\dfrac{1}{LW}$ b) $\dfrac{1}{abd}$ c) ef d) x^2

---

42. THE MULTIPLICATION AXIOM FOR EQUATIONS says this: If we multiply both sides of an equation by the same quantity, the new equation is equivalent to the original one. Here is the same principle in symbols:

If $\bigcirc = \square$,

then $(\triangle)(\bigcirc) = (\triangle)(\square)$.

Since a formula is an equation, this principle is true even when we multiply both sides of a formula by the same literal quantity. For example:

If $rs = m$, then $\left(\dfrac{1}{r}\right)(rs) = \left(\dfrac{1}{r}\right)(m)$.

If $A = LW$, then $\dfrac{1}{L}(A) = (\quad)(LW)$.

Answers to 41: a) $\dfrac{1}{x - d}$ b) $x + y$ c) $\dfrac{1}{3a(s + t)}$ d) $\dfrac{1}{w(R - r)}$

---

43. (a) If $2RF = W(R - r)$, then $\left(\dfrac{1}{R - r}\right)(2RF) = (\qquad)W(R - r)$.

(b) If $m = \dfrac{t}{c}$, then $c(m) = (\quad)\left(\dfrac{t}{c}\right)$.

Answer to 42: $\dfrac{1}{L}$

Answers to 43: a) $\left(\dfrac{1}{R - r}\right)$ b) (c)

9-4 IDENTIFYING TERMS IN FORMULAS

A formula is an equation. Just like an equation, each side of a formula contains one or more terms. In this section, we will review the types of terms and identify terms in formulas.

44. In a formula or equation, terms are separated by addition or subtraction symbols.

<u>Any single letter (with or without a subscript) is a term.</u>

In the formula: $I_E = I_C + I_B$

there is <u>one</u> term on the left side (I_E)
and <u>two</u> terms on the right side (I_C and I_B).

In the formula: $R = R_1 + R_2 + R_3$,

there are _______ terms on the right side.

45. <u>Any series of factors is one term</u>, no matter how many factors appear in the series.

F_1d_1 is a shorthand way of writing $(F_1)(d_1)$.

Though F_1d_1 is a series of <u>two</u> factors, it is <u>one</u> term.

$P_1V_1T_2$ is a shorthand way of writing $(P_1)(V_1)(T_2)$.

Though $P_1V_1T_2$ is a series of <u>three</u> factors, it is <u>one</u> term.

In $V_1T_2 = T_1V_2$: (a) There are _______ terms on the left side.
(b) There are _______ terms on the right side.

Answer to 44: three

46. In a formula or equation, terms are separated by addition ("+") or subtraction ("−") symbols.

In $T_j = T_a + \theta_T P_T$,
there are <u>two</u> terms (T_a and $\theta_T P_T$) on the right.
They are separated by a "+".

In $e = E - Ir$,
there are also <u>two</u> terms (E and Ir) on the right.
They are separated by a "−".

(a) In $F_1r_1 + F_2r_2 + F_3r_3 = 0$, there are _______ terms on the left.

(b) In $K = c_1a - c_2b$, there are _______ terms on the right.

Answers to 45:
a) one (V_1T_2)
b) one (T_1V_2)

Answers to 46:
a) three
b) two

47. A series of factors is one term, even if one of the factors is a grouping. (The "+" or "−" within a grouping does not separate two terms.) For example:

$a(b + c)$ is a two-factor expression. However, it is only one term.

In $\pi dR(F_1 - F_2)$: (a) There are how many factors? _______

(b) There are how many terms? _______

---

48. How many terms are contained in each of these expressions?

(a) $2t(a - t)(b - t_1)$ _______

(b) $t(V_2 - V_1)$ _______

(c) $tV_2 - tV_1$ _______

(d) $M[(V_1)^2 - (V_2)^2]$ _______

a) Four

b) One

---

49. Any single fraction is one term, no matter how complicated its numerator or denominator is.

All of the following fractions are one term:

$$\frac{F_1}{F_2} \qquad \frac{V_2 - V_1}{t} \qquad \frac{\pi dR(F_1 - F_2)}{33{,}000}$$

In $\frac{a + b}{c} + d$, "d" is not part of the fraction. Therefore, the expression $\frac{a + b}{c} + d$ contains how many terms? _______

a) One c) Two

b) One d) One

---

50. Draw a box around each term in the following expressions:

(a) $V_2 - at$

(b) $\frac{1}{D} + \frac{1}{d}$

(c) $F + g(r - r_1)$

(d) $\frac{T(V_1 - V_2)}{V_1(P_2 - P_1)}$

Two

---

51. In each of the following formulas, draw a box around each term:

(a) $e = E - Ir$

(b) $A = \frac{1}{2}bh$

(c) $M = \frac{WLX}{2} - \frac{WX^2}{2}$

(d) $H = ms(t_2 - t_1)$

a) $\boxed{V_2} - \boxed{at}$

b) $\boxed{\frac{1}{D}} + \boxed{\frac{1}{d}}$

c) $\boxed{F} + \boxed{g(r - r_1)}$

d) $\boxed{\frac{T(V_1 - V_2)}{V_1(P_2 - P_1)}}$

---

52. Draw a box around each term:

(a) $df + Df = Dd$

(b) $C_tC_2 = C_1C_2 - C_1C_t$

(c) $R(P + Q) = PQ$

(d) $m = \frac{1}{c} + t$

a) $\boxed{e} = \boxed{E} - \boxed{Ir}$

b) $\boxed{A} = \boxed{\frac{1}{2}bh}$

c) $\boxed{M} = \boxed{\frac{WLX}{2}} - \boxed{\frac{WX^2}{2}}$

d) $\boxed{H} = \boxed{ms(t_2 - t_1)}$

Answer to Frame 52: a) $\boxed{df} + \boxed{Df} = \boxed{Dd}$ b) $\boxed{C_tC_2} = \boxed{C_1C_2} - \boxed{C_1C_t}$ c) $\boxed{R(P + Q)} = \boxed{PQ}$ d) $\boxed{m} = \boxed{\frac{1}{c}} + \boxed{t}$

SELF-TEST 1 (Frames 44-52)

Draw a box around each term in the following formulas:

1. $P = A + T$
2. $R_t = aR_1 + bR_2$
3. $R_t = I(aR_1 + bR_2)$
4. $\frac{R_t}{I} = aR_1 + bR_2$
5. $a = \frac{v_2 - v_1}{t}$
6. $at = v_2 - v_1$
7. $4x = 12 - 7x$
8. $0.24I = \frac{E_o - e}{t}$

ANSWERS:

1. $\boxed{P} = \boxed{A} + \boxed{T}$
2. $\boxed{R_t} = \boxed{aR_1} + \boxed{bR_2}$
3. $\boxed{R_t} = \boxed{I(aR_1 + bR_2)}$
4. $\boxed{\frac{R_t}{I}} = \boxed{aR_1} + \boxed{bR_2}$
5. $\boxed{a} = \boxed{\frac{v_2 - v_1}{t}}$
6. $\boxed{at} = \boxed{v_2} - \boxed{v_1}$
7. $\boxed{4x} = \boxed{12} - \boxed{7x}$
8. $\boxed{0.24I} = \boxed{\frac{E_o - e}{t}}$

9-5 REARRANGING FORMULAS WITH ONE NON-FRACTIONAL TERM ON EACH SIDE

All of the following formulas contain one non-fractional term on each side:

$$A = LW \qquad P_1V_1 = P_2V_2 \qquad H = ms(t_2 - t_1)$$

In this section, we will rearrange formulas of this type in order to solve for one of the variables.

53. Here is a simple equation and a simple formula:

$$\boxed{11 = 7x} \qquad \boxed{A = LW}$$

To solve for "x" or "W", we use the multiplication axiom.

To solve for "x" in $11 = 7x$, we multiply both sides by $\frac{1}{7}$, the reciprocal of the coefficient of x.

To solve for "W" in $A = LW$, we multiply both sides by $\frac{1}{L}$, the reciprocal of the coefficient of W.

The two solutions are shown below:

$$\boxed{11 = 7x} \qquad \boxed{A = LW}$$

$$\frac{1}{7}(11) = \frac{1}{7}(7x) \qquad \frac{1}{L}(A) = \frac{1}{L}(LW)$$

$$\frac{11}{7} = 1\,x \qquad \frac{A}{L} = 1\,W$$

$$\frac{11}{7} = x \text{ or } x = \frac{11}{7} \qquad \frac{A}{L} = W \text{ or } W = \frac{A}{L}$$

We say that "W" is "solved for" in the bottom formula on the right <u>because "W" appears alone on one side.</u>

Go to next frame.

54. Formulas which contain one non-fractional term on each side are very similar to simple equations in one letter. In such formulas, we solve for a letter by multiplying both sides by the reciprocal of the coefficient of that letter.

Let's review the idea of factors and coefficients:

(a) In the term "2xy", there are <u>three</u> factors. They are _____, _____, and _____.

(b) In the term "m(p + q)", there are <u>two</u> factors. They are _______ and _______.

---

a) 2, x, and y

b) m and (p + q)

55. List the factors in each of the following terms:

(a) 3abc ______________ (c) 3vs(t + 1) ______________

(b) 2x(a − b) ______________ (d) a(b + c)(d − f) ______________

---

a) 3, a, b, and c

b) 2, x, and (a − b)

c) 3, v, s, and (t + 1)

d) a, (b + c), and (d − f)

56. Because multiplication is commutative, the order in which we multiply factors <u>does</u> <u>not</u> <u>matter</u>. For example:

2xy = 2yx abc = acb or bca, etc.

Complete: (a) x(y + 1) = (y + 1)(__)

(b) ab(d + c) = (a)(_____)(b)

(c) x(t + 1)(s + 5) = (__)(_____)(t + 1)

---

a) (x)

b) (d + c)

c) (x)(s + 5)

57. We can write 2xy as 2x(y). We then say that 2x is the <u>coefficient</u> of (y).

We can write 2xy as 2y(x). We then say that 2y is the ______________ of (x).

---

coefficient

58. (a) Since mpq = mp(q), we say that mp is the ______________ of (q).

(b) Since mpq = mq(p), the coefficient of (p) is _______.

(c) In mpq, what is the coefficient of (m)? _______

---

a) coefficient

b) mq

c) pq

59. In any term which contains a series of factors, the <u>coefficient</u> of any <u>one</u> factor is the remaining factor or factors.

In x(y + z): the coefficient of the factor (y + z) is (x); the coefficient of the factor (x) is (y + z).

In dt(x + y): (a) The coefficient of (x + y) is _________.

(b) The coefficient of (d) is _________.

(c) The coefficient of (t) is _________.

---

a) dt

b) t(x + y)

c) d(x + y)

60. In 2m(L + d): (a) The coefficient of (L + d) is _______.

(b) The reciprocal of this coefficient is _______.

---

a) 2m

b) $\frac{1}{2m}$

61. In $mg(s+t)$: (a) The coefficient of (g) is ________.

(b) The reciprocal of this coefficient is ________.

Answers: a) $m(s+t)$ b) $\frac{1}{m(s+t)}$

62. In $3xyz$: (a) The coefficient of (x) is ________.

(b) The reciprocal of this coefficient is ________.

Answers: a) $3yz$ b) $\frac{1}{3yz}$

63. In $2(F-t)(m+g)$: (a) The coefficient of $(m+g)$ is ________.

(b) The reciprocal of this coefficient is ________.

Answers: a) $2(F-t)$ b) $\frac{1}{2(F-t)}$

64. (Area of a Rectangle) $\boxed{A = LW}$

To solve for "L", we must multiply both sides by the reciprocal of the coefficient of L. Do so in the space on the right.

$L =$ ________

Answer: $\frac{A}{W} = L$ or $L = \frac{A}{W}$

65. (Boyle's Law For Gases) $\boxed{P_1V_1 = P_2V_2}$

To solve for V_1, we multiply both sides by $\left(\frac{1}{P_1}\right)$, the reciprocal of the coefficient of V_1. We did so on the right.

$$\frac{1}{P_1}(P_1V_1) = \frac{1}{P_1}(P_2V_2)$$

$$1\ V_1 = \frac{P_2V_2}{P_1}$$

$$V_1 = \frac{P_2V_2}{P_1}$$

(a) To solve for P_2 in the same formula, we must multiply both sides by ________.

(b) Do so on the right.

$\boxed{P_1V_1 = P_2V_2}$

$P_2 =$ ________

Answers: a) $\frac{1}{V_2}$ b) $P_2 = \frac{P_1V_1}{V_2}$

66. (Charles' Law for Gases) $V_1T_2 = V_2T_1$

(a) To solve for V_1, we must multiply both sides by the reciprocal of the coefficient of V_1, which is ______.

(b) Do so on the right.

$V_1 =$ ______

---

67. (Velocity, Acceleration, and Distance) $v^2 = 2as$

(a) To solve for "s", we must multiply both sides by ______.

(b) Do so on the right.

s = ______

Answers to 66: a) $\frac{1}{T_2}$ b) $V_1 = \frac{V_2T_1}{T_2}$

---

68. In the equation below, the coefficient of "a" is "2b". The reciprocal of this coefficient is $\frac{1}{2b}$.

(a) To solve for "a", we must multiply both sides by ______.

(b) Do so in the space at the right.

$2ab = c(d - f)$

a = ______

Answers to 67: a) $\frac{1}{2a}$ b) $s = \frac{v^2}{2a}$

---

69. In the equation below, the coefficient of "d" is (x - y). The reciprocal of this coefficient is $\frac{1}{x - y}$.

(a) To solve for "d", we must multiply both sides by ______.

(b) Do so:

$2mt = d(x - y)$

d = ______

Answers to 68: a) $\frac{1}{2b}$ b) $a = \frac{c(d - f)}{2b}$

Answers to 69: a) $\frac{1}{x - y}$ b) $d = \frac{2mt}{x - y}$

70. $\boxed{2RF = W(R - r)}$ (Differential Pulley Formula)

(a) Solve for W: (b) Solve for F:

W = ________ F = ________

---

71. (Fundamental Heat Equation) $\boxed{H = ms(t_2 - t_1)}$

(a) The coefficient of (s) is ________.

(b) The reciprocal of the coefficient of (s) is ________.

(c) Solve for s in the space on the right.

s = ________

a) $W = \frac{2RF}{R - r}$

b) $F = \frac{W(R - r)}{2R}$

---

72. (a) The coefficient of (b) is ________. $\boxed{a = 0.45b}$

(b) The reciprocal of the coefficient of (b) is ________.

(c) Solve for b in the space on the right.

b = ________

a) $m(t_2 - t_1)$

b) $\frac{1}{m(t_2 - t_1)}$

c) $\frac{H}{m(t_2 - t_1)} = s$

---

73. (Pressure at Various Depths of Water) $\boxed{P = 0.433h}$

Solve for h:

h = ________

a) 0.45

b) $\frac{1}{0.45}$

c) $b = \frac{a}{0.45}$

---

74. (Heat Produced by Electric Current) $\boxed{H = 0.24I^2Rt}$

(a) The coefficient of (R) is ________.

(b) Solve for R:

R = ________

$h = \frac{P}{0.433}$

---

a) $0.24I^2t$

b) $\frac{H}{0.24I^2t} = R$

9-6 A PREFERRED FORM FOR FRACTIONAL SOLUTIONS WHICH CONTAIN A NUMBER

In the last few frames, we obtained solutions of the following type:

$$h = \frac{P}{0.433} \qquad R = \frac{H}{0.24I^2t}$$

In each of these solutions, there is a number in the denominator. Mathematicians usually prefer to have numbers in the numerator rather than the denominator because the formula is then easier to work with. We will briefly show how we can eliminate numbers in denominators in this section.

75. In the formula on the left below, there is a number in the denominator of the fraction. Since the fraction stands for a division, we can convert it to a multiplication by multiplying P by the reciprocal of 0.433. We have done so on the right below.

$$\boxed{h = \frac{P}{0.433}} \quad \text{or} \quad h = P\left(\frac{1}{0.433}\right)$$

$\frac{1}{0.433}$, of course, is a division. You can perform this division by the long method, or on your slide rule.

Since $\frac{1}{0.433} = 2.31$, $h = P\left(\frac{1}{0.433}\right)$

$$h = P(2.31)$$

$$h = 2.31P$$

Have we eliminated the number in the denominator? _______

Answer: Yes.

76. Let's try another one: $\boxed{m = \frac{a}{0.45}}$ or $m = a\left(\frac{1}{0.45}\right)$

(a) Divide: $\frac{1}{0.45} =$ _______ (b) Therefpre, m = _______.

Answer: a) 2.22 b) m = a(2.22) or m = 2.22a

77. To eliminate a number in the denominator, we must be able to isolate a pure numerical fraction. In the last two frames, we did so by converting division to multiplication. In the formula below, we do so by factoring:

$$R = \frac{H}{0.24I^2t}$$

$$R = \frac{1H}{0.24I^2t} = \left(\frac{1}{0.24}\right)\left(\frac{H}{I^2t}\right)$$

Since $\frac{1}{0.24} = 4.17$, $R = \left(\frac{1}{0.24}\right)\left(\frac{H}{I^2t}\right) = (\quad)\left(\frac{H}{I^2t}\right) =$ _________

Answer: $R = (4.17)\left(\frac{H}{I^2t}\right) = \frac{4.17H}{I^2t}$

78. The formula for "capacitive reactance" in electronics is given below. Though "π" is a Greek letter, it really stands for a number. The number is approximately 3.14. Therefore:

$$X_c = \frac{1}{2\pi fC}$$

$$X_c = \frac{1}{2(3.14)fC}$$

$$X_c = \frac{1}{6.28fC}$$

Watch how we can factor to isolate a pure numerical fraction:

$$X_c = \frac{(1)(1)}{6.28fC} = \left(\frac{1}{6.28}\right)\left(\frac{1}{fC}\right)$$

(a) Divide: $\frac{1}{6.28} =$ ________ (b) Therefore, $X_c =$ ________

Answer to Frame 78: a) 0.159 b) $\frac{0.159}{fC}$

In general, we can eliminate a number in the denominator of a complicated fraction by some manipulation which makes it part of a pure numerical fraction. This simplification is not always done. You will have to learn the conventions for particular formulas as you meet them in your study of technology and science.

SELF-TEST 2 (Frames 53-78)

1. Solve for f: $X_L = 2\pi fL$

   f = ________

2. Solve for h: $P = 62.4h$

   h = ________

3. Solve for d_1: $F_1d_1 = F_2d_2$

   d_1 = ________

4. Solve for t: $K = mt(x - r)$

   t = ________

ANSWERS: 1. $f = \frac{X_L}{2\pi L}$ 2. $h = \frac{P}{62.4}$ or $h = 0.016P$ 3. $d_1 = \frac{F_2d_2}{F_1}$ 4. $t = \frac{K}{m(x - r)}$

9-7 FORMULAS WITH ONE TERM ON EACH SIDE CONTAINING ONE FRACTION

Each of the following formulas has one term on each side. In each case, one of the terms is a fraction.

$$P = \frac{W}{t} \qquad H = \frac{\pi dR(F_1 - F_2)}{33,000}$$

We will rearrange formulas of this type in this section.

79. Our first step in solving fractional equations was "clearing the fractions." We did so by using the multiplication axiom. We will use the same first step when rearranging formulas which contain fractions.

(a) To clear the fraction in $a = \frac{b}{3}$, we multiply both sides by ______.

(b) To clear the fraction in $a = \frac{b}{c}$, we multiply both sides by ______.

80. Clear the fraction in each formula below and write the resulting non-fractional equation:

(a) $v = \frac{s}{t}$ (b) $P = \frac{Fs}{t}$ (c) $Ft = \frac{mv}{g}$

______ ______ ______

Answers to 79: a) 3 b) c

81. Clear the fraction in each formula below and write the resulting equation:

(a) $H = \frac{AKT(t_2 - t_1)}{L}$ (b) $H = \frac{M[(V_1)^2 - (V_2)^2]}{1100gt}$

______ ______

Answers to 80: a) $vt = s$ b) $Pt = Fs$ c) $Ftg = mv$

Answers to 81: a) $HL = AKT(t_2 - t_1)$ b) $1100gtH = M[(V_1)^2 - (V_2)^2]$

82. Clear the fraction and write the resulting equation:

(a) $H = \frac{D^2N}{2.5}$ (b) $H = \frac{\pi dR(F_1 - F_2)}{33,000}$ (c) $X_c = \frac{1}{2\pi fC}$

________ ____________ ________

---

a) $2.5H = D^2N$

b) $33,000H = \pi dR(F_1 - F_2)$

c) $2\pi fCX_c = 1$

83. When a formula includes a fraction, the first step is "clearing the fraction." Then we solve for any letter in the usual way. Do the following:

(Power Formula) (Uniform Velocity Formula)

(a) Solve for W: $P = \frac{W}{t}$ (b) Solve for t: $v = \frac{s}{t}$

W = ______ t = ______

---

a) $W = Pt$

b) $t = \frac{s}{v}$

84. (Power Formula) (Momentum Formula)

(a) Solve for s: $P = \frac{Fs}{t}$ (b) Solve for F: $Ft = \frac{mv}{g}$

s = ______ F = ______

---

a) $s = \frac{Pt}{F}$ b) $F = \frac{mv}{gt}$

85. (Centripetal Force) (Heat Transfer Formula)

(a) Solve for r: $F_c = \frac{mv^2}{r}$ (b) Solve for T: $H = \frac{AKT(t_2 - t_1)}{L}$

r = ________ T = ________

86. (Momentum Formula) (Turbine Horsepower)

(a) Solve for v: $Ft = \frac{mv}{g}$ (b) Solve for M: $H = \frac{M[(V_1)^2 - (V_2)^2]}{1100gt}$

v = ________ M = ________

a) $r = \frac{mv^2}{F_c}$

b) $T = \frac{HL}{AK(t_2 - t_1)}$

87. (Centripetal Force) (Turbine Horsepower)

(a) Solve for m: $F_c = \frac{mv^2}{r}$ (b) Solve for t: $H = \frac{M[(V_1)^2 - (V_2)^2]}{1100gt}$

m = ________ t = ________

a) $v = \frac{Ftg}{m}$

b) $M = \frac{1100gtH}{(V_1)^2 - (V_2)^2}$

88. (Engine Horsepower) (Brake Horsepower)

(a) Solve for N: $H = \frac{D^2N}{2.5}$ (b) Solve for R: $H = \frac{\pi dR(F_1 - F_2)}{33,000}$

N = ________ R = ________

a) $m = \frac{F_c r}{v^2}$

b) $t = \frac{M[(V_1)^2 - (V_2)^2]}{1100gH}$

---

89. (Coulomb's Law) (Capacitive Reactance)

(a) Solve for q_2: $F = \frac{Kq_1q_2}{s^2}$ (b) Solve for C: $X_C = \frac{1}{2\pi fC}$

q_2 = ________ C = ________

a) $N = \frac{2.5H}{D^2}$

b) $R = \frac{33,000H}{\pi d(F_1 - F_2)}$

---

90. The formula for "induced voltage" is given on the left below. Since "t x 10^8" means "t times 10^8", we can rewrite the formula as we have done on the right below.

$E = \frac{N\varphi}{t \times 10^8}$ $E = \frac{N\varphi}{t(10^8)}$

(a) Solve for φ: (b) Solve for t:

φ = ________ t = ________

a) $q_2 = \frac{Fs^2}{Kq_1}$

b) $C = \frac{1}{2\pi fX_C}$

91. Below we have written two equivalent forms of the same formula. They are equivalent since: $\frac{1}{2}bh = \frac{1}{2}(bh) = \frac{bh}{2}$

$$A = \frac{1}{2}bh \qquad A = \frac{bh}{2}$$

On the left, $\frac{1}{2}$ is a factor in a three-factor multiplication. On the right, all three of the original factors are part of a single fraction.

Students frequently have difficulty "clearing the fraction" when faced with the form on the left. They have no difficulty in doing so with the form on the right. Therefore, <u>when faced with a formula in which a numerical fraction like "$\frac{1}{2}$" is a factor in a term, perform the multiplication to obtain a single fraction.</u>

$s = \frac{1}{2}at^2$ should be written as $s = \frac{at^2}{2}$

$E = \frac{1}{2}mv^2$ should be written as E = ______

Answers: a) $\varphi = \frac{Et(10^8)}{N}$ b) $t = \frac{N\varphi}{E(10^8)}$

---

92. (Area of a Triangle)

To solve for "b" or "h" in: $A = \frac{1}{2}bh$

Step 1: We write the formula in its equivalent form: $A = \frac{bh}{2}$

Step 2: We clear the fraction: $2A = bh$

Therefore: (a) b = ______ (b) h = ______

Answer: $E = \frac{mv^2}{2}$

---

93. (Formula Relating Distance, Acceleration, and Time) (Kinetic Energy)

(a) Solve for a: $s = \frac{1}{2}at^2$ (b) Solve for m: $E = \frac{1}{2}mv^2$

a = ______ m = ______

Answers: a) $b = \frac{2A}{h}$ b) $h = \frac{2A}{b}$

Answers: a) $a = \frac{2s}{t^2}$ b) $m = \frac{2E}{v^2}$

9-8 FORMULAS WITH ONE TERM ON EACH SIDE CONTAINING TWO FRACTIONS

In the formulas below, there is one fractional term on each side:

$$\frac{P_1}{P_2} = \frac{V_2}{V_1} \qquad \frac{I_1}{I_2} = \frac{(d_2)^2}{(d_1)^2}$$

We will rearrange formulas of this type in this section.

---

94. (Boyle's Law)

$$\boxed{\frac{P_1}{P_2} = \frac{V_2}{V_1}}$$

To solve for any variable in the formula on the right, we must "clear the fractions" first. To do so, we must multiply both sides by P_2 and V_1 at the same time.

Notice the steps:

$$(P_2V_1)\frac{P_1}{P_2} = (P_2V_1)\frac{V_2}{V_1}$$

$$\frac{P_2V_1P_1}{P_2} = \frac{P_2V_1V_2}{V_1}$$

$$\left(\frac{P_2}{P_2}\right)V_1P_1 = \left(\frac{V_1}{V_1}\right)P_2V_2$$

$$(1)(V_1P_1) = (1)(P_2V_2)$$

$$\_\_\_\_\_\_ = \_\_\_\_\_\_$$

---

95. (a) To clear both fractions in one step in $\frac{V_1}{V_2} = \frac{T_1}{T_2}$, we multiply both sides by ________.

(b) To clear both fractions in one step in $\frac{I_1}{I_2} = \frac{(d_2)^2}{(d_1)^2}$, we multiply both sides by ________.

Answer (94): $V_1P_1 = P_2V_2$ or $P_1V_1 = P_2V_2$

---

96. Clear the fractions in each formula below:

(a) $\frac{P_1V_1}{T_1} = \frac{P_2V_2}{T_2}$

(b) $\frac{I_1}{I_2} = \frac{(d_2)^2}{(d_1)^2}$

________ ________

Answers (95): a) V_2T_2 b) $I_2(d_1)^2$

Answers (96): a) $T_2P_1V_1 = T_1P_2V_2$ b) $(d_1)^2I_1 = I_2(d_2)^2$

97. (Charles' Law)

(a) Solve for T_1: $\dfrac{V_1}{V_2} = \dfrac{T_1}{T_2}$

T_1 = ____________

(General Gas Law)

(b) Solve for V_1: $\dfrac{P_1V_1}{T_1} = \dfrac{P_2V_2}{T_2}$

V_1 = ____________

98. (Boyle's Law)

(a) Solve for P_2: $\dfrac{P_1}{P_2} = \dfrac{V_2}{V_1}$

P_2 = ____________

(Light Intensity)

(b) Solve for I_1: $\dfrac{I_1}{I_2} = \dfrac{(d_2)^2}{(d_1)^2}$

I_1 = ____________

a) $T_1 = \dfrac{V_1T_2}{V_2}$

b) $V_1 = \dfrac{T_1P_2V_2}{T_2P_1}$

99. (a) Solve for T_2: $\dfrac{P_1V_1}{T_1} = \dfrac{P_2V_2}{T_2}$

T_2 = ____________

(b) Solve for I_2: $\dfrac{I_1}{I_2} = \dfrac{(d_2)^2}{(d_1)^2}$

I_2 = ____________

a) $P_2 = \dfrac{P_1V_1}{V_2}$

b) $I_1 = \dfrac{I_2(d_2)^2}{(d_1)^2}$

a) $T_2 = \dfrac{T_1P_2V_2}{P_1V_1}$ b) $I_2 = \dfrac{I_1(d_1)^2}{(d_2)^2}$

100. In the following frames, we will review various types of solutions:

(a) Solve for P: $W_S = P(V_1 - V_2)$

(b) Solve for m_1: $F = \frac{Gm_1m_2}{s^2}$

P = ________

m_1 = ________

Answers: a) $P = \frac{W_S}{V_1 - V_2}$ b) $m_1 = \frac{Fs^2}{Gm_2}$

---

101. (a) Solve for A: $H = \frac{AKT(t_2 - t_1)}{L}$

(b) Solve for t: $H = 0.24I^2Rt$

A = ________

t = ________

Answers: a) $A = \frac{HL}{KT(t_2 - t_1)}$ b) $t = \frac{H}{0.24I^2R}$

---

102. (a) Solve for d: $H = \frac{\pi dR(F_1 - F_2)}{33,000}$

(b) Solve for F_2: $\frac{F_1}{F_2} = \frac{d_1}{d_2}$

d = ________

F_2 = ________

Answers: a) $d = \frac{33,000H}{\pi R(F_1 - F_2)}$ b) $F_2 = \frac{F_1d_2}{d_1}$

SELF-TEST 3 (Frames 79-102)

1. Solve for t: $a = \frac{V_2 - V_1}{t}$	2. Solve for e: $Y = \frac{FL}{Ae}$	3. Solve for A_1: $\frac{F_1}{F_2} = \frac{A_1}{A_2}$
4. Solve for T_2: $\frac{P_1}{T_1} = \frac{P_2}{T_2}$	5. Solve for E: $W = \frac{EI}{S}$	6. Solve for g: $H = \frac{M[(V_1)^2 - (V_2)^2]}{1100gt}$
7. Solve for P_1: $\frac{P_1V_1}{T_1} = \frac{P_2V_2}{T_2}$	8. Solve for d: $H = \frac{\pi dR(F_1 - F_2)}{33,000}$	9. Solve for R_a: $I_a = \frac{E_i - E_o}{R_a}$

(Answers are on following page.)

ANSWERS: 1. $t = \frac{V_2 - V_1}{a}$ 2. $e = \frac{FL}{AY}$ 3. $A_1 = \frac{A_2F_1}{F_2}$ 4. $T_2 = \frac{P_2T_1}{P_1}$ 5. $E = \frac{SW}{I}$ 6. $g = \frac{M[(V_1)^2 - (V_2)^2]}{1100Ht}$ 7. $P_1 = \frac{P_2V_2T_1}{V_1T_2}$ 8. $d = \frac{33,000H}{\pi R(F_1 - F_2)}$ 9. $R_a = \frac{E_i - E_o}{I_a}$

9-9 FORMULAS WHICH CONTAIN MORE THAN ONE TERM ON ONE SIDE

Each formula below contains more than one term on one side.

$$I_E = I_C + I_B \qquad T_j = T_a + \theta_T P_T \qquad F_1r_1 + F_2r_2 + F_3r_3 = 0$$

Up to this point, we have avoided formulas which contain more than one term on one side. In this section, we will solve for variables in formulas of this type. Before doing so, however, we must review some basic principles.

---

103. The new step we need to rearrange formulas with more than one term on a side is "isolating a term." We will begin by reviewing the principles which are necessary for this step.

<u>Two quantities are a pair of opposites if their sum is "0".</u>

Write the opposites of each of the following:

(a) -14 ______ (c) -r ______ (e) 3VT ______

(b) P ______ (d) AB ______ (f) -4mq ______

Answers: a) +14 b) -P c) +r d) -AB e) -3VT f) +4mq

104. We have used the concept of opposites in <u>converting subtraction problems to addition form.</u>

a - b = a + (the opposite of b) = a + (-b)

Convert the following subtractions to addition form:

(a) $I_c - I_a$ = ______ (b) $T_1 - T_2$ = ______ (c) $F_1r_1 - F_2r_2$ = ______

Answers: a) $I_c + (-I_a)$ b) $T_1 + (-T_2)$ c) $F_1r_1 + (-F_2r_2)$

105. <u>Here is the "Addition Axiom for Equations" in symbol form:</u>

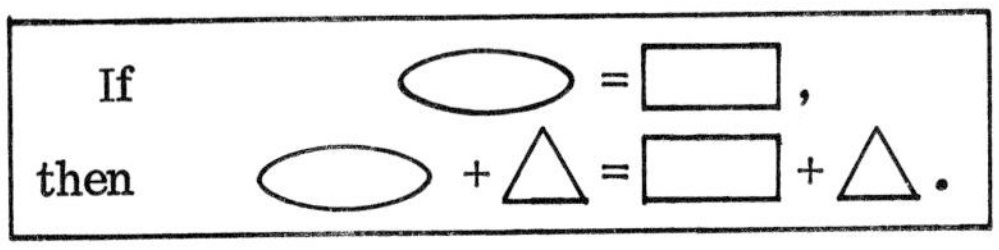

That is: If (3x + 2) = [2x + 3]

then (3x + 2) + △-2 = [2x + 3] + △-2

When using the addition axiom to eliminate a term from one side of an equation, we add the ______________ of the term to both sides.

Answer: opposite

106. There are two terms on the right side of the equation below:

$$m = p + q$$

To isolate one of these two terms, we use the addition axiom. For example, to isolate the term "p", we must eliminate the term "q". To do so, we use the addition axiom, adding the opposite of q (which is -q) to both sides.

What must we do to isolate the term "q" in the original equation?

____

Answer to 106: Add (-p) to both sides.

107. (a) In $I_E = I_C + I_B$, to isolate the term "I_B", we must add _______ to both sides.

(b) In $R_t = R_1 + R_2 + R_3$, to isolate the term "R_2", we must add _______ and _______ to both sides.

Answers to 107: a) $(-I_C)$, the opposite of I_C b) $(-R_1)$ and $(-R_3)$

108. (a) In $T_j = T_a + \theta_T P_T$, to isolate the term "T_a", we must add _______ to both sides.

(b) In $F_1r_1 + F_2r_2 + F_3r_3 = 0$, to isolate the term "$F_1r_1$", we must add _______ and _______ to both sides.

Answers to 108: a) $(-\theta_T P_T)$, the opposite of $\theta_T P_T$ b) $(-F_2r_2)$ and $(-F_3r_3)$

109. Let's isolate "b" in the equation on the right by using the addition axiom. To do so, we add "-c" to both sides. Notice the steps:

$$\boxed{a = b + c}$$

$$a + (-c) = b + \underbrace{c + (-c)}$$

$$a + (-c) = b + \quad 0$$

$$a + (-c) = b \quad \text{or} \quad b = a + (-c)$$

Isolate I_B in the formula on the right:

$$\boxed{I_E = I_C + I_B}$$

$I_B =$ ______________

Answer to 109: $I_B = I_E + (-I_C)$

110. Here are the two solutions we obtained in the last frame:

$$b = a + (-c) \qquad I_B = I_E + (-I_C)$$

On the right side in each case, we have an addition in which the second term is negative. In formula rearrangement, we traditionally convert additions of this type back to subtractions to eliminate the negative terms.

Just as a subtraction can be converted to addition by "adding the opposite," an addition can be converted back to subtraction by "subtracting the opposite." For example:

Just as $a - c = a + (\text{the opposite of } c)$
$= a + (-c)$

$a + (-c) = a - (\text{the opposite of } -c)$
$= a - (+c)$
$= a - c$

Convert the addition on the right side below to subtraction:

$$I_B = I_E + (-I_C)$$

$$I_B = \_\_ - \_\_\_\_$$

Answer: $I_B = I_E - I_C$

---

111. Convert each of the following additions to subtraction form:

(a) $T_1 + (-T_2) =$ ________

(b) $F_1r_1 + (-F_2r_2) =$ ________

(c) $R_t + (-R_1) + (-R_2) =$ ________

(d) $(-F_1r_1) + (-F_2r_2) =$ ________

Answers:

a) $T_1 - T_2$

b) $F_1r_1 - F_2r_2$

c) $R_t - R_1 - R_2$

d) $-F_1r_1 - F_2r_2$

(Note: The first term is still negative.)

---

112. When a variable in a formula is solved for, the solution should not be left in a form which contains the addition of a negative term. The preferred form, in this case, is subtraction. Therefore, we expect you to convert such additions to subtractions. For example:

Instead of $m = p + (-q)$, we write $m = p - q$.

(Transistor Current)

(a) Solve for I_C: $\boxed{I_E = I_C + I_B}$

(Series Resistance)

(b) Solve for R_1: $\boxed{R_t = R_1 + R_2 + R_3}$

$I_C =$ ________

$R_1 =$ ________

113. (Series Resistance) (Transistor Temperature)

(a) Solve for R_3: $R_t = R_1 + R_2 + R_3$ (b) Solve for T_a: $T_j = T_a + \theta_T P_T$

$R_3 =$ ______ $T_a =$ ______

a) $I_C = I_E - I_B$

[not $I_C = I_E + (-I_B)$]

b) $R_1 = R_t - R_2 - R_3$

[not $R_1 = R_t + (-R_2) + (-R_3)$]

---

114. In the last frame, we solved for T_a in: $T_j = T_a + \theta_T P_T$

Since T_a had no coefficient shown, the solution was completed when the "T_a" term was isolated.

If we wanted to solve for P_T in the same formula, we would first isolate the term in which P_T appears, and obtain:

$$T_j - T_a = \theta_T P_T$$

or

$$\theta_T P_T = T_j - T_a$$

Since P_T has a coefficient in this isolated term, one more step is required. We must use the multiplication axiom to eliminate this coefficient.

(a) To solve for P_T, we must multiply both sides by ______.

(b) Do so:

$P_T =$ ______

a) $R_3 = R_t - R_1 - R_2$

b) $T_a = T_j - \theta_T P_T$

---

a) $\frac{1}{\theta_T}$

b) $P_T = \frac{T_j - T_a}{\theta_T}$

115. $\boxed{H = E + PV}$ (Thermodynamics Formula)

(a) Solve for E: (b) Solve for P:

E = ______ P = ______

a) $E = H - PV$

b) $P = \dfrac{H - E}{V}$

9-10 THE OPPOSITING PRINCIPLE APPLIED TO FORMULAS

When isolating a term in a formula, you occasionally obtain an equation like this:

$$-t = a - b$$

Though the "t" term has been isolated, there is a "-" in front of the "t". Therefore, "t" is not yet solved for. To do so, we must eliminate the "-" in front of it. In this section, we will show you how to eliminate the "-" in such equations by using the oppositing principle for equations.

116. Let's review <u>the oppositing principle for equations.</u> It says this:

If we replace each side of an equation by its opposite, the new equation is equivalent to the original one.

<u>For example:</u> $-5x = 15$ and $5x = -15$ are equivalent equations since the root of each is "-3".

When faced with a formula like: $-t = a - b$

We can eliminate the "-" in front of the "t" by replacing each side with its opposite. We get:

$$t = b - a$$

In the next frame, we will show that the opposite of $a - b$ is $b - a$. That is, to get the opposite of $a - b$ we merely interchange the two terms.

117. We have seen that we obtain the opposite of an addition by replacing each term with its opposite. That is:

The opposite of $5 + 3$ is $(-5) + (-3)$.
The opposite of $3x + 7$ is $(-3x) + (-7)$.
The opposite of $a + (-b)$ is $(-a) + b$.

Let's show that the opposite of a subtraction can be obtained by merely interchanging the two terms:

Step 1: The opposite of $a - b$ = the opposite of $a + (-b)$
Step 2: $= (-a) + b$
Step 3: $= b + (-a)$
Step 4: $= b - a$

In Step 1, we converted the subtraction to addition.
In Step 2, we wrote the opposite of the addition.
In Step 3, we commuted the two terms.
In Step 4, we converted the addition back to subtraction.

---

118. It is easy to justify numerically that we can obtain the opposite of a subtraction by interchanging the two terms.

$(7 - 11)$ is the opposite of $(11 - 7)$, since:

$7 - 11 = -4$
$11 - 7 = +4$ (and $(-4) + (+4) = 0$)

$7 - 11$ and $11 - 7$ are a pair of opposites since their sum is ______.

Answer: Go to next frame.

---

119. Let's justify another example numerically:

$(1 - 2x)$ is the opposite of $(2x - 1)$, since:

If we plug in "3" for "x": $1 - 2x = 1 - 2(3) = 1 - 6 = -5$

$2x - 1 = 2(3) - 1 = 6 - 1 = +5$

and -5 and $+5$ are a pair of ______________.

Answer: 0

---

120. We can justify another example by performing the addition:

$(a - b)$ is the opposite of $(b - a)$, since:

$$(a - b) + (b - a) = a - b + b - a$$
$$= a + (-b) + b + (-a)$$
$$= \underbrace{a + (-a)} + \underbrace{b + (-b)}$$
$$= \quad 0 \quad + \quad 0 \quad = \_\_\_\_$$

Answer: opposites

---

121. You should be convinced by now that we can obtain the opposite of any subtraction by interchanging the two terms.

Write the opposite of each of the following:

(a) $T - S$ ______ (b) $ab - c$ ______ (c) $3x - 7$ ______

Answer: 0

Answers:
a) $S - T$
b) $c - ab$
c) $7 - 3x$

122. The root of the following equation is +4.

$$2x - 5 = 3$$

Let's apply the oppositing principle for equations to the equation above. If we replace each side with its opposite, we get:

$$5 - 2x = -3$$

This new equation is equivalent to the original one since its root is _____.

---

123. $3x - 14 = -8$ (The root is 2.)

(a) What new equation do you get if you take the opposite of each side? ______________

(b) Is 2 the root of this new equation? _______

Answer: +4

---

124. Use the oppositing axiom to complete each of these:

(a) If $-x = 2 - y$, $x =$ _______

(b) If $-T = S - V$, $T =$ _______

(c) If $-Ir = e - E$, $Ir =$ _______

Answers:
a) $14 - 3x = 8$
b) Yes

---

125. $\boxed{-y = a - 2b}$

In formula rearrangement, you can encounter an equation like the one above when trying to solve for a letter like y. The above equation has -y, the opposite of y. To solve for y, you merely use the oppositing principle for equations:

$$y = 2b - a$$

Solve for t in each of the following:

(a) $-t = 2a - 3b$ $t =$ ____________

(b) $-t = 3RF - P_1$ $t =$ ____________

Answers:
a) $y - 2$
b) $V - S$
c) $E - e$

---

126. (a) Solve for I: $\boxed{-I = E_1 - E_2}$ $I =$ __________

(b) Solve for m: $\boxed{-m = R - d}$ $m =$ __________

Answers:
a) $t = 3b - 2a$
b) $t = P_1 - 3RF$

---

127. $\boxed{d = e - f}$

Let's solve for f. We change the subtraction to addition first:

$$d = e + (-f)$$

(a) To eliminate e from the right side, we add "-e" to both sides and get: ______________

(b) Now what should you do to solve for f?

____

(c) Solve for f: $f =$ ______________

Answers:
a) $I = E_2 - E_1$
b) $m = d - R$

128. (Centigrade and Kelvin Conversion) $\boxed{T_k - T_c = 273°}$

Let's solve for T_c: $T_k + (-T_c) = 273°$

$-T_c = 273° + (-T_k)$

$-T_c = 273° - T_k$

(a) To solve for T_c, you now do what? ____________

(b) Solve for T_c: $T_c =$ ____________

a) $d + (-e) = -f$ or $d - e = -f$

b) Use the oppositing principle.

c) $f = e - d$

---

129. (Series Resistance) (Terminal Voltage Formula)

(a) Solve for R_1: $\boxed{R_t - R_1 = R_2}$

(b) Solve for E: $\boxed{e = E - Ir}$

$R_1 =$ ____________

$E =$ ____________

a) Use the oppositing principle.

b) $T_c = T_k - 273°$

---

130. (Acceleration Formula)

Clearing the fraction in: $\boxed{a = \frac{v_2 - v_1}{t}}$

we get: $at = v_2 - v_1$

Solve for v_1:

$v_1 =$ ____________

a) $R_1 = R_t - R_2$

b) $E = e + Ir$

---

131. $\boxed{a = \frac{v_2 - v_1}{t}}$

(a) Solve for v_2:

(b) Solve for t:

$v_2 =$ ____________

$t =$ ____________

$v_1 = v_2 - at$

132. To solve for "I" or "r" in the formula below, we isolate their term as follows:

$$\boxed{e = E - Ir}$$

(1) Convert the subtraction to addition: $e = E + (-Ir)$

(2) Add (-E) to both sides: $e - E = -Ir$

(3) Take the opposite of both sides: $E - e = Ir$

(a) Solve for I: (b) Solve for r:

$I =$ ________ $r =$ ________

Answers: a) $v_2 = at + v_1$ b) $t = \dfrac{v_2 - v_1}{a}$

---

133. $\boxed{F_1r_1 + F_2r_2 + F_3r_3 = 0}$ (Equilibrium of Forces)

To solve for F_1, we isolate its term as follows:

(1) Add $(-F_2r_2)$ and $(-F_3r_3)$ to both sides and get:

$$F_1r_1 = (-F_2r_2) + (-F_3r_3)$$

(2) Change the addition to subtraction form:

$$F_1r_1 = (-F_2r_2) - F_3r_3$$

or

$$F_1r_1 = -F_2r_2 - F_3r_3$$

(a) What must you do to complete the solution? ____________________

____

(b) Complete the solution:

$F_1 =$ ________

Answers: a) $I = \dfrac{E - e}{r}$ b) $r = \dfrac{E - e}{I}$

Answers: a) Eliminate the coefficient of F_1 by multiplying both sides by $\dfrac{1}{r_1}$.

b) $F_1 = \dfrac{-F_2r_2 - F_3r_3}{r_1}$

134. Solve for r_2: $F_1r_1 + F_2r_2 + F_3r_3 = 0$

$$r_2 = \frac{-F_1r_1 - F_3r_3}{F_2}$$

9-11 CONTRASTING COEFFICIENTS AND TERMS

Some students fail to recognize the difference between "two terms" and "one term with a coefficient." This failure is disastrous because a term is eliminated by the addition axiom whereas a coefficient is eliminated by the multiplication axiom. We will contrast the two in this section.

135. Let's compare the following two expressions:

(1) $X_L - X_C$ (2) $X_L(-X_C)$

In (1), there are two terms. The "-" symbol is the symbol for subtraction.

In (2), there is one term. The parentheses signify a multiplication.

Which of the following indicates a multiplication? ________

(a) $F_1 - F_2$ (b) $F_1(-F_2)$

136. The expression $E_1 - E_0$ means which of these? ________

(a) Multiply (E_1) and ($-E_0$) (b) Subtract E_0 from E_1

Answer to 135: (b) $F_1(-F_2)$, the one with the parentheses

137. In the expression $c - d$, is "c" a term or is it the coefficient of "-d"? ________________

Answer to 136: (b) $E_1 - E_0$ is a subtraction.

138. In which of the following expressions is "V_0" a coefficient? ________

(a) V_0V_1 (b) $V_0 - V_1$ (c) $V_0(-V_1)$

Answer to 137: It is a term in a subtraction.

Answer to 138: Both (a) and (c)

139. To solve for "X_C" in the formula below, we must eliminate "X_L". Since "X_L" is a term, we use the addition axiom (adding the opposite of "X_L" to both sides).

Solve for X_C: $\boxed{X = X_L - X_C}$

X_C = ________

Answer: $X_C = X_L - X$

140. The point we want to make is this: In $\boxed{I = E_1 - E_0}$,

E_1 is not the coefficient of "$-E_0$". It is a term. Therefore, it is eliminated by the addition axiom, and not by the multiplication axiom (multiplying by $\frac{1}{E_1}$).

Solve for E_0 in the formula above:

E_0 = ________

Answer: $E_0 = E_1 - I$

SELF-TEST 4 (Frames 103-140)

1. Solve for X_C: $X - X_L = -X_C$	2. Solve for H_r: $D = H_r + A_t$	3. Solve for R: $F - R = N$
4. Solve for d: $h = H + ad$	5. Solve for V: $H = E - PV$	6. Solve for c_2: $K = c_1 a + c_2 b$
7. Solve for K_2: $aK_1 + bK_2 + cK_3 = 0$	8. Solve for E_o: $I_a = \frac{E_i - E_o}{R_a}$	9. Solve for w: $p = bh - 2w$

ANSWERS:

1. $X_C = X_L - X$
2. $H_r = D - A_t$
3. $R = F - N$
4. $d = \frac{h - H}{a}$
5. $V = \frac{E - H}{P}$
6. $c_2 = \frac{K - c_1 a}{b}$
7. $K_2 = \frac{-aK_1 - cK_3}{b}$
8. $E_o = E_i - I_a R_a$
9. $w = \frac{bh - p}{2}$

9-12 CASES IN WHICH THE SAME VARIABLE APPEARS IN MORE THAN ONE TERM

Here are two formulas in which the same variable appears in more than one term:

$$\boxed{\overset{\downarrow}{M}S = R - Q\overset{\downarrow}{S}} \qquad \boxed{E = \overset{\downarrow}{I}R_1 + \overset{\downarrow}{I}R_2}$$

In this section, we will show how to solve for variables of this type. To do so, you must understand exactly what a solution means.

141. When solving for a letter in a formula, we must rearrange the formula until:

(1) The letter is isolated on one side with no coefficient.
(2) The letter does not appear on the other side.

Here is an example which is not a solution for "W" because it violates (2) above:

$$W = \frac{P + TW}{Q}$$

It is not a solution because "W" appears on both sides of the equation.

Which of the following is a solution for "t"? _______

(a) $t = \dfrac{m + pq}{r}$ (b) $t = \dfrac{a + bt}{s}$

Only (a). In (b), "t" appears on both sides.

142. $$bf = p + cf$$

We want to solve for "f" in the equation above. Notice that "f" appears in two terms. Let's try to solve for "f" by isolating "bf" alone or "cf" alone.

(1) Isolating "bf" alone, we get: $bf = p + cf$

$$f = \frac{p + cf}{b}$$

(2) Isolating "cf" alone, we get: $cf = bf - p$

$$f = \frac{bf - p}{c}$$

Neither of the above leads to a solution for "f". Why not?

____

Because in each case, "f" appears on both sides.

143. In the last frame, we attempted to solve for "f" in the equation below by isolating the "bf" term alone or the "cf" term alone. Neither approach led to a solution for "f", since in each case, we ended up with "f" on both sides of the equation.

To solve for "f", we must do the following:

(1) Isolate both terms in which "f" appears on one side.

(2) Factor by the distributive principle on the left side (which reduces the two terms to one term).

(3) Multiply both sides by the reciprocal of the coefficient of "f".

$$\boxed{bf = p + cf}$$

$$bf - cf = p$$

$$f(b - c) = p$$

$$f(b - c)\left(\frac{1}{b - c}\right) = p\left(\frac{1}{b - c}\right)$$

$$f(1) = \frac{p}{b - c}$$

$$f = \frac{p}{b - c}$$

---

144. When solving for a letter which appears in two terms, we must always factor by the distributive principle during the solution.

Let's review "factoring by the distributive principle."

Here is an instance of multiplying by the distributive principle:

$$M(V + T) = MV + MT$$

Notice this: (1) The left side is one term.
(2) The right side is two terms.

When we multiply by the distributive principle, we increase the number of terms from one to two.

Here is an instance of factoring by the distributive principle:

$$pt - pq = p(t - q)$$

(a) The left side contains how many terms? _______

(b) The right side contains how many terms? _______

(c) When we factor by the distributive principle, we decrease the number of terms from _______ to _______.

Go to next frame.

---

145. Here is an instance of the two factors in the distributive principle:

$$C_1(C_2 - C_3)$$

"C_1" is one factor; "$C_2 - C_3$" is the second factor.

Since the coefficient of a factor is the other factor or factors in the term:

(a) The coefficient of "C_1" is _______.

(b) The coefficient of $(C_2 - C_3)$ is _______.

a) Two

b) One

c) two to one

---

a) $C_2 - C_3$

b) C_1

146. In the expression $V(S + T)$: (a) The coefficient of $(S + T)$ is _______. (b) The coefficient of "V" is _______.	
147. In $W(R + M)$: (a) The coefficient of "W" is _______. (b) The reciprocal of the coefficient of "W" is _______.	a) V b) S + T
148. To solve for "S" in the formula below, you must follow these steps: $\boxed{MS = R - QS}$ (a) Isolate both terms which contain "S" on one side. Do so: _______ (b) Factor by the distributive principle. Do so: _______ (c) Complete the solution by multiplying both sides by the reciprocal of the coefficient of S. S = _______	a) R + M b) $\frac{1}{R + M}$
149. (Voltage Drop Formula) $\boxed{E = IR_1 + IR_2}$ Notice that both "I" terms are already isolated on the right. (a) To solve for "I", what must your first step be? _______ (b) Do so and complete the solution: I = _______	a) $MS + QS = R$ b) $S(M + Q) = R$ c) $S = \frac{R}{M + Q}$
150. To clear the fraction in the equation on the right, we must multiply both sides by "a" and "b" and "c". We get: $\boxed{\frac{1}{a} + \frac{1}{b} = \frac{1}{c}}$ $abc\left[\frac{1}{a} + \frac{1}{b}\right] = abc\left(\frac{1}{c}\right)$ $abc\left(\frac{1}{a}\right) + abc\left(\frac{1}{b}\right) = abc\left(\frac{1}{c}\right)$ $bc + ac = ab$ In this non-fractional equation: (a) "c" appears in how many terms? _______ (b) "a" appears in how many terms? _______ (c) "b" appears in how many terms? _______	a) Factoring by the distributive principle. b) $I = \frac{E}{R_1 + R_2}$

151. The equation from the last frame is given on the right. Let's solve for "c":

$$\boxed{\frac{1}{a} + \frac{1}{b} = \frac{1}{c}}$$

Step 1: Clearing the fraction. We get: $bc + ac = ab$

Step 2: Factoring (distributive principle): $c(b + a) = ab$

Step 3: Solve for "c":

$c =$ ________

a) Two

b) Two

c) Two

---

152. In the last frame, the solution for "c" was: $c = \frac{ab}{b + a}$

Students are tempted to "cancel" in the fraction on the right. However, this fraction can only be reduced to lower terms if we can find a common factor in both the numerator and denominator. Then we can factor out an instance of $\frac{n}{n}$.

The two factors in the numerator are "a" and "b".

(a) Is either "a" or "b" a common factor in "b + a"? ________

(b) Can $\frac{ab}{b + a}$ be written in a simpler form? ________

(c) Can we "cancel" anything in $\frac{ab}{b + a}$? ________

$c = \frac{ab}{b + a}$ or $c = \frac{ab}{a + b}$

---

153. (Parallel Circuit Resistance)

$$\boxed{\frac{1}{R_t} = \frac{1}{R_1} + \frac{1}{R_2}}$$

(a) To clear the fractions in one step, you multiply both sides by ________________.

(b) Clear the fractions. What new equation do you get?

____

(c) Solve for R_t:

$R_t =$ ________

a) No

b) No

c) No

---

a) $(R_t)(R_1)(R_2)$

b) $R_1R_2 = R_tR_2 + R_tR_1$

c) $R_t = \frac{R_1R_2}{R_2 + R_1}$

154. (Lens Equation)

$$\frac{1}{D} + \frac{1}{d} = \frac{1}{f}$$

(a) Clear the fractions in the formula on the right. What new equation do you get?

____

(b) Solve for "f":

f = ______________

155. (Series Capacitance Formula)

a) df + Df = Dd

b) $f = \frac{Dd}{d + D}$

$$\frac{1}{C_1} = \frac{1}{C_t} - \frac{1}{C_2}$$

(a) Clear the fractions and write the resulting equation:

____

(b) Solve for C_1:

C_1 = ______________

a) $C_tC_2 = C_1C_2 - C_1C_t$

b) $C_1 = \frac{C_tC_2}{C_2 - C_t}$

156. (Voltage Drop Formula)

"I" appears in three terms in this formula. To solve for "I":

$$E = IR_1 + IR_2 + IR_3$$

(a) We factor the right side first. Do so: ________________

(b) Complete the solution:

I = __________

a) $E = I(R_1 + R_2 + R_3)$

b) $I = \dfrac{E}{R_1 + R_2 + R_3}$

9-13 ALTERNATE METHODS OF SOLUTION AND EQUIVALENT FORMS

In each formula below, there is an instance of the distributive principle on one side.

$$Q = C(T_1 - T_2) \qquad M = P(L - X)$$

Two different methods can be used to solve for any of the variables within the groupings above. These alternate methods lead to equivalent solutions which look different. It is important that you can recognize that the solutions are equivalent. In this section, we will discuss these alternate methods for solving and the equivalency of the solutions.

157. Let's examine the following formula: $Q = C(T_1 - T_2)$ (Refrigeration Formula)

$C(T_1 - T_2)$ is an instance of the two factors in the distributive principle. Both "T_1" and "T_2" appear inside the grouping. To solve for either of them, we can either multiply by the distributive principle first, or isolate the grouping first. Therefore, there are two legitimate methods for the solution. Let's solve for "T_1" in each of these two ways.

Method 1: Multiplying by the distributive principle first.

$$Q = C(T_1 - T_2)$$

$$Q = CT_1 - CT_2$$

$$CT_1 = Q + CT_2$$

$$T_1 = \frac{Q + CT_2}{C}$$

(Continued on following page.)

157. (Continued)

Method 2: Isolating the grouping first.

$$Q = C(T_1 - T_2)$$

Multiplying both sides by the reciprocal of "C", we can isolate the grouping and drop the parentheses. We get:

$$\frac{Q}{C} = T_1 - T_2$$

$$T_1 = \frac{Q}{C} + T_2$$

The two methods led to the following two solutions. Superficially, these two solutions look different. They are, however, equivalent.

$$T_1 = \frac{Q + CT_2}{C} \qquad\qquad T_1 = \frac{Q}{C} + T_2$$

(Go to the next frame for a discussion of these two solutions.)

---

158. Let's examine the two solutions in the preceding frame:

$$(1)\ T_1 = \frac{Q + CT_2}{C} \qquad\qquad (2)\ T_1 = \frac{Q}{C} + T_2$$

Solution #1 fits the pattern for the sum of two fractions. We can show that it equals solution #2 by breaking the pattern down into its simpler parts:

$$T_1 = \frac{Q + CT_2}{C} = \frac{Q}{C} + \frac{CT_2}{C} = \frac{Q}{C} + T_2$$

We can also show that solution #2 equals solution #1 by adding the fraction and non-fraction.

$$T_1 = \frac{Q}{C} + T_2 = \frac{Q}{C} + \frac{CT_2}{C} = \frac{Q + CT_2}{C}$$

Though the two solutions superficially look different, it should be obvious that they are really equivalent.

---

159. Let's take the same formula and solve for T_2 in two ways.

Go to next frame.

Method 1: Multiplying by the distributive principle first.

$$\boxed{Q = C(T_1 - T_2)}$$

Step 1: $Q = CT_1 - CT_2$

Step 2: $Q - CT_1 = -CT_2$

Step 3: $CT_1 - Q = CT_2$

Step 4: $\frac{CT_1 - Q}{C} = T_2$ or $T_2 = \frac{CT_1 - Q}{C}$

(Notice how we used the oppositing principle for equations in going from Step 2 to Step 3.)

(Continued on following page.)

159. (Continued)

Method 2: Isolating the grouping first. (We isolate the grouping by multiplying both sides by the reciprocal of C.)

$$\boxed{Q = C(T_1 - T_2)}$$

Step 1: $\frac{Q}{C} = T_1 - T_2$

Step 2: $\frac{Q}{C} - T_1 = -T_2$

Step 3: $T_1 - \frac{Q}{C} = T_2$ or $T_2 = T_1 - \frac{Q}{C}$

(Notice again how we used the oppositing principle for equations in going from Step 2 to Step 3.)

To justify that the two solutions are equivalent, show how $T_1 - \frac{Q}{C}$ can be converted to $\frac{CT_1 - Q}{C}$.

$T_1 - \frac{Q}{C} = \frac{CT_1}{C} - \frac{Q}{C} = \frac{CT_1 - Q}{C}$

---

160. When one side of a formula is an instance of the distributive principle, we use the multiplication axiom to isolate the grouping.

(a) To isolate $(L - X)$ in $M = P(L - X)$, we multiply both sides by ________.

(b) To isolate $(t_2 - t_1)$ in $HL = AKT(t_2 - t_1)$, we multiply both sides by ________.

---

a) $\frac{1}{P}$, the reciprocal of P

b) $\frac{1}{AKT}$, the reciprocal of AKT

---

161. (a) Isolate the grouping first. Then solve for L and X:

$\boxed{M = P(L - X)}$ (Bending Moment Formula)

L = ________ X = ________

(b) Multiply by the distributive principle on the right side. Then solve for L and X:

$\boxed{M = P(L - X)}$

L = ________ X = ________

162. $H = \frac{AKT(t_2 - t_1)}{L}$ (Heat Transfer Formula)

To solve for t_2 and t_1, clear the fraction first and get:

$LH = AKT(t_2 - t_1)$

(a) Isolate the grouping and solve for t_2 and t_1:

$t_2 =$ ______ $t_1 =$ ______

(b) Multiply $AKT(t_2 - t_1)$ using the distributive principle and then solve for t_2 and t_1:

$t_2 =$ ______ $t_1 =$ ______

a) $L = \frac{M}{P} + X$

$X = L - \frac{M}{P}$

b) $L = \frac{M + PX}{P}$

$X = \frac{PL - M}{P}$

---

163. In the following frames, whenever two methods are possible, use the one which you prefer.

Solve for V_1: $W_s = P(V_1 - V_2)$ (Energy Formula)

$V_1 =$ ______

a) $t_2 = \frac{LH}{AKT} + t_1$

$t_1 = t_2 - \frac{LH}{AKT}$

b) $t_2 = \frac{LH + AKTt_1}{AKT}$

$t_1 = \frac{AKTt_2 - LH}{AKT}$

$V_1 = \frac{W_s}{P} + V_2$

or

$V_1 = \frac{W_s + PV_2}{P}$

164. Solve for t_1: $\boxed{H = ms(t_2 - t_1)}$ (Fundamental Heat Equation)

$t_1 =$ ________________

165. Solve for m: $\boxed{H = ms(t_1 - t_2)}$

$m =$ ________________

Answer to 164: $t_1 = t_2 - \frac{H}{ms}$ or $t_1 = \frac{mst_2 - H}{ms}$

166. Solve for F_1: $\boxed{H = \frac{\pi dR(F_1 - F_2)}{33,000}}$

$F_1 =$ ________________

Answer to 165: $m = \frac{H}{s(t_1 - t_2)}$

167. Solve for R: $\boxed{H = \frac{\pi dR(F_1 - F_2)}{33,000}}$

$R =$ ________________

Answer to 166: $F_1 = \frac{33,000H}{\pi dR} + F_2$ or $F_1 = \frac{33,000H + \pi dRF_2}{\pi dR}$

168. $\boxed{P = p_1 + w(h - h_1)}$ (Hydraulics Formula)

If we add $(-p_1)$ to both sides, we can isolate $w(h - h_1)$ on the right side:

$$P - p_1 = w(h - h_1)$$

(a) Using this form, multiply both sides by the reciprocal of w and solve for h and h_1:

$h =$ ______________ $h_1 =$ ______________

(b) Using the same form, multiply $w(h - h_1)$ and then solve for h and h_1:

$h =$ ______________ $h_1 =$ ______________

$$R = \frac{33,000H}{\pi d(F_1 - F_2)}$$

---

169. $\boxed{E = IR_1 + IR_2 + IR_3}$

Let's solve for R_1 by isolating its term first. We get:

$$IR_1 = E - IR_2 - IR_3$$

and

$$R_1 = \frac{E - IR_2 - IR_3}{I}$$

Write the solution in an equivalent form: $R_1 =$ ______________

a) $h = \frac{P - p_1}{w} + h_1$

$h_1 = h - \frac{P - p_1}{w}$

b) $h = \frac{P - p_1 + wh_1}{w}$

$h_1 = \frac{wh - (P - p_1)}{w}$

$$R_1 = \frac{E}{I} - R_2 - R_3$$

170. (Beam Bending Formula)

$$\boxed{M = \frac{WLX}{2} - \frac{WX^2}{2}}$$

To clear the fractions, we multiply both sides by "2" and get:

$$2M = WLX - WX^2$$

We can solve for "L" by:

(1) Isolating its term.

$$WLX = 2M + WX^2$$

(2) Multiplying both sides by the reciprocal of its coefficient.

$$L = \frac{2M + WX^2}{WX}$$

Write this solution in an equivalent form: $L =$ __________

---

171. Here is the same formula:

$$\boxed{M = \frac{WLX}{2} - \frac{WX^2}{2}}$$

Clearing the fractions, we get:

$$2M = WLX - WX^2$$

(a) To solve for "W", what must you do next?

____

(b) Do so and complete the solution:

$W =$ __________

Answer to 170:

$$L = \frac{2M}{WX} + X$$

Answers to 171:

a) Factor by the distributive principle since W appears in two terms.

b) $W = \dfrac{2M}{LX - X^2}$

SELF-TEST 5 (Frames 141-171)

1. Solve for P_2:

$$P_1P_2 = I - RP_2$$

2. Solve for T:

$$\frac{1}{P} - \frac{1}{A} = \frac{1}{T}$$

3. Solve for s:

$$\alpha = sd_1 + sd_2 - sd_3$$

4. Solve for V_1:

$$R = \frac{r_1r_2(V_1 - V_2)}{I}$$

5. Solve for P_1:

$$H = \alpha + t(P_1 - P_2)$$

6. Solve for A:

$$T = \frac{AFG}{3} - \frac{AG^2}{3}$$

ANSWERS: 1. $P_2 = \frac{I}{P_1 + R}$ 3. $s = \frac{\alpha}{d_1 + d_2 - d_3}$ 5. $P_1 = \frac{H - \alpha + P_2 t}{t}$ or $P_1 = \frac{H - \alpha}{t} + P_2$

2. $T = \frac{PA}{A - P}$ 4. $V_1 = \frac{RI + r_1 r_2 V_2}{r_1 r_2}$ or $V_1 = \frac{RI}{r_1 r_2} + V_2$ 6. $A = \frac{3T}{FG - G^2}$ or $A = \frac{3T}{G(F - G)}$

9-14 A PREFERRED FORM FOR FRACTIONAL SOLUTIONS

When solving for a variable in a formula, we occasionally obtain a solution like the following:

$$T = \frac{-T_R}{E - 1}$$

Mathematicians do not like a fractional solution which contains a negative sign in front of a single term like T_R. In this section, we will show you a method for writing the solution above in a more preferred form. To do so, we must show that two fractions are equivalent if their numerators and denominators are opposites.

172. Since $\frac{8}{4} = +2$ and $\frac{-8}{-4} = +2$, $\boxed{\frac{8}{4} = \frac{-8}{-4}}$

(a) The two numerators (8 and -8) are a pair of ____________.

(b) The two denominators (4 and -4) are a pair of ____________.

173. (a) Does $\frac{16}{4} = \frac{-16}{4}$? (b) Does $\frac{16}{4} = \frac{16}{-4}$? (c) Does $\frac{16}{4} = \frac{-16}{-4}$?

________ ________ ________

Answers to 172: a) opposites b) opposites

174. The point we are trying to make is this: $\boxed{\frac{a}{b} = \frac{-a}{-b}}$

That is, two fractions are equivalent if:

(1) Their numerators are a pair of opposites, and
(2) Their denominators are a pair of opposites.

Using the principle above, write each of the following in an equivalent form:

(a) $\frac{17}{8} =$ ____ (b) $\frac{-5}{-9} =$ ____ (c) $\frac{R}{W} =$ ____ (d) $\frac{-m}{-t} =$ ____

Answers to 173: a) No $(4 \neq -4)$ b) No $(4 \neq -4)$ c) Yes

175. Even with more complicated fractions, it is true that you can get an equivalent fraction by taking the opposite of both the numerator and denominator.

For example: $\boxed{\frac{a}{b - c} = \frac{-a}{c - b}}$

Let's plug in numbers to check this.

If $a = 16$, $b = 8$, $c = 4$, we get:

$$\frac{16}{8 - 4} = \frac{-16}{4 - 8}$$

$$\frac{16}{4} = \frac{-16}{-4}$$

$$4 = 4$$

(Continued on following page.)

Answers to 174: a) $\frac{-17}{-8}$ b) $\frac{5}{9}$ c) $\frac{-R}{-W}$ d) $\frac{m}{t}$

175. (Continued)

Using the principle above, fill in the missing numerator or denominator:

(a) $\dfrac{V_1}{T_1 - T_2} = \dfrac{\square}{T_2 - T_1}$ (c) $\dfrac{-D}{A - B} = \dfrac{\square}{B - A}$

(b) $\dfrac{M}{N - P} = \dfrac{-M}{\square}$ (d) $\dfrac{-G}{F - S} = \dfrac{G}{\square}$

---

176. Since you can use alternate methods in formula rearrangement, it is possible to obtain either of the following equivalent solutions when rearranging a particular formula:

(1) $t = \dfrac{a}{b - c}$ (2) $t = \dfrac{-a}{c - b}$

Mathematicians prefer the first solution since it involves fewer negative signs. If you obtain the second solution, you can easily take the opposite of both the numerator and denominator to reduce the number of negative signs, as follows:

$$t = \frac{-a}{c - b} = \frac{a}{b - c}$$

Write each formula in the preferred way:

(a) $V = \dfrac{-T}{S - D} =$ ________   (b) $R = \dfrac{-wr}{2F - w} =$ ________

Answers to 175: a) $-V_1$ b) $P - N$ c) D d) $S - F$

---

177. Let's solve for T. $E = \dfrac{T - T_R}{T}$

Clearing the fraction we get: $ET = T - T_R$

Notice that T appears in two terms, one on each side of the "=" sign. This means that factoring will be needed. There are two methods of solving for T. We will show one in this frame and one in the next.

$$ET = T - T_R$$

(Isolating both terms containing the factor T on one side.) $ET - T = -T_R$

(Factoring the left side.) $T(E - 1) = -T_R$

(Completing the solution.) $T = \dfrac{-T_R}{E - 1}$

The solution $T = \dfrac{-T_R}{E - 1}$ is not written in preferred form. How should it be written? $T =$ ________

Answers to 176: a) $V = \dfrac{T}{D - S}$ b) $R = \dfrac{wr}{w - 2F}$

Answer to 177: $T = \dfrac{T_R}{1 - E}$

178. Here is the second method of solving for T: $E = \frac{T - T_R}{T}$

(Fraction cleared)	$ET = T - T_R$
(Isolating "T" terms)	$ET - T = -T_R$
(Oppositing principle for equations)	$T - ET = T_R$
(Factoring by the distributive principle)	$T(1 - E) = T_R$
(Completing the solution)	$T = \frac{T_R}{1 - E}$

Do we arrive at the preferred solution by this method? ______

---

179. We have shown two methods of solving for T in the formula below. Each method is equally good. However, only one method leads immediately to the preferred form of the solution. This fact presents no problem since it is easy to change the non-preferred form of the solution to the preferred form.

Answer: Yes

(a) Solve for T_R: $E = \frac{T - T_R}{T}$

$T_R =$ ______

(b) In $T_R = T - ET$, the right side can be factored. Mathematicians prefer the factored form. Therefore, you should write:

$T_R =$ ______

---

180. (a) What equation do you get when the fractions are cleared? $\frac{1}{R_t} = \frac{1}{R_1} + \frac{1}{R_2}$

Answers: a) $T_R = T - ET$ b) $T_R = T(1 - E)$

______

(b) Solve for R_2: (Note: R_2 appears in two terms.)

$R_2 =$ ______

a) $R_1R_2 = R_tR_2 + R_tR_1$

b) $R_2 = \dfrac{R_tR_1}{R_1 - R_t}$

181. Solve for D: $\boxed{\dfrac{1}{D} + \dfrac{1}{d} = \dfrac{1}{f}}$

D = ________

$D = \dfrac{df}{d - f}$

182. Solve for C_2: $\boxed{\dfrac{1}{C_1} = \dfrac{1}{C_t} - \dfrac{1}{C_2}}$

C_2 = ________

$C_2 = \dfrac{C_1C_t}{C_1 - C_t}$

183. $\boxed{\mu = \dfrac{Mm}{M + m}}$ (Nuclear Technology Formula)

To solve for "M", we clear the fraction first and get:

$$\mu(M + m) = Mm$$

Notice that "M" appears on both sides. On the left, it appears inside the parentheses.

To solve for "M", we must multiply by the distributive principle on the left in order to get "M" out of the parentheses. We get:

$$\mu M + \mu m = Mm$$

Now it is easy to complete the solution, even though "M" appears in two terms. Do so:

M = ________

$M = \dfrac{\mu m}{m - \mu}$

184. In the last frame, we obtained this solution: $\boxed{M = \dfrac{\mu m}{m - \mu}}$

Can this solution be written in a simpler, equivalent form? ______

No. It cannot be simplified.

185. (Transistor Formula) $\alpha = \frac{\beta}{\beta + 1}$

(a) Clearing the fraction, you get: ________________

(b) To solve for β, what should you do next?

____

(c) Solve for β:

$\beta =$ ____________

---

186. (Fundamental Feedback Formula) $A_f = \frac{A}{1 - BA}$

(a) Clearing the fraction, you get: ________________

(b) To solve for A, you must multiply by the distributive principle on the left side to get A out of the parentheses. Do so:

____

(c) Complete the solution:

A = ____________

Answers to 185:

a) $\alpha(\beta + 1) = \beta$

b) Multiply by the distributive principle on the left side to get "β" out of the parentheses.

c) $\beta = \frac{\alpha}{1 - \alpha}$

---

187. Solve for m: $\mu = \frac{Mm}{M + m}$

m = ____________

Answers to 186:

a) $A_f(1 - BA) = A$

b) $A_f - A_fBA = A$

c) $A = \frac{A_f}{1 + A_fB}$

Answer to 187:

$m = \frac{\mu M}{M - \mu}$

188. Solve for B: $A_f = \dfrac{A}{1 - BA}$

B = ______

Answer: $B = \dfrac{A_f - A}{A_f A}$

189. Let's write the solution from the last frame in an equivalent form:

$$B = \frac{A_f - A}{A_f A} = \frac{A_f}{A_f A} - \frac{A}{A_f A} = \_\_\_\_ - \_\_\_\_$$

Answer: $B = \dfrac{1}{A} - \dfrac{1}{A_f}$

190. To solve for T, you merely clear the fraction. Do so: $S = \dfrac{T}{2t(a - t)(b - t_1)}$

T = ______

Answer: $T = 2tS(a - t)(b - t_1)$

9-15 THE ADDITION AXIOM AND INSTANCES OF THE DISTRIBUTIVE PRINCIPLE

In the formula below, there are two terms on the right side. The second term is an instance of the distributive principle.

$$P = p_1 + w(h - h_1)$$

To solve for "p_1" we must eliminate "$w(h - h_1)$" from the right. It is easier to do so without multiplying by the distributive principle first. We merely use the addition axiom, adding the opposite of "$w(h - h_1)$" to both sides. We will show the method in this section.

191. Before applying the addition axiom to formulas like the preceding one, we must examine the meaning of opposites for two-factor terms like: (5)(4)

Since (5)(4) = 20, the opposite of (5)(4) must be an expression which equals "-20".

We can obtain the opposite of (5)(4) by replacing one (not both) of the factors by its opposite.

(-5)(4) and (5)(-4) } are opposites of (5)(4), since both equal -20.

If we replace both factors by their opposites, we get: (-5)(-4)

Is (-5)(-4) the opposite of (5)(4)? _______

192. Which of the following expressions are the opposite of (10)(2)? _______

(a) (-10)(2) (b) (-10)(-2) (c) (10)(-2)

Answer to 191: No, since both equal +20 and their sum is 40, not "0".

193. 5(3 + 2) is an instance of the distributive principle (over addition). It is also a two-factor term.

Since 5(3 + 2) = 25, its opposite must equal "-25".

We can again obtain its opposite by replacing one of the factors by its opposite.

$$(-5)(3 + 2) = (-5)(5) = -25$$

or

$$(5)[(-3) + (-2)] = (5)(-5) = -25$$

If we replace both factors by their opposites, we get: (-5)[(-3) + (-2)]

(a) (-5)[(-3) + (-2)] = _______

(b) Is (-5)[(-3) + (-2)] the opposite of 5(3 + 2)? _______

Answer to 192: (a) and (c), not (b)

194. 5(4 - 1) is an instance of the distributive principle (over subtraction). It is a two-factor term.

Since 5(4 - 1) = 15, its opposite must equal "-15".

(a) Replacing 5 by its opposite, we get: (-5)(4 - 1). Is this new expression the opposite of (5)(4 - 1)? _______

(b) Replacing (4 - 1) by its opposite, we get: 5(1 - 4). Is this new expression the opposite of 5(4 - 1)? _______

(c) Replacing both factors by their opposites, we get: (-5)(1 - 4). Is this new expression the opposite of (5)(4 - 1)? _______

Answer to 193: a) +25 b) No. Their sum is +50.

195. Here is another instance of the distributive principle (over addition): $3(x + 2)$

We can obtain its opposite by replacing one factor (not both) by its opposite. However, we will only demonstrate the case in which the simpler factor is replaced by its opposite. (Only this case is useful in rearranging formulas.)

$(-3)(x + 2)$ is the opposite of $3(x + 2)$, since:

$$(-3)(x + 2) + 3(x + 2) = (-3x) + (-6) + 3x + 6$$
$$= \underbrace{(-3x) + 3x}_{0} + \underbrace{(-6) + 6}_{0}$$
$$= 0 + 0 = 0$$

Write the opposite of $V(t + I)$ by replacing the simpler factor with its opposite: ________________

a) Yes, since: $(-5)(4 - 1) = (-5)(3) = -15$

b) Yes, since: $(5)(1 - 4) = (5)(-3) = -15$

c) No, since: $(-5)(1 - 4) = (-5)(-3) = +15$

---

196. Here is another instance of the distributive principle (over subtraction): $5(y - 3)$

$(-5)(y - 3)$ is the opposite of $5(y - 3)$, since:

$$(-5)(y - 3) + 5(y - 3) = (-5y) + 15 + 5y - 15$$
$$= \underbrace{(-5y) + 5y}_{0} + \underbrace{15 + (-15)}_{0}$$
$$= 0 + 0 = 0$$

Write the opposite of $m(p - q)$ by replacing the simpler factor with its opposite: ________________

$(-V)(t + I)$

---

197. When manipulating equations, we always obtain the opposite of an instance of the distributive principle by replacing the simpler factor with its opposite. Write the opposites of these:

(a) $3(y + 2)$ ________ (b) $S(t_1 + t_2)$ ________ (c) $a(d - c)$ ________

$(-m)(p - q)$

---

198. We can convert the following subtractions to additions by adding the opposite of the instances of the distributive principle:

$$3 - 4(x + 5) = 3 + [\text{the opposite of } 4(x + 5)]$$
$$= 3 + (-4)(x + 5)$$

$$g - d(b - c) = g + [\text{the opposite of } d(b - c)]$$
$$= \text{\_\_\_\_\_\_\_\_\_\_\_\_\_\_\_\_}$$

a) $(-3)(y + 2)$

b) $(-S)(t_1 + t_2)$

c) $(-a)(d - c)$

---

199. We can convert the following addition back to subtraction by subtracting the opposite of the instance of the distributive principle:

$$10 + (-5)(y + 2) = 10 - [\text{the opposite of } (-5)(y + 2)]$$
$$= 10 - 5(y + 2)$$

Convert the following addition to subtraction:

$V + (-c)(t - s) =$ ________________

$g + (-d)(b - c)$

$V - c(t - s)$

200. $\boxed{f = g + d(b - c)}$

To solve for g, we add the opposite of $d(b - c)$ to both sides. Its opposite is $(-d)(b - c)$. We get:

$$f + (-d)(b - c) = g + \underbrace{d(b - c) + (-d)(b - c)}_{0}$$

$$f + (-d)(b - c) = g + 0$$

$$f + (-d)(b - c) = g$$

We then change the addition on the left back to subtraction. Do so:

$g =$ ______________

Answer: $g = f - d(b - c)$

---

201. $\boxed{P = p_1 + w(h - h_1)}$ (Hydraulics Formula)

(a) To solve for p_1, what should you add to both sides? ______________

(b) Do so, and convert the addition to subtraction:

$p_1 =$ ______________

Answers: a) $(-w)(h - h_1)$ b) $p_1 = P - w(h - h_1)$

---

202. Solve for r_3: $\boxed{F_1r_1 + F_2r_2 + F_3r_3 = 0}$

$r_3 =$ ______________

Answer: $r_3 = \dfrac{-F_1r_1 - F_2r_2}{F_3}$

---

203. $\boxed{P = p_1 + w(h - h_1)}$

To solve for w, we add $-p_1$ to both sides and get:

$$P - p_1 = w(h - h_1)$$

(a) Now to solve for w, what would you do?

____

(b) Solve for w:

$w =$ ______________

204. $\boxed{I_c = \beta I_B + (\beta + 1)I_{co}}$ (Transistor Formula)

To solve for I_B, we must isolate the term (βI_B) in which I_B appears. To isolate that term, we must eliminate $(\beta + 1)I_{co}$ from the right side.

(a) What must you add to both sides to eliminate $(\beta + 1)I_{co}$ from the right side? ______________

(b) Do so and complete the solution:

I_B = ________________

Answers to 203:

a) Multiply both sides by $\frac{1}{h - h_1}$

b) $w = \frac{P - p_1}{h - h_1}$

---

205. Solve for I_{co}: $\boxed{I_c = \beta I_B + (\beta + 1)I_{co}}$

I_{co} = ________________

Answers to 204:

a) $(\beta + 1)(-I_{co})$, its opposite.

b) $I_B = \frac{I_c - (\beta + 1)I_{co}}{\beta}$

---

206. $\boxed{2RF = w(R - r)}$ (Differential Pulley Formula)

We want to solve for "R". It appears in two terms. Since it appears within the parentheses on the right, we must multiply by the distributive principle on the right first. Solve for R:

R = ________________

Answer to 205:

$I_{co} = \frac{I_c - \beta I_B}{\beta + 1}$

Answer to 206:

$R = \frac{wr}{w - 2F}$

207. $I_c = \beta I_B + (\beta + 1)I_{co}$

(a) β appears in two places. To solve for β, you must multiply $(\beta + 1)I_{co}$ first to get β out of the parentheses. Do so:

____

(b) To solve for β, what would you do next?

____

(c) Solve for β:

β = ______________________

208. $2RF = w(R - r)$ (Differential Pulley Formula)

(a) Solve for w:

(b) Solve for r:

w = ______________ r = ______________

209. $\alpha = \dfrac{L_2 - L_1}{L_1(t_2 - t_1)}$

(a) Clear the fraction. You get: ______________________

(b) Solve for L_2:

L_2 = ______________________

Answers to 207:

a) $I_c = \beta I_B + \beta I_{co} + I_{co}$

b) Add "$-I_{co}$" to both sides.

c) $\beta = \dfrac{I_c - I_{co}}{I_B + I_{co}}$

Answers to 208:

a) $w = \dfrac{2RF}{R - r}$

b) $r = \dfrac{Rw - 2RF}{w}$ or

$r = \dfrac{R(w - 2F)}{w}$ or

$r = R - \dfrac{2RF}{w}$

210. Solve for T: $B = \dfrac{T(V_1 - V_2)}{V_1(P_2 - P_1)}$ (Fluid Compression Formula)

T = ________

a) $\alpha L_1(t_2 - t_1) = L_2 - L_1$

b) $L_2 = \alpha L_1(t_2 - t_1) + L_1$

or

$L_2 = L_1[\alpha(t_2 - t_1) + 1]$

---

211. Let's solve this formula for L_1: $\alpha = \dfrac{L_2 - L_1}{L_1(t_2 - t_1)}$

Clearing the fraction, we get: $\alpha L_1(t_2 - t_1) = L_2 - L_1$

Multiplying the factors on the left, we get: $\alpha L_1 t_2 - \alpha L_1 t_1 = L_2 - L_1$

Isolating all the terms containing L_1 on one side, we get: $\alpha L_1 t_2 - \alpha L_1 t_1 + L_1 = L_2$

(a) What can you factor out on the left side? ________

(b) Do so and solve for L_1:

L_1 = ________

$$T = \frac{BV_1(P_2 - P_1)}{V_1 - V_2}$$

---

a) L_1

b) $L_1 = \dfrac{L_2}{\alpha t_2 - \alpha t_1 + 1}$

212. Solve for P_2: $B = \dfrac{T(V_1 - V_2)}{V_1(P_2 - P_1)}$

$P_2 =$ ____________

$$P_2 = \frac{TV_1 - TV_2 + BV_1P_1}{BV_1}$$

or

$$P_2 = \frac{T(V_1 - V_2) + BV_1P_1}{BV_1}$$

or

$$P_2 = \frac{T(V_1 - V_2)}{BV_1} + P_1$$

213. Solve for t_2: $\alpha = \dfrac{L_2 - L_1}{L_1(t_2 - t_1)}$

$t_2 =$ ____________

$$t_2 = \frac{L_2 - L_1}{\alpha L_1} + t_1$$

or

$$t_2 = \frac{L_2 - L_1 + \alpha L_1 t_1}{\alpha L_1}$$

SELF-TEST 6 (Frames 172-213)

1. Solve for α:

$$\beta = \frac{\alpha}{1 - \alpha}$$

2. Solve for R_1:

$$\frac{1}{R_t} = \frac{1}{R_1} + \frac{1}{R_2}$$

3. Solve for t_1:

$$\alpha = \frac{L_2 - L_1}{L_1(t_2 - t_1)}$$

4. Solve for μ:

$$M = \frac{\mu}{m - \mu}$$

5. Solve for d:

$$\frac{1}{D} + \frac{1}{d} = \frac{1}{f}$$

6. Solve for V_1:

$$B = \frac{T(V_1 - V_2)}{V_1(P_2 - P_1)}$$

7. Solve for t_1:

$$H = ms(t_2 - t_1)$$

8. Solve for A_f:

$$A = \frac{A_f}{1 + BA_f}$$

ANSWERS:

1. $\alpha = \dfrac{\beta}{\beta + 1}$

2. $R_1 = \dfrac{R_t R_2}{R_2 - R_t}$

3. $t_1 = \dfrac{\alpha L_1 t_2 - L_2 + L_1}{\alpha L_1}$

or $t_1 = t_2 - \dfrac{L_2 - L_1}{\alpha L_1}$

or $t_1 = t_2 + \dfrac{L_1 - L_2}{\alpha L_1}$

4. $\mu = \dfrac{Mm}{M + 1}$

5. $d = \dfrac{Df}{D - f}$

6. $V_1 = \dfrac{TV_2}{BP_1 - BP_2 + T}$

7. $t_1 = t_2 - \dfrac{H}{ms}$

or $t_1 = \dfrac{mst_2 - H}{ms}$

8. $A_f = \dfrac{A}{1 - AB}$

PRACTICE PROBLEMS: FORMULA REARRANGEMENT

All the types of formulas that you have learned to rearrange are represented here. Test your skill by solving for the indicated letter. Do your work on a separate sheet. Answers follow.

1. Solve for b: $A = \frac{1}{2}bh$
2. Solve for A_1: $\frac{F_1}{F_2} = \frac{A_1}{A_2}$
3. Solve for h_1: $P = p_1 + w(h - h_1)$
4. Solve for f: $\frac{1}{D} + \frac{1}{d} = \frac{1}{f}$
5. Solve for r: $e = E - Ir$
6. Solve for T_1: $Q = C(T_1 - T_2)$
7. Solve for f: $X_C = \frac{1}{2\pi fC}$
8. Solve for β: $\alpha = \frac{\beta}{\beta + 1}$
9. Solve for T_a: $T_j = T_a + \theta_T P_T$
10. Solve for m_1: $F = \frac{Km_1m_2}{s^2}$
11. Solve for T: $E = \frac{T - T_R}{T}$
12. Solve for V_2: $W_S = P(V_1 - V_2)$
13. Solve for g: $F + g(r - r_1) = S$
14. Solve for r_1: $F_1r_1 + F_2r_2 + F_3r_3 = 0$
15. Solve for F_1: $H = \frac{\pi dR(F_1 - F_2)}{33,000}$
16. Solve for W: $M = \frac{WLX}{2} - \frac{WX^2}{2}$
17. Solve for V_1: $a = \frac{V_2 - V_1}{t}$
18. Solve for P: $M = P(L - X)$
19. Solve for P_1: $B = \frac{T(V_1 - V_2)}{V_1(P_2 - P_1)}$
20. Solve for E: $\frac{1}{P} = \frac{1}{I} - \frac{1}{E}$

ANSWERS:

1. $b = \frac{2A}{h}$
2. $A_1 = \frac{F_1A_2}{F_2}$
3. $h_1 = \frac{p_1 + wh - P}{w}$
4. $f = \frac{Dd}{d + D}$
5. $r = \frac{E - e}{I}$
6. $T_1 = \frac{Q}{C} + T_2$
7. $f = \frac{1}{2\pi CX_C}$
8. $\beta = \frac{\alpha}{1 - \alpha}$
9. $T_a = T_j - \theta_T P_T$
10. $m_1 = \frac{Fs^2}{Km_2}$
11. $T = \frac{T_R}{1 - E}$
12. $V_2 = V_1 - \frac{W_S}{P}$
13. $g = \frac{S - F}{r - r_1}$
14. $r_1 = \frac{-F_2r_2 - F_3r_3}{F_1}$
15. $F_1 = \frac{33,000H}{\pi dR} + F_2$
16. $W = \frac{2M}{LX - X^2}$
17. $V_1 = V_2 - at$
18. $P = \frac{M}{L - X}$
19. $P_1 = \frac{BV_1P_2 - T(V_1 - V_2)}{BV_1}$
20. $E = \frac{PI}{P - I}$

Chapter 10 SYSTEMS OF EQUATIONS AND FORMULA DERIVATION

In this chapter, we will examine systems of two equations and systems of two formulas. The following major topics will be discussed:

(1) The meaning of a system of equations or formulas.
(2) Methods for solving systems of equations.
(3) Some applied situations in which a system of equations must be solved.
(4) Methods for eliminating a variable (or variables) from a system of formulas to derive a new formula.

Though the solution of a system of equations in applied situations frequently requires a substantial amount of computation, the amount of computation in this chapter will be kept at a minimum. The computation will be minimized so that the principles of solutions can be emphasized.

10-1 MEANING OF A SYSTEM OF EQUATIONS

1. Here are two examples of systems of equations:

System #1	System #2
(1) $x - y = 4$	(1) $3F_1 - 4F_2 = 8$
(2) $x + y = 10$	(2) $2F_1 + 3F_2 = 6$

Note: (1) There are two variables in each equation in each system.
(2) In each system, the same two variables appear in each equation.

(a) In System #1 the two common variables are _____ and _____.

(b) In System #2 the two common variables are _____ and _____.

Answers: a) x and y b) F_1 and F_2

2. In a system of equations, the two equations do not have to be written in any particular form. In the two systems below, various arrangements of the letter-terms appear.

System #1	System #2
(1) $m = 9t$	(1) $R = S + 7$
(2) $m + t = 50$	(2) $2R + 3S = 18$

(a) The two common variables in System #1 are _____ and _____.

(b) The two common variables in System #2 are _____ and _____.

Answers: a) m and t b) R and S

3. Two equations form a system of equations only if they contain the same two variables.

Which of the following pairs of equations form a system of equations? ________

(a) $x + y = 12$, $p - q = 10$ (b) $t_1 = t_2 + 5$, $t_1 + t_2 = 9$ (c) $y = 5x$, $y + 2z = 9$

---

4. Here is a system of two equations:

(1) $x - y = 4$
(2) $x + y = 10$

(a) Does $x = 5$ and $y = 4$ satisfy: Equation (1)? ______ Equation (2)? ______

(b) Does $x = 9$ and $y = 5$ satisfy: Equation (1)? ______ Equation (2)? ______

(c) Does $x = 6$ and $y = 4$ satisfy: Equation (1)? ______ Equation (2)? ______

(d) Does $x = 7$ and $y = 3$ satisfy: Equation (1)? ______ Equation (2)? ______

Answer: Only (b). The pairs of equations in (a) and (c) do not contain the same two variables.

---

5. In the last frame, the pair of values "$x = 7$, $y = 3$" satisfied both equations. Therefore, we call this pair of values the solution of the system of equations.

Ordinarily, we are interested in solving a system of equations. By "solving" we mean finding a pair of values for the two variables which satisfies both equations at the same time.

Here is another system of equations:

(1) $F_1 + F_2 = 14$
(2) $F_1 - F_2 = 2$

Only one of the following pairs of values is a solution of the system above. Which pair is it? ______

(a) $F_1 = 10$, $F_2 = 4$ (c) $F_1 = 7$, $F_2 = 7$

(b) $F_1 = 8$, $F_2 = 6$ (d) $F_1 = 6$, $F_2 = 8$

Answers:
a) (1) No. (2) No.
b) (1) Yes. (2) No.
c) (1) No. (2) Yes.
d) (1) Yes. (2) Yes.

---

6. Here is another system of equations:

(1) $p = q + 1$
(2) $p = 5 - q$

Equation (1)

p	q
1	0
2	1
3	2
6	5
11	10

Equation (2)

p	q
5	0
4	1
3	2
0	5
-5	10

As we saw in Chapter 7, there are an infinite number of solutions for any single two-variable equation. We can find various solutions for each individual equation above by plugging in values for "q" and computing the corresponding values of "p". We have made up a table of solutions for each equation. These two tables are shown above.

(Continued on following page.)

Answer: (b) $F_1 = 8$, $F_2 = 6$

6. (Continued)

As you can see from the tables, there are various solutions for Equation (1) which do not satisfy Equation (2). There are also various solutions for Equation (2) which do not satisfy Equation (1). However, there is only one pair of values in the tables which satisfies both equations. This pair of values is the solution of the system of equations. It is: p = ____, q = ____

---

7. We can attempt to solve a system of equations by the trial-and-error method. When using this method, we simply keep trying pairs of values until we find a pair which satisfies both equations.

Use the trial-and-error method to solve the system at the right.

(1)	$x + y = 5$
(2)	$y = 4x$

The solution is: x = ____, y = ____.

p = 3, q = 2

---

x = 1, y = 4

10-2 GRAPHICAL SOLUTIONS OF SYSTEMS OF EQUATIONS

The trial-and-error method of solving a system of equations is not very efficient, to say the least. Therefore, mathematicians have developed various other methods for solving systems. One alternate method is to find the solution graphically. We will discuss the graphical method in this section.

---

8. The graphical method for solving a system of equations is based on this fact: The coordinates of any point on a line satisfy the equation of the line.

On the set of axes below, we have graphed both equations in the system on the right:

(1)	$x - y = 1$
(2)	$x + y = 5$

(a) There is only one point which lies on both lines. It is the point of intersection. The coordinates of this point are:

x = ____, y = ____

(b) Do the coordinates of the point of intersection satisfy both equations?

____

(c) Therefore, the solution of this system is:

x = ____, y = ____

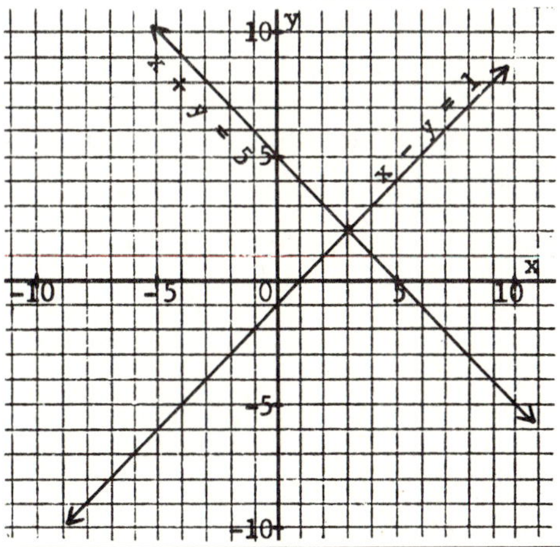

---

a) x = 3, y = 2

b) Yes.

c) x = 3, y = 2

9. On the axes below, we have graphed each equation from the system on the right.

(1)	$2x - y = 7$
(2)	$x + y = 2$

(a) The coordinates of point A are (6, 5). These coordinates satisfy equation #1. Do they also satisfy equation #2 ? ______

(b) The coordinates of point B are (-4, 6). These coordinates satisfy equation #2. Do they also satisfy equation #1 ? ______

(c) The coordinates of point C are (3, -1). Do these coordinates satisfy both equations? ______

(d) Therefore, the solution of this system is:

x = ________, y = ________

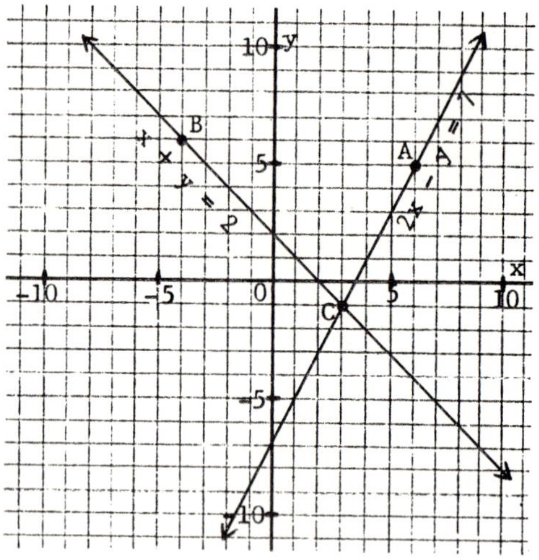

Answer to Frame 9: a) No. b) No. c) Yes. d) $x = 3$, $y = -1$

10. To solve a system of equations by the graphical method, we graph both equations on the same set of axes. We then look for points of intersection of the two graphs. The coordinates of any point of intersection are a solution of the system since its coordinates satisfy both equations. In fact, only the coordinates of a point of intersection are a solution since only intersection points lie on both graphs.

Whenever the graph of each equation is a straight line and the lines intersect:

(1) There is only one point of intersection.
(2) Therefore, there is only one solution to the system.

Let's solve the system on the right graphically. The graph of each equation is a straight line. Equation (1) is already graphed on the axes below. Complete the table for Equation (2) and then graph Equation (2) on the same axes.

(1)	$y = 3x$
(2)	$x + y = 4$

$x + y = 4$

x	y
-3	
0	
+2	
+3	

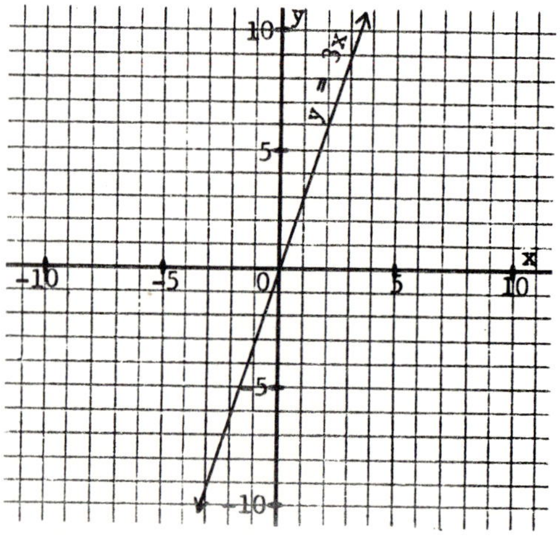

(a) There is only one point of intersection. Its coordinates are (,).

(b) Therefore, the only solution of the system is:

x = ______, y = ______

Answer to Frame 10: a) (1, 3)

b) x = 1, y = 3

x + y = 4

x	y
-3	7
0	4
+2	2
+3	1

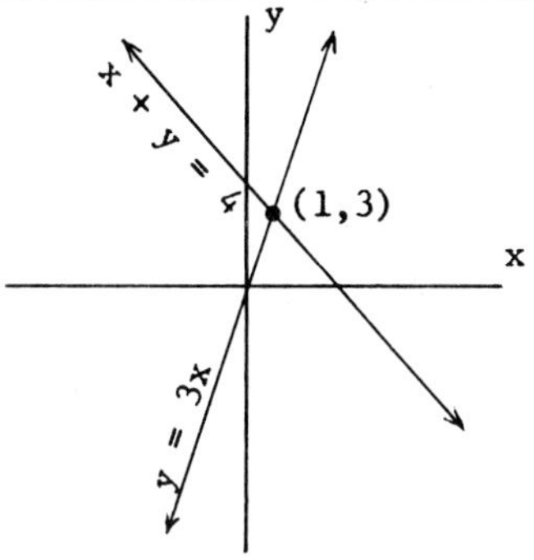

11. The graph of each equation in the system on the right is a straight line. Equation (2) is already graphed on the axes below. Complete the table for Equation (1) and then graph Equation (1) on the same axes:

(1) y = 10x
(2) y - x = 9

y = 10x

x	y
-2	
-1	
0	
+2	

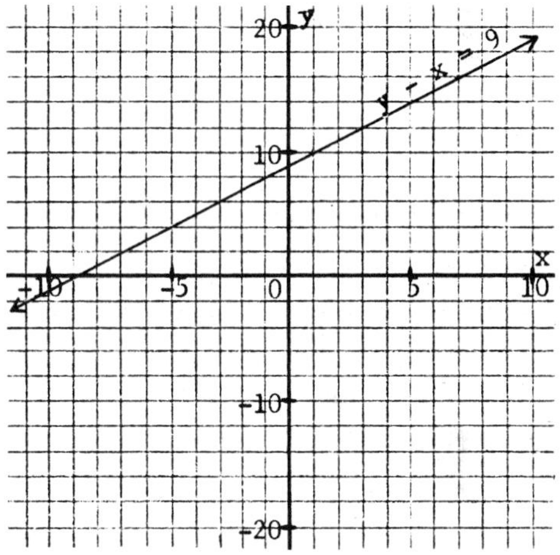

(a) The coordinates of the point of intersection are (,).

(b) Therefore, the solution of the system is:

x = _____, y = _____

Answer to Frame 11: a) (1, 10)

b) x = 1, y = 10

y = 10x

x	y
-2	-20
-1	-10
0	0
+2	20

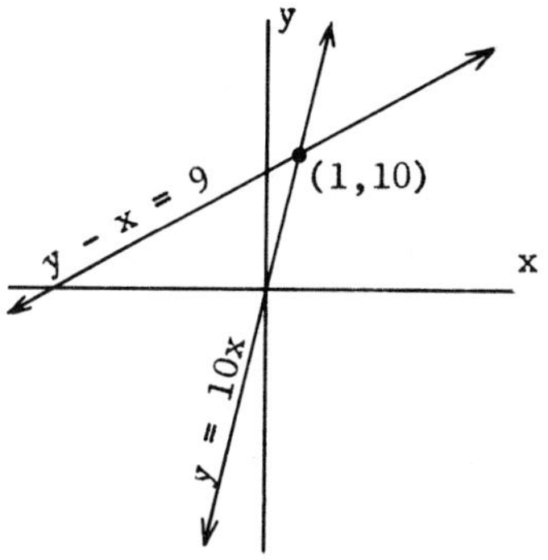

12. The graph of each equation in the system on the right is a straight line. Using the tables provided, graph both equations on the axes below. Plot "q" as the ordinate (on the vertical axis).

(1) $2p - q = 6$
(2) $p + 2q = 13$

(a) The coordinates of the point of intersection are (,).

(b) Therefore, the solution of the system is:

p = _____, q = _____

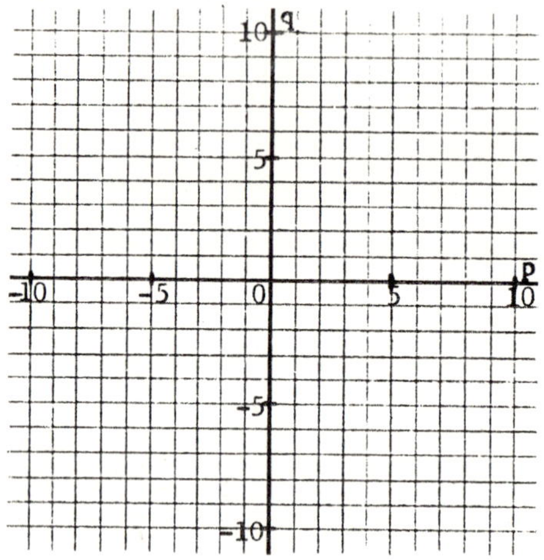

Equation (1)

p	q

Equation (2)

p	q

Answer to Frame 12: a) (5, 4)

b) p = 5, q = 4

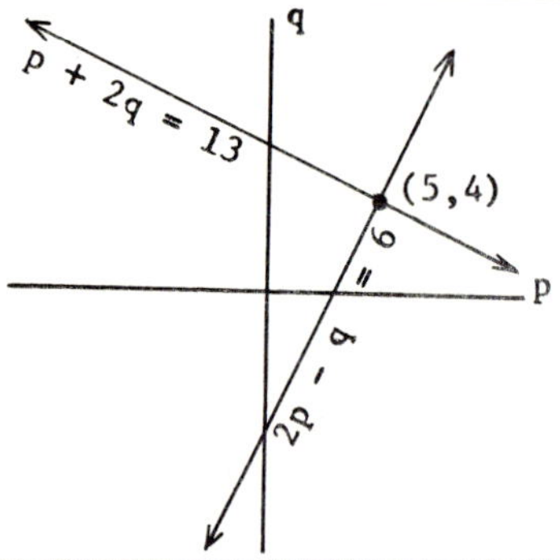

13. The two equations in the following system are graphed on the right.

(1) $3R - 2S = -9$
(2) $6R - 4S = 12$

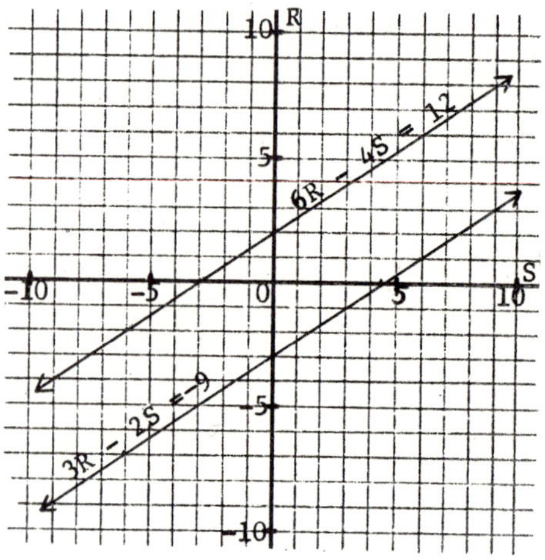

The two graphed lines are parallel. Since they are parallel:

(1) Is there a point of intersection? _____

(2) Does the system have a solution? _____

Answer to Frame 13: a) No b) No, since there is no point of intersection.

---

14. In the last frame, the graphed lines for the two equations were parallel. Therefore, there was no point on intersection and no solution. In applied situations, a system of equations without a solution is extremely rare. Therefore, we will not include any more systems of that type in this chapter.

The graphical method of solving a system of equations works well when the solution is a pair of whole numbers.

The solution of the system below, however, is not a pair of whole numbers:

$$x - y = 1.62$$
$$x + y = 6.78$$

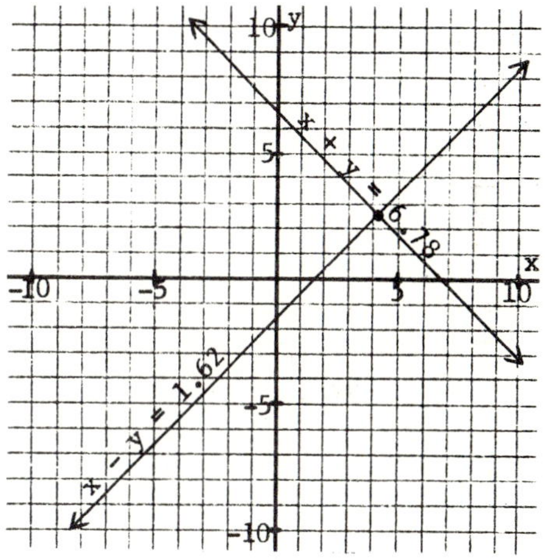

These two equations have been graphed at the right. Note that the coordinates of their intersection point are not whole numbers.

In obtaining the coordinates of the point of intersection, your readings on the two axes must clearly be estimates. The solution is close to:

$$x = 4.1 \text{ or } 4.2, \quad y = 2.6 \text{ or } 2.7$$

Since no pair of these values satisfy the two equations exactly, they are not the exact solutions. The exact solution is:

$$x = 4.2, \quad y = 2.58$$

To obtain the exact solution when both values are decimals, we will need a method which is more precise than the graphical method. The method needed is the algebraic method, which will be studied in the next section.

Go to Self-Test 1.

SELF-TEST 1 (Frames 1-14)

1. The two equations at the right are not a system of equations. Why not? ____________

$$2x - 3y = 8$$
$$4t - 5w = 3$$

2. Which one of the following pairs of values is a solution of the system at the right? ________

$$r + 2s = 7$$
$$5r - s = 2$$

(a) r = 3, s = 2 (b) r = -1, s = 4 (c) r = 1, s = 3 (d) r = 5, s = 1

Given this system:

$$x - y = 8$$
$$2x + y = 1$$

3. Graph both equations on the axes at the right.

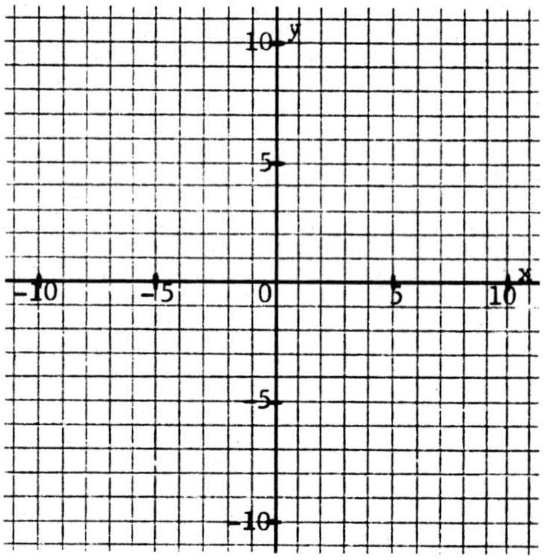

4. From the graph, the solution of the system is:

x = ______

y = ______

ANSWERS:

1. Because the equations do not have common variables.

2. (c) r = 1, s = 3

3. Your graph should look like the graph at the right.

4. x = 3, y = -5

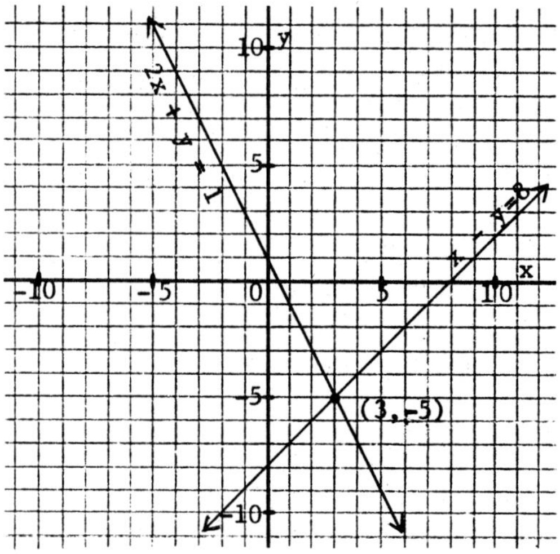

10-3 THE EQUIVALENCE METHOD OF SOLVING SYSTEMS OF EQUATIONS

Though the graphical method of solving systems of equations is occasionally used in applied situations, it has two drawbacks: (1) It is time-consuming; (2) It does not always give exact values for the pair of coordinates or solution.

Ordinarily, systems of equations are solved by algebraic manipulations. Four methods are possible: (1) the equivalence method, (2) the substitution method, (3) the addition-subtraction method, and (4) the determinant method. In this chapter, only the first two of these four methods will be discussed. The determinant method will be discussed in a later book. The addition-subtraction method will not be discussed in this series because it is not very suitable for solving systems arising from applied situations; it also does not generalize to systems of formulas and formula derivation.

In this section, we will discuss the equivalence method for solving systems of equations. Only very simple systems will be solved so that the overall procedure and the principle of "eliminating a variable" can be emphasized.

---

15. Every algebraic technique for solving systems of equations is based on some principle by which one of the variables can be eliminated. In the equivalence method, a variable is eliminated by the EQUIVALENCE PRINCIPLE.

> The EQUIVALENCE PRINCIPLE says this: If the same quantity equals two other quantities, those two quantities are equal.

This principle is stated symbolically on the left below. An example is given on the right.

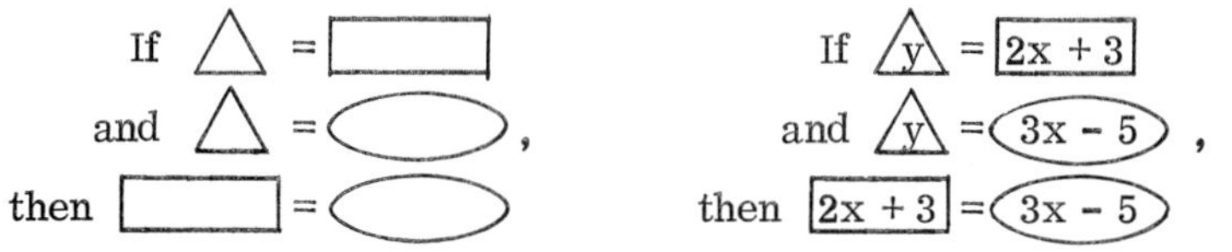

Using the equivalence principle, complete each of these:

(a) If $R = 3S$
and $R = S - 7$,

then _______ = _______

(b) If $p = 3q - 1$
and $p = 5q + 7$,

then _______ = _______

---

a) $3S = S - 7$

b) $3q - 1 = 5q + 7$

16. When using the equivalence principle to solve a system of equations, the same variable must be "solved for" in each equation.

In a two-variable equation, a variable is "solved for" if:

(1) It appears alone on one side without a numerical or literal coefficient.

(2) It does not also appear on the other side.

The variable "y" is "solved for" in each of these equations:

$y = 5x$ $\qquad$ $y = 2x - 1$

The variable "y" is not "solved for" in these equations because it has a coefficient in each:

$3y = 4x + 9$ $\qquad$ $ky = 7x$

The variable "y" is not "solved for" in these equations because it appears on both sides:

$y = 3x - y$ $\qquad$ $y = 3y + 5x$

In which of these equations is "m" solved for? _______

(a) $3m = 7t$ $\qquad$ (c) $tm = t + 2$

(b) $m = 5t + 1$ $\qquad$ (d) $m = 2t - 3m$

Only in (b)

17. If the same variable is "solved for" in each equation in a system, we can eliminate that variable by applying the equivalence principle. For example:

If $y = 10x$
and $y = x + 5$

then $10x = x + 5$

(a) How many variables are there in the new equation? _______

(b) What variable has been eliminated in the new equation? _______

---

a) One. Only "x"

b) "y"

18. The variable "T" is solved for in each equation in the system on the right:

(1) $T = 2R$
(2) $T = 5R - 6$

(a) What new equation do you get if you apply the equivalence principle to this system? _______ = _____________

(b) What variable is eliminated in this new equation? _______

---

a) $2R = 5R - 6$

b) "T"

19. To solve the system of equations on the right, we must find a pair of values for "x" and "y" which satisfy both equations. To do so, we need three steps.

$$y = 5x$$
$$y = 2x + 6$$

Step 1: Using the equivalence principle to eliminate "y", the "solved for" variable. We get:

$$5x = 2x + 6$$

Step 2: Finding the numerical value of "x", the "non-solved for" variable, by solving the new equation. We get:

$$5x = 2x + 6$$
$$3x = 6 \quad \text{or} \quad x = 2$$

Step 3: Finding the corresponding numerical value of "y", the "solved for" variable, by substituting in either of the original equations of the system.

Substituting "2" for "x" in the top equation, we get:

$$y = 5x = 5(2) = 10$$

Substituting "2" for "x" in the bottom equation, we get:

$$y = 2x + 6 = 2(2) + 6 = 4 + 6 = 10$$

Note: We obtained the same value "10" for "y" by substituting in either original equation.

Therefore, the solution of the system of equations is:

x = ________, y = ________

Answer: $x = 2$, $y = 10$

20. In the last frame, we obtained the solution ($x = 2$, $y = 10$) for the system of equations on the right. Check this solution in each of the two equations below.

$$y = 5x$$
$$y = 2x + 6$$

(a) $y = 5x$

(b) $y = 2x + 6$

Answers:

a) $y = 5x$
$10 = 5(2)$
$10 = 10$

b) $y = 2x + 6$
$10 = 2(2) + 6$
$10 = 4 + 6$
$10 = 10$

21. Let's solve this system: $\boxed{\begin{aligned} p &= 7q \\ p &= 4q - 6 \end{aligned}}$

(a) Eliminate "p" by applying the equivalence principle. Write the new equation.

_______ = ___________

(b) Find the numerical value of "q" by solving this new equation.

q = __________

(c) Plug this value for "q" in either original equation to find the corresponding value of "p".

p = __________

(d) The solution is: p = _______, q = _______

Answers: a) $7q = 4q - 6$ b) $q = -2$ c) $p = -14$ d) $p = -14,\ q = -2$

22. When the same variable is "solved for" in each equation in a system of two equations, the system can be solved in these three steps:

<u>Step 1</u>: Use the equivalence principle to eliminate the "solved for" variable.
<u>Step 2</u>: Find the numerical value of the "non-solved for" variable by solving this new equation.
<u>Step 3</u>: Find the corresponding value of the "solved for" variable by plugging the value of the "non-solved for" variable in either of the original equations.

(a) Using these three steps, solve this system: $\boxed{\begin{aligned} P &= 5E \\ P &= E + 4 \end{aligned}}$

E = _______, P = _______

(b) Check your solution in each original equation:

$P = 5E$ $\qquad\qquad$ $P = E + 4$

Answers: a) $E = 1,\ P = 5$

b)

$P = 5E$	$P = E + 4$
$5 = 5(1)$	$5 = 1 + 4$
$5 = 5$	$5 = 5$

10-4 REARRANGEMENTS NEEDED TO USE THE EQUIVALENCE METHOD

In every system solved in the last section, the same variable was already "solved for" in each equation in the system. Ordinarily, this will not be the case. Systems like those below are more common.

$$\boxed{\begin{aligned} 4d + R &= 9 \\ d + R &= 6 \end{aligned}} \qquad \boxed{\begin{aligned} 2y &= x - 5 \\ 3x + y &= 1 \end{aligned}}$$

To solve systems like those above by the equivalence method, we must solve for the same variable in each equation first so that the equivalence principle can be applied. We will discuss various types of equation rearrangements in this section.

23. Neither variable is solved for in the equation below.

$$x + y = 5$$

To solve for "x", we add "the opposite of y" or "-y" to both sides. We get:

$$x + \underbrace{y + (-y)}_{0} = 5 + (-y)$$

$$x + \quad 0 \quad = 5 + (-y)$$

$$x = 5 + (-y)$$

Ordinarily, we do not leave the solution in a form in which there is an addition of a negative quantity like "-y". We convert this addition to a subtraction. To make this conversion from addition to subtraction in the equation above, we subtract "the opposite of -y". We get:

$$x = 5 + (-y)$$

$$= 5 - (\text{the opposite of } -y)$$

$$= 5 - y$$

Solve for "y" in each equation below and write the solution in the preferred form:

(a) $x + y = 10$ (b) $y + 5x = 7$

y = ________ y = ________

a) $y = 10 - x$

b) $y = 7 - 5x$

24. To solve for "m" in the equation below, we add "-t" to both sides. We get:

$$t - m = 3$$

$$\underbrace{(-t) + t}_{0} - m = 3 + (-t)$$

$$0 \quad - m = 3 + (-t)$$

$$-m = 3 + (-t)$$

or

$$-m = 3 - t$$

The solution is not complete because there is a "-" in front of the "m". To eliminate this "-" we use the oppositing principle. That is, we replace each side with its opposite.

Since the opposite of "-m" is "m" and the opposite of "3 - t" is "t - 3", we get:

$$m = t - 3$$

To apply the oppositing principle to the equation above, you must remember that we get the opposite of a subtraction <u>by merely interchanging the terms.</u> That is:

The "opposite of 3 - t" is "t - 3".

Apply the oppositing principle to each of these equations:

(a) $-x = y - 5$ (b) $-R = 1 - 9S$

___ = ____ ___ = ____

Answers to Frame 24: a) $x = 5 - y$ b) $R = 9S - 1$

---

25. Solve for "S" in each equation below. Write each solution in the preferred form.

(a) $R - S = 7$ (b) $4E - S = 9$

S = ________ S = ________

Answers to Frame 25: a) $S = R - 7$ b) $S = 4E - 9$

---

26. Solve for "y" in each equation below.

(a) $7x + y = 5$ (b) $d - y = 1$ (c) $8t - y = 3$

y = ________ y = ________ y = ________

Answer to Frame 26: a) $y = 5 - 7x$ b) $y = d - 1$ c) $y = 8t - 3$

27. "y" is not solved for in the equation below because it has a numerical coefficient. To eliminate the coefficient "3", we multiply both sides by "$\frac{1}{3}$", the reciprocal of "3". We get:

$$3y = 2x + 7$$

$$\frac{1}{3}(3y) = \frac{1}{3}(2x + 7)$$

$$1\ y = \frac{2x + 7}{3}$$

$$y = \frac{2x + 7}{3}$$

Notice what we did on the right side. $\frac{1}{3}(2x + 7)$ is an instance of the distributive principle. Therefore:

$$\frac{1}{3}(2x + 7) = \frac{1}{3}(2x) + \frac{1}{3}(7) = \frac{2x}{3} + \frac{7}{3} = \frac{2x + 7}{3}$$

When solving systems of equations, it is easier if we write a solution as a single fraction rather than as an addition of fractions. That is:

We write $y = \frac{2x + 7}{3}$ instead of $y = \frac{2x}{3} + \frac{7}{3}$.

Solve for "t" in each of these. Write each solution in the preferred form.

(a) $5t = m + 7$ (b) $2t = 1 - 9p$

t = ________ t = ________

Answers: a) $t = \frac{m + 7}{5}$ b) $t = \frac{1 - 9p}{2}$

---

28. To solve for "V" in this equation: $2t + 8V = 7$

(1) We isolate its term and get: $8V = 7 - 2t$

(2) We multiply both sides by "$\frac{1}{8}$" and get: $V = \frac{7 - 2t}{8}$

Solve for "y" in each equation below:

(a) $x + 2y = 9$ (b) $7z + 5y = 1$

y = ________ y = ________

29. To solve for "k" in this equation: $5a - 3k = 4$

(1) We isolate the "k"-term and get: $-3k = 4 - 5a$

(2) We apply the oppositing principle and get: $3k = 5a - 4$

(3) We multiply both sides by $\frac{"1"}{3}$ and get: $k = \frac{5a - 4}{3}$

Solve for "x" in each equation below:

(a) $y - 7x = 3$ (b) $3y - 4x = 1$

x = ________ x = ________

a) $y = \frac{9 - x}{2}$

b) $y = \frac{1 - 7z}{5}$

---

30. Solve for "p" in each equation below:

(a) $3p + 5q = 9$ (b) $7t - 8p = 1$

p = ________ p = ________

a) $x = \frac{y - 3}{7}$

b) $x = \frac{3y - 1}{4}$

---

31. Neither variable is solved for in the equations in the system below. Solve for "y" in each equation. Write your rearranged equations in the lower box.

$$x + y = 7$$
$$4y = 2x - 1$$

y =

y =

a) $p = \frac{9 - 5q}{3}$

b) $p = \frac{7t - 1}{8}$

32. Solve for "t" in each equation below. Write your rearranged equations in the lower box.

$2m - t = 1$
$3t - 4m = 5$

$t =$
$t =$

$y = 7 - x$

$y = \frac{2x - 1}{4}$

---

33. Solve for F_2 in each equation below. Write your rearranged equations in the lower box.

$4F_2 = F_1 + 5$
$2F_1 - 7F_2 = 1$

$F_2 =$
$F_2 =$

$t = 2m - 1$

$t = \frac{4m + 5}{3}$

$F_2 = \frac{F_1 + 5}{4}$

$F_2 = \frac{2F_1 - 1}{7}$

10-5 FRACTIONAL EQUATIONS RESULTING FROM THE EQUIVALENCE PRINCIPLE

When we solve for the same variable in each equation in a system, the resulting equations frequently contain fractions. Here are two examples:

$$\boxed{\begin{aligned} V &= 2t + 1 \\ V &= \frac{t + 11}{2} \end{aligned}} \qquad \boxed{\begin{aligned} y &= \frac{x - 1}{5} \\ y &= \frac{3x + 4}{2} \end{aligned}}$$

If we apply the equivalence principle to eliminate the "solved for" variable in each system above, we get the following fractional equations:

$$2t + 1 = \frac{t + 11}{2} \qquad \frac{x - 1}{5} = \frac{3x + 4}{2}$$

In this section, we will review the method of solving fractional equations like those above.

---

34. At the right is a type of fractional equation which we can get by applying the equivalence principle to a system.

$$2t + 1 = \frac{t + 11}{2}$$

To clear the fraction, we multiply both sides by "2" and get:

$$2(2t + 1) = 2\left(\frac{t + 11}{2}\right)$$

$$4t + 2 = t + 11$$

Note: Since $2(2t + 1)$ is an instance of the distributive principle, both terms within the parentheses were multiplied by "2".

Find the root of the non-fractional equation. $t =$ ______

---

35. Write the non-fractional equation you get when you clear the fractions in each equation below. Do not solve the resulting equation.

Answer to 34: $t = 3$

(a) $x - 3 = \frac{3x + 5}{4}$ (b) $\frac{1 - 5t}{7} = 1 - t$

______ ______

---

a) $4x - 12 = 3x + 5$

b) $1 - 5t = 7 - 7t$

36. Find the root of each fractional equation below.

(a) $t + 5 = \frac{3t + 1}{4}$ (b) $\frac{5m - 7}{6} = 2m - 7$

t = ______ m = ______

Answers: a) $t = -19$ b) $m = 5$

---

37. At the right is another type of equation we can get by applying the equivalence principle to a system.

$$\frac{x - 1}{5} = \frac{3x + 4}{2}$$

We can clear the fractions by multiplying both sides by (5)(2). We get:

$$(5)(2)\left(\frac{x - 1}{5}\right) = (5)(2)\left(\frac{3x + 4}{2}\right)$$

Note: In the last step, we multiplied by the distributive principle on both sides.

$$\left(\frac{5}{5}\right)(2)(x - 1) = \left(\frac{2}{2}\right)(5)(3x + 4)$$

$$2(x - 1) = 5(3x + 4)$$

Find the root of the non-fractional equation:

$$2x - 2 = 15x + 20$$

x = ______

Answer: $x = -\frac{22}{13}$

---

38. When deciding on the multiplier to use to clear the fractions in a fractional equation, we examine the denominators. In cases where there are two denominators:

If they are both 5's, we multiply by 5.
If they are 3 and 4, we multiply by (3)(4).
If they are 4 and 12, we multiply by 12 alone, since 12 is a multiple of 4.

What multiplier would you use to clear the fractions in each equation below?

(a) $\frac{y - 6}{5} = \frac{y + 1}{3}$ (b) $\frac{m - 1}{7} = \frac{2m + 5}{7}$ (c) $\frac{d + 5}{3} = \frac{1 - 7d}{9}$

______ ______ ______

39. What non-fractional equation do you get when you clear the fractions in each equation below?

(a) $\frac{x+1}{5} = \frac{3x+4}{4}$ (b) $\frac{2t-1}{3} = \frac{5t-2}{3}$

__________ __________

a) (5)(3)

b) 7

c) 9, since 9 is a multiple of 3

---

40. What non-fractional equation do you get when you clear the fractions in each of these?

(a) $\frac{1-2p}{5} = \frac{3p+6}{10}$ (b) $\frac{3t-4}{7} = \frac{1-2t}{3}$

__________ __________

a) $4x + 4 = 15x + 20$

b) $2t - 1 = 5t - 2$

---

41. Solve each of the following fractional equations.

(a) $\frac{m-1}{2} = \frac{3m-5}{4}$ (b) $\frac{3-2y}{4} = \frac{y-4}{3}$

m = ______ y = ______

a) $2 - 4p = 3p + 6$

b) $9t - 12 = 7 - 14t$

---

a) $m = 3$

b) $y = \frac{5}{2}$

SELF-TEST 2 (Frames 15-41)

1. In the system at the right: $P = 5E$, $P = E + 4$

(a) Eliminate P by applying the equivalence principle to this system of equations. The resulting equation is: ____________

(b) The solution is: E = ____, P = ____

2. In the system at the right, rearrange each equation, solving each for i_2. Write each rearranged equation in the lower box. Do not solve the system.

$5i_1 - 2i_2 = 24$
$3i_1 - i_2 = 15$

$i_2 =$
$i_2 =$

3. Solve for x: $\frac{6 - 3x}{2} = 4x - 8$

x = ______

4. Solve for t: $\frac{4 - t}{6} = \frac{3t - 1}{2}$

t = ______

5. In solving the system at the right, the following solution was obtained: $F = 3.5$, $f = 0.5$. Check this solution by showing that it satisfies each equation in the system.

$F - 3f = 2$
$2F + 4f = 9$

ANSWERS:

1. (a) $5E = E + 4$
   (b) $E = 1$, $P = 5$

2. $i_2 = \frac{5i_1 - 24}{2}$
   $i_2 = 3i_1 - 15$

3. $x = 2$

4. $t = \frac{7}{10}$

5.
$F - 3f = 2$
$3.5 - 3(0.5) = 2$
$3.5 - 1.5 = 2$
$2 = 2$
Check

$2F + 4f = 9$
$2(3.5) + 4(0.5) = 9$
$7 + 2 = 9$
$9 = 9$
Check

10-6 REARRANGEMENTS AND THE EQUIVALENCE METHOD

When solving systems in an earlier section, we solved only those types in which the same variable was "solved for" in both original equations. In this section, we will solve types in which the same variable is not "solved for" in both original equations. To solve this latter type, the same procedure is used. However, we must begin by solving for the same variable in each equation so that the equivalence principle can be applied to eliminate a variable.

42. In the system on the right, neither variable is solved for in the original equations. We can solve this system in four steps:

$$4d + R = 9$$
$$d + R = 6$$

Step 1: Solve for the same variable in each equation. (We'll solve for "R".)

$$R = 9 - 4d$$
$$R = 6 - d$$

Step 2: Apply the equivalence principle to eliminate "R".

$$9 - 4d = 6 - d$$

Step 3: Find the numerical value of "d".

$$3 = 3d$$
$$d = 1$$

Step 4: Find the corresponding value of "R" by substituting "d = 1" into one of the earlier equations containing both "d" and "R".

$$R = 9 - 4d$$
$$= 9 - 4(1)$$
$$= 9 - 4 = 5$$

Note: It is easier to substitute "d = 1" into an equation in which "R" is solved for than to substitute it into one of the original equations.

or $R = 6 - d$
$= 6 - 1 = 5$

The solution is: $d = 1$, $R = 5$. Show that this solution satisfies each original equation. Do the work below:

(a) $4d + R = 9$ (b) $d + R = 6$

a) $4d + R = 9$
$4(1) + 5 = 9$
$4 + 5 = 9$
$9 = 9$

b) $d + R = 6$
$1 + 5 = 6$
$6 = 6$

43. On the right is the same system we solved in the last frame. The solution was: $d = 1$, $R = 5$. In the last frame, we began by solving for "R" and then eliminating "R". In this frame, we will begin by solving for "d" and then eliminating "d", to show that we get the same solution both ways. Here are the steps:

$$\boxed{\begin{array}{r} 4d + R = 9 \\ d + R = 6 \end{array}}$$

Step 1: Solving for "d" in each equation.

$$\boxed{\begin{array}{l} d = \dfrac{9 - R}{4} \\ d = 6 - R \end{array}}$$

Step 2: Applying the equivalence principle to eliminate "d".

$$\frac{9 - R}{4} = 6 - R$$

Step 3: Finding the value of "R".

$$4\left(\frac{9 - R}{4}\right) = 4(6 - R)$$
$$9 - R = 24 - 4R$$
$$3R = 15$$
$$R = 5$$

Step 4: Finding the corresponding value of "d" by substituting "R = 5" into one of the earlier equations.

$$d = \frac{9 - R}{4} = \frac{9 - 5}{4} = \frac{4}{4} = 1$$

or
$$d = 6 - R = 6 - 5 = 1$$

We obtained the solution: $d = 1$, $R = 5$. Is this the same solution we obtained in the last frame? _______

Yes

---

44. Here is a general summary of the equivalence method. We will use the system on the right as an example.

$$\boxed{\begin{array}{l} P - 2K = 16 \\ P + K = 1 \end{array}}$$

Step 1: Solve for the same variable in each equation. (We will solve for "P".)

$$\boxed{\begin{array}{l} P = 2K + 16 \\ P = 1 - K \end{array}}$$

Step 2: Apply the equivalence principle to eliminate the "solved for" variable "P".

$$2K + 16 = 1 - K$$

Step 3: Find the numerical value of "K".

$$3K = -15$$
$$K = -5$$

Step 4: Find the corresponding value of "P" by substitution. (Substitute "-5" for "K" in one of the equations in which "P" is "solved for.")

$$P = 2K + 16 = 2(-5) + 16 = (-10) + 16 = 6$$

Step 5: Check the solution ($K = -5$, $P = 6$) in both original equations. We have done so below.

$$P - 2K = 16$$
$$(6) - 2(-5) = 16$$
$$6 - (-10) = 16$$
$$6 + 10 = 16$$
$$16 = 16$$

$$P + K = 1$$
$$(6) + (-5) = 1$$
$$1 = 1$$

Go to next frame.

45. Solve the system at the right, using the procedures outlined in the last frame.

$$2y = x - 5$$
$$3x + y = 1$$

(a) The solution is: $x =$ _____, $y =$ _____

(b) Check your solution in each original equation.

---

46. Let's solve the following system:

$$2x - y = 4$$
$$2y - x = 1$$

(a) Solve for "x" in each equation.

$x =$

$x =$

(b) Apply the equivalence principle and write the non-fractional equation you get when the fraction is cleared.

____

(c) Find the value of "y".

$y =$ _______

(d) Find the corresponding value of "x".

$x =$ _______

(e) The solution of the system is: $x =$ _____, $y =$ _____

---

a) $x = 1$, $y = -2$

b)
$$2y = x - 5$$
$$2(-2) = 1 - 5$$
$$-4 = -4$$ Check

$$3x + y = 1$$
$$3(1) + (-2) = 1$$
$$1 = 1$$ Check

47. Let's solve this system:

$$3F_1 = 13 - 7F_2$$
$$5F_2 = 13 - 4F_1$$

(a) Solve for F_1 in each equation.

$F_1 =$

$F_1 =$

(b) Apply the equivalence principle and write the non-fractional equation you get when the fractions are cleared. ________

(c) Find the value of F_2.

$F_2 =$ ________

(d) Find the corresponding value of F_1.

$F_1 =$ ________

(e) The solution of the system is: $F_1 =$ ______, $F_2 =$ ______

Answers:

a) $x = \dfrac{y+4}{2}$, $x = 2y - 1$

b) $y + 4 = 4y - 2$

c) $y = 2$

d) $x = 3$

e) $x = 3, \ y = 2$

---

48. In the last frame, we solved the system on the right. The solution was $F_1 = 2$, $F_2 = 1$. Check this solution below by showing that it satisfies each of the original equations:

$$3F_1 = 13 - 7F_2$$
$$5F_2 = 13 - 4F_1$$

(a) $3F_1 = 13 - 7F_2$ (b) $5F_2 = 13 - 4F_1$

Answers:

a) $F_1 = \dfrac{13 - 7F_2}{3}$

$F_1 = \dfrac{13 - 5F_2}{4}$

b) $52 - 28F_2 = 39 - 15F_2$

c) $F_2 = 1$

d) $F_1 = 2$

e) $F_1 = 2, \ F_2 = 1$

---

a) $3F_1 = 13 - 7F_2$
$3(2) = 13 - 7(1)$
$6 = 6$

b) $5F_2 = 13 - 4F_1$
$5(1) = 13 - 4(2)$
$5 = 5$

SELF-TEST 3 (Frames 42-48)

1. Solve this system:

$$\begin{aligned} 5R + 2T &= 2 \\ 3R - T &= 10 \end{aligned}$$

2. Solve this system:

$$\begin{aligned} d - 3p &= 1 \\ 5d - p &= 12 \end{aligned}$$

ANSWERS: 1. $R = 2$, $T = -4$ 2. $d = \frac{5}{2}$, $p = \frac{1}{2}$

10-7 EQUATIONS WITH DECIMAL COEFFICIENTS

When you encounter a system of equations in applied situations, the coefficients of the variables will frequently be decimals. Since it is easier to work with whole numbers than with decimals, in this section we will show you a technique which enables you to convert the decimals to whole numbers.

49. Here is a simple decimal equation: $1.2x = 2.4$

By using the multiplication axiom with a suitable power of ten, we can eliminate the decimals. Since the last digit of each decimal is in the "tenths" position, we will multiply both sides by "10". We get:

$$1.2x = 2.4$$

$$(10)(1.2x) = (10)(2.4)$$

$$12x = 24$$

The root of $12x = 24$ is 2. Is "2" the root of $1.2x = 2.4$? _______

---

50. In the last frame, we saw that the following two equations have the same root.

$$1.2x = 2.4 \quad \text{and} \quad 12x = 24$$

When two equations have the same root, we call them _______________ equations.

Answer (49): Yes, since:

$$1.2(2) = 2.4$$

$$2.4 = 2.4$$

---

51. When we multiply both sides of an equation by some power of ten to eliminate the decimals, the new equation is equivalent to the original equation.

Here's another equation which contains decimals: $0.3t = 12.9$

(a) If we multiply both sides by "10", what new equation do we obtain? ____________

(b) The root of the new equation is _______.

(c) The root of the original equation must be _______.

Answer (50): equivalent

---

52. Here's another equation which contains decimals: $3.2R = 4.48$

The last digit of "3.2" is in the "tenths" position.
The last digit of "4.48" is in the "hundredths" position.

To eliminate both decimals in one step, we must multiply both sides by 10^2 (or 100). We get:

$$(10^2)(3.2R) = (10^2)(4.48)$$

or

$$320R = 448$$

(a) The root of $320R = 448$ (in decimal form) is _______.

(b) The root of $3.2R = 4.48$ must be _______.

Answer (51):

a) $3t = 129$

b) 43

c) 43

Answer (52):

a) 1.4

b) 1.4

53. When multiplying by a power of ten to clear the decimal numbers in an equation, the power of ten is determined by the number or coefficient with the most decimal places. If its last digit is:

(1) in the tenths place, we multiply by 10^1 (or 10)
(2) in the hundredths place, we multiply by 10^2 (or 100)
(3) in the thousandths place, we multiply by 10^3 (or 1,000)

What power of ten must you use to clear the decimal coefficients in each of the following equations?

(a) $5.2x = 6.95$ ______ (c) $1.95t = 4.7$ ______

(b) $1.17y = 0.465$ ______ (d) $0.3M = 19.875$ ______

---

a) 10^2 (or 100)
b) 10^3 (or 1,000)
c) 10^2 (or 100)
d) 10^3 (or 1,000)

54. This same principle can be extended to decimal equations that contain two letters. For example, let's clear the following equation of decimals:

$$1.2x + 0.3y = 2.5$$

In this equation, each decimal is carried to the "tenths" position. Therefore, it is necessary to multiply both sides by 10. We get:

$$(10)(1.2x + 0.3y) = (2.5)(10)$$

or

$$12x + 3y = 25$$

Here is another equation: $2.1x + 0.7y = 8.3$

(a) To clear this equation of decimals, what power of ten should you multiply by? ______

(b) Do so. The resulting equation is ____________.

---

a) 10^1 or 10
b) $21x + 7y = 83$

55. (a) To clear the equation on the right of decimals, what power of ten should be your multiplier? ______ $3.4F_1 + 0.07F_2 = 18.05$

(b) Do so. The resulting equation is : ____________

---

a) 10^2 or 100
b) $340F_1 + 7F_2 = 1,805$

56. Write the resulting equation after clearing the decimals in each of the following:

(a) $5.7p = 3.25q + 1.213$ ____________

(b) $14V = 1.2 - 0.9t$ ____________

(c) $4i_1 - 0.9i_2 = 1.2$ ____________

---

a) $5700p = 3250q + 1213$
b) $140V = 12 - 9t$
c) $40i_1 - 9i_2 = 12$

57. In the box on the right, the second equation was obtained by multiplying each side of the first equation by "10".

$0.2x + 0.1y = 0.5$

and

$2x + y = 5$

These two equations are <u>equivalent</u> because:

(1) <u>The same pairs of values satisfy each equation.</u>

Here is a partial table which shows some pairs of values which satisfy each equation:

x	y
-1	7
0	5
2	1

(2) Since the same pairs of values satisfy each equation, <u>their graphs are identical.</u> Here is the graph for each:

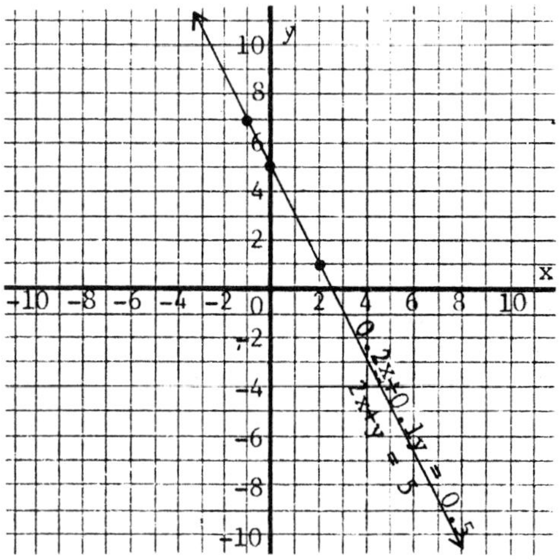

---

58. The bottom equation shown below was obtained by multiplying both sides of the top equation by 10^2 (or 100).

$$1.56F_1 + 2.7F_2 = 1.84$$

and

$$156F_1 + 270F_2 = 184$$

(a) Would the same pairs of values satisfy both equations? _______

(b) If we graphed both equations, would their graphed lines be identical or different? _______________

Go to next frame.

---

59. The point we are trying to make is this:

If a system of equations, like the one on the right, contains decimals,

(1) $1.5x + 1.0y = 2.8$
(2) $3.0x + 0.9y = 4.5$

we can use the multiplication axiom with powers of ten to obtain a new system which does not contain decimals.

(1) $15x + 10y = 28$
(2) $30x + 9y = 45$

Since the two systems contain equivalent equations, the solution of the "non-decimal" system is identical to the solution of the "decimal" system. The solution of the "non-decimal" system is: $x = 1.2$, $y = 1$.

(Continued on following page.)

a) Yes.

b) Identical.

59. (Continued)

In the space below, show that this solution satisfies each equation in the "decimal" system:

(a) $1.5x + 1.0y = 2.8$ (b) $3.0x + 0.9y = 4.5$

60.

$$1.5F_1 + 1.0F_2 = 4.0$$
$$2.1F_1 + 0.9F_2 = 2.1$$

(a) What new system of equations do you obtain when you eliminate the decimals in each of the equations in the system above?

(1)

(2)

(b) Solve the "non-decimal" system.

F_1 = ______, F_2 = ______

(c) Show that this solution satisfies both equations in the original "decimal" system:

a)
$$1.5x + 1.0y = 2.8$$
$$1.5(1.2) + 1.0(1) = 2.8$$
$$1.8 + 1.0 = 2.8$$
$$2.8 = 2.8$$

b)
$$3.0x + 0.9y = 4.5$$
$$3.0(1.2) + 0.9(1) = 4.5$$
$$3.6 + 0.9 = 4.5$$
$$4.5 = 4.5$$

61. Sometimes the numbers in the solution of a system are non-ending decimals. In this case, use your slide rule and report the numbers with three digits. Here is an example:

$$\boxed{\begin{array}{l} 6x + 7y = 15 \\ 11x - 3y = 9 \end{array}}$$

When "x" is eliminated, we eventually get this equation:

$$95y = 111 \quad \text{or} \quad y = \frac{111}{95} = 1.17 \quad \text{(Slide rule accuracy)}$$

Plugging this value in for "y" in the top equation, we get:

$$6x + 7(1.17) = 15$$
$$6x + 8.19 = 15$$
$$6x = 6.81 \quad \text{and} \quad x = \frac{6.81}{6} = 1.13 \quad \text{(Slide rule accuracy)}$$

The solution is approximately: $x = 1.13$, $y = 1.17$

Since there is a slight inaccuracy in the solution, there will also be a slight inaccuracy when we check the solution in the two original equations.

In the top equation, for example, we get:

$$6(1.13) + 7(1.17) \doteq 15$$
$$6.78 + 8.19 \doteq 15$$
$$14.97 \doteq 15$$

Note: The symbol "$\doteq$" means "is approximately equal to."

Check the solution in the bottom equation: $11x - 3y = 9$

a)
$$\boxed{\begin{array}{l} (1)\ 15F_1 + 10F_2 = 40 \\ (2)\ 21F_1 + 9F_2 = 21 \end{array}}$$

b) $F_1 = -2$, $F_2 = 7$

c)
$$1.5F_1 + 1.0F_2 = 4.0$$
$$1.5(-2) + 1.0(7) = 4.0$$
$$(-3) + 7 = 4$$
$$4 = 4$$

$$2.1F_1 + 0.9F_2 = 2.1$$
$$2.1(-2) + 0.9(7) = 2.1$$
$$(-4.2) + 6.3 = 2.1$$
$$2.1 = 2.1$$

Answer to Frame 61:

$$11x - 3y = 9$$
$$11(1.13) - 3(1.17) \doteq 9$$
$$12.43 - 3.51 \doteq 9$$
$$8.92 \doteq 9$$

SELF-TEST 4 (Frames 49-61)

1. Eliminate the decimals in the following system. Write the new system in the box at the right. (Note: Do not solve for v_1 and v_2.)

$$\boxed{\begin{array}{l} 2.73v_1 + 0.92v_2 = 18.4 \\ 0.05v_1 + 8.1v_2 = 7.32 \end{array}}$$

2. Eliminate the decimals in the following system. Write the new system in the box at the right. (Note: Do not solve for i_1 and i_2.)

$$\boxed{\begin{array}{l} 0.184i_1 - 0.27i_2 = 1.9 \\ 4.06i_1 + 0.365i_2 = 8 \end{array}}$$

ANSWERS:

1.
$$\boxed{\begin{array}{l} 273v_1 + 92v_2 = 1840 \\ 5v_1 + 810v_2 = 732 \end{array}}$$

2.
$$\boxed{\begin{array}{l} 184i_1 - 270i_2 = 1900 \\ 4060i_1 + 365i_2 = 8000 \end{array}}$$

10-8 SYSTEMS CONTAINING OTHER TYPES OF EQUATIONS

The typical equation in a system of two equations is a non-fractional equation containing three terms - one term for each variable, and a constant term. Here are two examples:

$$3F_1 + 5F_2 = 7 \qquad\qquad x - 9 = 2y$$

In practice, however, some equations which appear in systems do not fit this typical pattern because:

(1) They contain fractions: $\frac{m}{3} + \frac{n}{2} = 3$

(2) They contain an instance of the distributive principle: $4e_1 - 3(e_2 - 1) = 11$

In this section, we will examine systems which contain equations of these types.

---

62. The following system contains two fractional equations:

(1)	$\frac{p}{3} + \frac{q}{2} = 4$
(2)	$\frac{p}{4} - \frac{q}{3} = \frac{25}{12}$

The first step in solving a system of this type is to clear the fractions and obtain a system in which the typical non-fractional equations appear.

To clear the fractions in Equation (1), we multiply both sides by "3" and "2" at the same time. We get:

$$(3)(2)\left(\frac{p}{3} + \frac{q}{2}\right) = (3)(2)(4)$$

$$(3)(2)\left(\frac{p}{3}\right) + (3)(2)\left(\frac{q}{2}\right) = (6)(4)$$

$$\left(\frac{3}{3}\right)(2p) + \left(\frac{2}{2}\right)(3q) = 24$$

$$2p + 3q = 24$$

To clear the fractions in Equation (2), we multiply both sides by "12", since "12" is a multiple of both "4" and "3". We get:

$$12\left(\frac{p}{4} - \frac{q}{3}\right) = 12\left(\frac{25}{12}\right)$$

$$12\left(\frac{p}{4}\right) - 12\left(\frac{q}{3}\right) = \left(\frac{12}{12}\right)(25)$$

$$3p - 4q = 25$$

Using these two new equations, we obtain the new system shown on the right:

$$2p + 3q = 24$$
$$3p - 4q = 25$$

We then use the usual method to solve the new system. The solution will also satisfy the original system.

Go to next frame.

63. Clear the fractions in each equation in the system on the right. Write the non-fractional equations in the box below.

$$\frac{x}{5} - \frac{y}{3} = 1$$

$$\frac{x}{2} + 1 = \frac{y}{4}$$

---

64. In the system on the right, each equation contains an instance of the distributive principle:

(1) $3k - 2(k + t) = 15$

(2) $5t - 4(2k - t) = 11$

$3x - 5y = 15$
$2x + 4 = y$

To obtain a system with typical equations, we must multiply each instance of the distributive principle and then combine terms. Here is the method for Equation (1):

$$3k - [2(k + t)] = 15$$

$$3k - [2k + 2t] = 15$$

$$3k + (-2k) + (-2t) = 15$$

$$k - 2t = 15$$

Multiply by the distributive principle and combine terms in Equation (2):

$5t - 4(2k - t) = 11$

---

65. In the box below, write each of the equations in the following system in the typical form:

$$R + \frac{4t}{3} = 17$$

$$2R + 3(R + t) = 63$$

$9t - 8k = 11$

Solution:

$$5t - [4(2k - t)] = 11$$

$$5t - [8k - 4t] = 11$$

$$5t - [8k + (-4t)] = 11$$

$$5t + (-8k) + 4t = 11$$

$$9t - 8k = 11$$

66. In the box below, write each equation in the following system in the typical form. (Note: Eliminate the decimal coefficients in the top equation.)

$$W = 7.22Q + 1.09$$
$$5(W - Q) - 4Q = 0$$

Answer: $3R + 4t = 51$, $5R + 3t = 63$

---

67. In the box below, write each of the equations in the following system in the typical form:

$$x = 3(5 - y)$$
$$5x - 2(x - 4y) = 9$$

Answer: $100W = 722Q + 109$, $5W - 9Q = 0$

Answer: $x = 15 - 3y$, $3x + 8y = 9$

68. Let's solve the following system:

$$e - 5i = 0$$
$$4e - 3(e - i) = 8$$

(a) In the box on the right, write each equation in the typical form:

(b) Solve for "e" in each equation:

$e =$

$e =$

(c) Apply the equivalence principle and write the resulting equation: ______________

(d) Find the value of "i":

$i =$ ______________

(e) Substitute to find the corresponding value of "e":

$e =$ ______________

(f) The solution is: $i =$ _____, $e =$ _____

(g) Show that this solution satisfies each original equation:

$e - 5i = 0$ $\qquad\qquad$ $4e - 3(e - i) = 8$

Answer to Frame 68:

a) $e - 5i = 0$, $e + 3i = 8$

b) $e = 5i$, $e = 8 - 3i$

c) $5i = 8 - 3i$

d) $i = 1$

e) $e = 5$

f) $i = 1$, $e = 5$

g)

$$e - 5i = 0$$
$$5 - 5(1) = 0$$
$$0 = 0$$

$$4e - 3(e - i) = 8$$
$$4(5) - 3(5 - 1) = 8$$
$$20 - 3(4) = 8$$
$$8 = 8$$

SELF-TEST 5 (Frames 62-68)

Solve this system:

$$w - (r - 3) = r - 1$$
$$\frac{r-1}{2} = \frac{w-1}{3}$$

w = _______, r = _______

ANSWER: $w = 10$, $r = 7$

10-9 APPLIED PROBLEMS WHICH LEAD TO SYSTEMS OF EQUATIONS

In this section, we will show various applied problems in which a system of equations must be solved. The main purpose of this section is to show you that solving systems is sometimes necessary. In each case, we will set up the system of equations from the applied situation. Since we do not expect all of you to understand the scientific principles underlying the solution, don't worry if you cannot see how we obtained the original equations. We will also minimize the amount of work which you have to do.

69. Problem: One alloy contains 55% silver; a second alloy contains 80% silver. By mixing them, we want to obtain 40 ounces of an alloy that contains 75% silver. How many ounces of the two original alloys must we use in the mixture?

In this problem, there are two variables: (1) the number of ounces of the first alloy, (2) the number of ounces of the second alloy. Since there are two variables, we need two equations (a system) to obtain the solution.

(1) Let "f" = number of ounces of the first alloy.
"s" = number of ounces of the second alloy.

Since we want to obtain 40 ounces of the new alloy, we can set up this equation:

$$f + s = 40$$

(2) We also know the percent of silver in each of the three alloys. Therefore, this statement is true:

(55% of "f") + (80% of "s") = (75% of 40)

Converting the percents to decimals, we get an equivalent equation:

$$0.55f + 0.80s = 0.75(40)$$
or
$$0.55f + 0.80s = 30$$

We now have our system of two equations:

(1) $f + s = 40$
(2) $0.55f + 0.80s = 30$

The solution of this system is: $f = 8$ ounces, $s = 32$ ounces

Therefore: To obtain an alloy containing 75% silver, we must mix _____ ounces of an alloy with 55% silver and _____ ounces of an alloy containing 80% silver.

Answer to Frame 69: 8 32

70. Systems of equations arise in electrical work. For example, in the circuit on the right, it is desired to calculate the current in the 20-ohm resistor, in the 30-ohm resistor, and in the 40-ohm resistor.

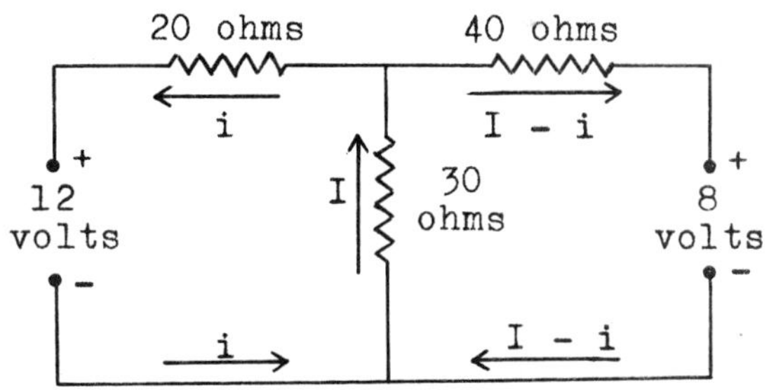

To find the currents, let I = current in 30-ohm resistor, in amperes
i = current in 20-ohm resistor, in amperes

Then I - i = current in 40-ohm resistor.

Using basic circuit principles, the following system of equations can be set up:

$$30I + 20i = 12$$
$$40(I - i) + 30I = 8$$

The solution of this system is: $I = 0.246$, $i = 0.231$

Therefore: (a) The current in the 20-ohm resistor is _______ ampere.

(b) The current in the 30-ohm resistor is _______ ampere.

(c) The current in the 40-ohm resistor is _______ ampere.

Answer to Frame 70: a) 0.231 amp. b) 0.246 amp. c) 0.015 amp. (from I - i)

71. The following formula occurs in the study of "heat." It contains four variables.

$H = m(t_2 - t_1)$, where H = amount of heat
m = mass
t_2 = final temperature
t_1 = original temperature

If you know the value of the same two variables in two situations, you can find the value of the other two variables by means of a system of equations. For example:

In one situation, if $m = 15$ and $t_2 = 50$, we get: $H = 15(50 - t_1)$

In another situation, if $m = 25$ and $t_2 = 40$, we get: $H = 25(40 - t_1)$

Here is the system of equations:

$$H = 15(50 - t_1)$$
$$H = 25(40 - t_1)$$

Notice that it is easy to eliminate "H" since it is "solved for" in each case. Find the solution of this system:

Answer to Frame 71: $H = 375$, $t_1 = 25$

---

72. A steel beam of uniform cross-section is 20 feet long, weighs 3,000 lbs., and is supported at each end. A load of 5,000 lbs. is applied to the beam 3 feet from its left end, and a load of 2,000 lbs. is applied to the beam 6 feet from its right end. Refer to the diagram at the right.

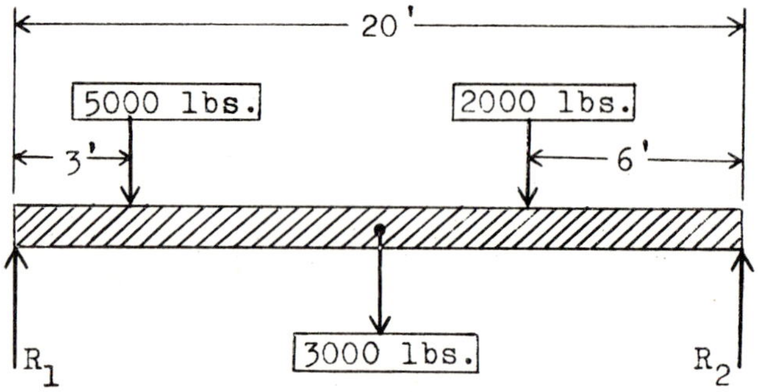

Problem: The forces exerted by the supports upon the beam are called "reactions" and are labeled R_1 and R_2 in the diagram. Find R_1 and R_2 in pounds.

(Continued on following page.)

72. (Continued)

Solution: Using the principles of translational and rotational equilibrium, the following two equations can be set up:

$$R_1 + R_2 = 10{,}000$$
$$10R_1 + (4)(2{,}000) = 10R_2 + (7)(5{,}000)$$

Simplifying this system and solving each equation for R_2, the following simpler equivalent system can be set up. Solve this system for R_1 and R_2:

$$R_2 = 10{,}000 - R_1$$
$$R_2 = R_1 - 2{,}700$$

R_1 = ________ lbs.

R_2 = ________ lbs.

Answer to Frame 72: $R_1 = 6{,}350$ lbs., $R_2 = 3{,}650$ lbs.

73. The electrical circuit shown at the right is called a "series-parallel" circuit. The applied voltage is 22 volts.

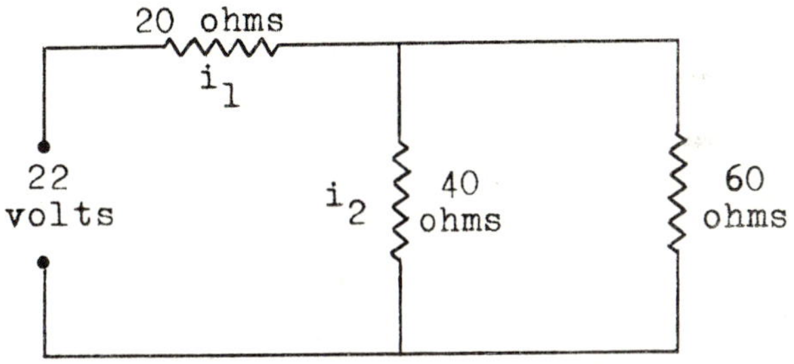

Problem: Find the current in the 20-ohm resistor, and the current in the 40-ohm resistor. These two currents are called i_1 and i_2, respectively.

Solution: Using basic circuit principles, the system of equations at the right can be set up.

$$20i_1 + 40i_2 = 22$$
$$60(i_1 - i_2) = 40i_2$$

Simplifying this system and solving each equation for i_2, the following simpler equivalent system can be set up. Solve this system for i_1 and i_2:

$$i_2 = \frac{11 - 10i_1}{20}$$
$$i_2 = \frac{3i_1}{5}$$

i_1 = ________ ampere

i_2 = ________ ampere

Answer to Frame 73: $i_1 = 0.50$ ampere, $i_2 = 0.30$ ampere

10-10 THE EQUIVALENCE METHOD AND FORMULA DERIVATION

In science, new formulas are frequently derived from a system of existing formulas. This process, called formula derivation, is very similar to the procedure for solving systems of equations. It is based on the possibility of "eliminating a variable" by means of the equivalence principle. However, its goal is the derivation of a new formula, and not finding a numerical solution.

In this section, we will define a system of formulas. Then we will show how the equivalence method can be used to perform formula derivations.

---

74. Two formulas form a system if they contain at least one common variable. Here is an example:

$$Pt = W$$
$$W = Fs$$

These two formulas form a system because they contain one common variable, namely _____.

---

75. Do the pairs of formulas in each box below form a system:

(a) $E = IR$, $F = ma$ ______ (b) $E = IR$, $P = EI$ ______

Answer to 74: W

---

76. The two formulas on the right form a system because they contain the common variable "W":

(1) $Pt = W$
(2) $W = Fs$

The first formula shows the relationship between three variables: "P", "t", and "W".

The second formula shows the relationship between a different set of three variables: "W", "F", and "s".

Since "W" can be eliminated from this system, we can derive a new formula showing a new relationship between variables. Here is the method:

Since in each formula "W" is solved for:

$$W = Pt$$
$$W = Fs$$

we can eliminate "W" by applying the equivalence principle.

$$Pt = Fs$$

Now we have a new relationship between four variables: "P", "t", "F", and "s".

(a) Which variable has been eliminated? ______

(b) Are any of the four variables "solved for" in the new formula? ______

Answers to 75:

a) No, because they do not contain a common variable.

b) Yes, because they contain two common variables, "E" and "I".

Answers to 76:

a) W

b) No

77. When we performed the formula derivation in the last frame, we obtained the following new formula: Pt = Fs. No variable is "solved" for in this formula.

Ordinarily, a formula is written with one of the variables "solved for." Therefore, after deriving a new formula, we solve for one of the variables. Any of the variables can be solved for.

In $\boxed{Pt = Fs}$: (a) Solve for "P".

P = ________

(b) Solve for "s".

s = ________

---

a) $P = \frac{Fs}{t}$

b) $s = \frac{Pt}{F}$

---

78. There are two parts to a formula derivation:

(1) Eliminating a common variable from the system.
(2) Solving for one of the variables in the new formula.

The common variable in the system on the right is "N". Let's eliminate "N" and derive a new formula.

(1) $S = 0.26DN$

(2) $T = \frac{FL}{N}$

Step 1: Solving for "N" in each formula, we get: →

(1) $N = \frac{S}{0.26D}$

(2) $N = \frac{FL}{T}$

Step 2: Applying the equivalence principle to eliminate "N", we get: → $\frac{S}{0.26D} = \frac{FL}{T}$

Step 3: Solving for one of the variables in the new formula. We will solve for "T": → $T = \frac{0.26DFL}{S}$

Go to next frame.

79. In a system of formulas, each common variable can be eliminated to obtain a new relationship.

In the system on the right, there are two common variables, "E" and "I". Each of the two can be eliminated, and therefore, we can derive two new relationships.

(1) $E = IR$
(2) $P = EI$

1. Deriving a new relationship by eliminating "E":

Solve for "E" in each formula:

$$E = IR$$
$$E = \frac{P}{I}$$

Apply the equivalence principle:
Note that the new formula does not contain "E".

$$IR = \frac{P}{I}$$

2. Deriving a new relationship by eliminating "I":

(a) Solve for "I" in each formula: I = ______

I = ______

(b) Apply the equivalence principle: ______ = ______

(c) Does "I" appear in the new relationship? ______

---

a) $I = \frac{E}{R}$ and $I = \frac{P}{E}$

b) $\frac{E}{R} = \frac{P}{E}$

c) No, it has been eliminated.

---

80. Here is the same system:

$E = IR$
$P = EI$

"R" cannot be eliminated from this system because "R" is not common to both formulas.

Which other variable cannot be eliminated from the system? ______

P

---

81. In the system on the right, both "F" and "m" are common variables. Sometimes, when one common variable is eliminated, a second common variable is also eliminated. It happens in this case.

$F = ma$

$F = \frac{GMm}{r^2}$

We can eliminate "F" by applying the equivalence principle. We get:

$$ma = \frac{GMm}{r^2}$$

Now, if we solve this new relationship, "m" will also be eliminated when we reduce to lowest terms. Let's solve for "a":

$$a = \frac{GMm}{r^2m} = \left(\frac{GM}{r^2}\right)\left(\frac{m}{m}\right) = \left(\frac{GM}{r^2}\right)(\quad) = \_\_\_\_\_\_$$

$$\left(\frac{GM}{r^2}\right)(1) = \frac{GM}{r^2}$$

82. Here is the same system. If we eliminate "m" and solve for "a", "F" is also eliminated.

$$F = ma$$

$$F = \frac{GMm}{r^2}$$

(a) Solve for "m" in each formula. Write your solution in the box on the right.

m =

m =

(b) Apply the equivalence principle to eliminate "m".

_____ = _____

(c) Solve this new formula for "a". Notice how "F" is eliminated when you reduce to lowest terms.

a = _____

Answers to frame 82:

a) $m = \frac{F}{a}$ $\quad m = \frac{Fr^2}{GM}$

b) $\frac{F}{a} = \frac{Fr^2}{GM}$

c) $a = \frac{GM}{r^2}$

---

83. Let's eliminate "g" from this system and solve for "v".

$$v = gt$$

$$d = \frac{1}{2}gt^2$$

(a) Solve for "g" in each formula, and then apply the equivalence principle to eliminate "g":

$\frac{v}{t} =$ _____

(b) To clear the fractions in the new relationship, you should multiply both sides by _____.

(c) Do so and solve for "v":

v = _____

84. Look closely at this system of formulas:

$$He = \frac{mV}{R}$$
$$V = RT$$

(a) Two variables can be eliminated, namely _____ and _____.

(b) Eliminate "V" from this system and solve for "e". (Be sure to write the formula for "e" in lowest terms.)

e = _______

(c) Was "R" also eliminated? _______

Answers:

a) $\frac{v}{t} = \frac{2d}{t^2}$

b) t^2

c) $vt = 2d$

$v = \frac{2d}{t}$

---

85. Here is a tougher problem:

$$E = A(N - BE)$$
$$E = GN$$

(a) Solve for "E" in the first equation. (Notice that "E" appears in two places. This is a case in which you must eventually factor by the distributive principle.)

E = __________

(b) Eliminate "E" and solve for "G":

G = __________

Answers:

a) V and R

b) $e = \frac{mT}{H}$

Here is the work:

$\frac{HeR}{m} = RT$

$HeR = mRT$

$e = \frac{mRT}{HR}$

$e = \frac{mT}{H}\left(\frac{R}{R}\right)$

$e = \frac{mT}{H}$

c) Yes.

86. Sometimes it is possible to eliminate two variables at one time. In the system on the right, we can eliminate both "W" and "m" at the same time because we can solve for the fraction $\frac{\text{"W"}}{\text{m}}$ in both formulas.

$$g = \frac{W}{m}$$
$$W = \frac{GMm}{r^2}$$

Solving for $\frac{\text{"W"}}{\text{m}}$ in each, we get:

$$\frac{W}{m} = g$$
$$\frac{W}{m} = \frac{GM}{r^2}$$

Applying the equivalence principle, we get:

$$g = \frac{GM}{r^2}$$

Solve this new formula for "M":

M = ______

a) $E = \frac{AN}{1 + AB}$

Here is the complete solution:

$E = A(N - BE)$

$E = AN - ABE$

$E + ABE = AN$

$E(1 + AB) = AN$

$E = \frac{AN}{1 + AB}$

b) $G = \frac{A}{1 + AB}$ Since:

$GN = \frac{AN}{1 + AB}$

$G = \frac{AN}{(1 + AB)N}$

$G = \left(\frac{A}{1 + AB}\right)\left(\frac{N}{N}\right)$

$G = \frac{A}{1 + AB}$

---

87. Here is another system in which we can eliminate two variables at one time because we can solve for "as" in both formulas.

$$W = mas$$
$$V_f^2 = V_o^2 + 2as$$

(a) Solve for "as" in both formulas.

as =

as =

(b) Apply the equivalence principle and solve for "W".

W = ______

$M = \frac{gr^2}{G}$

Answer to Frame 87: a) $as = \frac{W}{m}$

$as = \frac{V_f^2 - V_o^2}{2}$

b) $W = \frac{m(V_f^2 - V_o^2)}{2}$ or

$W = \frac{mV_f^2 - mV_o^2}{2}$ or

$W = \frac{1}{2}mV_f^2 - \frac{1}{2}mV_o^2$

10-11 THE SUBSTITUTION METHOD AND SYSTEMS OF EQUATIONS

Up to this point, we have used the equivalence method to solve systems of equations and perform derivations. Other methods are possible. One other useful method is called the "substitution method." It is called the "substitution" method because, when using it, we eliminate a variable by substitution rather than by the equivalence principle.

In this section, we will briefly show how the substitution method can be used to solve systems of equations. In the next section, we will show how it can be used to perform formula derivation.

88. In the system on the right, "y" is solved for in the top equation but not in the bottom equation. When using the equivalence method, we would eliminate "y" by solving for it in the bottom equation and then applying the equivalence principle.

$$y = 4x$$
$$x + y = 15$$

There is a second method we can use to eliminate "y". We can do so by substitution. We substitute the solution ("4x") for "y" in the top equation for the "y" in the bottom equation. We get:

$$x + y = 15$$
$$x + (4x) = 15$$

Is "y" eliminated in the new equation? ______

Yes

89. "p" is solved for in the bottom equation on the right.

$$t = p - 1$$
$$p = 2t$$

(a) Substitute the "2t" for the "p" in the top equation. ______

(b) Is "p" eliminated in this new equation? ______

a) $t = 2t - 1$

b) Yes

90. "m" is solved for in the top equation on the right. If we substitute "3R" for the "m" in the bottom equation and multiply, we get:

$$m = 3R$$
$$R + 2m = 7$$

$$R + 2(3R) = 7$$
$$R + 6R = 7$$

"x" is solved for in one of the equations in each system below. Substitute "5y" for the "x" in the other equation in each system, and simplify the resulting equation.

(a) $x = 5y$, $4x + 3y = 46$

(b) $2y - 3x = 13$, $x = 5y$

________________ ________________

a) $4(5y) + 3y = 46$
$20y + 3y = 46$

b) $2y - 3(5y) = 13$
$2y - 15y = 13$

91. Let's use the substitution method to solve the system on the right:

$$t = 2s$$
$$s + 3t = 7$$

Step 1: Substituting the "2s" for the "t" in the bottom equation, we get:

$$s + 3(2s) = 7$$
$$s + 6s = 7$$

Step 2: Solving this new equation for "s", we get:

$$7s = 7$$
$$s = 1$$

Step 3: Substituting "1" for "s" in either original equation, we can find the corresponding value of "t".

$$t = 2s$$
$$t = 2(1)$$
$$t = 2$$

The solution is: $s = 1$, $t = 2$. Show that this solution satisfies each of the original equations.

(a) $t = 2s$

(b) $s + 3t = 7$

a) $2 = 2(1)$
$2 = 2$

b) $1 + 3(2) = 7$
$1 + 6 = 7$
$7 = 7$

92. When using the equivalence method, we must solve for the same variable in both equations. When using the substitution method, we must solve for one of the variables in only one equation.

Neither variable is solved for in the equations in the system on the right.

$$\begin{array}{|l|}\hline x + y = 5 \\ 2x - y = 1 \\ \hline\end{array}$$

To use the substitution method, we must solve for one of the variables in one equation. If we solve for "x" in the top equation, we get: $x = 5 - y$. Substituting this solution for the "x" in the bottom equation, we get:

$$2(5 - y) - y = 1$$

or

$$10 - 2y - y = 1$$

In each system below, solve for "y" in the top equation and substitute the solution for the "y" in the bottom equation:

(a) $\begin{array}{|l|}\hline x + y = 7 \\ x + 5y = 1 \\ \hline\end{array}$ (b) $\begin{array}{|l|}\hline y - x = 9 \\ 3x - 2y = 6 \\ \hline\end{array}$

______________ ______________

Answers to frame 92:

a) $y = 7 - x$

$x + 5(7 - x) = 1$
$x + 35 - 5x = 1$

b) $y = 9 + x$

$3x - 2(9 + x) = 6$
$3x - 18 - 2x = 6$

---

93. Neither variable is solved for in the equations in the system on the right.

$$\begin{array}{|l|}\hline 3t - 2m = 0 \\ 2t + m = 7 \\ \hline\end{array}$$

If we solve for "t" in the top equation, we get:

$$3t - 2m = 0$$
$$3t = 2m$$
$$t = \frac{2m}{3}$$

Substituting this solution for the "t" in the bottom equation, we get:

$$2t + m = 7$$
$$2\left(\frac{2m}{3}\right) + m = 7$$
$$\frac{4m}{3} + m = 7$$

(Continued on following page.)

93. (Continued)

In each system below, solve for the "y" in the top equation and substitute the solution for the "y" in the bottom equation:

(a) $\begin{array}{r} 2y = 5x \\ 3y + 2x = 1 \end{array}$ (b) $\begin{array}{r} 5y - 4x = 0 \\ 3x - 2y = 9 \end{array}$

____________ ____________

a) $y = \frac{5x}{2}$

$3\left(\frac{5x}{2}\right) + 2x = 1$

$\frac{15x}{2} + 2x = 1$

b) $y = \frac{4x}{5}$

$3x - 2\left(\frac{4x}{5}\right) = 9$

$3x - \frac{8x}{5} = 9$

94. After applying the substitution process, the steps for solving a system of equations by the substitution method are identical to the steps in the equivalence method.

Let's solve this system: $\begin{array}{r} 2m - 3t = 0 \\ 2t - m = 2 \end{array}$

Step 1: Solving for "m" in the top equation, we get:

$$2m = 3t$$
$$m = \frac{3t}{2}$$

Step 2: Substituting this solution for the "m" in the bottom equation, we get:

$$2t - \frac{3t}{2} = 2$$

Step 3: Clearing the fraction and solving for "t", we get:

$$2\left(2t - \frac{3t}{2}\right) = 2(2)$$
$$2(2t) - 2\left(\frac{3t}{2}\right) = 4$$
$$4t - 3t = 4$$
$$t = 4$$

Step 4: Substituting "4" for "t" in the top equation, we can find the corresponding value of "m":

$$2m - 3t = 0$$
$$2m - 3(4) = 0$$
$$2m = 12$$
$$m = 6$$

The solution of the system is: t = _____, m = _____

t = 4, m = 6

95. Let's solve this system by the substitution method:

$$\begin{aligned} x + 2y &= 10 \\ 3x - y &= 2 \end{aligned}$$

(a) Solve for "x" in the top equation:

$x =$ ______

(b) Substitute this solution for the "x" in the bottom equation:

(c) Find the numerical value of "y" by solving this new equation:

$y =$ ______

(d) Substitute this value for "y" in either of the original equations to obtain the corresponding value of "x".

$x =$ ______

(e) The solution is: $x =$ _____, $y =$ _____.

a) $x = 10 - 2y$

b) $3(10 - 2y) - y = 2$
$30 - 6y - y = 2$
$30 - 7y = 2$

c) $y = 4$
(from $-7y = -28$)

d) $x = 2$

e) $x = 2$, $y = 4$

<u>Note</u>: The substitution method is easy to use with some systems. With others, it is more difficult to use because the substitution leads to a complex fraction. When you encounter such cases, <u>use the equivalence method</u>. In fact, the equivalence method is a perfectly satisfactory method in all cases.

10-12 THE SUBSTITUTION METHOD AND FORMULA DERIVATION

In this section, we will show how the substitution method can be used to perform a formula derivation. The procedure is similar to that introduced in the last section.

96. It is easy to use the substitution method to eliminate "E" from the system on the right because "E" is solved for in the top formula.

E = IR
P = EI

Substituting "IR" for the "E" in the bottom formula, we get:

$P = (IR)I$
$P = I^2R$

Solve for "R" in this new formula.

R = ______

---

97. To eliminate "I" from the same system by the substitution method, we must solve for "I" in one of the formulas first.

E = IR
P = EI

Answer to 96: $R = \frac{P}{I^2}$

(a) Solve for "I" in the top formula:

I = ______

(b) Substitute this solution for the "I" in the bottom formula:

(c) Solve for "R" in this new formula:

R = ______

Answers to 97:

a) $I = \frac{E}{R}$

b) $P = E\left(\frac{E}{R}\right)$

$P = \frac{E^2}{R}$

c) $R = \frac{E^2}{P}$

98. Let's eliminate "R" by the substitution method, and solve the new formula for "T".

$$He = \frac{mV}{R}$$
$$V = RT$$

(a) Solve for "R" in the top formula:

R = ______

(b) Substitute this solution for the "R" in the bottom formula:

(c) Solve for "T" in the new formula:

T = ______

(d) What other variable besides "R" was eliminated from the system? ______

---

99. Let's eliminate "X_L" and solve for "L" by the substitution method.

$$X = X_L - X_c$$
$$X_L = 2\pi fL$$

(a) "X_L" is solved for in the bottom formula. Substitute this solution for the "X_L" in the top formula:

______

(b) Now solve for "L" in the new formula:

______

Answers to 98:

a) $R = \frac{mV}{He}$

b) $V = \left(\frac{mV}{He}\right)T$

$V = \frac{mVT}{He}$

c) $T = \frac{He}{m}$

(Did you reduce to lowest terms?)

d) V

100. Let's eliminate "r" and solve for "q" by the substitution method.

$$p = q + r$$
$$r = 2p + t$$

(a) "r" is solved for in the bottom equation. Substitute this solution for the "r" in the top formula: ______

(b) Now solve for "q" in the new formula: ______

a) $X = 2\pi fL - X_c$

b) $L = \dfrac{X + X_c}{2\pi f}$

---

101. In the system on the right, we can eliminate by "E" and "I" at the same time since "EI" is solved for in the bottom formula.

$$J = EIt$$
$$P = EI$$

(a) Substituting "P" for "EI" in the top formula, we get: ______

(b) Solve this new formula for "t": ______

a) $p = q + (2p + t)$
$p = q + 2p + t$

b) $q = p - 2p - t$
$q = (-p) - t$
$q = -p - t$

---

102. In this system, let's eliminate "t" by substitution:

$$V = gt$$
$$d = \frac{1}{2}gt^2$$

(a) Solve the top formula for "t": $t =$ ______

(b) Substitute this value of "t" for the "t" appearing in the bottom formula: $d = \frac{1}{2}g(\quad)^2$

(c) Simplify this last result. You get: $d =$ ______

a) $J = (P)t$
$J = Pt$

b) $t = \dfrac{J}{P}$

103. We have used both the equivalence method and the substitution method in formula derivation. We will use both methods to eliminate "a" and solve for "t" in the system on the right.

$$v = at$$
$$v^2 = 2as$$

Equivalence Method:

(a) Solve for "a" in both formulas:

a =

a =

(b) Apply the equivalence principle and solve for "t":

t = ______

Substitution Method:

(c) Solve for "a" in the top formula:

a = ______

(d) Substitute this solution for the "a" in the bottom formula, and simplify:

______

(e) Now solve for "t":

t = ______

a) $t = \frac{V}{g}$

b) $d = \frac{1}{2}g\left(\frac{V}{g}\right)^2$

c) $d = \frac{V^2}{2g}$

Since:

$$d = \frac{1}{2}g\left(\frac{V}{g}\right)^2$$
$$= \frac{1}{2}g\left(\frac{V^2}{g^2}\right)$$
$$= \frac{gV^2}{2g^2}$$
$$= \left(\frac{V^2}{2g}\right)\left(\frac{g}{g}\right)$$
$$= \frac{V^2}{2g}$$

Answer to Frame 103: a) $a = \frac{v}{t}$, $a = \frac{v^2}{2s}$ b) $t = \frac{2s}{v}$ c) $a = \frac{v}{t}$ d) $v^2 = 2\left(\frac{v}{t}\right)s$, $v^2 = \frac{2vs}{t}$ e) $t = \frac{2s}{v}$

Note: The same statement can be made about using the substitution method for formula derivations as was made about using it for solving systems of equations. In some cases, it works easily. In others, it leads to complex fractions which are difficult to handle. In these latter cases, you will find it easier to use the equivalence method.

104. Eliminate "E" and solve for "R":

$$i = \frac{E}{R + r}$$

$$E = IR$$

R = ________

Answer: $R = \frac{ir}{I - i}$

105. Here are the same two formulas. This time eliminate "R" and solve for "E":

$$i = \frac{E}{R + r}$$

$$E = IR$$

E = ________

Answer: $E = \frac{Iir}{I - i}$

SELF-TEST 6 (Frames 74-105)

1. Using the equivalence method, eliminate "P" and solve for "V":

$$H = P + KV$$
$$V = A - P$$

V = ____________

2. Using the substitution method, solve this system:

$$d + 2t = 9$$
$$t = 2 - 3d$$

d = _____, t = _____

3. Using the substitution method, eliminate "w" and solve for "b":

$$r = \frac{bw}{s}$$
$$w = \frac{r}{2s}$$

b = __________

4. Using the substitution method, eliminate "N" and solve for "E":

$$AE - K = BN$$
$$EK + N = 0$$

E = __________

ANSWERS:

1. $V = \frac{H - A}{K - 1}$
2. $d = -1$, $t = 5$
3. $b = 2s^2$
4. $E = \frac{K}{A + BK}$

CHAPTER 10 PRACTICE EXERCISE

1. Solve:

$$5T - 2V = -16$$
$$3V - 4T = 17$$

2. Solve:

$$2(3R_1 + R_2) = 16$$
$$R_1 + 11 = 3R_2 - 3$$

3. Solve:

$$\frac{I_1}{6} + \frac{I_2}{10} = 2.5$$
$$I_1 + \frac{I_2}{5} = 13$$

4. Solve:

$$E_1 + E_2 = 5.6$$
$$1.3E_1 - 3.7E_2 = 5.28$$

5. Solve:

$$0.56T_1 + 0.42T_2 = 1.92$$
$$T_1 = 0.74T_2$$

6. Solve:

$$4.10(H - 2W) = 17.62$$
$$0.24H - 0.77W = 0$$

7. Eliminate "PV" and solve for "E":

$$H = E + PV$$
$$\frac{PV}{C(T_1 - T_2)} = R$$

8. Eliminate "H" and solve for "R":

$$\frac{HL}{t_2 - t_1} = AKT$$
$$\frac{H}{0.24} = IR$$

ANSWERS:

1) $T = -2$, $V = 3$

2) $R_1 = 1$, $R_2 = 5$

3) $I_1 = 12$, $I_2 = 5$

4) $E_1 = 5.2$, $E_2 = 0.4$

5) $T_1 = 1.7$, $T_2 = 2.3$

6) $H = 11.5$, $W = 3.6$

7) $E = H - RC(T_1 - T_2)$

8) $R = \frac{AKT(t_2 - t_1)}{0.24IL}$